Lecture Notes in Computer Science 1665

Edited by G. Goos, J. Hartmanis and J. van Leeuwen

T0217229

Springer

Berlin
Heidelberg
New York
Barcelona
Hong Kong
London
Milan
Paris
Singapore
Tokyo

Peter Widmayer Gabriele Neyer
Stephan Eidenbenz (Eds.)

Graph-Theoretic Concepts in Computer Science

25th International Workshop, WG' 99
Ascona, Switzerland, June 17-19, 1999
Proceedings

Springer

Series Editors

Gerhard Goos, Karlsruhe University, Germany
Juris Hartmanis, Cornell University, NY, USA
Jan van Leeuwen, Utrecht University, The Netherlands

Volume Editors

Peter Widmayer
Gabriele Neyer
Stephan Eidenbenz
Institute for Theoretical Computer Science, ETH Zürich
ETH Zentrum, 8092 Zürich, Switzerland
E-mail: {widmayer/neyer/eidenben}@inf.ethz.ch

Cataloging-in-Publication data applied for

Die Deutsche Bibliothek - CIP-Einheitsaufnahme

Graph theoretic concepts in computer science : 25th international
workshop ; proceedings / WG '99, Ascona, Switzerland, June 17 - 19,
1999. Peter Widmayer ... (ed.). - Berlin ; Heidelberg ; New York ;
Barcelona ; Hong Kong ; London ; Milan ; Paris ; Singapore ; Tokyo
: Springer, 1999
  (Lecture notes in computer science ; Vol. 1665)
  ISBN 3-540-66731-8

CR Subject Classification (1998): F.2, G.2.2, F.1.2-3, F.3-4, E.1, I.3.5

ISSN 0302-9743
ISBN 3-540-66731-8 Springer-Verlag Berlin Heidelberg New York

© Springer-Verlag Berlin Heidelberg 1999
Printed in Germany

Typesetting: Camera-ready by author
SPIN: 10704274   06/3142 – 5 4 3 2 1 0   Printed on acid-free paper

Preface

The 25th International Workshop on Graph-Theoretic Concepts in Computer Science (WG'99) celebrated the anniversary of the workshop series at the Centro Stefano Franscini on Monte Verità, Ascona, Switzerland, from June 17 to 19, 1999. It was organized by ETH Zürich, sponsored by ETH, the Centro Stefano Franscini, the Swiss National Science Foundation, and Swissphone.

The workshop looks back on a remarkable tradition of a quarter century, with predecessors organized at various places in Europe. WG'99 has been an inspiring mix of questions from theory and practice, and of 74 young scientists and established researchers from all over the world, including Bangladesh, Belgium, Canada, the Czech Republic, France, Germany, Israel, Italy, Japan, Korea, the Netherlands, Russia, the Slovak Republic, Spain, Sweden, Switzerland, the UK, and the USA.

Four invited lectures reflect the history, theory, practice and this year's setting. Hartmut Noltemeier as the only one of four founding members of the WG series who served on the program committee throughout the entire history of WG talked about past achievements and future challenges of WG and of the field. Susanne Albers lectured on the theory of online algorithms. Thomas Lengauer presented practical packing problems from the textile and car manufacturing industries, as well as their solutions. A workshop held at the Centro Stefano Franscini traditionally offers a talk that is open to the general (Italian speaking) public – Nicola Santoro delivered this Stefano Franscini talk by reporting on computer science, school, and community, as seen from the eye of the storm.

At the age of 25, WG is healthy and strong. The number of submissions continues to be high, and the selection process continues to stricktly filter out the highest quality papers. Out of 64 submitted papers from two dozen countries, the program committee selected 33 for presentation at the workshop. The program committee consisted of

H. Bodlaender, Utrecht (NL)	A. Brandstädt, Rostock (D)
M. Habib, Montpellier (F)	J. Hromkovič, Aachen (D)
M. Kaufmann, Tübingen (D)	L. Kučera, Prague (CR)
A. Marchetti-Spaccamela, Rome (I)	E. Mayr, Munich (D)
R. Möhring, Berlin (D)	M. Nagl, Aachen (D)
H. Noltemeier, Würzburg (D)	S. Olariu, Norfolk (USA)
F. Parisi Presicce, Roma (I)	O. Sykora, Bratislava (SK)
G. Tinhofer, Munich (D)	D. Wagner, Konstanz (D)
P. Widmayer, Zurich (CH)	

and worked as a wonderful team: It punctually produced at least four reviews per submission, and it discussed quite a few papers in great detail. Comments and discussion at the workshop have been taken into account by the authors in the papers in this volume.

It is my pleasure to thank many people for their contribution in making WG'99 a memorable event: the authors of papers for submitting their work and for presenting all accepted papers at the workshop; the invited speakers for delivering fascinating lectures; all members of the program committee and all reviewers for their careful and timely evaluations, and in particular Andreas Brandstädt, Juraj Hromkovič, and Dorothea Wagner for coming to Zürich for the final discussion; Stephan Eidenbenz and Gabriele Neyer for organizing the workshop autonomously and with full responsibility (since I managed to escape into a sabbatical at the right time), and for taking care of the logistics, before, during and after the workshop, including the preparation of this volume; the Centro Stefano Franscini, represented by Christian Stamm, Claudia Lafranchi, and the team at the Monte Verità for hosting us; ETH Zürich, the Swiss National Science Foundation, and Swissphone for sponsoring the workshop; and, obviously, the participants for their active interest that continues to shape our thriving field.

Zürich, August 1, 1999   Peter Widmayer

The 25 WGs and Their Chairs

WG '75	U. Pape, Berlin
WG '76	H. Noltemeier, Göttingen
WG '77	M. Mühlbacher, Linz
WG '78	M. Nagl, H.J. Schneider, Schloss Feuerstein near Erlangen
WG '79	U. Pape, Berlin
WG '80	H. Noltemeier, Bad Honnef
WG '81	J. Mühlbacher, Linz
WG '82	H.J. Schneider, H. Göttler, Neunkirchen near Erlangen
WG '83	M. Nagl, J. Perl, Haus Ohrbeck near Osnabrück
WG '84	U. Pape, Berlin
WG '85	H. Noltemeier, Schloss Schwanenberg near Würzburg
WG '86	G. Tinhofer, G. Schmidt, Stift Bernried near Munich
WG '87	H. Göttler, H.J. Schneider, Schloss Banz near Bamberg
WG '88	J. van Leeuwen, Amsterdam
WG '89	M. Nagl, Schloss Rolduc near Aachen
WG '90	R.H. Möhring, Johannesstift Berlin
WG '91	G. Schmidt, R. Berghammer, Richterheim Fischbachau, Munich
WG '92	E.W. Mayr, W.-Kempf-Haus, Wiesbaden-Naurod
WG '93	J. van Leeuwen, Sports Center Papendal near Utrecht
WG '94	G. Tinhofer, E.W. Mayr, G. Schmidt, Herrsching near Munich
WG '95	M. Nagl, Haus Eich at Aachen
WG '96	G. Ausiello, A. Marchetti-Spaccamela, Cadenabbia
WG '97	R.H. Möhring, Bildungszentrum am Müggelsee, Berlin
WG '98	J. Hromkovič, O. Sýkora, Smolenice Castle near Bratislava
WG '99	P. Widmayer, Centro Stefano Franscini, Monte Verità, Ascona

List of Reviewers

R. Ahlswede
L. Becchetti
S. Bezrukov
H.-J. Böckenhauer
U. Brandes
T. Calamoneri
M. Chladny
S. Cornelsen
E. Dahlhaus
S. De Agostino
S. Dobrev
F.F. Dragan
P. Ďuriš
J. Ebert
S. Eidenbenz
G. Fertin
I. Finocchi
P. Franciosa
W. von Gudenberg
Y. Guo
D. Handke
B. Haverkort
R. Heckel
P. Horak
F. Imhoff
R. Klasing
B. Klinz
T. Kloks
E. Köhler
J. Kratochvíl
D. Kratsch
V.B. Le
A. Liebers
F. Malvestuto
A. Massini
R. McConnell
A. Monti

M. Moscarini
H. Müller
M. Naatz
K. Nakano
G. Neyer
R. Niedermeier
W. Oberschelp
D. Pardubska
C. Paul
S. Perennes
U. Quernheim
T. Roos
A. Schürr
P. Scheffler
K. Schlude
H. Schröder
S. Seibert
J.F. Sibeyn
R. Silvestri
M. Simeoni
M. Skutella
O. Spaniol
L. Staiger
C. Stamm
L. Stacho
D. Stefankovic
F. Stork
R. Szelepcsenyi
D.M. Thilikos
O. Togni
R. Ulber
W. Unger
M. Veldhorst
I. Vrťo
E. Wanke
K. Weihe
G.M. Ziegler

Table of Contents

Silver Graphs:
Achievements and New Challenges

Hartmut Noltemeier

Department of Mathematics and Computer Science,
University of Wuerzburg,
Am Hubland, D-97074 Wuerzburg, Germany

Abstract. The following text is a written version of the dinner speech, which was given by the author at the occasion of the 25th Workshop on Graphtheoretic Concepts in Computer Science (WG '99), Ascona, June 18, 1999.

Dear participants, dear friends and especially dear Peter Widmayer,
thank you for inviting me and giving me the opportunity to talk a little bit about the history, the achievements and challenges of this series of 25 workshops, we can celebrate today.
After listening to a great variety of exciting talks and hiking in the beautiful Valle Verzasca, I am sure, you will be very hungry – as me too –, thus, I promise you, I will be very short.

First let me give you some remarks on the origin and starting events of these workshops. The first workshop took place in Berlin 1975 and was organized by Uwe Pape; to be more precise: originally it was the third meeting of the German chapter of the ACM, devoted to Graph-Languages and it was a combination of a tutorial on "Graph Languages", presented by Victor Basili, and a workshop on "Graph Manipulation Systems", extending this tutorial.

The real starting point for this successful series of workshops however was, when Jörg Mühlbacher, Uwe Pape and me met at a conference in Bad Homburg near Frankfurt in autumn 1975. As there seemed to be very promising problems and applications in the field of Algorithmic Graph Theory, I suggested to install a series of workshops on "Graphtheoretic Concepts in Computer Science" and we agreed to try to realize this idea with enthusiasm and much effort.
Thus, as a first consequence, the ACM-meeting was additionally defined to be the first workshop in our series, too, and the proceedings got the title "Graph-Languages and Algorithms on Graphs". The publication was done by Carl Hanser-Publishing company in Munich, which had some connection to the German chapter of the ACM.

Widmayer et al. (Eds.): WG'99, LNCS 1665, pp. 1–9, 1999.
© Springer-Verlag Berlin Heidelberg 1999

Now, initiating such a series of workshops, of course we had some *goals* in mind. First, we wanted to realize a real *workshop* within a familiar atmosphere, which allows a lot of social contacts, and with some special social events, too. (I remember very well some exciting sports events, e.g. a basketball competition (1985), a soccer tournament (1989), or some very fine concerts: Miriam Makeba, Berlin Philharmonic Orchestra, organ concert etc.).

Secondly we would like to have this events not twice at the same location, but everytime at a new beautiful place - as here in Ascona on top of Monte Verita, for example.

Third and our *main goal*:
we wanted to have a high-level event, which may cover many facets of this fascinating field, but should focus on most promising lines and potential applications of Graphtheoretic Concepts and Graph-Algorithms.

Additionally we soon got aware of another goal, namely we should try to ensure some kind of independence, independence from societies as well as from specific institutions and industries and take the burden of preparing and organizing the workshops and take care for the publications within a small group of responsible, very interested colleagues.

To give you an idea of this topic, let me mention only the following:
We asked people from the ACM and the German Society for Computer Science (GI) for cooperation. Immediately there were at least five special interest groups, which would like to incorporate our activity or to participate at least in the program selection process.

This showed us, that we were focusing on a central issue of widespread interest, but on the other hand showed us, too, that we would be immediately a minority within this setting and probably could not realize our specific idea of these events.

Thus the next workshop was the first workshop following *our specific ideas*; it was organized by me at Göttingen University in 1976. Jörg Mühlbacher continued with the third Workshop in Linz/Austria in 1977.

At that time we had no explicit program committee. Thus the three organizers were responsible for the quality of the contributions, at least in the following sense:
we asked people to present recent work and submit a short abstract and we made a preselection, which was not very tough. But after the conference, we asked all participants to submit full papers within a couple of weeks, which should of course incorporate suggestions, stimulations and results of discussions of the workshop. All these papers were carefully reviewed by us and a lot of well-known experts, too.

The 25 WGs and their Chairs

WG'75	U. Pape	Berlin
WG'76	H. Noltemeier	Goettingen
WG'77	J. Mühlbacher	Linz
WG'78	M. Nagl, H. J. Schneider	Castle Feuerstein near Erlangen
WG'79	U. Pape	Berlin-Lichtenrade
WG'80	H. Noltemeier	Bad Honnef / Aachen
WG'81	J. Mühlbacher	Linz
WG'82	H. J. Schneider, H. Göttler	Neunkirchen near Erlangen
WG'83	M. Nagl, J. Perl	Osnabrueck
WG'84	U. Pape	European Academy Berlin-Grunewald
WG'85	H. Noltemeier	Castle Schwanberg near Wuerzburg
WG'86	G. Tinhofer, G. Schmidt	Stift Bernried near Munich
WG'87	H. Göttler, H. J .Schneider	Monastery Banz near Bamberg
WG'88	J. van Leeuwen	CWI / Amsterdam
WG'89	M. Nagl	Castle Rolduc near Aachen
WG'90	R. H. Möhring	Johannesstift Berlin
WG'91	G. Schmidt, R. Berghammer	Fischbachau near Munich
WG'92	E. W. Mayr	W.-Kempf-Haus, Naurod / Frankfurt
WG'93	J. van Leeuwen	Sports Center Papendal near Utrecht
WG'94	G. Tinhofer, E. W. Mayr, G. Schmidt	Herrsching near Munich
WG'95	M. Nagl	Haus Eich at Aachen
WG'96	G. Ausiello, A. Marchetti-Spaccamela	Foundation Konrad Adenauer, Cadenabbia / Lake Como
WG'97	R. H. Möhring	Müggelsee / Berlin
WG'98	O. Sýkora, J. Hromkovič	Smolenice Castle, Slovak Republic
WG'99	P. Widmayer	Monte Verita / Ascona

Now, as the first three workshops attracted a lot of people and attention, we were happy to enlarge the set of organizers, and by this Hans-Jürgen Schneider and with him later on Manfred Nagl entered our board and strengthened our basis with respect to Graph Grammars and their applications. Both were organizing the fourth workshop at Burg Feuerstein near Erlangen, a wonderful place, too.

In 1979 we entered the next cycle and Uwe Pape again did organize this workshop in Berlin, but at a very different place.

In 1980 I did it again, this time at Bad Honnef near Bonn, when I was affiliated with the Technical University of Aachen.

As I of course remember this event very well, I would like to use it as an example to give some more detailed comments and to point to some problems in general.

Contents of LNCS 100 (WG'80)

First of all, I have to point out, that it was the first time we published our proceedings within the Lecture Notes by Springer and we were very proud to get the jubilee number vol. 100. (At that time – you must know – there were not more than a dozen volumes each year within the LNCS-series).

Thus it was a special acknowledgement of our work and a great honor to us.

Additionally it was the first time EATCS (the European Association for Theoretical Computer Science) was – at least ideally – supporting the workshop. And – I guess – both facts especially in combination with the high-level-contents of the volume were in some sense essential that this WG-series got to be well-known world wide.

More specifically let us take a closer look at the contents of this proceedings:

You can observe a lot of presentations/talks/papers on

- Data Structures
- Hypergraphs and Data Bases
- Graph Grammars
- Complexity
- Petri-Nets/Concurrency
- Optimization
- Graph Embeddings
- Connection of Graph Theory, Data Structures and Computational Geometry
- Partial Graphs and Programs

Thus we had a wide range of very interesting talks and of course the question did arise, should we enlarge the whole workshop to be a full-week conference or should we maintain our approach and focus and concentrate on specific topics. This question of course was and still is alive all the time (since the beginning).

For me today a list of *promising fields*, which is far from being complete, may contain topics like

- Network Design
  Network Upgrade and Improvement
- Massive Graphs
  in Telecommunication: Call Graphs, External Memory Graph
  Algorithms and Data Structures, Routing;
  Scheduling (crew, flight, train, ...)
- Mobile Computing
  Channel Assignment, Tracking, Ad-Hoc Networks
- Distributed Systems
  On-Line Algorithms, Reactive Systems, Information Retrieval
  (Web-Search,...)
- Random Graph Algorithms
- Sequencing, Structure Prediction in Biology

- Graph-Transformation-based Applications
    Evolutionary Trees, Constraint Solving Tools
- Animation/Visualization of Graph-Concepts and Graph-Algorithms
    Algorithm Engineering, User Interface, Drawings

This list is far from being complete; you will certainly miss some standard topics, which I have left out to present only one slide.

More importantly this selection is more or less application oriented and of course there is a big challenge in *Structural Graph Theory*, too, although there have been a lot of exciting breakthroughs within the past 25 years in this field (see for instance the survey on Graph Classes, recently published by A. Brandstädt, V. B. Le and J. P. Spinrad).

Meanwhile a lot of special events, devoted to exactly one of the topics mentioned above, have been developed (Dagstuhl- and DIMACS-seminars, Graph Grammar conferences, workshops on Mobile Computing, Petri-Nets etc.).

To me it is essential:

We cannot cover all interesting topics, but we should maintain the focus on concepts with promising longstanding potential, while we should not loose contact to developing specialized fields and conferences.

And we do this, and shall continue, at least by inviting people from this neighbouring or more specialized fields! And I am sure:

as long as we have colleagues, who are fascinated by our field and take the burden to organize such exciting events, we will continue the success-story of this series. And we had remarkable success!

To conclude let me *thank* all the former organizers and their staffs for their wonderful engaged work, all the participants (there were about a thousand within this series) and thanks to the authors of about 2000 papers, which were submitted.

Many thanks to the referees and members of program committees, which usually had a very hard time with respect to the tough refereeing schedule, and thanks to the publishing companies, especially to Hanser and Springer, and to numerous sponsors, too.

Last, but not least, I would like to thank Peter Widmayer and his team very cordially for selecting such a beautiful place and organizing this workshop in an excellent way.

Dear friends, please raise your glasses with me and drink a toast to Peter Widmayer and his team, and to the *further successful workshops* of the next century

- to the future -

CHEERS!

Appendix: Proceedings of the Workshops on Graph-Theoretic Concepts in Computer Science WG '75-WG '98

U. Pape (Ed.):
Graph Languages and Algorithms (Proc. WG '75),
Hanser, Munich, 1976, 236 pages, ISBN 3-446-12215-X

H. Noltemeier (Ed.):
Graphs, Algorithms, Data Structures (Proc. WG '76),
Graphtheoretic Concepts in Computer Science. Hanser, Munich, 1977, 336 pages,
ISBN 3-446-12330-X

J. Mühlbacher (Ed.):
Data Structures, Graphs, Algorithms (Proc. WG '77),
Hanser, Munich, 1978, 368 pages, ISBN 3-446-12526-4

M. Nagl, H. J. Schneider(Eds.):
Graphs, Data Structures, Algorithms (Proc. WG '78),
Hanser, Munich, 1979, 320 pages, ISBN 3-446-12748-8

U. Pape (Ed.):
Discrete Structures and Algorithms (Proc. WG '79),
Hanser, Munich, 1980, 270 pages, ISBN 3-446-13135-3

H. Noltemeier (Ed.):
Graphtheoretic Concepts in Computer Science (Proc. WG '80),
Lecture Notes in Computer Science 100, 403 pages,
Springer, Berlin/New York, 1981, ISBN 0-387-10291-4

J. Mühlbacher (Ed.):
Proceedings WG '81
Hanser, Munich, 1982, 355 pages, ISBN 3-446-13538-3

H. J. Schneider, H. Göttler (Eds.):
Proceedings WG '82
Hanser, Munich, 1983, 280 pages, ISBN 3-446-13778-5

M. Nagl, J. Perl (Eds.):
Proceedings WG '83
Trauner, Linz, 1984, 397 pages, ISBN 3-85320-311-6

U. Pape (Ed.):
Proceedings WG '84
Trauner, Linz, 1985, 381 pages, ISBN 3-85320-334-5

H. Noltemeier (Ed.):
Proceedings WG '85
Trauner, Linz, 1986, 443 pages, ISBN 3-85320-357-4

G. Tinhofer,G. Schmidt (Eds.):
Proceedings WG '86
Lecture Notes in Computer Science 246,
Springer, Berlin/New York, 1987, 305 pages, ISBN 0-387-17218-1

H. Göttler, H. J. Schneider (Eds.):
Proceedings WG '87
Lecture Notes in Computer Science 314,
Springer, Berlin/New York, 1988, 254 pages, ISBN 0-387-17218-1

J. van Leeuwen (Ed.):
Proceedings WG '88
Lecture Notes in Computer Science 344,
Springer, Berlin/New York, 1989, ISBN 0-387-50728-0

M. Nagl (Ed.):
Proceedings WG '89
Lecture Notes in Computer Science 411,
Springer, Berlin/New York, 1990, ISBN 0-387-52292-1

R. H. Möhring (Ed.):
Proceedings WG '90
Lecture Notes in Computer Science 484,
Springer, Berlin/New York, 1991, ISBN 0-387-53832-1

G. Schmidt, R. Berghammer (Eds.):
Proceedings WG '91
Lecture Notes in Computer Science 570,
Springer, 1992, ISBN 0-387-55121-2

E. W. Mayr (Ed.):
Proceedings WG '92
Lecture Notes in Computer Science 657,
Springer, 1993, ISBN 0-387-56402-0

J. van Leeuwen (Ed.):
Proceedings WG '93
Lecture Notes in Computer Science 790,
Springer, 1994, ISBN 0-387-57899-4

E. W. Mayr, G. Schmidt, G. Tinhofer (Eds.):
Proceedings WG '94
Lecture Notes in Computer Science 903,
Springer, 1995, ISBN 3-540-59071-4

M. Nagl (Ed.):
Proceedings WG '95
Lecture Notes in Computer Science 1017,
Springer, 1995, ISBN 3-540-60618-1

F. d'Amore, P. G. Franciosa, A. Marchetti-Spaccamela (Eds.):
Proceedings WG '96
Lecture Notes in Computer Science 1197,
Springer, 1997, ISBN 3-540-62559-3

R. H. Möhring (Ed.):
Proceedings WG '97
Lecture Notes in Computer Science 1335,
Springer, 1997, ISBN 3-540-63757-5

J. Hromkovič, O. Sýkora (Eds.):
Proceedings WG '98
Lecture Notes in Computer Science 1517,
Springer, 1998, ISBN 3-540-65195-0

Online Algorithms:
A Study of Graph-Theoretic Concepts

Susanne Albers

Max-Planck-Institut für Informatik,
Im Stadtwald, 66123 Saarbrücken, Germany.
albers@mpi-sb.mpg.de
www.mpi-sb.mpg.de/albers/

Abstract. In this paper we survey results on the design and analysis of online algorithms, focusing on problems where graphs and graph-theoretic concepts have proven particularly useful in the formulation or in the solution of the problem. For each of the problems addressed, we also present important open questions.

1 Introduction

The traditional design and analysis of algorithms assumes that an algorithm, which generates output, has complete knowledge of the entire input. However, this assumption is often unrealistic in practical applications. Many of the algorithmic problems that arise in practice are *online*. In these problems the input is only partially available because some relevant input arrives only in the future and is not accessible at present. An online algorithm must generate output without knowledge of the entire input. Online problems arise in areas such as resource allocation in operating systems, data structuring, robotics, distributed computing and scheduling. We give some illustrative examples.

Caching: In a two-level memory system, consisting of a small fast memory and a large slow memory, a caching algorithm has to keep actively referenced pages in fast memory not knowing which pages will be requested next.

Data structures: In a given data structure, we wish to access elements at low cost. The goal is to keep the data structure in a good state, not knowing which elements will have to be accessed in the future.

Robot exploration: A robot is placed in an unknown environment and has to construct a complete map of the environment using a path that is as short as possible. The environment is explored only as the robot travels around.

Distributed data management: In a network of processors we wish to dynamically re-allocate files so that accesses can be served with low communication cost. Future accesses are unknown.

We will address these problems in more detail in the following sections.

During the last ten years *competitive analysis* has proven to be a powerful tool for analyzing the performance of online algorithms. In a competitive analysis, introduced by Sleator and Tarjan [50], an online algorithm A is compared to an

Widmayer et al. (Eds.): WG'99, LNCS 1665, pp. 10–26, 1999.

optimal offline algorithm OPT. An optimal offline algorithm knows the entire input in advance and can process it optimally. Given an input I, let $C_A(I)$ and $C_{OPT}(I)$ denote the costs incurred by A and OPT in processing I. Algorithm A is called c-competitive if there exists a constant a such that

$$C_A(I) \leq c \cdot C_{OPT}(I) + a,$$

for all inputs I. Note that a competitive algorithm must perform well on *all* input sequences.

2 Caching

In this section, we first investigate uniform caching, also known as paging. For this problem, access graphs are very useful in the modeling of realistic request sequences. In the second part of this section we discuss caching problems that arise in large networks such a the world-wide web.

2.1 Paging

We first define the paging problem formally. In the two-level memory system, all memory pages have the same size. We assume that the fast memory can store k pages at any time. A paging algorithm is presented with a *request sequence* $\sigma = \sigma(1), \ldots, \sigma(m)$, where each request $\sigma(t)$, $1 \leq t \leq m$, specifies a page in the memory system. A request can be served immediately, if the requested page is in fast memory. If a requested page is not in fast memory, a *page fault* occurs. Then the missing page has to be copied from slow memory to fast memory so that the request can be served. At the same time the paging algorithm has to evict a page from fast memory in order to make room for the incoming page. Typically, a paging algorithm has to work online, i.e. the decision which page to evict on a fault must be made without knowledge of any future requests. The goal is to minimize the number of page faults.

We note that an online paging algorithm A is called c-competitive if, for all request sequences σ, the number of faults incurred by A is at most c times the number of faults incurred by an optimal offline algorithm OPT.

There are two very well-known online algorithms for the paging problem.

Least-Recently-Used (LRU): On a fault evict the page that has been requested least recently.

First-In First-Out (FIFO): On a fault evict the page that has been in fast memory longest.

An optimal offline algorithm was presented by Belady [16]. The algorithm is called MIN and has a polynomial running time.

MIN: On a fault evict the page whose next request is farthest in the future.

Sleator and Tarjan [50] analyzed the performance of LRU and FIFO and showed that on any request sequence the number of page faults incurred by the algorithms is bounded by k times the number of faults made by OPT. They also showed that this is optimal.

Theorem 1. [50] *LRU and FIFO are k-competitive.*

Theorem 2. [50] *No deterministic online algorithm for the paging problem can achieve a competitive ratio smaller than k.*

While Theorem 1 gives a strong worst-case analysis of LRU and FIFO, the competitive ratio of k is not very meaningful from a practical point of view. A fast memory can often hold several hundreds or thousands of pages. In fact, the competitive ratio of k is much higher than observed in practice. In an experimental study presented by Young [53], LRU achieves a competitive ratio between 1 and 2. Also, in practice, the performance of LRU is much better than that of FIFO. This does not show in the competitive analysis.

The high competitive ratio is due to the fact that a competitive analysis allows arbitrary request sequences whereas in practice only restricted classes of request sequences occur. Request sequences generated by real programs typically exhibit locality of reference: Whenever a page is requested, the next request is usually to a page that comes from a very small set of associated pages. Locality of reference can be modeled by *access graphs*, introduced by Borodin *et al.* [27]. In an access graph, the nodes represent the memory pages. Whenever a page p is requested, the next request can only be to a page that is adjacent to p in the access graph.

Formally, let $G = (V, E)$ be an unweighted graph. As mentioned above, V represents the set of memory pages. E is a set of edges. A request sequence $\sigma = \sigma(1), \ldots, \sigma(m)$, is *consistent with* G if $(\sigma(t), \sigma(t+1)) \in E$ for all $t = 1, \ldots, m-1$. We say that an online algorithm A is c-competitive on G if there exists a constant a such that $C_A(\sigma) \leq c \cdot C_{OPT}(\sigma) + a$ for all σ consistent with G. The competitive ratio of A on G, denoted by $R(A, G)$, is the infimum of all c such that A is c-competitive on G. Let

$$R(G) = \min_A R(A, G)$$

be the best competitive ratio achievable on G. Borodin *et al.* [27] showed that LRU achieves the best possible competitive ratio on access graphs that are trees. Trees represent the access graphs for many data structures. For any tree T, let $l(T)$ denote the number of leaves of T. Let

$$\mathcal{T}_n(G) = \{T \mid T \text{ is an } n\text{-node subtree of } G\}.$$

Theorem 3. [27] *For any graph G consisting of at least $k + 1$ nodes,*

$$R(G) \geq \max_{T \in \mathcal{T}_{k+1}} \{l(T) - 1\}.$$

Theorem 4. [27] *Let G be a tree. Then $R(LRU, G) = \max_{T \in \mathcal{T}_{k+1}} \{l(T) - 1\}$.*

Borodin *et al.* [27] also analyzed $R(LRU, G)$ on arbitrary graphs. They showed that the ratio depends on the number of *articulation nodes* of G. An articulation node is a node whose removal disconnects the graph. Another important result, due to Chrobak and Noga [26], is that LRU is never worse than FIFO on access graphs. Moreover, Borodin *et al.* [27] showed that there exist graphs for which the competitive ratio of FIFO is much higher than that of LRU.

Theorem 5. [26] *For any graph G, $R(LRU, G) \leq R(FIFO, G)$.*

Theorem 6. [27] *For any connected graph consisting of at least $k + 1$ nodes, $R(FIFO, G) \geq (k + 1)/2$.*

Borodin *et al.* [27] also presented an optimal online algorithm for any access graph.

FAR: The algorithm is a marking strategy working in phases. Whenever a page is requested, it is marked. If there is a fault at a request to a page p, then FAR evicts an unmarked page from fast memory that has the largest distance to a marked page in the access graph. A phase ends when all pages in fast memory are marked and a fault occurs. At this point all marks are erased and a new phase is started.

Irani *et al.* [39] showed that this algorithm achieves the best possible competitive ratio, up to a constant factor, for all access graphs.

Theorem 7. [39] *For any graph G, $R(FAR, G) = O(R(G))$.*

Fiat and Karlin [30] presented randomized online paging algorithms for access graphs that achieve an optimal competitive ratio.

A disadvantage of the algorithm FAR is that the access graph has to be known in advance. Fiat and Rosen [31] proposed a scheme that grows a dynamic weighted access graph over time. Whenever two pages p and q are requested successively for the first time, an edge (p, q) of weight 1 is inserted into the graph. Every time p and q are requested successively again, the weight w of the edge is decreased to $\min\{\alpha w, 1\}$ for some $\alpha < 1$. After each round of γk requests, all weights are increased by β, where $\beta, \gamma > 1$ are some fixed chosen constants. Fiat and Rosen [31] proposed the following variant of the algorithm FAR, called FARL: If there is a fault, the algorithm evicts the page that has the largest distance from the page requested just before the fault. Fiat and Rosen presented an experimental study in which FARL incurs fewer page faults than even LRU.

So far we have addressed undirected access graphs. An initial investigation of directed access graph was presented by Irani *et al.* [39], who considered structured program graphs.

Open Problem: Develop online paging algorithms for general directed access graphs.

2.2 Generalized Caching

Caching problems that arise in large networks such as the world-wide web differ from ordinary caching in two aspects. Pages may have different sizes and may incur different costs when loaded into fast memory. The pages or *documents* to be cached may be text files, pictures or web pages; the cost of loading a missing page into fast memory may depend on the size of the page and on the distance to the nearest node in the network holding the page. In generalized caching we have again a two-level memory system consisting of a fast and a slow memory. (In the network setting, the fast memory is the memory of a given node. The slow memory is the memory of the remaining network.) We assume that the fast memory has a capacity of K. For any page p, let $\text{SIZE}(p)$ be the size and $\text{COST}(p)$ be the cost of p. The total size of the pages in fast memory may never exceed K. The goal is to serve a sequence of requests so that the total loading cost is as small as possible. Various cost models have been proposed in the literature.

1. **The Bit Model [38]:** For each page p, we have $\text{COST}(p) = \text{SIZE}(p)$. (The delay in bringing the page into fast memory depends only upon its size.)
2. **The Fault Model [38]:** For each page p, we have $\text{COST}(p) = 1$ while the sizes can be arbitrary.
3. **The Cost Model:** For each page p, we have $\text{SIZE}(p) = 1$ while the costs can be arbitrary.
4. **The General Model:** For each page p, both the cost and size can be arbitrary.

In the Bit Model, and hence in the General Model, computing an optimal offline service schedule for a given request sequence is NP-hard. The problem is polynomially solvable in the Cost Model [25]. In fact, caching in the Cost Model is also known as *weighted caching*, a special instance of the k-server problem. In the Fault Model the complexity is unknown.

Open Problem: Determine the complexity of caching in the Fault Model.

Young [54] gave a k-competitive online algorithm for the General Model.

Landlord: For each p in fast memory, the algorithm maintains a variable $\text{CREDIT}(p)$ that takes values between 0 and $\text{COST}(p)$. If a requested page p is already in fast memory, then $\text{CREDIT}(p)$ is reset to any value between its current value and $\text{COST}(p)$. If the requested page is not in fast memory, then the following two steps are executed until there is enough room to load p. (1) For each page q in fast memory, decrease $\text{CREDIT}(q)$ by $\Delta \cdot \text{SIZE}(q)$, where $\Delta = \min_{q \in F} \text{CREDIT}(q)/\text{SIZE}(q)$ and F is the set of pages in fast memory. (2) Evict any page q from fast memory with $\text{CREDIT}(q) = 0$. When there is enough room, load p and set $\text{CREDIT}(p)$ to $\text{COST}(p)$.

Theorem 8. [54] *Landlord is k-competitive in the General Model.*

For the Bit and the Fault Model, Irani presented $O(\log^2 k)$-competitive online algorithms and $O(\log k)$-approximation offline algorithms. Here k is the ratio of

K to the size of the smallest page requested. For the offline problem, Albers *et al.* [2] gave constant factor approximation algorithms using only a small amount of additional space in fast memory, say $O(1)$ times the largest page size. Note that the largest page size is typically a very small fraction of the total size of the fast memory, say 1%. The approach is to formulate the caching problems as integer linear programs and then solve a relaxation to obtain a fractional optimal solution. The *integrality gap* of the linear programs is unbounded, but nevertheless one can show the following.

Theorem 9. [2] *There is a polynomial-time algorithm that, given any request sequence, finds a solution of cost $c_1 \cdot \mathrm{OPT_{LP}}$, where $\mathrm{OPT_{LP}}$ is the cost of the fractional solution (with fast memory K). The solution uses $K + c_2 \cdot S$ memory, where S is the size of the largest page in the sequence. The values of c_1 and c_2 are as follows for the various models. Let ϵ and δ be real numbers with $\epsilon > 0$ and $0 < \delta \leq 1$.*

1. *$c_1 = 1/\delta$ and $c_2 = \delta$ for the Bit Model,*
2. *$c_1 = (1 + \epsilon)/\delta$ and $c_2 = \delta(1 + 1/(\sqrt{1 + \epsilon} - 1))$ for the Fault Model,*
3. *$c_1 = 4 + \epsilon$ and $c_2 = 6/\epsilon$ for the General Model.*

The c_1, c_2 values in the above theorem express trade-offs between the approximation ratio and the additional memory needed. For example, in the Bit Model, we can get a solution with cost $\mathrm{OPT_{LP}}$ using at most S additional memory. In the Fault model, we can get a solution with cost $4\mathrm{OPT_{LP}}$ using at most $2S$ additional memory. The approximation ratio can be made arbitrarily close to 1 by using $c_2 S$ additional memory for a large enough c_2. In the General Model we obtain a solution of $5\mathrm{OPT_{LP}}$ using $6S$ additional memory, but we can achieve approximation ratios arbitrarily close to 4.

Open Problem: Are there constant factor approximation algorithms that do not require extra space in fast memory?

3   Data Structures

Many online problems arise in the area of data structures. We consider the list update problem which is among the most extensively studied online problems.

The list update problem is to maintain a dictionary as an unsorted linear list. Consider a set of items that is represented as a linear linked list. We receive a request sequence σ, where each request is one of the following operations. (1) It can be an *access* to an item in the list, (2) it can be an *insertion* of a new item into the list, or (3) it can be a *deletion* of an item. To access an item, a list update algorithm starts at the front of the list and searches linearly through the items until the desired item is found. To insert a new item, the algorithm first scans the entire list to verify that the item is not already present and then inserts the item at the end of the list. To delete an item, the algorithm scans the list to search for the item and then deletes it.

In serving requests a list update algorithm incurs cost. If a request is an access or a delete operation, then the incurred cost is i, where i is the position of the requested item in the list. If the request is an insertion, then the cost is $n+1$, where n is the number of items in the list before the insertion. While processing a request sequence, a list update algorithm may rearrange the list. Immediately after an access or insertion, the requested item may be moved at no extra cost to any position closer to the front of the list. These exchanges are called *free exchanges*. Using free exchanges, the algorithm can lower the cost on subsequent requests. At any time two adjacent items in the list may be exchanged at a cost of 1. These exchanges are called *paid exchanges*.

With respect to the list update problem, we require that a c-competitive online algorithm has a performance ratio of c *for all size lists*. More precisely, a deterministic online algorithm for list update is called c-competitive if there is a constant a such that for all size lists and all request sequences σ, $C_A(\sigma) \leq c \cdot C_{OPT}(\sigma) + a$.

Linear lists are one possibility to represent a dictionary. Certainly, there are other data structures such as balanced search trees or hash tables that, depending on the given application, can maintain a dictionary in a more efficient way. In general, linear lists are useful when the dictionary is small and consists of only a few dozen items [19]. Furthermore, list update algorithms have been used as subroutines in algorithms for computing point maxima and convex hulls [18,32]. Recently, list update techniques have been very successfully applied in the development of data compression algorithms [6,20,24].

There are three well-known deterministic online algorithms for the list update problem.

Move-To-Front: Move the requested item to the front of the list.

Transpose: Exchange the requested item with the immediately preceding item in the list.

Frequency-Count: Maintain a frequency count for each item in the list. Whenever an item is requested, increase its count by 1. Maintain the list so that the items always occur in nonincreasing order of frequency count.

The formulations of list update algorithms generally assume that a request sequence consists of accesses only. It is obvious how to extend the algorithms so that they can also handle insertions and deletions. On an insertion, the algorithm first appends the new item at the end of the list and then executes the same steps as if the item was requested for the first time. On a deletion, the algorithm first searches for the item and then just removes it.

In the following, we discuss the algorithms Move-To-Front, Transpose and Frequency-Count. We note that Move-To-Front and Transpose are *memoryless* strategies, i.e., they do not need any extra memory to decide where a requested item should be moved. Thus, from a practical point of view, they are more attractive than Frequency-Count. Sleator and Tarjan [50] analyzed the competitive ratios of the three algorithms.

Theorem 10. [50] *The Move-To-Front algorithm is 2-competitive.*

Proposition 1. *The algorithms Transpose and Frequency-Count are not c-competitive, for any constant c.*

Albers [1] presented another deterministic online algorithm for the list update problem. The algorithm belongs to the Timestamp(p) family of algorithms that were introduced in the context of randomized online algorithms and are defined for any real number $p \in [0, 1]$, see [1]. For $p = 0$, the algorithm is deterministic and can be formulated as follows.

Timestamp(0): Move the requested item, say x, in front of the first item in the list that precedes x and that has been requested at most once since the last request to x. If there is no such item or if x has not been requested so far, then leave the position of x unchanged.

Theorem 11. [1] *The Timestamp(0) algorithm is 2-competitive.*

Note that Timestamp(0) is not memoryless. We need information on past requests in order to determine where a requested item should be moved. Timestamp(0) is interesting because it has a better overall performance than Move-To-Front. The algorithm achieves a competitive ratio of 2, as does Move-To-Front. However, Timestamp(0) is considerably better than Move-To-Front on request sequences that are generated by probability distributions [6]. For any probability distribution, the asymptotic expected cost incurred by TS(0) is at most 1.5 times the asymptotic expected cost incurred by an optimal offline algorithm. The corresponding bound for Move-To-Front is not better than $\pi/2$.

Karp and Raghavan [42] developed a lower bound on the competitiveness that can be achieved by deterministic online algorithms. This lower bound implies that Move-To-Front and Timestamp(0) have an optimal competitive ratio.

Theorem 12. [42] *Let A be a deterministic online algorithm for the list update algorithm. If A is c-competitive, then $c \geq 2$.*

An important question is whether the competitive ratio of 2 can be improved using randomization. We analyze randomized online algorithms problem against oblivious adversaries [17]. An oblivious adversary has to construct the entire request sequence in advance and is not allowed to see the random choices made by an online algorithm.

Many randomized online algorithms for list update have been proposed [1,7,35,36,49]. We present the two most important algorithms. Reingold et al. [49] gave a very simple algorithm, called Bit.

Bit: Each item in the list maintains a bit that is complemented whenever the item is accessed. If an access causes a bit to change to 1, then the requested item is moved to the front of the list. Otherwise the list remains unchanged. The bits of the items are initialized independently and uniformly at random.

Theorem 13. [49] *The Bit algorithm is 1.75-competitive against any oblivious adversary.*

Interestingly, it is possible to combine the algorithms Bit and Timestamp(0), see Albers *et al.* [7]. This combined algorithm achieves the best competitive ratio that is currently known for the list update problem.

Combination: With probability 4/5 the algorithm serves a request sequence using Bit, and with probability 1/5 it serves a request sequence using Timestamp(0).

Theorem 14. [7] *The algorithm Combination is 1.6-competitive against any oblivious adversary.*

Teia [51] presented a lower bound for randomized list update algorithms.

Theorem 15. [51] *Let A be a randomized online algorithm for the list update problem. If A is c-competitive against any oblivious adversary, then $c \geq 1.5$.*

A slightly better lower bound of 1.50084 was presented recently by Ambühl *et al.* [8]. However, the lower bound only holds in the partial cost model where the cost of serving a request to the i-th item in the list incurs a cost of $i - 1$ rather then i.

Open Problem: Give tight bounds on the competitive ratio achieved by randomized online algorithms against any oblivious adversary.

4   Robot Exploration

In robot exploration problems, a robot has to construct a complete map of an *unknown environment* using a path that is as short as possible. Many geometric and graph-theoretic problems have been studied in the past [3,13,22,28,29,33,34,46]. A general problem setting was introduced by Deng *et al.* [28]. The robot is placed in a room with obstacles. The exterior wall of the room as well as the obstacles are modeled by simple polygons. Figure 4 shows an example in which the room is a rectangle and all obstacles are rectilinear. The robot has 360° vision. Its task is to move through the scene so that it sees all parts of the room. More precisely, every point in the room must be visible from some point on the path traversed.

Given a scene S, let $L_A(S)$ be the length of the path traversed by algorithm A to explore S. Since A does not know S in advance it is also referred to as an *online algorithm*. Let $L_{OPT}(S)$ be the length of the path of an optimum algorithm that *knows the scene in advance*. We call an online exploration algorithm A *c-competitive* if for all scenes S, $L_A(S) \leq c \cdot L_{OPT}(S)$.

Exploration algorithms achieving a constant competitive ratio were given for rooms without obstacles [28,33,34,46]. Note that the exploration problem is non-trivial even in rooms without obstacles because the room might be a general polygon. Deng *et al.* [28] gave an $O(n)$-competitive algorithm for exploring rectilinear rooms with n rectilinear obstacles. Albers and Kursawe [5] showed that no exploration algorithm in rooms with n obstacles can be better than $\Omega(\sqrt{n})$-competitive. This lower bound holds even if the obstacles are rectangles.

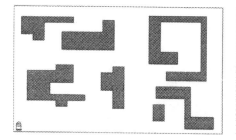

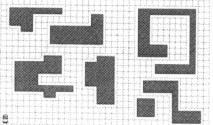

Fig. 1. A sample scene   **Fig. 2.** A sample scene with a grid

4.1 Exploration of Grid Graphs

In the scenario described above it is assumed that the robot can see an infinite range as long as no obstacle or exterior wall blocks the view. However, in practice, a robot's sensors can often scan only a distance of a few meters. This situation can be modeled by adding a grid to the scene, as shown in Figure 4, and requiring that the robot moves on the nodes and edges of the grid. A node in the grid models the vicinity that the robot can see at a given point. Now the robot has to explore all nodes and edges of the grid using as few edge traversals as possible. A node is explored when it is *visited* for the first time and an edge is explored when it is *traversed* for the first time. At any node the robot knows its global position and the directions of the incident edges. Note that using a depth-first strategy, the graph can be explored using $O(m)$ edge traversals, which is optimal. Here m denotes the total number of edges of the graph.

Betke *et al.* [22] introduced an interesting, more complicated variant of this problem where an additional *piecemeal constraint* has to be satisfied, i.e, the robot has to return to a start node s every so often. These returns might be necessary because the robot has to refuel or drop samples collected on a trip. Betke *et al.* developed two algorithms for piecemeal exploration of grids with rectangular obstacles. The algorithms, called *Wavefront* and *Ray*, need $O(m)$ edge traversals. The *Wavefront* algorithm implements a breadth-first strategy while the *Ray* algorithm implements a simple and elegant depth-first strategy.

Theorem 16. [22] *A grid with rectangular obstacles can be explored in a piecemeal fashion using $O(m)$ edge traversals.*

Albers and Kursawe [5] present an algorithm that explores a grid with arbitrary (rectilinear) obstacles using $O(m)$ edge traversals, which is optimal. The algorithm is a generalization of the *Ray* algorithm by Betke, Rivest and Singh. In the original *Ray* algorithm it is required that the robot always knows a path back to the start node whose length is most the radius of the graph. When exploring grids with arbitrary obstacles, this constraint cannot be satisfied. Albers and Kursawe [5] solve this problem by presenting an efficient strategy for exploring the boundary of arbitrary obstacles.

Theorem 17. [5] *A grid with arbitrary rectilinear obstacles can be explored in a piecemeal fashion using $O(m)$ edge traversals.*

4.2 Exploration of General Graphs

The graph-theoretic abstraction of a scene can be taken even further. Suppose that the environment is modeled by a strongly connected graph $G = (V, E)$. G can be directed or undirected. Such a general, graph-theoretic modeling of a scene allows us to neglect geometric features of the environment and to concentrate on combinatorial aspects of the exploration problem. Let n denote the number of nodes and m denote the number of edges of G.

Awerbuch *et al.* [13] consider piecemeal exploration of arbitrary undirected graphs and give a nearly optimal algorithm. The algorithm explores the graphs in strips, where each strip is explored using a breadth-first strategy.

Theorem 18. *An undirected graph can be explored in a piecemeal fashion using $O(m + n^{1+o(1)})$ edge traversals.*

Open Problem: Is there an algorithm that achieves an optimal bound of $O(m + n)$ on the number of traversals?

The most general graph-theoretic exploration problem was formulated by Deng and Papadimitriou [29]. The environment is now modeled by a strongly connected directed graph. At any point during the exploration process the robot knows (1) all visited nodes and edges and can recognize them when encountered again; and (2) the number of unvisited edges leaving any visited node. The robot does not know the head of unvisited edges leaving a visited node or the unvisited edges leading into a visited node. At each point in time, the robot visits a *current node* and has the choice of leaving the current node by traversing a specific known or an arbitrary (i.e. given by an adversary) unvisited outgoing edge. An edge can only be traversed from tail to head, not vice versa. As usual, the goal is to minimize the total number T of edge traversals. A piecemeal constraint does not have to be satisfied here.

If the graph is Eulerian, $2m$ edge traversals suffice [29]. For a non-Eulerian graph, let the *deficiency* d be the minimum number of edges that have to be added to make the graph Eulerian. Deng and Papadimitriou [29] suggested to study the dependence of T on m and d and showed the first upper and lower bounds. They gave a graph such that any algorithm needs $\Omega(d^2m/\log d)$ edge traversals. This lower bound was improved by Koutsoupias [46].

Theorem 19. [46] *There exist graphs for which every exploration algorithm needs $\Omega(d^2m)$ edge traversals.*

Deng and Papadimitriou gave an exponential upper bound.

Theorem 20. [29] *There is an algorithm that explores a graph with deficiency d using $d^{O(d)}m$ edge traversals.*

Deng and Papadimitriou asked the question whether the exponential gap between the upper and lower bound can be closed. The paper by Albers and Henzinger [3] is a first step in this direction: They give an algorithm that is sub-exponential in d, namely it achieves an upper bound of $d^{O(\log d)}m$. Albers and Henzinger also show that several exploration strategies based on greedy, depth-first and breadth-first approaches do not work well. There are graphs for which these strategies need $2^{\Omega(d)}m$ traversals.

We sketch the basic idea of the sub-exponential algorithm. At any time, the algorithm tries to explore new edges that have not been visited so far. That is, starting at some visited node x with unvisited outgoing edges, the robot explores new edges until it gets *stuck* at a node y, i.e., it reaches y on an unvisited incoming edge and y has no unvisited outgoing edge. Since the robot is not allowed to traverse edges in the reverse direction, an adversary can always force the robot to visit unvisited nodes until it finally gets stuck at a visited node.

The robot then relocates, using visited edges, to some visited node z with unexplored outgoing edges and continues the exploration. The *relocation* to z is the only step where the robot traverses visited edges. To minimize T one has to minimize the total number of edges traversed during all relocations. It turns out that a locally greedy algorithm that tries to minimize the number of traversed edges during each relocation is not optimal. Instead, the algorithm uses a divide-and-conquer approach. The robot explores a graph with deficiency d by exploring d^2 subgraphs with deficiencies $d/2$ each and uses the same approach recursively on each of the subgraphs. To create subgraphs with small deficiencies, the robot keeps track of visited nodes that have more visited outgoing than visited incoming edges. Intuitively, these nodes are *expensive* because the robot, when exploring new edges, can get stuck there. The relocation strategy tries to keep portions of the explored subgraphs "balanced" with respect to their expensive nodes. If the robot gets stuck at some node, then it relocates to a node z such that "its" portion of the explored subgraph contains the minimum number of expensive nodes.

Theorem 21. [3] *There is an algorithm that explores a graph with deficiency d using $d^{O(\log d)}m$ edge traversals.*

Open Problem: Is there an exploration algorithm for directed graphs that achieves an upper bound on the number of edge traversals that is polynomial in d?

5   Online Problems in Networks

Many online problems also arise in the area of distributed computing. We describe only a few problems here. Consider a network of processors each of which has its own local memory. Such a network can be modeled by a weighted undirected graph. The nodes of the graph represent the processor in the network and the edges represent the communication links. Let n be the number of nodes and m be the number of edges of the graph.

5.1 Migration and Replication Problems

First we address a problem in distributed data management, known as the *file allocation problem*. The goal is to dynamically re-allocate files in the network so that a sequence read and write requests to files an be served at low communication costs. The configuration of the system can be changed by *migrating* and *replicating* files, i.e., a file is moved resp. copied from one local memory to another.

In the investigation of the problem, we generally concentrate on one particular file in the system. We say that *a node v has the file* if the file is contained in v's local memory. *A request at a node v* occurs if v wants to read or write the file. Immediately after a request, the file may be migrated or replicated from a node holding the file to another node in the network. We use the cost model introduced by Bartal *et al.* [15] and Awerbuch *et al.* [12]. (1) If there is a read request at v and v does not have the file, then the incurred cost is $dist(u, v)$, where u is the closest node with the file. (2) The cost of a write request at node v is equal to the cost of communicating from v to all other nodes with a file replica. (3) Migrating or replicating a file from node u to node v incurs a cost of $d \cdot dist(u, v)$, where d is the file size factor. (4) A file replica may be erased at 0 cost.

Theorem 22. [12,15] *There exist deterministic and randomized online algorithms for the file allocation problem that achieve competitive ratios of $O(\log n)$.*

The randomized solution, due to Bartal *et al.* [15], is very simple and elegant.

Coinflip: If there is a read request at node v and v does not have the file, then with probability $1/d$, replicate the file to v. If there is a write request at node v, then with probability $1/\sqrt{3}d$, migrate the file to v and erase all other file replicas.

The *file migration* problem is a restricted version of the file allocation problem where we keep only one copy of each file in the entire system. If a file is writable, this avoids the problem of keeping multiple copies of a file consistent. For this problem, constant competitive algorithms are known, see [12,14,55]. In the *file replication* problem, files are assumed to be read-only and we have to determine which local memories should contain copies of the read-only files. Constant competitive algorithms are known for specific network topologies such as uniform networks, trees and rings [4,23]. A uniform network is a complete graph in which all edges have the same length.

All of the solutions mentioned above assume that the local memories of the processors have infinite capacity. Bartal *et al.* [15] showed that if the local memories have finite capacity, then no online algorithm for file allocation can be better than $\Omega(N)$-competitive, where N is the total number of files that can be accommodated in the system. They also presented an $O(N)$-competitive algorithm for uniform networks.

Open Problem: Is there an $O(N)$-competitive algorithm for arbitrary network topologies when the nodes have limited memory capacity?

5.2 Routing Problems

Many different online routing problems have been studied in the literature, see [47] for a survey. In the *virtual circuit routing problem* each communication link e in the network has a given maximum capacity c_e. The input consists of a sequence σ of communication requests, where each request $\sigma(t)$ can be describe by a 5-tupel $(u_t, v_t, r_t, d_t, b_t)$. Here u_t and v_t are the nodes to be connected, r_t is the bandwidth requirement of the request, d_t is its duration and b_t is a certain benefit. In response to each request we wish to establish a virtual circuit on a path connecting u_t and v_t with the given bandwidth. The benefit parameter is only specified in problems where calls may also be rejected. A benefit is obtained if the call is indeed routed.

Aspnes *et al.* [9] considered the problem variant when connection requests have unlimited duration and every call has to be routed. The goal is to minimize the maximum load on any of the links. The idea of their algorithm is to assign with every edge in the network a cost that is exponential in the fraction of the capacity of the edge assigned to on-going circuits. Let $\mathcal{P} = \{P_1, \ldots, P_t\}$ be the routes assigned by the online algorithm to the first t requests. Similarly, let $\mathcal{P}^{OPT} = \{P_1^{OPT}, \ldots, P_t^{OPT}\}$ be the routes assigned by the optimal offline algorithm. For every edge e in the network, we define a *relative load* after t requests,

$$l_e(t) = \sum_{\substack{s:e \in P_s \\ s \le t}} r_s/c_e.$$

The online algorithm given below assumes knowledge of a value Λ which is an estimate on the maximum load obtained by an optimal offline algorithm when all the requests are routed. Such a value can be obtained using a doubling strategy. Whenever the current guess turns out to bee too small, it is doubled.

Assign-Route: Let a be a constant and let (u, v, r) be the current request to be routed. Set $\bar{r} = r/\Lambda$ and $\bar{l}_e = l_e/\Lambda$ for all $e \in E$. Let

$$cost_e = a^{\bar{l}_e + \bar{r}/c_e} - a^{\bar{l}_e}$$

for all $e \in E$. Let P be a shortest path from u to v in the graph with respect to costs $cost_e$. Route the request along P and set $l_e = l_e + r/c_e$ for all edges on P.

Aspnes *et al.* [9] show that for any sequence of requests that can be routed using the given edge capacities, the maximum load achieved by Assign-Route is at most $O(\log n)$ times as large as the maximum load of an optimal solution.

Theorem 23. [9] *Assign-Route is an $O(\log n)$-competitive algorithm for the problem of minimizing the maximum load on the links.*

The virtual circuit routing problem has also been studied in its throughput version. In this variant, called the *call control problem*, a benefit is associated with every request. Requests can be accepted or rejected while link capacities may not be exceeded. Awerbuch *et al.* [11] examined the case that each call has a limited duration and showed the following result based on an algorithm similar to Assign-Route.

Theorem 24. [11] *There is an $O(\log nT)$-competitive algorithm for the problem of maximizing throughput. T denotes the maximum duration of a call.*

The bound given in Theorem 24 is tight.

6   Concluding Remarks

There are many online problems related to graphs that we have not addressed in this survey. A classical problem is *online coloring*: The nodes of a graph arrive online and we wish to color them using as few colors as possible. There is a significant body of work on this problem, see e.g. [37,45,48,52]. In *online matching*, a newly arriving node can be matched to a node already present. Unweighted and weighted versions of this problem have been considered [43,41,44]. The *generalized Steiner tree problem* is an extensively studied problem where we have to construct a minimum weight tree in a graph such that certain connectivity requirements are satisfied. In the online variant nodes as well as connectivity requirements arrive online [10,21,40,56].

References

1. S. Albers. Improved randomized on-line algorithms for the list update problem. *SIAM Journal on Computing*, 27:670–681, 1998.
2. S. Albers, S. Arora and S. Khanna. Page replacement for general caching problems. *Proc. 10th Annual ACM-SIAM Symp. on Discrete Algorithms*, 31–40, 1999.
3. S. Albers and M. Henzinger. Exploring unknown environments. *Proc. 29th Annual ACM Symposium on Theory of Computing*, 416–425, 1997.
4. S. Albers and H. Koga. New on-line algorithms for the page replication problem. *Journal of Algorithms*, 27:75–96, 1998.
5. S. Albers and K. Kursawe. Exploring unknown environments with obstacles. *Proc. 10th Annual ACM-SIAM Symp. on Discrete Algorithms*, 842–843, 1999.
6. S. Albers and M. Mitzenmacher. Average case analyses of list update algorithms, with applications to data compression. *Algorithmica*, 21:312–329, 1998.
7. S. Albers, B. von Stengel and R. Werchner. A combined BIT and TIMESTAMP algorithm for the list update problem. *Information Processing Letters*, 56:135–139, 1995.
8. C. Ambühl, B. Gärtner and B. von Stengel. Towards new lower bounds for the list update problem. To appear in *Theoretical Computer Science*.
9. J. Aspnes, Y. Azar A. Fiat, S. Plotkin and O. Waarts. On-line load balancing with applications to machine scheduling and virtual circuit routing. *Proc. 25th ACM Annual ACM Symp. on the Theory of Computing*, 623–631, 1993.
10. B. Awerbuch, Y. Azar and. Y. Bartal. On-line generalized Steiner tree problem. *Proc. 7th ACM-SIAM Symposium on Discrete Algorithms*, 68–74, 1996.
11. B. Awerbuch, Y. Azar and S. Plotkin. Throughput-competitive online routing. *34th IEEE Symp. on Foundations of Computer Science*, 32–40, 1993.
12. B. Awerbuch, Y. Bartal and A. Fiat. Competitive distributed file allocation. *Proc. 25th Annual ACM Symp. on Theory of Computing*, 164–173, 1993.

13. B. Awerbuch, M. Betke, R. Rivest and M. Singh. Piecemeal graph learning by a mobile robot. *Proc. 8th Conference on Computational Learning Theory*, 321–328, 1995.

14. Y. Bartal, M. Charikar and P. Indyk. On page migration and other relaxed task systems. *Proc. of the 8th Annual ACM-SIAM Symp. on Discrete Algorithms*, 43–52, 1997.

15. Y. Bartal, A. Fiat and Y. Rabani. Competitive algorithms for distributed data management. *Proc. 24th Annual ACM Symp. on Theory of Computing*, 39–50, 1992.

16. L.A. Belady. A study of replacement algorithms for virtual storage computers. *IBM Systems Journal*, 5:78–101, 1966.

17. S. Ben-David, A. Borodin, R.M. Karp, G. Tardos and A. Wigderson. On the power of randomization in on-line algorithms. *Algorithmica*, 11:2-14,1994.

18. J.L. Bentley, K.L. Clarkson and D.B. Levine. Fast linear expected-time algorithms for computing maxima and convex hulls. *Proc. of the 1st Annual ACM-SIAM Symposium on Discrete Algorithms*, 179–187, 1990.

19. J.L. Bentley and C.C. McGeoch. Amortized analyses of self-organizing sequential search heuristics. *Communication of the ACM*, 28:404–411, 1985.

20. J.L. Bentley, D.S. Sleator, R.E. Tarjan and V.K. Wei. A locally adaptive data compression scheme. *Communication of the ACM*, 29:320–330, 1986.

21. P. Berman and C. Coulston. Online algorithms for Steiner tree problems. *Proc. 29th Annual ACM Symposium on Theory of Computing*, 344–353, 1997.

22. M. Betke, R. Rivest and M. Singh. Piecemeal learning of an unknown environment. *Proc. 5th Conference on Computational Learning Theory*, 277–286, 1993.

23. D.L. Black and D.D. Sleator. Competitive algorithms for replication and migration problems. Technical Report Carnegie Mellon University, CMU-CS-89-201, 1989.

24. M. Burrows and D.J. Wheeler. A block-sorting lossless data compression algorithm. DEC SRC Research Report 124, 1994.

25. M. Chrobak, H. Karloff, T. Paye and S. Vishwanathan. New results on the server problem. *SIAM Journal on Discrete Mathematics*, 4:172–181, 1991.

26. M. Chrobak and J. Noga. LRU is better than FIFO. *Proc. 9th Annual ACM-SIAM Symposium on Discrete Algorithms*, 78–81, 1998.

27. A. Borodin, S. Irani, P. Raghavan and B. Schieber. Competitive paging with locality of reference. *Journal on Computer and System Sciences*, 50:244–258, 1995.

28. X. Deng, T. Kameda and C. H. Papadimitriou. How to learn an unknown environment. *Journal of the ACM*, 45:215–245, 1998.

29. X. Deng and C. H. Papadimitriou. Exploring an unknown graph. *Proc. 31st Symposium on Foundations of Computer Science*, 356–361, 1990.

30. A. Fiat and A. Karlin. Randomized and multipointer paging with locality of reference. *Proc. 27th Annual ACM Symposium on Theory of Compting*, 626–634, 1995.

31. A. Fiat and Z. Rosen. Experimental studies of access graph based heuristics: Beating the LRU standard. *Proc. 8th Annual ACM-SIAM Symp. on Discrete Algorithms*, 63–72, 1997.

32. M.J. Golin. PhD thesis, Department of Computer Science, Princeton University, 1990. Technical Report CS-TR-266-90.

33. F. Hoffmann, C. Icking, R. Klein and K. Kriegel. A competitive strategy for learning a polygon. *Proc. 8th ACM-SIAM Symposium on Discrete Algorithms*, 166–174, 1997.

34. F. Hoffmann, C. Icking, R. Klein and K. Kriegel. The polygon exploration problem:A new strategy and a new analysis technique. *Proc. 3rd International Workshop on Algorithmic Foundations of Robotics*, 1998.

35. S. Irani. Two results on the list update problem. *Information Processing Letters*, 38:301–306, 1991.

36. S. Irani. Corrected version of the SPLIT algorithm. Manuscript, January 1996.

37. S. Irani. Coloring inductive graphs on-line. *Algorithmica*, 11(1):53–62, 1994.

38. S. Irani. Page replacement with multi-size pages and applications to Web caching. *Proc. 29th Annual ACM Symp. on Theory of Computing*, 701–710, 1997.

39. S. Irani, A.R. Karlin and S. Phillips. Strongly competitive algorithms for paging with locality of reference. *Proc. 3rd Annual ACM-SIAM Symposium on Discrete Algorithms*, 228–236, 1992.

40. M. Imase and B.M. Waxman. Dynamic Steiner tree problems. *SIAM Journal on Discrete Mathematics*, 4:349–384, 1991.

41. B. Kalyanasundaram and K. Pruhs. Online wieghted matching. *Journal of Algorithms*, 14:139–155, 1993.

42. R. Karp and P. Raghavan. From a personal communication cited in [49].

43. S. Khuller, S.G. Mitchell and V.V. Vazirani. On-line weighted bipartite matching. *Proc. 18th International Colloquium on Automata, Languages and Programming (ICALP)*, Springer LNCS, Vol. 510, 728–738, 1991.

44. R. Karp, U. Vazirani and V. Vazirani. An optimal algorithm for online bipartite matching. *Proc. 22nd ACM Symp. on Theory of Computing*, 352–358, 1990.

45. H.A. Kierstead. The linearity of First-Fit coloring of interval graphs. *SIAM Journal on Discrete Mathematics*, 1:526–530, 1988.

46. J. Kleinberg. On-line search in a simple polygon. *Proc. 5th Annual ACM-SIAM Symposium on Discrete Algorithms*, 8–15, 1994.

47. S. Leonardi. On-line network routing. *Online Algorithms: The State of the Art*, Edited by A. Fiat and G. Woeginger, Springer LNCS, Volume 1442, 242-267, 1998.

48. L. Lovasz, M. Saks and M. Trotter. An online graph coloring algorithm with sublinear performance ratio. *Discrete Mathematics*, 75:319–325, 1989.

49. N. Reingold, J. Westbrook and D.D. Sleator. Randomized competitive algorithms for the list update problem. *Algorithmica*, 11:15–32, 1994.

50. D.D. Sleator and R.E. Tarjan. Amortized efficiency of list update and paging rules. *Communications of the ACM*, 28:202–208, 1985.

51. B. Teia. A lower bound for randomized list update algorithms. *Information Processing Letters*, 47:5–9, 1993.

52. S. Vishwanathan. Randomized online graph coloring. *Journal of Algorithms*, 13:657–669, 1992.

53. N. Young. The k-server dual and loose competitiveness for paging. *Algorithmica*, 11:525–541, 1994.

54. N.E. Young. Online file caching. *Proc. 9th Annual ACM-SIAM Symp. on Discrete Algorithms*, 82–86, 1998.

55. J. Westbrook. Randomized algorithms for the multiprocessor page migration. *SIAM Journal on Computing*, 23:951–965, 1994.

56. J. Westbrook and D.C. Yan. Lazy and greedy: On-line algorithms for Steiner problems. *Proc. Workshop on Algorithms and Data Structures*, Springer LNCS Volume 709, 622–633, 1993.

Discrete Optimization Methods for Packing Problems in Two and Three Dimensions — With Applications in the Textile and Car Manufacturing Industries

Thomas Lengauer

Institute for Algorithms and Scientific Computing (SCAI)
GMD - German National Research Center for Information Technology GmbH
Sankt Augustin, Germany
and
Department of Computer Science
University of Bonn

Abstract. This talk surveys two- and three-dimensional packing problems in the textile and leather manufacturing industry as well as in the automobile industry and presents methods for solving such problems. We solve these problems with methods from combinatorial optimization. Our research leads to software that is used in the respective industrial branches.

On leather, the problem is to place stencils (for objects such as sofas, car seats, shoes etc.) on a leather hide such as to minimize the waste and to obey certain restrictions pertaining to leather quality. Since runtime requirements are tough, we are using a greedy placement strategy here that employs a shape fitting method based on a variant of geometric hashing. Work on leather is joint with Jörg Heistermann, Ralf Heckmann, and Birgit Fromme.

On textiles, the problem is analogous to the leather case. Here, the fabric is rectangular, rotations of the stencils are restricted, and quality restrictions are replaced with a wide variety of constraints pertaining to patterns on the fabric. Cutting images can be reused in this case, and therefore more runtime is allowed. The most successful algorithmic variant for this case employs a fast randomized strategy for generating cutting images. This strategy is based on Minkowski sums. We can generate up to 1 Mio images per hour, the best of which is postoptimized with simulated annealing or a linear-programming method and then offered as the solution. On fabric, we can also compute lower bounds. For this purpose, we extend a branch-and-bound technique by Milenkovic to compute bounds that are quite tight (below 1 percentage point of waste) for pants and less tight (within several percentage points) on jackets. We also developed an algorithm that uses a database of cutting images to generate cutting images based on their similarity to already computed high-quality cutting images. By continually extending the database with cutting images computed by the algorithm, the algorithm is able to learn. Work on textiles is joint with Ralf Heckmann at GMD-SCAI. Research

Widmayer et al. (Eds.): WG'99, LNCS 1665, pp. 27–28, 1999.
© Springer-Verlag Berlin Heidelberg 1999

on textiles has been partially supported by the German National Science Foundation (DFG).
Arrangement problems in the car manufacturing industry take many forms and are generally not yet very accessible to automation. The reason is the multitude of constraints on these problems (geometric, electromagnetic, thermal, maintenance, ergonomic) and the elusiveness of the cost function. We have chosen an approach in stages to make these problem more amenable to automation. In stage 1, we extended our approach to chip layout to three dimensions, in order to pack boxes with electronic modules in cars. Here the shapes are brick-like. In stage 2, we develop methods based on combinatorial optimization that pack more general - even nonconvex - 3D shapes. We use a two-phase method here. In the first phase we generate variants of the relative position and orientation of these shapes. In the second stage we apply a local compaction method that is able to change the orientation of the objects within limits. Here, we extend a method by Milenkovic to three dimensions. In the third stage, we integrate the methods that have been developed and propose scenarios for their application in the automobile industry. Packing electronic parts, packing parts in trunks of cars, and inserting modules like dynamos etc. into preplaced engine spaces are examples of such scenarios. Work on automobiles is in progress and is joint with Mike Schäfer at the Department of Computer Science, University of Bonn. Research on automobiles has been partially supported by the German Federal Ministry for Education and Research (BMBF).

References

1. J. Heistermann, T. Lengauer. The Nesting Problem in the Leather Manufacturing Industry. *Annals of Operations Research* 57 (1995) 147-173
2. R. Heckmann, T. Lengauer. Computing Upper and Lower Bounds on Textile Nesting Problems. *European Journal of Operational Research* 108,3 (1998) 473-489
3. V. Milenkovic. Translational Polygon Containment and Minimal Enclosure using Linear Programming Based Restriction. *Proc. 28th Annual ACM Symposium on Theory of Computing* (1996) 109-118
4. V. Milenkovic. Rotational Polygon Overlap Minimization. *Thirteenth Annual ACM Symposium on Computational Geometry* (1997) 334-343

Informatica, Scuola, Communità: Uno Sguardo dall' Occhio del Ciclone

N. Santoro

Carleton University, Ottawa

Abstract. Le trasformazioni in corso costituiscono una ri-voluzione, una rottura con le premesse culturali che ci hanno guidato finora, uno strappo dalle promesse che politologi e futurologi ci hanno prodigato nel tempo. Ogni continuita' col passato e' solo apparente, le dinamiche in re-alta' sono caotiche, e le divergenze sono radicali. L'informatica, uno degli enzimi di questa trasformazione, si manifesta come ibrido scientifico-filosofico ed agisce come motore sia tecnico che culturale di questo processo di ri-definizione della vita sociale. La scuola diventa il nuovo centro di trasformazione del sociale, il momento di fissione delle regole attive della nuova societa'. Al tempo stesso, il sostrato su cui queste trasformazioni sono costrette ad operare e' la comunita'.

Uno sguardo, dal centro della trasformazione, a queste tre componenti e la loro interazione.

Widmayer et al. (Eds.): WG'99, LNCS 1665, pp. 29–29, 1999.

Proximity-Preserving Labeling Schemes and Their Applications

David Peleg[*]

Department of Applied Mathematics and Computer Science,
The Weizmann Institute, Rehovot 76100, Israel.
peleg@wisdom.weizmann.ac.il

Abstract. This paper considers informative labeling schemes for graphs. Specifically, the question introduced is whether it is possible to label the vertices of a graph with short labels in such a way that the *distance* between any two vertices can be inferred from inspecting their labels. A labeling scheme enjoying this property is termed a *proximity-preserving* labeling scheme. It is shown that for the class of n-vertex weighted trees with M-bit edge weights, there exists such a proximity-preserving labeling scheme using $O(M \log n + \log^2 n)$ bit labels. For the family of all n-vertex unweighted graphs, a labeling scheme is proposed that using $O(\log^2 n \cdot \kappa \cdot n^{1/\kappa})$ bit labels can provide *approximate* estimates to the distance, which are accurate up to a factor of $\sqrt{8\kappa}$. In particular, using $O(\log^3 n)$ bit labels the scheme can provide estimates accurate up to a factor of $\sqrt{2 \log n}$. (For weighted graphs, one of the $\log n$ factors in the label size is replaced by a factor logarithmic in the network's diameter.) In addition to their theoretical interest, proximity-preserving labeling systems seem to have some relevance in the context of communication networks. We illustrate this by proposing a potential application of our labeling schemes to efficient distributed connection setup in circuit-switched networks.

1 Introduction

Most traditional approaches to the problem of graph representation are based on storing the adjacency information using some kind of a data structure, e.g., an adjacency matrix. Such representation enables one to decide, given the names of two vertices, whether or not they are adjacent in the graph, simply by looking at the appropriate entry in the table. However, note that this decision cannot be made in the absence of the table. That is, the names themselves contain no useful information, and they serve only as "place holders," or pointers to entries in the table.

In contrast, one may consider using more "informative" naming schemes for graphs. The idea is to associate with each vertex a label to be used as its name,

---

[*] Supported in part by grants from the Israel Science Foundation and from the Israel Ministry of Science and Art.

and select the labels in a way that will allow us to infer the adjacency of two vertices *directly* from their labels, without using any additional memory.

Obviously, labels of unrestricted size can be used to encode any desired information in them. Specifically, it is possible to encode the entire row i in the adjacency matrix of the graph in the label chosen for vertex i. As another example, labeling systems of general graphs based on Hamming distances are studied by Breuer and Folkman [4,5]. An (m, T)-labeling system is based on labeling each vertex of the graph with an m-bit label, such that two vertices are adjacent iff their labels are at Hamming distance T or less of each other. In [5] it is shown that every n-vertex graph has a $(2n\Delta, 4\Delta - 4)$-labeling, where Δ is the maximum vertex degree in the graph.

It is clear, however, that a labeling scheme is most useful if it uses relatively *short* labels (say, of length polylogarithmic in n), and yet allows us to deduce adjacencies efficiently (say, within polylogarithmic time). This leads to the following definition.

Definition 1. *A family $\mathcal{F}$ of graphs has an $l(n)$ adjacency-labeling scheme if there is a function Label labeling the vertices of each n-vertex graph in $\mathcal{F}$ with distinct labels of up to $l(n)$ bits, and there exists a polynomial time algorithm that given two labels of vertices in a graph from $\mathcal{F}$, decides the adjacency of these vertices.*

It is shown in [10] how to construct $O(\log n)$ adjacency labeling schemes for a number of graph families, including trees, bounded arboricity graphs (including, in particular, graphs of bounded degree and graphs of bounded genus, e.g., planar graphs), various intersection-based graphs such as interval graphs, and c-decomposable graphs.

The ability to decide adjacency is only one of the basic properties a representation may be required to possess. This paper makes one step further along this line of study, and addresses the more general question of retrieving information about arbitrary (i.e., possibly non-adjacent) vertices.

In particular, a natural property we may require a labeling scheme to possess is the ability to determine, or at least estimate, the *distance* between two vertices efficiently (say, within polylogarithmic time) given their labels. In this paper we introduce the notion of *proximity-preserving* labeling schemes, which are scheme possessing this ability.

More specifically, we introduce the following definitions. Recall that in a weighted graph $G(V, E, \omega)$, the length of a path is the combined weight of the edges composing it, and the distance between two vertices u, v, denoted $dist(u, v, G)$, is defined as the length of the shortest path connecting them.

Definition 2. *A family $\mathcal{F}$ of weighted graphs has an $l(n)$ distance labeling scheme if there is a function Label labeling the vertices of each n-vertex graph in $\mathcal{F}$ with distinct labels of up to $l(n)$ bits, and there exists a polynomial time algorithm that given two labels of vertices u, v in a graph G from $\mathcal{F}$, computes the distance between these vertices in the graph.*

It is clear that distance labeling schemes with short ($O(\log n)$-bit) labels are available for highly regular graph classes, such as rings, meshes, tori, hypercubes, and the like. It is less clear whether more general graph classes can be labeled in this fashion. In Sect. 2 we give one such example: it is shown that the family of n-vertex weighted trees enjoys an $O(M \log n + \log^2 n)$ distance labeling scheme, where M denotes the maximum number of bits required for representing an edge weight in the graph. The same approach extends to handle the class of n-vertex c-decomposable graphs for constant c, which includes also the classes of series-parallel graphs and k-outerplanar graphs, with $c = 2k$ (cf. [6]).

For larger classes of graphs, it seems harder to capture precise distance information using short labels. As one particular example, we do not know whether the class of planar graphs enjoys a distance labeling scheme with polylogarithmic length labels. Nevertheless, note that for *very* large graph classes the problem becomes easy again: for a family of n-vertex graphs with $\Omega(\exp(n^{1+\epsilon}))$ non-isomorphic graphs, any labeling scheme must use labels whose total combined length is $\Omega(n^{1+\epsilon})$, hence at least one label must be of $\Omega(n^{\epsilon})$ bits.

Instead of insisting on labeling schemes capturing precise distance information, we may settle for labeling schemes that efficiently provide a decent *estimate* on the distance between any two given vertices.

Definition 3. *A family $\mathcal{F}$ of weighted graphs has an $(l(n), R)$ approximate-distance labeling scheme (for some fixed $R > 1$) if there is a function Label labeling the vertices of each n-vertex graph in $\mathcal{F}$ with distinct labels of up to $l(n)$ bits, and there exists a polynomial time algorithm that given two labels of vertices u, v in a graph G from $\mathcal{F}$, provides an estimate $\tilde{D}(u, v)$ for the distance between these vertices in the graph, such that*

$$\frac{1}{R} \cdot \tilde{D}(u,v) \le dist(u, v, G) \le R \cdot \tilde{D}(u,v).$$

To illustrate this notion, in Sect. 3 it is shown that the family of all n-vertex weighted graphs enjoys an $(O(\Lambda \log n \cdot \kappa \cdot n^{1/\kappa}), \sqrt{8\kappa})$ approximate-distance labeling scheme for any integer $\kappa \ge 1$. (Here $\Lambda = \lceil \log Diam(G) \rceil$, where $Diam(G)$ denotes the weighted *diameter* of the graph G, i.e., $\max_{u,v \in V}(dist(u, v, G))$.) In particular, the family of weighted graphs enjoys also an $(O(\Lambda \log^2 n), \sqrt{2 \log n})$ approximate-distance labeling scheme.

Finally, we point out that the labeling problem studied in this paper seems to have some relevance in the context of communication networks. This fact is illustrated in Section 4 by proposing a potential application of our labeling schemes, namely, a fast and memory-efficient best-service (shortest-path, optimal setup time) distributed connection setup procedure for establishing virtual channels in circuit-switched networks.

2   Distance-Preserving Labelings

This section describes a distance labeling scheme for the family of weighted trees. The construction makes use of tree separators.

Definition 4. *Given a tree T, a separator is a vertex v_0 whose removal breaks T into disconnected subtrees of at most $n/2$ vertices each.*

We rely on the following well-known fact regarding tree separators.

Lemma 1. *Every n-vertex tree T, $n \geq 2$, has a separator (which can be found in linear time).*

2.1   The Labeling System

The vertices of a given n-vertex weighted tree T are labeled as follows. As a preprocessing step, arbitrarily prelabel each vertex v of T with a distinct integer $I(v)$ from $[1, n]$. The actual labeling is constructed by a recursive procedure PROC$_1$, defined next.

The procedure recursively partitions the tree by finding a separator. For example, in the tree T depicted in Fig. 1, vertex 31 separates T into T_1 and T_2, which are in turn partitioned by the separator vertices 21 and 22, respectively, and so on. Thus each vertex belongs to a unique subtree on each level of this partitioning, up to the level in which it is selected as the separator. For a vertex v which is internal to some subtrees on $p - 1$ levels, and becomes the separator on level p, its label consists of p triples, $Label(v) = \mathcal{J}_1(v) \circ \ldots \circ \mathcal{J}_p(v)$, where each triple $\mathcal{J}_j(v)$ gives the index $I(w)$ of the separator of v's tree on the $(j-1)$st level, the distance from v to that separator, and the index of the subtree to which v belongs on the jth level.

Applied to a subtree T', Procedure PROC$_1$ operates as follows. If T' contains a single vertex v_0 then it is labeled by $Label(v_0) = (I(v_0), 0, 0)$. Otherwise, the vertices of T' are labeled by performing the following steps. First, find a separator v_0 for T'. The removal of v_0 breaks T into disconnected subtrees $T_1, \ldots, T_k$, each with at most $n/2$ vertices. Recursively apply procedure PROC$_1$ to label each vertex v in each subtree T_i by $Label_i(v)$. Now consider a vertex $v \in T_i$. Let $\mathcal{J}(v) = (I(v_0), dist(v, v_0, T), i)$, and label v by the concatenation of the new triple $\mathcal{J}(v)$ and $Label_i(v)$, the label of v in T_i, i.e.,

$$Label(v) \ = \ \mathcal{J}(v) \circ Label_i(v).$$

Finally label the separator v_0 itself by $Label(v_0) = (I(v_0), 0, 0)$. Figure 1 illustrates the resulting labeling for a tree T.

2.2   Computing the Distances

Let us next describe a recursive procedure PROC$_2$ for computing the distance between two vertices u, v in T.

Consider two vertices u, v in T, with labels

$$Label(u) = \mathcal{J}_1(u) \circ \ldots \circ \mathcal{J}_q(u)$$

and

$$Label(v) = \mathcal{J}_1(v) \circ \ldots \circ \mathcal{J}_p(v),$$

respectively. Procedure PROC$_2$ considers the following cases.

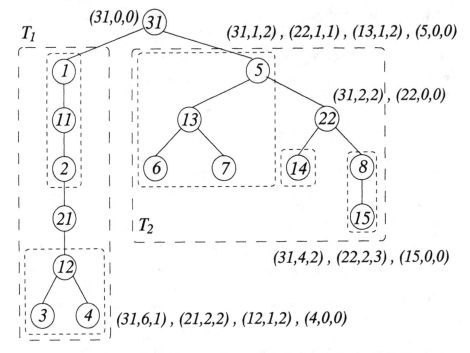

Fig. 1. A tree T and its recursive partitioning and distance labeling.

1. $p = 1$: /* v is the separator */   Return the second field in $\mathcal{J}_1(u)$.
2. $q = 1$: /* u is the separator */   Return the second field in $\mathcal{J}_1(v)$.
3. $p, q > 1$: Let

$$\mathcal{J}_1(u) = (I(w) , \; dist(u, w, T) , \; i)$$

$$\mathcal{J}_1(v) = (I(w) , \; dist(v, w, T) , \; j)$$

for some i, j. There are two subcases to consider.

(a) $i \neq j$: /* u and v belong to different subtrees */
Return the sum of the second fields in $\mathcal{J}_1(u)$ and $\mathcal{J}_1(v)$.

(b) $i = j$: /* u and v belong to the same subtree */
Then do the following:

   i. Peel off the first triple $\mathcal{J}_1(u)$ from $Label(u)$, and the triple $\mathcal{J}_1(v)$ from $Label(v)$, remaining with

$$Label_i(u) = \mathcal{J}_2(u) \circ \ldots \circ \mathcal{J}_q(u)$$

$$Label_i(v) = \mathcal{J}_2(v) \circ \ldots \circ \mathcal{J}_p(v) ;$$

   ii. Invoke procedure PROC$_2$ recursively on $Label_i(u)$ and $Label_i(v)$ to compute $dist(u, v, T_i)$;

   iii. Return this value.

In the example of Fig. 1, the distance between the vertices 15 and 31 is decided by case 1, the distance between 15 and 4 is decided by case 3(a), while the distance between 15 and 6 falls under case 3(b), and thus requires a recursive invocation of the procedure, for computing this distance within subtree T_2.

2.3   Analysis

Lemma 2. *The labeling scheme uses $O(M \log n + \log^2 n)$ bit labels, where M denotes the maximum number of bits required for representing an edge weight in the graph.*

Proof. On each level of the recursion, the sublabel $\mathcal{J}(v)$ contains $O(M + \log n)$ bits. As the maximum tree size is halved in each application of the recursive labeling procedure, there are at most $\log n$ levels, hence the lemma follows.   □

Lemma 3. *For every weighted tree T and vertices u, v, the output of the algorithm is $dist(u, v, T)$.*

Proof. Consider an arbitrary pair of vertices $u, v \in V$ in T, with labels

$$Label(u) = \mathcal{J}_1(u) \circ \ldots \circ \mathcal{J}_q(u)$$

and

$$Label(v) = \mathcal{J}_1(v) \circ \ldots \circ \mathcal{J}_p(v),$$

respectively. Let us examine the cases considered by the procedure one by one. If $p = 1$, then v is the chosen separator of T, hence $Label(v) = (I(v) , 0 , 0)$ and $\mathcal{J}_1(u) = (I(v) , dist(u, v, T) , i)$ for some i. Consequently, the second field in $\mathcal{J}_1(u)$ indeed gives the required distance. The case $q = 1$ is symmetric to the previous case.

Now suppose both $p, q > 1$, with $\mathcal{J}_1(v) = (I(w) , dist(v, w, T) , i)$ and $\mathcal{J}_1(u) = (I(w) , dist(u, w, T) , j)$ for some i, j. The first subcase considered by the procedure is where $i \neq j$. In this case, the unique path connecting v to u in T goes through w, and therefore $dist(v, u, T) = dist(v, w, T) + dist(u, w, T)$.

The second and final subcase is when $i = j$. In this case, both v and u belong to the same subtree T_i in the partition induced by the separator w, and hence the path connecting them in T (and determining their distance) is contained in its entirety in T_i. Therefore $dist(u, v, T)$ is equal to $dist(v, u, T_i)$, which is the value computed by the recursive invocation of the procedure.   □

We conclude the following.

Theorem 1. *There exists an $O(M \log n + \log^2 n)$ distance labeling scheme for the class of all n-vertex weighted trees with M-bit edge weights.*   □

The same approach extends to handle the class of n-vertex c-decomposable graphs (for constant c), where c affects the constant in the bound on the length of the resulting label.

Theorem 2. *There exists an $O(M \log n + \log^2 n)$ distance labeling scheme for the class of all n-vertex c-decomposable graphs with M-bit edge weights.*   □

3   Distance-Approximating Labelings

We next describe a approximate-distance labeling scheme for arbitrary weighted graphs. The construction makes use of the concept of tree covers, introduced in [1]. A *tree cover* for a graph G is a special collection of trees in the graph, containing all vertices of G. Intuitively, a tree cover captures locality by obeying the following property. For every two vertices that are not too far apart in the graph, the tree cover should contain a tree common to both, on which their distance is close to optimal. Naturally, it is desirable that the tree cover be also sparse. Formally, we have the following definition. The ℓ-*neighborhood* of a vertex $v \in V$ is the collection of vertices at distance ℓ or less from it in G, $\Gamma_\ell(v) = \{w \mid dist(w,v,G) \leq \ell\}$.

Definition 5. *Given a weighted graph $G = (V, E, \omega)$, an ℓ-tree cover is a collection TC of trees in G, with the property that for every vertex $v \in V$, there is a tree $T \in TC$ that spans its entire ℓ-neighborhood, namely, $\Gamma_\ell(v) \subseteq V(T)$. The depth of a tree cover TC is*

$$Depth(TC) = \max_{T \in TC}\{Depth(T)\},$$

and the overlap of TC is the maximum, over all vertices v, of the number of different trees containing v,

$$\texttt{Overlap}(TC) = \max_{v \in V}|\{T \in TC \mid v \in V(T)\}|.$$

We make use of the following result of [1].

Theorem 3. [1] *For every weighted graph $G = (V, E, \omega)$, $|V| = n$, and integers $\kappa, \ell \geq 1$, it is possible to construct an ℓ-tree cover $TC = TC_{\kappa,\ell}$ for G, with $Depth(TC) \leq (2\kappa - 1)\ell$ and $\texttt{Overlap}(TC) \leq \lceil 2\kappa \cdot n^{1/\kappa} \rceil$.* □

3.1   The Labeling System

Recall that $\Lambda = \lceil \log Diam(G) \rceil$. For every $1 \leq i \leq \Lambda$, construct a 2^i-tree-cover $TC_i = TC_{\kappa,2^i}$ for G as in Theorem 3. Separately in each tree cover TC_i, assign a distinct tag (from 1 to n) to each of the trees in TC_i. Each vertex v is now given a label composed of the concatenation of Λ tuples, $Label(v) = (T_1(v), \ldots, T_\Lambda(v))$, where the tuple $T_i(v)$ consists of the tags of all the trees in TC_i containing v.

3.2   The Distance Estimation Algorithm

The distance between vertices is estimated using the following algorithm. Given two labels $Label(v)$ and $Label(u)$, compare corresponding tuples in these labels one by one (starting with $T_1(v)$ and $T_1(u)$ and going upwards), until reaching the first level j such that $T_j(v)$ and $T_j(v)$ contain a common tag. Return $\tilde{D}(u,v) = \sqrt{2\kappa} \cdot 2^j$ as the estimate for $dist(u,v,G)$.

3.3   Analysis

Lemma 4. *The labeling scheme uses $O(\Lambda \log n \cdot \kappa \cdot n^{1/\kappa})$ bit labels.*

Proof. First note that the tags attached to the trees in the tree covers require at most $O(\log n)$ bits each, since there are at most $\lceil 2\kappa \cdot n^{1/\kappa}\rceil \cdot n$ trees in each tree cover. Each vertex v occurs on at most $\mathtt{Overlap}(\mathcal{TC}_i)$ different trees in $\mathcal{TC}_i$, hence its ith tuple $T_i(v)$ contains at most this many tags. The result now follows by the bound of Theorem 3 on $\mathtt{Overlap}(\mathcal{TC}_{\kappa,2^i})$.   □

Lemma 5. *For every weighted graph $G = (V, E, \omega)$ and vertices u, v, the estimate returned by the algorithm satisfies $\frac{1}{R} \cdot \tilde{D}(u,v) \le dist(u,v,G) \le R \cdot \tilde{D}(u,v)$ for $R = \sqrt{8\kappa}$.*

Proof. Consider an arbitrary pair of vertices $u, v \in V$. Suppose that $2^{i-1} < dist(u,v,G) \le 2^i$. Also suppose that the algorithm returned $\tilde{D}(u,v) = \sqrt{2\kappa} \cdot 2^j$ as the estimate for this distance. That is, j was the first level on which u and v shared a common tree in $\mathcal{TC}_j$.

By definition, the tree cover $\mathcal{TC}_i = \mathcal{TC}_{\kappa,2^i}$ constructed for the graph contains a tree T such that $\Gamma_{2^i}(u) \subseteq V(T)$. Hence both $u, v \in V(T)$. Therefore necessarily $j \le i$, and hence

$$dist(u,v,G) \ge 2^{i-1} \ge 2^{j-1} = \frac{1}{\sqrt{8\kappa}} \cdot (\sqrt{2\kappa} \cdot 2^j) = \frac{1}{R} \cdot \tilde{D}(u,v).$$

For the other direction, let T' be the tree common to u and v in $\mathcal{TC}_j$. By Theorem 3, $Depth(T') \le (2\kappa - 1) \cdot 2^j$. Consequently, the tree T provides a path of length at most $2 \cdot (2\kappa - 1) \cdot 2^j$ connecting u and v, hence necessarily

$$dist(u,v,G) \le 4\kappa \cdot 2^j = \sqrt{8\kappa} \cdot (\sqrt{2\kappa} \cdot 2^j) = R \cdot \tilde{D}(u,v).$$ □

Theorem 4. *For any integer $\kappa \ge 1$, there exists an $(O(\Lambda \log n \cdot \kappa \cdot n^{1/\kappa}), \sqrt{8\kappa})$ approximate-distance labeling scheme for the class of all n-vertex graphs.*   □

Corollary 1. *For any integer $\kappa \ge 1$, there exists an $(O(\log^2 n \cdot \kappa \cdot n^{1/\kappa}), \sqrt{8\kappa})$ approximate-distance labeling scheme for the class of all n-vertex unweighted graphs.*   □

Setting $\kappa = \log n$, we have

Corollary 2.   *1. There exists an $(O(\Lambda \log^2 n), \sqrt{2 \log n})$ approximate-distance labeling scheme for the class of all n-vertex graphs.*
2. There exists an $(O(\log^3 n), \sqrt{2 \log n})$ approximate-distance labeling scheme for the class of all n-vertex unweighted graphs.   □

4   Application to Efficient Distributed Connection Setup

The labeling problem studied in this paper seems to have some relevance in the context of communication networks. In this section we illustrate this by proposing a potential application of our labeling schemes to connection setup in circuit-switched networks. For the purpose of this discussion we model a communication network as an unweighted graph $G(V, E)$, whose vertices represent the network switches and whose edges represent the network's communication channels.

A number of architectures for communication networks rely on the concept of *virtual channels (VC's)*. A virtual channel $VC(x, y)$ is a logical connection set up between a pair of switches x, y in the network, such that all traffic between those switches can be transmitted along it. The virtual channel $VC(x, y)$ is constructed by a *connection setup (CS)* procedure in response to a request by one of the two endpoints. Such mechanisms are used in *circuit-switched* telecommunication networks, but also in fast and modern integrated network architectures, such as fiber-optic based ATM networks.

There are two natural "quality-of-service" (QoS) properties commonly associated with connection setup. First, note that a virtual channel is typically set up with the intention of using it for a relatively long period of time. Therefore, it is important that it is as efficient as possible. In particular, it is commonly required that the virtual channel $VC(x, y)$ uses a (hop-wise) shortest route between its endpoints x and y.

The second natural QoS requirement that is often made is that the virtual channel $VC(x, y)$ is established *as fast as possible*, once the request is issued at one of the two endpoints. The speed of connection setup is measured by the usual distributed time complexity measure, namely, the amount of time elapsing in the worst case from the time the connection request was made by x or y until the virtual channel $VC(x, y)$ is established, assuming that each message transmission along an edge takes at most one time unit. Since sending a message along a path of k edges requires at most k time units in the worst case, it is clear that a virtual channel $VC(x, y)$ connecting two switches x and y in the network cannot be set up any faster than the (unweighted) distance between them, $dist(x, y, G)$. Hence it is natural to concentrate on CS procedures that set up a connection in time *exactly $dist(x, y, G)$*.

In view of the above discussion, we concentrate in what follows on *best-service* connection setup procedures, namely, CS procedures providing a virtual channel $VC(x, y)$ of length $dist(x, y, G)$ within time $dist(x, y, G)$. Under these QoS requirements, one may consider a number of complexity measures for evaluating the efficiency of best-service connection-setup procedures. (Here we talk about "system-internal" measures, that are of little of no consequence to the client of the connection service, but affect the overall system performance.) In particular, for a given connection-setup procedure CS, we define the following measures:

1. Quiescence time: the time it takes the procedure CS to terminate (assuming that it might continue to operate even after time $dist(x, y, G)$, by which the virtual channel $VC(x, y)$ is already set up), denoted $T_{CS}(x, y)$.
2. Memory requirements per switch, denoted M_{CS}.

Let us first examine two extreme approaches for the design of a best-service procedure for setting up virtual channels, which differ on their quiescence time and memory requirements. The first method, called the *full tables (FT)* method, is based on storing full routing tables in the network. The quiescence time of this method is just $T_{FT}(x, y) = dist(x, y, G)$. However, its memory requirements are rather large, namely, $M_{FT} = \Theta(n \log n)$ bits per switch.

The other method is based on searching for the shortest path for the virtual channel from scratch upon each request, by applying a shortest-path algorithm (say, by flooding). Call this method the *Bellman-Ford (BF)* method. Here, the memory requirements M_{BF} are minimal (as no special information needs to be stored in advance for this protocol). However, note that the process may continue to work in other parts of the network, even after the connection is set up. Hence the entire process will terminate only after time $T_{BF} = Diam(G)$.

In view of these two extreme methods, it is natural to ask whether there exists an "intermediate" method which achieves close-to-optimal performance in both parameters.

We now propose a new approach for designing a best-service CS procedure, using restricted flooding based on a distance-preserving labeling of the network. This method, named the *label-based (LB)* method, operates as follows. Start by assigning a distance-approximating labeling to the vertices as in Part 2 of Cor. 2. Upon a request for setting up a virtual channel between two vertices x, y, use BF-style flooding for connection setup, but limit the flooding to distance $\hat{d} = \tilde{D}(x, y) \cdot \sqrt{2 \log n}$ (by not allowing the level counter to go beyond this value), where $\tilde{D}(x, y)$ is the estimate obtained out of the labels of x and y to $dist(x, y, G)$. Clearly, the shortest path should be found within this range, given that the $\tilde{D}$ upper bounds the actual distance. Moreover, since $\hat{d}/(2 \log n)$ lower bounds the actual distance, it follows that $T_{LB}(x, y) = dist(x, y, G) \cdot \log n$. The memory requirements are only $M_{LB} = O(\log^3 n)$, as each vertex x needs to store only its own label, $Label(x)$. (Of course, in order to send a message to vertex y, the vertex x must first obtain y's label, $Label(y)$.)

Theorem 5. *The label-based procedure LB achieves best-service connection setup. Its quiescence time and memory requirements are both within a poly-logarithmic factor of the optimum, i.e., $T_{LB}(x, y) = dist(x, y, G) \cdot \log n$ and $M_{LB} = O(\log^3 n)$.*

5   Discussion

This paper introduces the concept of proximity-preserving labeling schemes, and presents an exact scheme for the class of n-vertex weighted trees with M-bit edge weights using $O(M \log n + \log^2 n)$ bit labels, and an approximate scheme for the family of all n-vertex graphs, that using $O(\log^3 n)$ bit labels can provide estimates to the distance, which are accurate up to a factor of $\sqrt{2 \log n}$.

Very recently, the tightness of the labeling scheme presented herein for trees has been established in [7], by proving that any distance labeling scheme for the class of all n-vertex trees with M-bit edge weights must use $\Omega(M \log n + \log^2 n)$ bit labels for some vertices. Some lower bounds have also been established in [7] on the size of the labels in a distance-approximating labeling scheme, where the distance is approximated to a multiplicative factor s. Specifically, it is shown that there exists graphs with n nodes that require labels of size $\Omega(n)$ for every factor $s < \sqrt{3}$.

Our results for general graphs are apparently not optimal in terms of the resulting label size and approximation ratio, and the complexity of extracting the distance (or estimated distance) from two given labels. In particular, we are currently studying an alternative approach for constructing proximity-preserving labeling schemes, based on low-distortion embeddings of grneral metrics in low-dimensional Euclidean spaces [3,11]. It appears that this embedding-based approach can be used to derive labeling systems with properties similar to or even slightly better than the ones presented herein [8], but it also has some limitations. In particular, the best approximation ratio that can be achieved for general graphs using this approach appears to be $\Omega(\log n)$, whereas the method of Section 3 allows any desired approximation ratio $\kappa \leq \log n$. Moreover, the embedding algorithm of [11] is randomized, and while it can probably be derandomized (cf. [9]), it is likely to yield random-looking labels, that reveal little or nothing on the structure of the network. In contrast, the method proposed herein is more direct, and the resulting labels capture a considerable amount of information on the network topology.

This last point may prove particularly significant when considering additional potential applications for proximity-preserving labeling schemes. In particular, an interesting direction for future research, suggested by Richard Tan [12], involves using proximity-preserving labeling schemes for *memory-free* routing. This is a rather strict variant of the extensively studied *compact routing* problem (cf. [2]), which requires us to label the vertices of the network with routing labels in such a way that a message can be routed between any two vertices relying solely on their labels (and the labels of the intermediate vertices along the route), without the need to store any additional information in any of the vertices.

Evidently, since a memory-free routing scheme allows each node to store its own label, it is not really "memory-free." In fact, *every* routing scheme $\mathcal{R}$ can be transformed into a memory-free scheme simply by associating with each vertex v a label $Label(v)$ consisting of a string which contains all the information that needs to be stored at v by the scheme $\mathcal{R}$. Hence in essence, the problem is to come up with routing scheme that require very little memory *per vertex*, rather than just having low total memory requirements.

Indeed, the labeling scheme for trees presented in Sect. 2 can be easily modified to support memory-free routing along shortest paths, still using only $O(\log^2 n)$ bit labels. Similarly, for general graphs, the labeling scheme of Sect. 3

can be modified to support memory-free routing along near-shortest paths, i.e., paths longer than optimal by a prescribed factor. (In fact, as indicated earlier, a result of the latter type can also be derived in a straightforward way using, say, the compact routing schemes of [2].) The development of tighter and more efficient memory-free routing schemes is left for future study.

Acknowledgements

I am grateful to Cyril Gavoille, Yehuda Hassin and Richard Tan for helpful discussions.

References

1. Baruch Awerbuch, Shay Kutten, and David Peleg. On buffer-economical store-and-forward deadlock prevention. In *Proc. INFOCOM*, pages 410–414, 1991.
2. Baruch Awerbuch and David Peleg. Routing with polynomial communication-space trade-off. *SIAM J. on Discr. Math.*, pages 151–162, 1992.
3. J. Bourgain. On Lipschitz embeddings of finite metric spaces in Hilbert spaces. *Israel J. Math.*, pages 46–52, 1985.
4. Melvin A. Breuer. Coding the vertexes of a graph. *IEEE Trans. on Information Theory*, IT-12:148–153, 1966.
5. Melvin A. Breuer and Jon Folkman. An unexpected result on coding the vertices of a graph. *J. of Mathematical Analysis and Applications*, 20:583–600, 1967.
6. Greg N. Frederickson and Ravi Janardan. Space-efficient message routing in *c*-decomposable networks. *SIAM J. on Computing*, pages 164–181, 1990.
7. Cyril Gavoille, David Peleg, Stéphane Pérennes, and Ran Raz. Distance labeling in graphs. In preparation, 1999.
8. Y. Hassin. Private communication, 1999.
9. P. Indyk and R. Motwani. Approximate nearest neighbors: towards removing the curse of dimensionality. In *Proc. 30th ACM Symp. on Theory of Computing*, 1998.
10. Sampath Kannan, Moni Naor, and Steven Rudich. Implicit representation of graphs. In *Proc. 20th ACM Symp. on Theory of Computing*, pages 334–343, May 1988.
11. N. Linial, E. London, and Y. Rabinovich. The geometry of graphs and some of its algorithmic applications. *Combinatorica*, 15:215–245, 1995.
12. R.B. Tan. Private communication, 1997.

Euler Is Standing in Line
Dial-a-Ride Problems with Precedence-Constraints

D. Hauptmeier[1], S. O. Krumke[1*], J. Rambau[1], and H.-C. Wirth[2]

[1] Konrad-Zuse-Zentrum für Informationstechnik Berlin
Department Optimization, Takustr. 7, 14195 Berlin-Dahlem, Germany.
{hauptmeier,krumke,rambau}@zib.de
[2] University of Würzburg
Department of Computer Science, Am Hubland, 97074 Würzburg, Germany.
wirth@informatik.uni-wuerzburg.de

Abstract In this paper we study algorithms for "Dial-a-Ride" transportation problems. In the basic version of the problem we are given transportation jobs between the vertices of a graph and the goal is to find a shortest transportation that serves all the jobs. This problem is known to be NP-hard even on trees. We consider the extension when precedence relations between the jobs with the same source are given. Our results include a polynomial time algorithm on paths and approximation algorithms general graphs and trees with performances of 9/4 and 5/3, respectively.

1 Introduction and Overview

Transportation problems where objects are to be transported between given sources and destinations in a metric space are classical problems in combinatorial optimization. Applications include the routing of pick-up-and-delivery vehicles, the control of automatic storage systems and scheduling of elevators. This leads to the following optimization problem (DARP): We are given transportation jobs between the vertices of a graph and the goal is to find a shortest transportation that serves all the jobs.

A natural extension of DARP is the addition of precedence constraints between the jobs that start at the same vertex. This variant which we call FIFO-DARP is motivated by applications in which first-in-first-out waiting lines are present at the sources of the transportation jobs. In this case, jobs can be served only according to their order in the line. Examples with first-in-first-out lines are cargo elevator systems where at each floor conveyor belts deliver the goods to be transported. Elevators also motivate the restriction of DARP to paths, i.e., to the case where the underlying graph forms a path.

This paper is organized as follows. In Section 2 we formally state the problem FIFO-DARP. We also show that FIFO-DARP can be equivalently formulated as a graph augmentation problem. This key observation will be used to design

* Research supported by the German Science Foundation (DFG, grant Gr 883/5-2)

our algorithms. In Section 4 we prove structural facts about Eulerian cycles in a graph that respect a given "FIFO-order" on the arcs (A formal definition of FIFO-orders appears in Section 2.2. The class of FIFO-orders contains as a special case partial orders which model first-in-first-out waiting lines).

Section 5 contains a polynomial algorithm for FIFO-DARP when restricted to paths. In Section 6 we present an approximation algorithms for general graphs with performance of $(\rho_{\text{TSP}} + 3)/2$, where ρ_{TSP} is the performance of the best approximation for the TSP with triangle inequality. An improved algorithm for trees with performance 5/3 is given in Section 7.

2   Preliminaries and Problem Formulations

A *mixed graph* $G = (V, E, A)$ consists of a set V of vertices, a set E of undirected edges, and a set A of directed arcs (parallel arcs allowed). An edge with endpoints u and v will be denoted by $[u, v]$, an arc from u to v by (u, v). We denote by $n := |V|$, $m_E := |E|$ and $m_A := |A|$ the number of vertices, edges and arcs, respectively. For $v \in V$ we let A_v be the set of arcs in A emanating from v.

If $X \subseteq E \cup A$, then we denote by $G[X]$ the *subgraph of G induced by X*, that is, the subgraph of G consisting of the arcs and edges in X together with their endpoints. Throughout the paper we assume that $G[E]$ is connected and contains all endpoints of arcs from A. The *out-degree* of a vertex v in G, denoted by $d_G^+(v)$, equals the number of arcs in G leaving v. Similarly, the *in-degree* $d_G^-(v)$ is defined to be the number of arcs entering v. If $X \subseteq A$, we briefly write $d_X^+(v)$ and $d_X^-(v)$ instead of $d_{G[X]}^+(v)$ and $d_{G[X]}^-(v)$.

A graph G is called *degree balanced* if $d_G^+(v) = d_G^-(v)$ for all vertices $v \in V$. A *closed walk* in a mixed graph is a cycle which may visit vertices, edges and arcs multiple times. A *directed spanning tree rooted towards $o \in V$* is a subgraph of a directed graph $H = (V, R)$ which is a tree and which has the property that for each $v \in V$ it contains a directed path from v to o.

Since most of the problems under study are NP-hard, we are interested in approximation algorithms for them. Let Π be a minimization problem. A polynomial-time algorithm A is said to be a *ρ-approximation algorithm* for Π, if for every problem instance I of Π with optimal solution value $\text{OPT}(I)$ the solution of value $\text{A}(I)$ returned by the algorithm satisfies $\text{A}(I) \leq \rho \cdot \text{OPT}(I)$.

2.1   Basic Problem

Definition 1 (Dial-a-Ride Problem (Darp)). *The input for DARP consists of a finite mixed graph $G = (V, E, A)$, an origin vertex $o \in V$ and a nonnegative weight function $c \colon E \cup A \to \mathbb{R}_{\geq 0}$. It is assumed that for any arc $a = (u, v) \in A$ its cost $c(a)$ equals the length of a shortest path from u to v in $G[E]$.*

The goal of DARP is to find a closed walk in G of minimum cost which starts in o and traverses each arc in A.

The problem DARP is also known as the Stacker-Crane-Problem. In [7] it is shown that the problem is NP-hard even on trees, i.e., if the graph $G[E]$ is a tree. In [8] the authors present a 9/5-approximation algorithm for the problem on general graphs. An improved algorithm for trees with performance 5/4 is given in [7]. On paths DARP can be solved in polynomial time [3].

An important observation is that DARP can be equivalently formulated as a *graph augmentation problem*. Let $A(E)$ be the set of arcs such that for each undirected edge $e \in E$ the set $A(E)$ contains an anti-parallel pair of arcs between the endpoints of e. We can then extend the cost function c to $A(E)$ in a natural way by defining the cost of an arc in $A(E)$ to be the cost of the corresponding undirected edge in E.

It is straightforward to see that any feasible solution for DARP, i.e., a closed walk that starts in o and traverses each arc in A, corresponds to an augmenting multiset S of arcs from $A(E)$ such that $G[A \cup S]$ is Eulerian and contains o, and vice versa. This enables us to reformulate DARP equivalently as the problem of finding a multiset $S \subseteq A(E)$ minimizing the weight $c(A \cup S)$ subject to the constraint that $G[A \cup S]$ is Eulerian and includes o.

2.2   Precedence Constraints

FIFO-DARP is an extension of DARP. For each vertex $v \in V$ we are additionally given a partial order $\prec_v$ on the arcs in A_v and we require that, whenever $a \prec_v a'$, then a must be traversed before a' in any feasible solution.

A partial order $\prec$ on the arc set of a graph is a *FIFO-order*, if it satisfies: $r \prec r'$ implies that r and r' have the same source. The partial order $\prec$ on the arc set A of $G = (V, E, A)$ resulting from the disjoint union of the partial orders $\prec_v$ is clearly a FIFO-order. In the sequel $\prec$ is extended to $A \cup A(E)$ by defining that arcs from $A(E)$ are incomparable to each other and to those of A.

It is easy to give examples where an optimal FIFO-respecting transportation is strictly longer than the optimal transportation neglecting the precedences. In the special case that for a given FIFO-order $\prec$ the restriction of $\prec$ to each arc set A_v is total, then $\prec$ models first-in-first-out waiting lines.

Again, FIFO-DARP can be reformulated as a graph augmentation problem. To do this, we need some additional notations:

Definition 2 ($\prec$-respecting Eulerian Cycle, $\prec$-Eulerian). *Let $H = (V, R)$ be a directed graph, $\prec$ be a FIFO-order on the arcs R, and $o \in V$. A $\prec$-respecting Eulerian cycle in H with start o is a Eulerian Cycle C in G such that $a \prec a'$ implies that in the walk from o along C the arc a appears before a'. The graph H is then called $\prec$-Eulerian with start o.*

Notice that in contrast to the case of classical Eulerian cycles, for $\prec$-respecting Eulerian cycles it is meaningful to specify a start node explicitly.

Definition 3 (Graph Augmentation Version of Fifo-Darp). *An instance of the problem FIFO-DARP consists of the same input as for DARP and additionally a FIFO-order $\prec$ on the arc set A. The goal is to find a multiset S of arcs from $A(E)$ minimizing the weight $c(A \cup S)$ such that $G[A \cup S]$ is $\prec$-Eulerian with start o and to determine a $\prec$-respecting Eulerian cycle in $G[A \cup S]$.*

In the sequel we consider FIFO-DARP as a graph augmentation problem. We use S^* to denote an optimal solution and OPT $:= c(A \cup S^*)$ to denote its cost. Notice that if C^* is a $\prec$-respecting Eulerian cycle in $G[A \cup S^*]$ with start o, then the length of C^* is equal to OPT.

Since FIFO-DARP generalizes DARP, it follows from [7] that FIFO-DARP is NP-hard even on trees. However, we can strengthen the hardness result of [7] and show that DARP is hard on caterpillar graphs. A caterpillar graph is a tree of maximum degree three, where all non-leaf nodes lie on the same path. Due to lack of space we omit the proof of the following theorem.

Theorem 4. DARP *and* FIFO-DARP *on caterpillars are NP-hard. This continues to hold, if the transportation jobs are restricted to have sources and targets only in the feet (i.e., leaves) of the caterpillar. Furthermore, all hardness results for* FIFO-DARP *remain true if the FIFO-ordering is restricted to be total.* □

3 Basic Observations and Balancing

We first start with some technical assumptions about input instances $(G = (V, E, A), c, o, \prec)$ of FIFO-DARP. While all these assumptions are without loss of generality they greatly simplify the presentation of our algorithms.

Assumption 5 (Technical assumption for Fifo-Darp on trees). Each vertex of degree one or two in $G[E]$ is either the origin o or incident to at least one arc from A.

This assumption is indeed no restriction: Let $v \neq o$ be not adjacent to any arc in A. If v is of degree two, replace the two adjacent edges by one edge of cost equal to the sum of the two edges. If v is a leaf, it can be removed without affecting the optimal solution (cf. [7] for DARP on trees).

If $G[E]$ is a path it is easy to see that we can make an even stronger assumption without loss of generality (cf. [3] for DARP on paths):

Assumption 6 (Technical assumption for Fifo-Darp on paths). Each vertex $v \in V$ is incident to at least one arc from A.

Assumption 7 (Technical assumption for Fifo-Darp on general graphs).
  (i) Each vertex $v \in V$ is incident to at least one arc from A.
  (ii) $G[E]$ is complete and the cost function c obeys the triangle inequality, i.e., for any edge $[u, v] \in E$ the cost $c(u, v)$ does not exceed the length of a shortest path in $G[E]$ between u and v.

Note that Assumption 7 can be enforced without increasing the value of an optimal solution. If the start vertex o is not incident to any arc from A we can add a new vertex o', a new arc (o, o') and a new edge $[o, o']$, each of cost zero. The new vertex o' is joined by undirected edges to all neighbors v of o. The cost of an edge $[o', v]$ is set to $c(o, v)$. We can then remove vertices which are not incident on an arc. For every pair u and v of vertices we insert new edges of cost equal to the shortest path in $G[E]$ between u and v. (cf. [8] for DARP).

Notice that Assumption 7 can not be made without loss of generality for FIFO-DARP on trees, since removing vertices and later completing the graph as described in general destroys the "tree-property".

An important concept for tackling FIFO-DARP on paths and trees is that of *balancing*.

Definition 8 (Balancing set). *Let $G = (V, E, A)$ be a mixed graph. A multiset $B \subseteq A(E)$ of arcs is called a balancing set if in $H = G[A \cup B]$ we have $d_H^+(v) = d_H^-(v)$ for all vertices v of H.*

Suppose that $G[E]$ is a tree and that Assumption 5 is satisfied. Any edge $[x, y] \in E$ defines a partition $V = X \cup Y$ of the node set. The cut (X, Y) must be traversed by a closed walk W traversing each arc the same number of times in each direction. Denote by $\phi(X, Y) := |\{ (x, y) \in A \mid x \in X, y \in Y \}|$ the number of arcs emanating from X. Hence, W must traverse edge $[x, y]$ from x to y at least $b(x, y)$ times, where

$$b(x, y) := \begin{cases} 1 & \text{if } \phi(X, Y) = \phi(Y, X) = 0 \\ \phi(Y, X) - \phi(X, Y) & \text{if } \phi(Y, X) > \phi(X, Y) \\ 0 & \text{otherwise.} \end{cases}$$

This observation has the following consequence for the graph augmentation version: If $B \subseteq A(E)$ is a multiset of arcs such that B contains exactly $b(x, y)$ copies of the directed arc (x, y), then there is at least one optimal solution S^* such that $B \subseteq S^*$. This yields the following lemma which is proved in [3,7].

Lemma 9. *Let (G, c, o) be an instance of DARP such that $G[E]$ is a tree. Then in time $\mathcal{O}(nm_A)$ one can find a balancing set $B \subseteq A(E)$ such that $B \subseteq S^*$ for some optimal solution S^*.*   □

Notice also that Lemma 9 remains valid even in the presence of FIFO-orders. As is also shown in [3,7] the time bound of $\mathcal{O}(nm_A)$ can be improved to $\mathcal{O}(n + m_A)$ by allowing balancing arcs to be from $V \times V$ instead of just $A(E)$ (which does not change the problem: the cost function c is extended from $A(E)$ to $V \times V$ by the length of shortest paths).

4   Euler Tours Respecting FIFO-Orders

Let C be an Eulerian cycle starting at o in a connected directed graph. We define the *set of last arcs* of C, denoted by L, to contain for each vertex $v \in V$ the unique arc emanating from v which is traversed last by C. Observe that L contains a directed spanning tree rooted towards o.

Let $\prec$ be a FIFO-order. We denote the set of maximal elements with respect to $\prec$ by $M_\prec$, that is, $M_\prec := \{ a \in A : \text{there is no arc } a' \text{ such that } a \prec a' \}$.

Definition 10 (Possible set of last arcs). *Let $H = (V, R)$ be a directed graph and $o \in V$ be a distinguished vertex. A set $L \subseteq R$ is called a* possible set of last arcs, *if it satisfies the following conditions:*

    (i) $d_L^+(v) = 1$ for all $v \in V$, and
    (ii) for each $v \in V$ there is a path from v to o in $H[L]$.

Theorem 11. *Let $H = (V, R)$ be a directed Eulerian graph with distinguished vertex $o \in V$ and let $L \subseteq R$ be a possible set of last arcs.*

    (i) There exists an Eulerian cycle C in H such that for each vertex $v \in V$ the (unique) arc from L emanating from v is traversed last at v by C.
    (ii) Let $\prec$ be any FIFO-order with $L \subseteq M_\prec$. Then there exists a $\prec$-respecting Eulerian cycle with start o in H. This cycle can be found in time $\mathcal{O}(|V| + |R|)$.

Proof. We first show (i). Color the arcs from L red and the arcs in $R \setminus L$ blue. We claim that by the following procedure we construct an Eulerian cycle C in H with the desired properties. Start with current vertex o. If possible, choose an arbitrary (but yet untraversed) blue arc emanating from the current vertex, otherwise choose the red arc. Traverse the arc, let its target be the new current vertex, and repeat the iteration. Stop, if there is no untraversed arc emanating from the current vertex. Call the resulting path of traversed arcs C. Since H is Eulerian by assumption, for each vertex its in-degree equals its out-degree. Therefore, C must end in the origin o and forms in fact a cycle.

We show that there is no arc in H which is not traversed by C. For a node $v \in V$, let dist(v, o) be the distance (i.e., the number of arcs) on the shortest path from v to o in the subgraph $H[L]$. We show by induction on dist(v, o) that all arcs emanating from v are contained in C.

If dist$(v, o) = 0$ then $v = o$. Since our procedure stopped, all arcs emanating from o are contained in C. This proves the induction basis. Assume that the claim holds true for all vertices with distance $t \geq 0$ and let $v \in V$ with dist$(v, o) = t+1$. Let $a = (v, w)$ be the unique red arc emanating from v. Then dist$(w, o) = t$ and by the induction hypothesis all arcs emanating from w are contained in C. For $d_H^+(w) = d_H^-(w)$, it follows that all arcs entering w, in particular arc a, are also contained in C. Since red arc a is chosen last by our procedure, all other arcs emanating from v must be contained in C. This completes the induction. Hence, C is actually an Eulerian cycle with the claimed properties.

We proceed to show (ii). Analogously to (i) construct a Eulerian cycle with the sole difference that at each node v we choose the next arc according to the $\prec$-constraint at v. Since by assumption $L \subseteq M_\prec$ this yields a valid $\prec$-respecting Eulerian cycle with start o.   □

Corollary 12. *Let $H = (V, R)$ be a graph, $o \in V$ and $\prec$ a FIFO-order. Then the following two statements are equivalent:*

1. H is $\prec$-Eulerian with start o.

2. *H is Eulerian and the set $M_\prec$ of maximal elements with respect to $\prec$ contains a possible set of last arcs.*   □

Observe that Corollary 12 in fact implies a polynomial time algorithm for deciding whether a given graph H is $\prec$-Eulerian with start o. Provided H is Eulerian it suffices to check whether the subgraph formed by the arcs from $M_\prec$ contains a directed spanning tree D rooted towards o (which can be done in linear time). Adding to D an arbitrary arc from $A_o \cap M_\prec$ then yields indeed a possible set of last arcs.

5   A Polynomial Time Algorithm for Fifo-Darp on Paths

Let $G = (V, E, A)$ be a mixed graph such that $G[E]$ is a path. We assume throughout this section that Assumption 6 holds.

Lemma 13. *The set $B \cup N$ returned by Algorithm Alg-Path is a feasible solution for* FIFO-DARP, *i.e., $G[A \cup B \cup N]$ is $\prec$-Eulerian with start o.*

Proof. Since $G[A \cup B]$ is degree balanced and N consists of pairs of anti-parallel arcs, even $G[A \cup B \cup N]$ is degree balanced. By construction, it contains a directed spanning tree rooted towards o. Hence it is strongly connected and Eulerian.

The set L of arcs determined in Step 6 is clearly a possible set of last arcs. The claim now follows from Theorem 11 (ii).   □

Theorem 14. *Algorithm Alg-Path finds an optimal solution.*

Proof. Let S^* be an optimal solution such that $B \subseteq S^*$. By feasibility of S^* the graph $G[A \cup S^*]$ is $\prec$-Eulerian with start o.

Consider the multi-set $Z := (A \cup S^*) \setminus (A \cup B) = S^* \setminus B$. Since $G[A \cup B]$ and $G[A \cup S^*] = G[A \cup B \cup Z]$ are degree balanced and $Z \cap (A \cup B) = \emptyset$, we can decompose the set Z into arc disjoint cycles. Since Z consists of (multiple copies of) arcs from $A(E)$ and $G[E]$ is a tree it follows that $r \in Z$ implies that $r^{-1} \in Z$.

Let C be a $\prec$-respecting Eulerian cycle in $G[A \cup S^*]$ and let L be its last set of arcs. Notice that $L \subseteq B \cup M_\prec \cup Z$, where $M_\prec$ is defined in Step 2 of the algorithm. The set L must contain a directed spanning tree D' rooted towards o. We partition D' into the sets $D'_{B \cup M_\prec} := D' \cap (B \cup M_\prec)$ and $D'_Z := D' \cap Z$. Thus, $c'(D'_{B \cup M_\prec}) = 0$ and $c'(D'_Z) = c(D'_Z)$. Since we have seen that for each arc $r \in Z$ also its anti-parallel version $r^{-1} \in Z$ (and D'_Z does not contain a pair of anti-parallel arcs) we get that

$$c(Z) \geq 2c(D'_Z) = 2c'(D'_Z) + 2c'(D'_{B \cup M_\prec}) = 2c'(D') \geq 2c'(D). \qquad (1)$$

Here, D is the directed spanning tree of minimum weight computed in Step 4. The set N computed in Step 5 has cost

$$c(N) = 2c(D \setminus (B \cup M_\prec)) = 2c'(D \setminus (B \cup M_\prec)) = 2c'(D) \overset{(1)}{\leq} c(Z). \qquad (2)$$

Input: A mixed graph $G = (V, E, A)$, such that $G[E]$ is a path, a cost function c on E, an initial vertex $o \in V$, and a FIFO-order $\prec$
1 Compute a balancing set $B \subseteq A(E)$ such that $B \subseteq S^*$ for some optimal solution S^*.
2 Let $M_\prec$ be the set of maximal elements with respect to $\prec$.
3 Set $H = G[B \cup M \cup A(E)]$ with cost function c' on the arcs defined by

$$c'(r) = \begin{cases} 0 & \text{if } r \in B \cup M_\prec \\ c(r) & \text{if } r \in A(E) \setminus (B \cup M_\prec) \end{cases}$$

4 Compute a directed spanning tree D rooted towards o of minimum weight $c'(D)$ in $G[B \cup M_\prec \cup A(E)]$.
5 Set $N := \emptyset$. For each directed arc $r \in D$ which is not in $B \cup M_\prec$, add r and its anti-parallel r^{-1} to N.
6 Define $L := D \cup \{r\}$, where r is an arbitrary arc from $A_o \cap (M_\prec \cup B)$
   {Notice that such an arc must exist since o is source or target of at least one job and $G[A \cup B]$ is degree-balanced.}
7 Use the method from Theorem 11 to find a $\prec$-respecting Eulerian cycle C with start o in $G[A \cup B \cup N]$ such that L is the last set of arcs of C.
8 **return** the set $B \cup N$ and the cycle C.

Algorithm 1: Algorithm Alg-Path for FIFO-DARP on paths.

Using this result yields that

$$c(A \cup B \cup N) = c(A \cup B) + c(N) = c(A \cup (S^* \setminus Z)) + c(N)$$
$$\overset{(2)}{\leq} c(A \cup (S^* \setminus Z)) + c(Z) = c(A \cup S^*).$$

Thus, $B \cup N$ is an optimal solution as claimed.   □

   We briefly comment on the running time of Algorithm Alg-Path. A balancing set B can be found in time $\mathcal{O}(nm_A)$ by techniques as shown in [3]. As noted before this time bound can be improved to $\mathcal{O}(n + m_A)$ by allowing balancing arcs to be from $V \times V$ instead of just $A(E)$. A rooted spanning tree of minimum weight in a graph with n vertices and m arcs can be computed in time $\mathcal{O}(\min\{m \log n, n^2\})$ by the algorithm from [12]. Thus Algorithm Alg-Path can be implemented to run in time $\mathcal{O}(n + m_A + \min\{(m_A + n) \log n, n^2\})$.

6   An Approximation Algorithm for General Graphs

In this section we present our approximation algorithm for FIFO-DARP on general graphs. The algorithm uses ideas similar to the ones in [8]. In this section we will assume tacitly that Assumption 7 is satisfied.

   Our algorithm actually consists of *two* different sub-algorithms, Alg-TSP and Alg-Last-Arcs, which are run both and the best solution is picked. The first sub-algorithm, Alg-TSP, is extremely simple: It computes a shortest tour which visits

---

Input: A mixed graph $G = (V, E, A)$, a cost function c on E, an initial
     vertex $o \in V$, and a FIFO-order $\prec$
1 Let V_s be the set of vertices which are sources of arcs from A.
2 Compute a complete undirected auxiliary graph U with vertex set V_s. The weight
   $d(v, w)$ of edge $[v, w]$ is set to be the length of a shortest path in $G[E]$ from v
   to w.
3 Find an approximately shortest TSP tour P in U starting and ending in o. Let the
   order in which the vertices of V are visited by P be $v_0 = o, v_1, \ldots, v_{|V_s|}, v_{|V_s|+1} =$
   o.
4 Construct a feasible tour C for FIFO-DARP as follows:
5 Start with the empty tour C.
6 **for** $i := 0, \ldots, |V_s|$ **do**
7   Let $\{a_1, \ldots, a_k\} \leftarrow A_{v_i}$. Set $C \leftarrow C + (a_1, p_1, \ldots, a_k, p_k)$, where p_j is a shortest
      path in $G[E]$ from the endpoint of a_j to v_i.
8   Append to C the shortest path in $G[E]$ from v_i to v_{i+1}.
9 **end for**
10 Let S be the multi-set of directed edges used in C which are not contained in A.
11 **return** the set S and the cycle C.

Algorithm 2: TSP-based Approximation Algorithm Alg-TSP for FIFO-DARP.

Input: A mixed graph $G = (V, E, A)$, a cost function c on E, an initial
     vertex $o \in V$, and a FIFO-order $\prec$
1 Compute a balancing multiset $B \subseteq A(E)$ of minimum cost.
2 Follow steps 2 to 7 of Algorithm Alg-Path to compute a set N of arcs and a
   $\prec$-respecting Eulerian cycle C with start o.
3 **return** the set $B \cup N$ and the cycle C

Algorithm 3: Algorithm Alg-Last-Arcs "mimicking" the algorithm for paths.

each vertex from which emanates an arc at least once. Then, it uses this TSP-tour to obtain a feasible solution for FIFO-DARP in the most obvious way. The algorithm is displayed in Algorithm 2.

Lemma 15. *If in Step 3 a ρ_{TSP}-approximation algorithm for computing a TSP-tour is employed, then Alg-TSP finds a solution of cost at most $\rho_{\mathrm{TSP}} \cdot OPT + 2c(A)$.*

Proof. Let S^* be an optimum augmenting set and C^* be a $\prec$-respecting Eulerian cycle in $G[A \cup S^*]$ starting at o. Since C^* visits all vertices from V_s, the length of C^* (which equals OPT) is at least that of a shortest TSP-tour on the vertices V_s. Thus, the tour computed in Step 3 will have length at most $\rho_{\mathrm{TSP}} \cdot OPT$. The additional cost incurred in Step 7 is not greater than $2c(A)$, since each path added has weight not greater than the corresponding arc from A. □

Our second algorithm, Alg-Last-Arcs, is based on similar ideas as the algorithm from Section 5 for paths and is shown in Algorithm 3.

By a proof similar to Lemma 13 it follows that the set $B \cup N$ found by Algorithm Alg-Last-Arcs is indeed a feasible solution.

Lemma 16. *The balancing set B found in Step 1 of algorithm Alg-Last-Arcs has cost at most $OPT - c(A)$. Step 1 can be accomplished in the time needed for one minimum cost flow computation on a graph with n vertices and $2m_E$ arcs.*

Proof. Let $S^* \subseteq A(E)$ be an optimal solution, i.e., an augmenting (multi-) set of arcs from $A(E)$ with minimum cost. Then the graph $G[A \cup S^*]$ is $\prec$-Eulerian with start o. Thus, the addition of the arcs from S^* turns G Eulerian, in particular degree balanced. Thus, the cost $c(S^*) = OPT - c(A)$ is at least that of a minimum cost set $B \subseteq A(E)$ which achieves the degree balance.

Step 1 can be carried out by performing a minimum cost flow computation in the auxiliary graph $F = (V, A(E))$. A vertex v has charge $d_G^-(v) - d_G^+(v)$ and the cost of sending one unit of flow over arc $r \in A(E)$ equals its cost $c(r)$. We then compute an integral minimum cost flow in F. If the flow on an arc r is $t \in \mathbb{N}$, we add t copies of arc r to the set B. □

Lemma 17. *The cost of N computed by Alg-Last-Arcs is at most $2(OPT - c(A))$.*

Proof. The proof of the lemma is similar to the one for Theorem 14. The major difference is that in general we can not assure that the balancing set B computed in Step 1 is a subset of an optimal solution.

Let S^* be again an optimal augmenting set and L be the set of last arcs of a $\prec$-respecting Eulerian cycle in $G[A \cup S^*]$. We can find a directed spanning tree rooted towards o in L. The only arcs from A that L can contain are those from the set $M_\prec$. Thus $L \setminus (A \cup B) = L \setminus (M_\prec \cup B)$. Similar to Theorem 14 we can now conclude that

$$OPT - c(A) = c(S^*) \geq c(L \setminus (A \cup B)) = c(L \setminus (M_\prec \cup B)) = c'(L) \geq c(N)/2.$$

This shows the claim. □

Corollary 18. *Alg-Last-Arcs finds a solution of cost at most $3 \cdot OPT - 2c(A)$.*

We are now ready to combine our algorithms into one with an improved performance guarantee. The combined algorithm Alg-Combine simply runs both algorithms and picks the better solution.

Theorem 19. *Algorithm Alg-Combine has a performance of $\frac{1}{2}(\rho_{\mathrm{TSP}} + 3)$.*

Proof. Let $\beta := \frac{4}{3 - \rho_{\mathrm{TSP}}}$. If $OPT \leq \beta c(A)$, then the solution returned by Alg-TSP has cost at most

$$(\rho_{\mathrm{TSP}} + 2/\beta)\, OPT = \left(\rho_{\mathrm{TSP}} + 2\,\frac{3 - \rho_{\mathrm{TSP}}}{4}\right) OPT = \frac{\rho_{\mathrm{TSP}} + 3}{2} \cdot OPT.$$

If OPT $> \beta c(A)$, then the cost of the solution found by Alg-Last-Arcs is bounded from above by

$$(3 - 2/\beta)\,\mathrm{OPT} = \left(3 - 2\,\frac{3 - \rho_{\mathrm{TSP}}}{4}\right)\mathrm{OPT} = \frac{\rho_{\mathrm{TSP}} + 3}{2} \cdot \mathrm{OPT}.$$

This shows the claim of the theorem.   □

Using Christofides' algorithm [4] with $\rho_{\mathrm{TSP}} = 3/2$ results in a performance guarantee of $3/4 + 3/2 = 9/4$ for algorithm Alg-Combine.

Corollary 20. *There is an approximation algorithm for* FIFO-DARP *with performance* 9/4. *This algorithm can be implemented to run in time* $\mathcal{O}(\max\{n^3 + m_A m_E + m_A n \log n, m_E^2 \log n + m_E n \log^2 n\})$.

Proof. The performance has already been proved. The running time of Algorithm Alg-TSP is dominated by that of Christofides' algorithm, which can be implemented to run in time $\mathcal{O}(n^3)$, and the time needed for the addition of the paths in Step 7 which can be done in total time $\mathcal{O}(m_A m_E + m_A n \log n)$. The running time of Alg-Last-Arcs is dominated by the minimum cost flow computation which can be accomplished in time $\mathcal{O}(m_E^2 \log n + m_E n \log^2 n)$ by using Orlin's enhanced capacity scaling algorithm [1].   □

7   Improved Approximation Algorithm on Trees

For graph classes where the TSP can be approximated within a factor better than 3/2, the performance improves over the one stated in Corollary 20. In particular, for trees where the TSP can be solved in polynomial time Theorem 19 already implies a 2-approximation algorithm. However, we can still improve this performance guarantee.

Theorem 21. *There exists a polynomial time approximation algorithm for* FIFO-DARP *on trees with performance* 5/3. *This algorithm can be implemented to run in time* $\mathcal{O}(nm_A + n^2 \log n)$.

Proof. Our algorithm for trees uses a modified version of Alg-Last-Arcs. We defer removal of the vertices in V which are neither start nor endpoint of an arc from A and the completion of G via shortest paths until after the (modified) balancing step. The balancing step Step 1 of Alg-Last-Arcs is modified so that we find a balancing subset $B \subseteq S^*$ as in Lemma 9. After the balancing we remove all vertices which are not incident to the arcs in $A \cup B$ and continue with Alg-Last-Arcs from Step 2 on.

Let $I = (G = (V, E, A), c, o, \prec)$ be the original instance given such that $G[E]$ is a tree. We can consider the instance $I' = (G = (V, E, A \cup B), c, o, \prec)$ of FIFO-DARP (still on a tree) which results from adding the balancing arcs B as new transportation jobs. Since any feasible solution to I will have to use the arcs from B anyway (cf. Lemma 9), we get that $\mathrm{OPT}(I) = \mathrm{OPT}(I')$.

Now look at the instance I'' of FIFO-DARP which is obtained by removing vertices and completing G along shortest paths as in our algorithm. It is easy to see that $\mathrm{OPT}(I'') = \mathrm{OPT}(I')$. Notice also that we can transform any feasible solution of I'' to a feasible solution of I' (by replacing arcs not in $A(E)$ by shortest paths). Let S^* and S'' be optimal solutions for I and I'', respectively. Define $Z := S^* \setminus B$ and $Z'' := S'' \setminus B$. Since $\mathrm{OPT}(I) = c(A \cup B) + c(Z) = \mathrm{OPT}(I'') = c(A \cup B) + c(Z'')$, we have that $c(Z) = c(Z'')$.

Let $A \cup B \cup N$ be the solution found by the modified version of Alg-Last-Arcs. Then, using the arguments of Lemma 17 we get that

$$
\begin{aligned}
c(A \cup B \cup N) &= c(A \cup B) + c(N) = c(S^*) - c(Z) + c(N) \\
&\le c(S^*) - c(Z) + 2c(Z'') \le c(S^*) + c(Z) \\
&= 2 \cdot \mathrm{OPT}(I) - c(A).
\end{aligned}
$$

As noted before, Alg-TSP finds a solution of cost at most $\mathrm{OPT} + 2c(A)$, since we can solve the TSP on the tree $G[E]$ in polynomial time. We can estimate the cost of the best of the two solutions returned by Alg-TSP and the modified Alg-Last-Arcs by the techniques from Theorem 19 where this time $\beta = 3$. This yields a performance of $5/3$ as claimed.

The time bound for the algorithm is derived as follows: We can solve the TSP on the metric space induced by $G[E]$ in time $\mathcal{O}(n)$. We then root the tree $G[E]$ at an arbitrary vertex. With $\mathcal{O}(n)$ preprocessing time, the least common ancestor of any pair of vertices can be found in constant time (see [10,11]). Thus, we can implement Alg-TSP in such a way that the invocations of Step 7 take total time $\mathcal{O}(nm_A)$. This means that Alg-TSP can be implemented to run in time $\mathcal{O}(nm_A)$.

The balancing in the modified version of Alg-Last-Arcs can be accomplished in time $\mathcal{O}(n + m_A)$. Completion of the graph by computing all-pairs shortest paths can be done in time $\mathcal{O}(nm_E + n^2 \log n) = \mathcal{O}(n^2 \log n)$ [5,1]. All other steps can be carried out in time $\mathcal{O}(n^2)$ where again the algorithm from [12] is employed for computing a minimum weight directed spanning tree. □

References

1. R. K. Ahuja, T. L. Magnanti, and J. B. Orlin, *Networks flows*, Prentice Hall, Englewood Cliffs, New Jersey, 1993.
2. N. Ascheuer, M. Grötschel, S. O. Krumke, and J. Rambau, *Combinatorial online optimization*, Proceedings of the International Conference of Operations Research (OR'98), Springer, 1998, pp. 21–37.
3. M. J. Atallah and S. R. Kosaraju, *Efficient solutions to some transportation problems with applications to minimizing robot arm travel*, SIAM Journal on Computing **17** (1988), no. 5, 849–869.
4. N. Christofides, *Worst-case analysis of a new heuristic for the traveling salesman problem*, Tech. report, Graduate School of Industrial Administration, Carnegie-Mellon University, Pittsburgh, PA, 1976.
5. T. H. Cormen, C. E. Leiserson, and R. L. Rivest, *Introduction to algorithms*, MIT Press, 1990.

6. G. N. Frederickson and D. J. Guan, *Preemptive ensemble motion planning on a tree*, SIAM Journal on Computing **21** (1992), no. 6, 1130–1152.

7. G. N. Frederickson and D. J. Guan, *Nonpreemptive ensemble motion planning on a tree*, Journal of Algorithms **15** (1993), no. 1, 29–60.

8. G. N. Frederickson, M. S. Hecht, and C. E. Kim, *Approximation algorithms for some routing problems*, SIAM Journal on Computing **7** (1978), no. 2, 178–193.

9. D. J. Guan, *Routing a vehicle of capacity greater than one*, Discrete Applied Mathematics **81** (1998), no. 1, 41–57.

10. D. Harel and R. E. Tarjan, *Fast algorithms for finding nearest common ancestors*, SIAM Journal on Computing **13** (1984), no. 2, 338–355.

11. B. Schieber and U. Vishkin, *On finding lowest common ancestors: Simplification and parallelization*, SIAM Journal on Computing **17** (1988), no. 6, 1253–1262.

12. R. E. Tarjan, *Finding optimum branchings*, Networks **7** (1977), 25–35.

Lower Bounds for
Approximating Shortest Superstrings
over an Alphabet of Size 2

Sascha Ott

Lehrstuhl für Informatik I
(Algorithmen und Komplexität)
RWTH Aachen, Ahornstr. 55,
52056 Aachen, Germany
sott@i1.informatik.rwth-aachen.de

Abstract. The shortest common superstring problem (SCS) is known
to be NP-hard and APX-hard. The APX-hardness was proved for the
SCS in [BJLTY94], but the reduction used in that paper produces instances with arbitrarily large alphabets. We show that the problem is
APX-hard even if the size of the alphabet is 2.
A lot of results concerning approximation algorithms have been published. We use our result to establish the first explicit inapproximability
results for the SCS.

Keywords: Superstrings, approximability, APX-hardness, lower bounds

1 Introduction

The shortest common superstring problem (SCS) is to compute a shortest string
that contains every string in a given set of strings as a substring. The SCS is
NP-hard even if strong restrictions are made, for example:
(i) all strings are primitive (i.e. no character appears more than once in a string)
and of length $H \geq 3$ [GMS80],
(ii) all strings have length 3 and the maximal orbit size is 8 (i.e. no character
appears more than 8 times) [Midd94],
(iii) all strings are of the form 10^p10^q, $p, q \in \mathbb{N}$ [Midd98].

Furthermore, the problem was shown to be APX-hard in [BJLTY94]. The
L-reduction constructed in that paper produces SCS instances with arbitrarily
large alphabets. More precisely, a given instance of a special version of the TSP
(see below) with n vertices is transformed to an SCS instance over an alphabet
with $2n + 1$ characters.

We prove that the SCS is APX-hard even if the alphabet contains just 2
characters.

This result is of special interest since the SCS has important applications in
DNA sequencing [Li90]. In this context, an alphabet of size 4 is used, namely
$\{A, C, G, T\}$. The characters represent four different nucleotides, of which the

Widmayer et al. (Eds.): WG'99, LNCS 1665, pp. 55–64, 1999.
© Springer-Verlag Berlin Heidelberg 1999

DNA molecules consist. Strings over this alphabet represent pieces of DNA molecules, which have to be put together. The difficulty in DNA sequencing is that the pieces are from several copies of the DNA molecule of interest such that they may overlap. Furthermore, one has usually no idea from which part of the molecule the pieces are.

The SCS also has applications in data compression as discussed in [Stor88]. A large amount of research work has been spent on approximation algorithms for the SCS. We name only two examples for such results, which give the best known upper bounds to date. [Swee99] gives an 2.5-approximation algorithm for shortest superstrings w.r.t. the length measure (defined below). An $\frac{63}{38}$-approximation algorithm w.r.t. the compression measure can be found in [KPS94].

Building on the work of [Eng99], our APX-hardness results imply the following inapproximability results. W.r.t. the length measure, the lower bound $1 + \frac{1}{17245}$ is established, and w.r.t. the compression measure the lower bound $1 + \frac{1}{11216} - \varepsilon$ (for every $\varepsilon > 0$) is proved.

The paper is organized as follows. We give some basic definitions in Section 2. We prove the APX-hardness result in Section 3 and derive the lower bounds in Section 4.

2   Basic Definitions

Let Σ be an alphabet. For $s \in \Sigma^+$, we use $|s|$ to denote the length of s. Given $S \subset \Sigma^+$, we define the length of S as $\|S\| = \sum_{i=1}^{n} |s_i|$.

The SCS is the following NPO problem. The input consists of a non-empty set Σ (the alphabet), $n \in \mathbb{N}_{\geq 1}$ and a finite set $S = \{s_1, \ldots, s_n\} \subset \Sigma^+$. The task is to compute a string $w \in \Sigma^+$ such that each $s_i \in S$ is a substring of w, i.e. $w = x^\wedge s_i{}^\wedge y$ for some strings x, y. (We use the symbol $^\wedge$ to denote the concatenation.) A solution w for S can be measured in two different ways: the length measure and the compression measure. Let $opt(S)$ denote the length of shortest superstrings for S and let $comp(S)$ abbreviate $\|S\| - opt(S)$. If the length measure is used, one considers $\frac{|w|}{|w_{opt}|}$ as the approximation ratio of w, using the compression measure means to consider $\frac{comp(S)}{\|S\| - |w|}$ as the approximation ratio[1].

Given strings s and t, we use $pref(s,t)$ to denote the shortest string x such that $s = x^\wedge u$ and $t = u^\wedge v$ for some strings u, v. $|pref(s,t)|$ is called the *distance from s to t*. The *overlap of s and t* (in this order), denoted by $ov(s,t)$, is defined as the longest string x such that $s = u^\wedge x$ and $t = x^\wedge v$ for some strings u, v.

The edge-weighted directed graph $G_O(S) = (V, E, o)$ with $V = \{1, \ldots, |S|\}$, $E = V^2 - I$ and $o : E \to \mathbb{N}$ defined by $o(i,j) = |ov(s_i, s_j)|$ is called the *overlap graph of S*.

Let $s_1, \ldots, s_n$ be strings. Defining $\langle s_1, \ldots, s_n \rangle$ as

$$pref(s_1, s_2)^\wedge pref(s_2, s_3)^\wedge \ldots {}^\wedge pref(s_{n-1}, s_n)^\wedge s_n$$

---

[1] W.l.o.g. we assume $\|S\| - |w| > 0$.

is a simple method of constructing superstrings. Note that for any SCS input $S = \{s_1, \ldots, s_n\}$ there exists a permutation $\pi : \{1, \ldots, n\} \to \{1, \ldots, n\}$ such that $\langle s_{\pi(1)}, \ldots, s_{\pi(n)} \rangle$ is an optimal solution for S. Similarly to the definition of overlap graphs based on $ov(\cdot, \cdot)$, one can define *distance graphs* based on $pref(\cdot, \cdot)$. Distance graphs fulfill the triangle inequality. Thus, optimal solutions for the asymmetric TSP with triangle inequality yield solutions for the SCS, which are at most $min\{|s_i| \mid s_i \in S\}$ characters longer than $opt(S)$. In [Turn89], a relation between the SCS and the path version of the TSP is shown.

We need the notion of AP-reducibility in this paper. We use the one of [CKST96]. Let $A = (I_A, sol_A, m_A, type_A)$, $B = (I_B, sol_B, m_B, type_B)$ be two NPO problems where I_A denotes the set of instances of A, $sol_A(x)$ $(x \in I_A)$ is the set of feasible solutions, $m_A(x, y)$ $(y \in sol_A(x))$ is the positive integer measure of y, and $type_A \in \{max, min\}$.

A is said to be *AP-reducible* to B, if functions f, g, t_f, and t_g and a positive constant α exist such that:

1. For any $x \in I_A$ and for any $r > 1$, $f(x, r) \in I_B$ is computable in time $t_f(|x|, r)$.
2. For any $x \in I_A$, for any $r > 1$, and for any $y \in sol_B(f(x, r))$, $g(x, y, r) \in sol_A(x)$ is computable in time $t_g(|x|, |y|, r)$.
3. For any fixed r, both $t_f(\cdot, r)$ and $t_g(\cdot, \cdot, r)$ are bounded by a polynomial.
4. For any fixed n, both $t_f(n, \cdot)$ and $t_g(n, n, \cdot)$ are non-increasing functions.
5. For any $x \in I_A$, for any $r > 1$, and for any $y \in sol_B(f(x, r))$, $R_B(f(x, r), y) \leq r$ implies $R_A(x, g(x, y, r)) \leq 1 + \alpha(r - 1)$. ($R_A$ and R_B denote the approximation ratios.)

The triple (f, g, α) is said to be an *AP-reduction from A to B*.

3   APX-Hardness

The goal of this section is to prove the APX-hardness of the SCS over alphabets of size 2. We need the following definition.

Definition 1. $(TSP_Z(1, 2))$
The TSP(1,2) is the special case of the TSP where all distances are either 1 or 2. We use $TSP_Z(1, 2)$ to denote the TSP(1,2) for directed graphs with the following modifications:
(1) Every vertex is incident to at most 12 edges[2] with the weight 1.
(2) For every vertex, there is at least 1 edge with weight 1 out of the vertex.
(3) The input graph has at least Z vertices $(Z \in \mathbb{N}_{\geq 2})$.[3]
(4) The task is to compute a shortest Hamiltonian path.
W.l.o.g. we assume that $TSP_Z(1, 2)$ instances do not have reflexive edges.

---

[2] We have to ensure that this definition gives an APX-hard problem. The APX-hardness has been proved in [PY93] for undirected graphs with a degree of at most 6. By representing undirected edges with pairs of anti-parallel directed edges, we can bound the degree in the directed case by 12.

[3] This will be useful for the inapproximability results.

Theorem 1. *The SCS is APX-hard both w.r.t. the length measure and w.r.t. the compression measure, even if the alphabet has size 2 and every string is of the form $10^m1^n01^m0^{n+4}10$ or $01^m0^n10^p1^q01^m0^n10^r1^s01$ with $m,n,p,q,r,s \in \mathbb{N}_{\geq 2}$. (We call this restricted version SCS_2.)*

Proof. The $TSP(1,2)$ is known to be APX-hard even for the version $TSP_Z(1,2)$ [PY93,BJLTY94]. We modify the L-reduction from [BJLTY94], which produces SCS instances with arbitrarily large alphabets, to get an AP-reduction from the $TSP_Z(1,2)$ to the SCS_2.

The following definition provides the transformation from $TSP_Z(1,2)$ instances to SCS_2 instances. The idea of this transformation is explained in the example following the definition.

Definition 2. *(trans(·))*
We specify an $TSP_Z(1,2)$ instance I by a directed graph G, which contains precisely the edges with weight 1. All edges of I not contained in G have weight 2. Let $G = (V,E)$ be such an input for the $TSP_Z(1,2)$, $V = \{v_1,\dots,v_n\}$. An ordering for the edges out of a certain vertex can be achieved using the subscripts of the destination vertices. We will use w_j^v $(1 \leq j \leq out(v))$ to denote the jth successor of v in this order where $out(v)$ denotes the outdegree of v.

We use the alphabet $\{0,1\}$. We will use $\underline{v_i}$ $(v_i \in V)$ to abbreviate $0^{n-i+2}1^{i+1}$ and v_i' to abbreviate $1^{n-i+2}0^{i+5}$.

We define $trans(G)$ as

$$\{1\underline{v_i}0v_i'10 \mid v_i \in V\} \cup \{0v_i'1w_j^{v_i}0v_i'1w_{j+1}^{v_i}01 \mid v_i \in V, 1 \leq j \leq out(v_i)\} ,$$

where subscript arithmetic is modulo $out(v_i)$.[4]

For every vertex $v_i \in V$ we have a string $1\underline{v_i}0v_i'10$ (called connector or $conn(v_i)$) and $out(v_i)$ strings (called edge-strings) representing the edges out of v_i.

Observe that there is exactly one appearance of 101 in every connector and there are exactly two appearances of 101 in every edge-string of $trans(G)$. Furthermore, $trans(G)$ is substring free, i.e. no string in $trans(G)$ is a substring of another string in $trans(G)$.

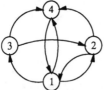

Figure 1 An $TSP_Z(1,2)$ instance. The graph contains only the edges with weight 1

Example 1.
Let $G = (V,E)$ denote the graph represented in Figure 1, $V = \{v_1, v_2, v_3, v_4\}$. The connectors are:
$conn(v_1) = 10000011011111100000010$ (string 1)
$conn(v_2) = 10000111011111000000010$ (string 2)

---
[4] We use the following modified notation throughout the paper: $x \bmod x = x$,
$(x+1) \bmod x = 1$

$conn(v_3) = 1000111101110000000010$ (string **3**)
$conn(v_4) = 1001111101100000000010$ (string **4**)
The edge-strings of v_1 are:
0111110000001**1000111**01111**10000001000111101** (string **5**)
0111110000001**1000111**10111**110000001001111101** (string **6**)
0111110000001**1001111**10111**11000000010000111101** (string **7**)
(Letters of successors of v_1 ($w_j^{v_1}$, $j = 1, \ldots, 3$) are written bold.)

The following figure shows $G_O(trans(G))$. The vertices 8 to 12 correspond to the edge-strings of v_2, v_3, and v_4. The vertices of connectors are drawn big.

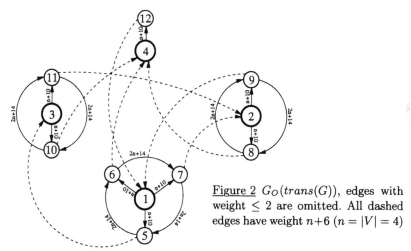

Figure 2 $G_O(trans(G))$, edges with weight ≤ 2 are omitted. All dashed edges have weight $n+6$ ($n = |V| = 4$)

The idea is that a superstring for $trans(G)$ is short, if all strings related to a vertex of the $TSP_Z(1,2)$ instance appear successively in the string. Disturbing overlaps are avoided with the following method. There are only two positions in every connector and four positions in every edge-string with "fast changes", i.e. positions where 101 or 010 appear as substrings.

In the figure above, the following order of vertices defines a Hamiltonian path with maximal total weight: $(1, 6, 7, 5, 3, 10, 11, 2, 9, 8, 4, 12)$. This path corresponds to the optimal solution $(1, 3), (3, 2), (2, 4)$ for the $TSP_Z(1,2)$ instance specified by G.

Lemma 1. *Let* $G = (V, E)$ *specify an instance of* $TSP_Z(1, 2)$.

1. *Given a Hamiltonian path of weight* $|V| - 1 + k$ *for the instance specified by* G, *a superstring for* $trans(G)$ *of length* $2|V|^2 + 2|V||E| + (13 + k)|V| + 12|E| + 5k + 6$ *(we use* $f(k)$ *to abbreviate this) is computable in polynomial time.*

2. *Given a superstring for* $trans(G)$ *of length* $f(k)$ *($k \in \mathbb{N}$), a Hamiltonian path of weight at most* $|V| - 1 + k$ *for the instance specified by* G *is computable in polynomial time.*

3. *Given a superstring for* $trans(G)$ *shorter than* $f(k)$ *($k \in \mathbb{N}$), a Hamiltonian path of weight at most* $|V| - 2 + k$ *is computable in polynomial time.*

Proof of Lemma 1.

1.) W.l.o.g. we assume $(v_1, v_2), \ldots, (v_{n-1}, v_n)$ is a Hamiltonian path with weight $|V| - 1 + k$ ($V = \{v_1, \ldots, v_n\}$). Thus, exactly k edges of weight 2 are used in this path. Let $i \in \mathbb{N}$ with $1 \le i \le n$. If $i < n$ and $(v_i, v_{i+1}) \in E$ (i.e. the edge has weight 1), then let $j \in \mathbb{N}$ such that v_{i+1} is the jth successor of v_i. Otherwise, let $j = 1$. Obviously, the string

$$\langle 1\underline{v_i}0\underline{v_i'}10,\ 0\underline{v_i'}1w_j^{v_i}0\underline{v_i'}1w_{j+1}^{v_i}01,\ 0\underline{v_i'}1w_{j+1}^{v_i}0\underline{v_i'}1w_{j+2}^{v_i}01, \ldots, 0\underline{v_i'}1w_{j-1}^{v_i}0\underline{v_i'}1w_j^{v_i}01\rangle \ ,$$

is a superstring of the strings of $trans(G)$ which are related to vertex v_i. (Subscript arithmetic is modulo $out(v_i)$.) We call this string $string(v_i)$. By adding up the distances from the left to the right and adding the length of the last string, we see that the length of this string is $(n + 4) + (out(v_i) - 1)(2 + (n + 3) + (n + 7)) + (6 + 2(n + 3) + 2(n + 7)) = (2 \cdot out(v_i) + 3)|V| + 12 \cdot out(v_i) + 18$. Thus, constructing such a string for each vertex and concatenating these strings yields a string, which is a superstring for $trans(G)$ and has the length $\sum_{i=1}^{n}((2 \cdot out(v_i) + 3)|V| + 12 \cdot out(v_i) + 18) = 3|V|^2 + 2|V||E| + 18|V| + 12|E|$.

For every edge (v_i, v_{i+1}) with weight 1, since $v_{i+1} = w_j^{v_i}$ (j as above) and $string(v_{i+1})$ starts with $conn(v_{i+1})$, $\langle string(v_i), string(v_{i+1})\rangle$ is shorter than $string(v_i)^{\wedge}string(v_{i+1})$ by $|V| + 6$ letters.

For every edge (v_i, v_{i+1}) with weight 2, since the overlap of an edge-string and a connector (in this order!) has length $|V| + 6$ or 1, $\langle string(v_i), string(v_{i+1})\rangle$ is shorter than $string(v_i)^{\wedge}string(v_{i+1})$ by 1 letter. No edge-string of v_i fits to get the overlap $|V| + 6$.

Thus,

$$|\langle string(v_1), \ldots, string(v_n)\rangle| =$$
$$2|V|^2 + 2|V||E| + (13 + k)|V| + 12|E| + 5k + 6 = f(k) \ .$$

Obviously, this string can be computed in polynomial time, which completes the proof of 1.

Similarly to [BJLTY94], we denote strings of the form

$$\langle string(v_{\pi(1)}), \ldots, string(v_{\pi(n)})\rangle$$

for some permutation $\pi : \{1, \ldots, n\} \to \{1, \ldots, n\}$ and arbitrary cyclic shifts of the edge-strings as *standard superstrings*.

2.) Let u be a superstring for $trans(G)$ with $|u| = f(k)$. For every string $s \in trans(G)$, there is a leftmost occurence of s in u. This defines an ordering, since two different strings $s, t \in trans(G)$ cannot have occurences in u starting at the same position, because $trans(G)$ is substring free. Thus, superstrings for $trans(G)$ with length $A \in \mathbb{N}$ correspond to Hamiltonian paths in $G_O(trans(G))$ with weight $\|trans(G)\| - A$ and vice versa. Let H denote the Hamiltonian path in the graph $G_O(trans(G))$, which corresponds to u, and $w(H)$ its weight.

At first, we take a look at $G_O(trans(G))$. As noticed above, in every connector there is exactly one occurence of 101 and in every edge-string of $trans(G)$ there are exactly two occurences of 101. We observe that the overlap of different connectors has length 2. Two edge-strings, which are related to different vertices, have an overlap of length 2. Moreover, the overlap of $conn(v_i)$ and an edge-string (in this order!), which is not related to v_i, has length 1. In case of an edge-string, which is related to v_i, this overlap has length $n + 10$. Finally, the overlap of an edge-string and a connector (in this order!) has length $n + 6$ or 1.

We will show that u can be transformed to a standard superstring which is not longer than u.

If u is not a standard superstring, then we have an $1 \leq i \leq n$ such that, for every cyclic shift of the edge-strings in $string(v_i)$, $string(v_i)$ is not a substring of u.

Let H' denote the remainder of H after deleting every vertex (and the incident edges), which is not a vertex of an edge-string of v_i. Let $k \in \mathbb{N}$ be the number of connected components of H'.
(cf. Example 1: If $i = 1$, then in Figure 2 the vertices of H' are 5, 6, and 7.)

We cut the subpaths of H, which correspond to the connected components of H', out of H and we delete the edge out of the vertex of $conn(v_i)$ (if there is one). After this, we delete all edges in these subpaths and connect the vertices of strings related to v_i in the way their strings are merged in $string(v_i)$ (arbitrary cyclic shift of the edge-strings). Finally, we connect the resulting paths arbitrarily to get a new Hamiltonian path P in $G_O(trans(G))$.
We have to prove $w(P) \geq w(H)$.

At first, let us assume that there is no edge in H out of the vertex of $conn(v_i)$ to the beginning of one of the subpaths. Then, cutting out a subpath deletes edges of weight at most $n + 8$. If there is an edge out of the vertex of $conn(v_i)$, its weight is at most 2. We added an edge from the vertex of $conn(v_i)$ to a vertex of an edge-string of v_i and we obtained $(k - 1)$ more edges connecting vertices of edge-strings of v_i than there were before. Summarizing, $w(P) - w(H) \geq (n + 10) + (k - 1) \cdot (2n + 14) - 2 - k(n + 8) = k(n + 6) - n - 6 \geq 0$.

If there is an edge in H out of the vertex of $conn(v_i)$ to an edge-string of v_i, then this edge is deleted and such an edge (possibly the same one) is added again in the procedure above. Thus, we get analogously $w(P) - w(H) \geq (k - 1)(2n + 14) - (k - 1) \cdot (n + 8) - (n + 6) \geq 0$ in the case $k \geq 2$. If $k = 1$, then the choice of i implies that there is at least one edge connecting vertices of edge-strings of v_i with weight 2. Therefore, we have $w(P) - w(H) \geq (2n + 14) - 2 - (n + 6) \geq 0$.

We obtained a new Hamiltonian path, which does not have a lower weight than H, and the corresponding superstring has $string(v_i)$ as a substring. Note that if for some $1 \leq j \leq n$ the vertices related to v_j form a subpath of H and they are in the order of their strings in $string(v_j)$ (with any cyclic shift of the edge-strings), then this also holds for P.

Thus, repeating this step at most $|V|$ times yields a solution for $trans(G)$, which is a standard superstring of length at most $f(k)$.

This standard superstring corresponds to an order of the vertices of V. In this order, we must add at most k edges to G to form a Hamiltonian path. Thus, we have a solution for the $TSP_Z(1, 2)$ instance specified by G with a weight of at most $|V| - 1 + k$.

Obviously, this solution can be computed in polynomial time.

3.) The proof of 2 gives a transformation from superstrings for $trans(G)$ to Hamiltonian paths for the instance specified by G, which are induced by standard superstrings. We call this transformation $trans'(\cdot, \cdot)$.

Given a superstring w for $trans(G)$ shorter than $f(k)$ for some $k \in \mathbb{N}$, the standard superstring computed by $trans'(\cdot, \cdot)$ is also shorter than $f(k)$. Since every standard superstring has length $f(l)$ for some $l \in \mathbb{N}$, this one has length at most $f(k-1)$. Thus, $trans'(G, w)$ is a solution for the $TSP_Z(1, 2)$ instance specified by G of length at most $|V| - 2 + k$. $\qquad\square$

Proof of Theorem 1. *(continued)*
At first, we consider the length measure. We prove that $(trans(\cdot), trans'(\cdot, \cdot), 15 + \frac{96}{Z})$ is an AP-reduction from $TSP_Z(1, 2)$ to SCS_2. Let $G = (V, E)$ specify an instance of $TSP_Z(1, 2)$, $r > 1$, and s be a superstring for $trans(G)$ such that $\frac{|s|}{opt(trans(G))} \leq r$ holds. Let $k \in \mathbb{N}$ be such that optimal solutions for the instance specified by G have the weight $|V| - 1 + k$. Lemma 1 implies $opt(trans(G)) = f(k)$.

Let $l \in \mathbb{N}$ be such that $f(l) \leq |s| < f(l+1)$. This implies $\frac{f(l)}{opt(trans(G))} \leq r$, which implies $l - k \leq \frac{(r-1) \cdot f(k)}{|V|+5}$, and $w(trans'(G, s)) \leq |V| - 1 + l$. Thus,

$$\frac{w(trans'(G, s))}{opt(G)} \leq \frac{|V| - 1 + l}{|V| - 1 + k} = 1 + \frac{l - k}{|V| - 1 + k} \leq 1 + \frac{(r-1) \cdot f(k)}{(|V| - 1 + k) \cdot (|V| + 5)}.$$

Therefore, we have to bound $\frac{f(k)}{(|V| - 1 + k) \cdot (|V| + 5)}$ with a constant. With $|E| \leq 6 \cdot |V|$ (see Definition 1), we have $\frac{f(k)}{(|V| - 1 + k) \cdot (|V| + 5)} \leq 15 + \frac{96}{Z}$, which completes the proof w.r.t. the length measure.

Similarly, one can prove that $(trans(\cdot), trans'(\cdot, \cdot), 12 + \frac{85}{Z})$ is an AP-reduction from the $TSP_Z(1, 2)$ to the SCS_2 w.r.t. the compression measure. $\qquad\square$

4   Inapproximability Results

The result of [Eng99] combined with our reduction in Section 3 allow to derive lower bounds for the SCS_2.

In this section, we will use a more restricted version of the TSP(1,2). Let $TSP_Z^*(1, 2)$ denote the $TSP_Z(1, 2)$, where the vertices are incident to at most 4 edges with the weight 1.

Theorem 2. *The SCS_2 is not approximable within $1 + \frac{1}{17245}$ w.r.t. the length measure (unless $P = NP$).*

Proof. For every $\varepsilon > 0$, it is NP-hard to approximate the asymmetric TSP(1,2) within $\frac{2805}{2804} - \varepsilon$ [Eng99]. We show that this result also holds for the $TSP_{Z(\varepsilon)}^*(1, 2)$ for a sufficiently large $Z(\varepsilon)$ (depending only on ε). We have to check the four modifications made in Definition 1.
(1) The definition of the $TSP_Z^*(1, 2)$ does not allow vertices that are incident with more than 4 edges of weight 1. In [Eng99], a reduction from E2-Lin(3) mod 2 to the TSP(1,2) is developed.[5] All vertices of the graphs constructed by this

---

[5]   *E2-Lin(3) mod 2 is the problem to maximize the number of satisfied equations in a system of linear equations mod 2 with exactly two variables in each equation and exactly three occurrences of each variable.*

reduction are incident with at most 4 edges of weight 1.

(2) The reduction by Engebretsen produces graphs such that for every vertex there is at least one edge with weight 1 out of the vertex.

(3) The reduction in [Eng99] produces graphs with $25 \cdot n$ vertices, given an E2-Lin(3) mod 2 instance with $2 \cdot n$ variables and $3 \cdot n$ equations.

(4) Let $G = (V, E)$ specify an $TSP_Z^*(1,2)$ instance. Let P_{opt} denote an optimal Hamiltonian path for this instance, C_{opt} an optimal Hamiltonian cycle, let $w(P_{opt})$ resp. $w(C_{opt})$ denote their weights, and let P be a Hamiltonian path satisfying $w(P) \leq (1 + \varepsilon_1) \cdot w(P_{opt})$ for some $\varepsilon_1 > 0$.

Obviously, the following inequalities hold:

$w(C_{opt}) \geq w(P_{opt}) + 1, \; w(C_{opt}) \geq |V|.$

We have:

$w(P) + 2 \leq (1 + \varepsilon_1) \cdot w(P_{opt}) + 2 \leq (1 + \varepsilon_1 + \frac{1}{|V|}) \cdot w(C_{opt}).$

Thus, an $(1 + \varepsilon_1)$-approximation for a shortest Hamiltonian path leads to an $(1 + \varepsilon_1 + \frac{1}{|V|})$-approximation for a shortest Hamiltonian cycle, and $\frac{1}{|V|} \leq \frac{1}{Z}$.

We have: For every $\varepsilon > 0$, it is NP-hard to approximate the $TSP_Z^*(1,2)$ within $\frac{2805}{2804} - \frac{1}{Z} - \varepsilon$.

Using the bound 4, one can prove analogously to the proof of Theorem 1 that $(trans(\cdot), trans'(\cdot, \cdot), 6.15 + \frac{48}{Z})$ is an AP-reduction from the $TSP_Z^*(1,2)$ to the SCS_2 w.r.t. the length measure. Thus, for every $\varepsilon > 0$ an $(1 + \varepsilon)$-approximation for the SCS_2 leads to an $(1 + (6.15 + \frac{48}{Z}) \cdot \varepsilon)$-approximation for the $TSP_Z^*(1,2)$. Therefore, it is NP-hard to approximate the SCS_2 within $1 + \frac{1}{17245}$ w.r.t. the length measure. Just choose Z^* such that $\frac{1}{Z^*} < \varepsilon^*$ (and therefore $\frac{48}{17245 \cdot Z^*} < \varepsilon^*$) with $\varepsilon^* = \frac{1}{2} \cdot (\frac{1}{2804} - \frac{6.15}{17245})$. $\qquad \square$

Theorem 3. *For every $\varepsilon > 0$, the SCS_2 is not approximable within $1 + \frac{1}{11216} - \varepsilon$ w.r.t. the compression measure (unless $P = NP$).*

Proof. Similarly to the proof of Theorem 1, one can prove that $(trans(\cdot), trans'(\cdot, \cdot), 4 + \frac{29}{Z})$ is an AP-reduction from the $TSP_Z^*(1,2)$ to the SCS_2 w.r.t. the compression measure. Thus, for every $\varepsilon > 0$, an $(1 + \varepsilon)$-approximation for the SCS_2 leads to an $(1 + (4 + \frac{29}{Z}) \cdot \varepsilon)$-approximation for the $TSP_Z^*(1,2)$.

We have: For every $\varepsilon > 0$, it is NP-hard to approximate the SCS_2 within $1 + \frac{1}{11216} - \varepsilon$ w.r.t. the compression measure. $\qquad \square$

5   Conclusion

We have proved that the SCS is APX-hard even if the alphabet contains just two characters. We used this result to derive explicit lower bounds. Any improvement of the lower bound in [Eng99] would immediately imply an improvement of these lower bounds.

The main open problem is the large remaining gap between the upper bounds 2.5 resp. $\frac{63}{38}$ and our lower bounds $1 + \frac{1}{17245}$ resp. $1 + \frac{1}{11216} - \varepsilon$.

Acknowledgement

The author would like to thank Juraj Hromkovič for helpful discussions about the SCS, Hans-Joachim Böckenhauer for very carefully reading the manuscript and helping to improve the readability, and Sebastian Seibert for helpful comments. Furthermore, I would like to thank the anonymous referees.

References

BJLTY94. A. Blum, T. Jiang, M. Li, J. Tromp, M. Yannakakis (1994), "Linear approximation of shortest superstrings", *Journal of the ACM*, Vol. **41**, No. 4, pp. 630-647.

CKST96. P. Crescenzi, V. Kann, R. Silvestri, L. Trevisan (1996), "Structure in Approximation Classes", *Electronic Colloquium on Computational Complexity*, TR96-**066**.

Eng99. L. Engebretsen (1999), "An explicit lower bound for TSP with distances one and two", *Proc.* **16th** *Annual Symposium on Theoretical Aspects of Computer Science*.

GMS80. J. Gallant, D. Maier, J. A. Storer (1980), "On finding minimal length superstrings", *Journal of Computer and System Sciences*, Vol. **20**, pp. 50-58.

KPS94. S. R. Kosaraju, J. K. Park, C. Stein (1994), "Long tours and short superstrings", *Proc. of the* **35th** *IEEE Symposium on Foundations of Computer Science*, pp. 166-177.

Li90. M. Li (1990), "Towards a DNA Sequencing Theory", *Proc. of the* **31st** *IEEE Symposium on Foundations of Computer Science*, pp. 125-134.

Midd94. M. Middendorf (1994), "More on the complexity of common superstring and supersequence problems", *Theoretical Computer Science*, Vol. **125**, pp. 205-228.

Midd98. M. Middendorf (1998), "Shortest common superstrings and scheduling with coordinated starting times", *Theoretical Computer Science*, Vol. **191**, pp. 205-214.

PY93. C. Papadimitriou, M. Yannakakis (1993), "The traveling salesman problem with distances one and two", *Mathematics of Operations Research*, Vol. **18**(1), pp. 1-11.

Stor88. J. Storer (1988), "Data compression: methods and theory", *Computer Science Press*.

Swee99. Z Sweedyk (1999), "A $2\text{-}\frac{1}{2}$ approximation algorithm for shortest superstring", *SIAM Journal on Computing*, to appear.

Turn89. J. S. Turner (1989), "Approximation algorithms for the shortest common superstring problem", *Information and Computation*, Vol. **83**, pp. 1-20.

Complexity Classification of Some Edge Modification Problems

Assaf Natanzon, Ron Shamir, and Roded Sharan

Department of Computer Science, Tel Aviv University
Tel-Aviv, Israel
{natanzon,shamir,roded}@math.tau.ac.il.

Abstract. In an edge modification problem one has to change the edge set of a given graph as little as possible so as to satisfy a certain property. We prove in this paper the NP-hardness of a variety of edge modification problems with respect to some well-studied classes of graphs. These include perfect, chordal, chain, comparability, split and asteroidal triple free. We show that some of these problems become polynomial when the input graph has bounded degree. We also give a general constant factor approximation algorithm for deletion and editing problems on bounded degree graphs with respect to properties that can be characterized by a finite set of forbidden induced subgraphs.

1   Introduction

Problem Definition: Edge modification problems call for making small changes to the edge set of an input graph in order to obtain a graph with a desired property. These include completion, deletion and editing problems. Let Π be a family of graphs. In the Π-*Editing* problem the input is a graph $G = (V, E)$, and the goal is to find a minimum set $F \subseteq V \times V$ such that $G' = (V, E \triangle F) \in \Pi$, where $E \triangle F$ denotes the symmetric difference between E and F. In the Π-*Deletion* problem only edge deletions are permitted, i.e., $F \subseteq E$. The problem is equivalent to finding a maximum subgraph of G with property Π. In the Π-*Completion* problem one is only allowed to add edges, i.e., $F \cap E = \emptyset$. Equivalently, we seek a minimum supergraph of G with property Π. In this paper we analyze edge modification problems with respect to some well-studied graph properties.

Motivation: Graph modification problems are fundamental in graph theory. Already in 1979, Garey and Johnson mentioned 18 different types of vertex and edge modification problems [11, Section A1.2]. Edge modification problems have applications in several fields, including molecular biology and numerical algebra. In many application areas a graph is used to model experimental data, and then edge modifications correspond to correcting errors in the data: Adding an edge corrects a false negative error, and deleting an edge corrects a false positive error. We summarize below some of these applications. Definitions of the graph classes are given in Section 3.

Interval modification problems have important applications in physical mapping of DNA (see [5,8,12,14]). Depending on the biotechnology used and the kind

Widmayer et al. (Eds.): WG'99, LNCS 1665, pp. 65–77, 1999.

of experimental errors, completion, deletion and editing problem arise, both for interval graphs and for unit interval graphs.

The chordal completion problem, which is also called the *minimum fill-in problem*, arises when numerically performing a Gaussian elimination on a sparse symmetric positive-definite matrix [30].

Chordal deletion problems occur when trying to solve the CLIQUE problem. Some heuristics for finding a large clique (see, e.g., [33]) aim to find a maximum chordal subgraph of the input graph, on which a maximum clique can be found in polynomial time.

Previous Results: Strong negative results are known for *vertex* deletion problems: Lewis and Yannakakis [24] showed that for any property which is non-trivial and hereditary, the maximum induced subgraph problem is NP-complete. Furthermore Lund and Yannakakis [26] proved that for any such property, and for every $\epsilon > 0$, the maximum induced subgraph problem cannot be approximated with ratio $2^{\log^{1/2-\epsilon} n}$ in quasi-polynomial time, unless $\tilde{P} = \tilde{N}P$. (Throughout we use n and m to denote the number of vertices and edges, respectively, in a graph).

For edge modification problems no such general results are known, although some attempts have been made to go beyond specific graph properties [3,10,2]. Most of the results obtained so far concerning edge modification problems are NP-hardness ones. (For simplicity we shall often refer to the decision version of the optimization problems). Chain Completion and Chordal Completion were shown to be NP-complete in [34]. As noted in [12], the NP-completeness of Interval Completion and Unit Interval Completion also follows from [34]. Interval Completion was directly shown to be NP-complete in [11, problem GT35] and [23]. Deletion problems on interval graphs and unit interval graphs were proven to be NP-complete in [12]. Cograph Completion and Cograph Deletion were shown to be NP-complete in [10]. Threshold Completion and Threshold Deletion were shown to be NP-complete in [27]. Comparability Completion was shown to be NP-complete in [17] and Comparability Deletion was shown to be NP-complete in [35].

Much fewer results are known for editing problems: Chordal Editing was proven to be NP-complete in [4]. The connected bipartite interval (caterpillar) editing problem was proven to be NP-complete in [8]. Split Editing was shown to be polynomial in [19].

Several authors studied variants of the completion problem, motivated by DNA mapping, in which the input graph is pre-colored and the required supergraph also obeys the coloring (see [5] and references thereof). Other biologically motivated problems, called *sandwich* problems, seek a supergraph satisfying a given property which does not include (pre-defined) forbidden edges. Polynomial algorithms or NP-hardness results are known for many sandwich problems [16,15,18,21]. Several results on the parametric complexity of completion problems were also obtained [22,7].

Approximation algorithms exist for several problems. In [28] an $8k$ approximation algorithm is given for the minimum fill-in problem, where k denotes the size of an optimum solution. In [1] an $O(m^{1/4} \log^{3.5} n)$ approximation algorithm is given for the minimum chordal supergraph problem (where one wishes to minimize the total number of edges in the resulting graph). For the minimum interval supergraph problem an $O(\log^2 n)$ approximation algorithm was given in [29]. In [8] it was shown that the minimum number of edge editions needed in order to convert a graph into a caterpillar cannot be approximated in polynomial time to within an additive term of $O(n^{1-\epsilon})$, for $0 < \epsilon < 1$, unless P=NP.

Contribution of this paper: In this paper we study the complexity of edge modification problems on some well-studied classes of graphs. We show, among other results, that deletion problems are NP-hard for perfect, chain, chordal, split and asteroidal triple free graphs; and that editing problems are NP-hard for perfect and comparability graphs. We also show that it is NP-hard to approximate comparability modification problems to within a factor of 18/17. The reader is referred to Figure 1 which summarizes the complexity results for the (decision version of) modification problems that we considered.

Positive complexity results are given for bounded degree input graphs: We give a simple, general constant factor approximation algorithm for the deletion and editing problems w.r.t. any hereditary property that is characterized by a finite set of forbidden induced subgraphs. We also show that Chain Deletion and Editing, Split Deletion and Threshold Deletion and Editing become polynomial when the input degrees are bounded.

Organization of the paper: Section 2 contains simple basic results that show connections between the complexity of related modification problems. Section 3 contains the main hardness results. Section 4 gives the positive results on bounded degree graphs. For lack of space, some proofs are omitted and many corollaries are only alluded to in Figure 1.

2   Basic Results

In this section we summarize some easy observations on modification problems, which will help us deduce complexity results from results on related graph families, and concentrate on those modification problems which are meaningful.

Definitions and Notation: All graphs in this paper are simple and contain no self-loops. Let $G =^{'} (V, E)$ be a graph. We denote its set V of vertices also by $V(G)$. We denote by $\overline{G}$ the *complement graph* of G, i.e., $\overline{G} = (V, \bar{E})$, where $\bar{E} = (V \times V) \setminus E$. (Throughout, we abuse notation for the sake of brevity, and for a set S we use $S \times S$ to denote $\{(s_1, s_2) : s_1, s_2 \in S, s_1 \neq s_2\}$.) If $G = (U, V, E)$ is bipartite then its *bipartite complement* is the bipartite graph $\overline{G} = (U, V, \bar{E})$, where $\bar{E} = (U \times V) \setminus E$. For a subset $A \subseteq V$ we denote by G_A the subgraph induced on the vertices of A. For a vertex $v \in V$ we denote by $N(v)$ the set of vertices adjacent to v in G. For a vertex $v \notin V$ we denote by $G \cup v$ the graph obtained by adding v to G as an isolated vertex. We denote by $G + v$ the graph obtained from G by adding v and connecting it to every other vertex of G. For a graph property Π the notation $G \in \Pi$ indicates that G satisfies Π. For basic

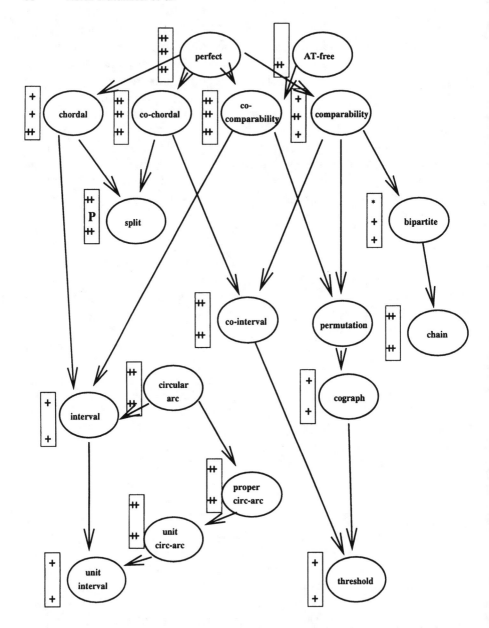

Fig. 1. The complexity status of edge modification problems for some graph classes. A→B indicates that class A contains class B. The box to the left of each class contains the status of the completion (top), editing (middle) and deletion (bottom) problems. +: NP-hard, previously known; ++: NP-hard, new result; P: polynomial; *: not meaningful.

definitions of graph properties and much more on the graph classes discussed here see, e.g., [13,6].

Let Π be a graph property. If F is a set of non-edges such that $G' = (V, E \cup F) \in \Pi$ and $|F| \leq k$, then F is called a k-completion set w.r.t. Π. k-deletion set and k-editing set are similarly defined.

Basic Results: A graph property Π is called *hereditary* if when a graph G satisfies Π every induced subgraph of G satisfies Π. Π is called *hereditary on subgraphs* if when G satisfies Π, every subgraph of G satisfies Π. Π is called *ancestral* if when G satisfies Π, every supergraph of G satisfies Π.

Proposition 1. *If property Π is hereditary on subgraphs then Π-Deletion and Π-Editing are polynomially equivalent, and Π-Completion is not meaningful.*

Proposition 2. *If Π is an ancestral graph property then Π-Completion and Π-Editing are polynomially equivalent, and Π-Deletion is not meaningful.*

Proposition 3. *If Π and Π' are graph properties such that for every graph G and a vertex $v \notin V(G)$, G satisfies Π iff $G \cup v$ satisfies Π', then Π-Deletion is polynomially reducible to Π'-Deletion. If in addition, Π' is a property such that $G \in \Pi'$ implies $G \cup v \in \Pi'$, then Π-Completion (Π-Editing) is polynomially reducible to Π'-Completion (Π'-Editing).*

Proposition 4. *If Π and Π' are graph properties such that for every graph G and a vertex $v \notin V(G)$, G satisfies Π iff $G + v$ satisfies Π', then Π-Completion is polynomially reducible to Π'-Completion. If in addition, Π' is a property such that $G \in \Pi'$ implies $G + v \in \Pi'$, then Π-Deletion (Π-Editing) is polynomially reducible to Π'-Deletion (Π'-Editing).*

For a graph property Π, we define the *complementary property* $\overline{\Pi}$ as follows: For every graph G, G satisfies $\overline{\Pi}$ iff $\overline{G}$ satisfies Π. Some well known examples are co-chordality and co-comparability.

Proposition 5. *For every graph property Π, Π-Deletion and $\overline{\Pi}$-Completion are polynomially equivalent.*

Proposition 6. *For every graph property Π, Π-Editing and $\overline{\Pi}$-Editing are polynomially equivalent.*

3 NP-Hard Modification Problems

3.1 Chain Graphs

A bipartite graph $G = (P, Q, E)$ is called a *chain graph* if there is an ordering π of the vertices in P, $\pi : \{1, \ldots, |P|\} \to P$, such that $N(\pi(1)) \subseteq N(\pi(2)) \subseteq \ldots \subseteq N(\pi(|P|))$. Yannakakis introduced this class of graphs and proved that Chain

Completion is NP-complete [34]. He also showed that G is a chain graph iff it does not contain an *independent pair of edges* (an induced $2K_2$). In this section we prove that Chain Deletion is NP-complete. This result will be the starting point to many of our subsequent reductions.

Lemma 1. *The bipartite complement of a chain graph is a chain graph.*

Proof. The claim follows from the observation that the chain containment order is reversed for the bipartite complement of a chain graph.   □

Corollary 1. *Chain Deletion is NP-complete.*

3.2   Perfect Graphs

A graph $G = (V, E)$ is called *perfect* if for any induced subgraph H of G, $\chi(H) = \omega(H)$, where $\chi(H)$ denotes the chromatic number of H, and $\omega(H)$ denotes the size of a maximum clique in H. It is easy to see that a perfect graph contains no induced cycle of odd length.

Theorem 1. *Perfect Completion is NP-hard.*

Proof: Reduction from Chain Completion. Let $< G = (P, Q, E), k >$ be an instance of Chain Completion. Since Chain Completion is NP-hard even when the input graph is connected [34], we can assume that the partition (P, Q) is known. We build the following instance $< P(G) = (N, E'), k >$ of Perfect Completion: Define $N = P \cup Q \cup C$, where

$$C = \{v^1_{q_1,q_2,i}, v^2_{q_1,q_2,i}, v^3_{q_1,q_2,i} : (q_1, q_2) \in Q \times Q, 1 \le i \le k + 1\},$$

and $E' = E \cup (P \times P) \cup E_1$, where

$$E_1 = \{ (q_1, v^1_{q_1,q_2,i}), (v^1_{q_1,q_2,i}, v^2_{q_1,q_2,i}), (v^2_{q_1,q_2,i}, v^3_{q_1,q_2,i}), (v^3_{q_1,q_2,i}, q_2) : \\ (q_1, q_2) \in Q \times Q, 1 \le i \le k + 1 \}.$$

We now prove the validity of the reduction.

⇒ Suppose that F is a chain k-completion set for G, that is $G' = (P, Q, E \cup F)$ is a chain graph. We claim that F is also a perfect k-completion set for $P(G)$. Let $K = (N, E' \cup F)$ and let $H = (V_H, E_H)$ be any induced subgraph of K. We have to show that $\omega(H) = \chi(H)$. If $E_H = \emptyset$ then H is trivially perfect, since $\chi(H) = \omega(H) = 1$. We therefore assume that $E_H \neq \emptyset$. Let $V_1 = P \cap V_H$ and let $V_2 = V_H \setminus V_1$. If $|V_1| = 0$ we can color H with two colors and $\omega(H) = \chi(H)$. Otherwise, there are two cases to examine:

   1. Suppose there is a vertex in V_2 which is adjacent to all vertices in V_1. Then $w(H) \ge |V_1| + 1$. We can color H with $|V_1| + 1$ colors in the following way:

     (a) Color the vertices of V_1 with $|V_1|$ colors.

(b) Color the vertices of Q with color number $|V_1| + 1$.

(c) Color all vertices of type $v^2_{q_1,q_2,i}$ with color number $|V_1| + 1$.

(d) Color all vertices of types $v^1_{q_1,q_2,i}$ and $v^3_{q_1,q_2,i}$ with color number $|V_1|$.

Hence, $\chi(H) \leq \omega(H)$ and the claim follows (since always $\omega(H) \leq \chi(H)$).

2. If no vertex in V_2 is adjacent to all vertices in V_1, then $\omega(H) \geq |V_1|$ and since G' is a chain graph, there is a vertex $p \in V_1$ such that no vertex in $V_2 \cap Q$ is adjacent to p. We can color the vertices of H using $|V_1|$ colors as follows:

(a) Color the vertices of V_1 with $|V_1|$ colors.

(b) Color the vertices of $V_2 \cap Q$ with the color of p.

(c) Color the vertices of type $v^2_{q_1,q_2,i}$ with the color of p.

(d) Color the vertices of types $v^1_{q_1,q_2,i}$ and $v^3_{q_1,q_2,i}$ with any other color.

If $|V_1| > 1$ we used $|V_1|$ colors. If $|V_1| = 1$ we used two colors. In any case, $\chi(H) = \omega(H)$.

$\Leftarrow$ Suppose that F is a perfect k-completion set. Let $F' = F \cap (P \times Q)$. We will show that $G' = (P, Q, E \cup F')$ is a chain graph. Suppose to the contrary that G' contains a pair of independent edges $(p_1, q_1), (p_2, q_2)$ such that $p_1, p_2 \in P$ and $q_1, q_2 \in Q$. Since $|F| \leq k$, there exists some $1 \leq i \leq k + 1$ such that the five edges $(q_1, v^2_{q_1,q_2,i}), (q_1, v^3_{q_1,q_2,i}), (v^1_{q_1,q_2,i}, v^3_{q_1,q_2,i}), (v^1_{q_1,q_2,i}, q_2)$ and $(v^2_{q_1,q_2,i}, q_2)$ are not in F. Hence, $(N, E' \cup F)$ contains an induced cycle of odd length: If $(q_1, q_2) \in F$ then $\{q_1, v^1_{q_1,q_2,i}, v^2_{q_1,q_2,i}, v^3_{q_1,q_2,i}, q_2\}$ induce a cycle of length 5. Otherwise, $\{p_1, q_1, v^1_{q_1,q_2,i}, v^2_{q_1,q_2,i}, v^3_{q_1,q_2,i}, q_2, p_2\}$ induce a cycle of length 7. In any case we arrive at a contradiction. $\square$

The perfect graph theorem by Lovasz [25] states that the complement of a perfect graph is perfect. Hence, we conclude that Perfect Deletion is also NP-hard.

Theorem 2. *Perfect Editing is NP-hard.*

Proof. Reduction from Chain Completion. Let $< G = (P, Q, E), k >$ be an instance of Chain Completion. We build the following instance $< P(G) = (N, E'), k >$ of Perfect Editing: Define $N = P \cup Q \cup C \cup D$, where

$$C = \{v^1_{q_1,q_2,i}, v^2_{q_1,q_2,i}, v^3_{q_1,q_2,i} : (q_1, q_2) \in Q \times Q, 1 \leq i \leq k + 1\},$$

$$D = \{w^1_{p,q,i}, w^2_{p,q,i}, w^3_{p,q,i} : (p, q) \in (P \times P) \cup E, 1 \leq i \leq k + 1\},$$

and $E' = E \cup (P \times P) \cup E_1 \cup E_2$, where

$$E_1 = \{ (q_1, v^1_{q_1,q_2,i}), (v^1_{q_1,q_2,i}, v^2_{q_1,q_2,i}), (v^2_{q_1,q_2,i}, v^3_{q_1,q_2,i}), (v^3_{q_1,q_2,i}, q_2) : (q_1, q_2) \in Q \times Q, 1 \leq i \leq k + 1\},$$

$$E_2 = \{ (p, w^1_{p,q,i}), (q, w^1_{p,q,i}), (p, w^2_{p,q,i}), (w^2_{p,q,i}, w^3_{p,q,i}), (w^3_{p,q,i}, q) : (p, q) \in E \cup (P \times P), 1 \leq i \leq k + 1\}.$$

The validity proof is similar to that of Theorem 1 and is omitted. The additional edges of E_2 "protect" the edges in $E \cup (P \times P)$ and prevent their removal. $\square$

3.3   Chordal Graphs

A graph is called *chordal* if it contains no induced cycle of length greater than 3. We show in this section that Chordal Deletion is NP-complete.

Theorem 3. *Chordal Deletion is NP-complete.*

Proof: The problem is in NP since chordal graphs can be recognized in linear time [31]. We prove NP-hardness by reduction from Chain Deletion. Let $< G = (P, Q, E), k >$ be an instance of Chain Deletion. Build the following instance $< C(G) = (V', E'), k >$ of Chordal Deletion: Define $V' = P \cup Q \cup V_P \cup V_Q$, where $V_P = \{v_1, \ldots, v_k\}$ and $V_Q = \{v_{k+1}, \ldots, v_{2k}\}$. Define $E' = E \cup (P \times P) \cup (Q \times Q) \cup (P \times V_P) \cup (Q \times V_Q)$. We show that the Chordal Deletion instance has a solution iff the Chain Deletion instance has a solution.

$\Rightarrow$ Suppose that F is a chain k-deletion set. We claim that F is also a chordal k-deletion set. Let $H = (V', E' \setminus F)$. Suppose to the contrary that H is not chordal, and let C be an induced cycle of length greater than 3 in H. If C contains any vertex $v \in V_P$ then it must contain at least two vertices from P, a contradiction. The same holds for V_Q. Hence, $C \cap V_P = C \cap V_Q = \emptyset$. Since P and Q are cliques, C must be of the form $\{p_1, p_2, q_1, q_2\}$, where $p_1, p_2 \in P$ and $q_1, q_2 \in Q$. But then (p_1, q_2) and (p_2, q_1) are independent edges in $(P, Q, E \setminus F)$, a contradiction.

$\Leftarrow$ Suppose that F is a chordal k-deletion set. We will prove that $F \cap E$ is a chain k-deletion set. Let $G' = (P, Q, E \setminus F)$. If G' is not a chain graph then it contains a pair of independent edges $(p_1, q_1), (p_2, q_2)$, where $p_1, p_2 \in P$ and $q_1, q_2 \in Q$. In $C(G)$, p_1, p_2 and also q_1, q_2 were connected by an edge and k edge-disjoint paths of length 2. Hence, both pairs are still connected in $H = (V', E' \setminus F)$ and p_1, q_1, q_2 and p_2 are on an induced cycle of length at least 4 in H, a contradiction. $\square$

3.4   Split Graphs

A graph G is called a *split graph* if there is a partition (K, I) of $V(G)$, so that K induces a clique and I induces an independent set. We prove that Split Deletion is NP-complete. Since the complement of a split graph is a split graph, this result implies that Split Completion is also NP-complete.

Theorem 4. *Split Deletion is NP-complete.*

Proof. Membership in NP is trivial. We prove NP-hardness by reduction from CLIQUE. Let $< G = (V, E), k >$ be an instance of CLIQUE. Build the following instance $< G' = (V', E'), k_2 = n^2(n - k + 1) - 1 >$ of Split Deletion: Define $V' = V \cup W$, where $W = \{w_1, \ldots, w_{n^2+1}\}$, and define $E' = E \cup (V \times W)$. If G has a clique K of size at least k, then denote $K' = K \cup \{w_1\}$ and partition V' into $(K', V' \setminus K')$. The number of edges that should be deleted from G' so that it becomes a split graph w.r.t. this partition is at most $n^2(n - k) + \binom{n-k}{2} < n^2(n - k + 1)$. On the other hand, suppose that G' has a k_2-deletion set, resulting in a split partition (K, I). If $|K \cap V| < k$ then at least $n^2(n - (k - 1)) > k_2$ edges in $(V \setminus K) \times (W \setminus K)$ should have been deleted from G', a contradiction. $\square$

3.5   AT-Free Graphs

An *asteroidal triple* is a set of three independent vertices such that there is a path between every pair of vertices which avoids the neighborhood of the third vertex. G is called *Asteroidal Triple free*, or *AT-free*, if G contains no asteroidal triple. Several families of graphs are asteroidal triple free, e.g., interval and comparability graphs. For characterizations of AT-free graphs see cf. [9]. We prove here that AT-free Deletion is NP-complete.

Theorem 5. *AT-free Deletion is NP-complete.*

Proof. The problem is clearly in NP. The hardness proof is by reduction from Chain Deletion. Let $< G = (U, V, E), k >$ be an instance of Chain Deletion. Build the following instance $< (V', E'), k >$ of AT-free Deletion: Define $V' = U \cup V \cup V_q \cup V_w \cup V_z$, where $V_q = \{q_1, \ldots, q_k\}, V_w = \{w_1, \ldots, w_{k+1}\}$ and $V_z = \{z_1, \ldots, z_{k+1}\}$. Define $E' = E \cup (U \times U) \cup (U \times V_q) \cup (U \times V_w) \cup ((V_w \cup V_z) \times (V_w \cup V_z))$. The validity proof is omitted.   □

3.6   Comparability Graphs

A graph is called a *comparability* graph if it has a transitive orientation of its edges, that is, an orientation F for which $(a, b), (b, c) \in F$ implies $(a, c) \in F$. We show below that Comparability Editing is NP-complete. We also prove that it is NP-hard to approximate comparability modification problems to within a factor of 18/17.

Theorem 6. *Comparability Editing is NP-complete.*

Proof: Membership in NP is trivial. The hardness proof is by reduction from MAX-CUT. Given a MAX-CUT instance $< G = (V, E), k >$ we build a Comparability Editing instance $< C(G) = (N, E'), k_2 = |E| - k >$ as follows: Define $N = V \cup \{e^1_{u,v}, e^2_{u,v} : (u, v) \in E\} \cup W$, where $W = \{w^v_i : v \in V, 1 \le i \le 2k_2 + 1\}$. Also define $E' = E_1 \cup E_2$, where

$$E_1 = \{(v, w^v_i) : v \in V, w^v_i \in W\},$$
$$E_2 = \{(v, e^1_{v,w}), (e^1_{v,w}, e^2_{v,w}), (e^2_{v,w}, w) : (v, w) \in E\}.$$

(for each $(v, w) \in E$ the choice of which vertex to connect to $e^1_{v,w}$ is arbitrary). In other words, we attach $2k_2 + 1$ private neighbors to each original vertex, and replace each edge by a path of length three. The validity proof follows.

$\Rightarrow$ Suppose that (V_1, V_2) is a cut of weight at least k in G, i.e., $|E \cap (V_1 \times V_2)| \ge k$. For each edge $e = (v, w) \in ((V_1 \times V_1) \cup (V_2 \times V_2)) \cap E$ we remove the edge $(e^1_{v,w}, e^2_{v,w})$ from its corresponding path in $C(G)$. In total, we remove k_2 edges. We now give a transitive orientation to the resulting graph, thus proving that it is a comparability graph. Orient each edge incident on $v \in V_1$ out of v, and each edge incident on $w \in V_2$ into w. For each edge $(v, w) \in (V_1 \times V_2) \cap E$, orient $(e^1_{v,w}, e^2_{v,w})$ from $e^2_{v,w}$ to $e^1_{v,w}$.

⇐ Suppose that F is a solution to the comparability instance, and let $H = (N, E' \triangle F)$ be the modified comparability graph. Let R be a transitive orientation of H. For each vertex $v \in V$ its private neighbors in $N(v) \cap W$ ensure that either all edges incident on v are directed in R into v, or they are all directed out of v. Define a partition (V_1, V_2) of V, in which $v \in V_1$ iff all edges incident on v are directed into v. We shall prove that the weight of this cut is at least k. Since we modified at most $|E| - k$ edges, there are at least k paths in H of the form $\{v, e^1_{v,w}, e^2_{v,w}, w\}$, for some $(v, w) \in E$, such that no edge in F is incident on any of those paths. For each such path, its corresponding edge must be across the cut, as otherwise R could not have been transitive.   □

A slight modification of the above reduction shows that if Comparability Editing can be approximated with ratio $1 + \theta$ then MAX-CUT can be approximated with ratio $1/(1-\theta)$. In [32,20] it is shown that approximating MAX-CUT to within a factor of $17/16$ is NP-hard. We conclude:

Corollary 2. *It is NP-hard to approximate Comparability Editing to within a factor of* $18/17$.

We comment that our reduction from MAX-CUT applies also to Comparability Completion and Comparability Deletion. Hence, it is also NP-hard to approximate the completion and deletion problems to within a factor of $18/17$.

4   Positive Results on Bounded Degree Graphs

We present below a constant factor approximation algorithm for the deletion and editing problems on bounded degree graphs. The result applies to any hereditary family which can be characterized by a finite set of forbidden induced subgraphs. Examples include cographs and claw-free graphs. An analogous result for vertex deletion problems was given by Yannakakis and Lund [26]. We also show that for bounded degree graphs Chain Deletion and Editing, Split Deletion and Threshold Deletion and Editing are polynomial.

Let Π be an hereditary graph property that can be characterized by a finite set $\mathcal{F}$ of forbidden induced subgraphs. Let $G = (V, E)$ be the input graph. We assume that each forbidden subgraph contains at most t vertices and that G has maximum degree d. In the following we further assume that no forbidden subgraph contains an isolated vertex. The approximation algorithm follows.

1) $A \leftarrow \emptyset$
2) **While** $G_{V \setminus A}$ contains an induced subgraph H isomorphic to some $F \in \mathcal{F}$,
          do: $A \leftarrow A \cup V(H)$.
3) Remove all edges $\{(v, w) \in E : v \in A, w \in V\}$ from G.

The algorithm is clearly polynomial since finding a forbidden induced subgraph with at most t vertices can be done in $O(n^t)$ time.

Theorem 7. *The algorithm approximates Π-Deletion and Π-Editing to within a factor of td.*

Proof. **Correctness**: After Step 2 is completed, $G_{V \setminus A}$ contains no forbidden induced subgraph. After Step 3 is completed, all vertices in A become isolated. Since no forbidden induced subgraph contains an isolated vertex, at the end of the algorithm G satisfies Π.

Approximation ratio: Let F be an optimum solution of size k. For any forbidden induced subgraph H found at Step 2 of the algorithm, F must contain an edge incident on H. Hence, at the end of the algorithm $|A| \leq kt$, and at most ktd edges are deleted from G.   □

It can be shown that our result extends for all hereditary properties that can be characterized by a finite set of forbidden induced subgraphs.

In the following we give some polynomial results for edge modification problems on bounded degree graphs. These results are derived by observing that for the properties in question the search space becomes bounded when the problem is restricted to bounded degree graphs.

Theorem 8. *Chain Deletion and Chain Editing are polynomially solvable on bounded degree graphs.*

Theorem 9. *Split Deletion is polynomially solvable on bounded degree graphs.*

Theorem 10. *Threshold Deletion and Threshold Editing are polynomially solvable on bounded degree graphs.*

5   Concluding Remarks

Most of the results obtained here and previously on edge modification problems are hardness results. Proving a general hardness result similar to that obtained for vertex deletion problems [24], is a challenging open problem.

The study of bounded-degree edge modification problems is still very preliminary. Such restriction is motivated by some real applications (see, e.g., [21]). Other realistic restrictions may be appropriate for particular problems. Studying the parameterized complexity of the NP-hard problems is also of interest.

Like every attempt to organize a body of results into a table or a diagram, Figure 1 immediately identifies numerous open problems. Many of those have not been investigated yet, and we are in the process of studying some of them.

References

1. A. Agrawal, P. Klein, and R. Ravi. Cutting down on fill using nested dissection: provably good elimination orderings. In A. George, J. R. Gilbert, and J. W. H. Liu, editors, *Graph Theory and Sparse Matrix Computation*, pages 31–55. Springer, 1993.

2. T. Asano. An application of duality to edge-deletion problems. *SIAM Journal on Computing*, 16(2):312–331, 1987.
3. T. Asano and T. Hirata. Edge-deletion and edge-contraction problems. In *Proceedings of the Fourteenth Annual ACM Symposium on Theory of Computing*, pages 245–254, San Francisco, California, 5–7 May 1982.
4. A. Ben-Dor. Private communication, 1996.
5. H. Bodlaender and B. de Fluiter. On intervalizing k-colored graphs for DNA physical mapping. *Discrete Applied Math.*, 71:55–77, 1996.
6. A. Brandstädt, V. B. Le, and J. P. Spinrad. *Graph Classes - a Survey*. SIAM, Philadelphia, 1999. SIAM Monographs in Discrete Mathematics and Applications.
7. L. Cai. Fixed-parameter tractability of graph modification problems for hereditary properties. *Information Processing Letters*, 58:171–176, 1996.
8. K. Cirino, S. Muthukrishnan, N. Narayanaswamy, and H. Ramesh. Graph editing to bipartite interval graphs: exact and asymptotic bounds. Technical report, Bell Laboratories Innovations, Lucent Technologies, 1996.
9. D. G. Corneil, S. Olariu, and L. Stewart. The linear structure of graphs: asteroidal triple-free graphs. In *Proc. 19th Int. Workshop (WG '93), Graph-Theoretic Concepts in Computer Science*, pages 211–224. Springer-Verlag, 1994. LNCS 790.
10. El-Mallah and Colbourn. The complexity of some edge deletion problems. *IEEE Transactions on Circuits and Systems*, 35(3):354–362, 1988.
11. M. R. Garey and D. S. Johnson. *Computers and Intractability: A Guide to the Theory of NP-Completeness*. W. H. Freeman and Co., San Francisco, 1979.
12. P. W. Goldberg, M. C. Golumbic, H. Kaplan, and R. Shamir. Four strikes against physical mapping of DNA. *Journal of Computational Biology*, 2(1):139–152, 1995.
13. M. C. Golumbic. *Algorithmic Graph Theory and Perfect Graphs*. Academic Press, New York, 1980.
14. M. C. Golumbic, H. Kaplan, and R. Shamir. On the complexity of DNA physical mapping. *Advances in Applied Mathematics*, 15:251–261, 1994.
15. M. C. Golumbic, H. Kaplan, and R. Shamir. Graph sandwich problems. *Journal of Algorithms*, 19:449–473, 1995.
16. M. C. Golumbic and R. Shamir. Complexity and algorithms for reasoning about time: A graph-theoretic approach. *J. ACM*, 40:1108–1133, 1993.
17. S. L. Hakimi, E. F. Schmeichel, and N. E. Young. Orienting graphs to optimize reachability. *Information Processing Letters*, 63(5):229–235, 1997.
18. P. L. Hammer, T. Ibaraki, and U. N. Peled. Threshold numbers and threshold completions. In P. Hansen, editor, *Studies on Graphs and Discrete Programming*, pages 125–145. North-Holland, 1981.
19. P. L. Hammer and B. Simeone. The splittance of a graph. *Combinatorica*, 1:275–284, 1981.
20. J. Håstad. Some optimal inapproximability results. In *Proc. 29th STOC*, pages 1–10, 1997. Full version: E-CCC Report number TR97-037.
21. H. Kaplan and R. Shamir. Physical maps and interval sandwich problems: Bounded degrees help. In *Proc. ISTCS*, pages 195–201, 1996.
22. H. Kaplan, R. Shamir, and R. E. Tarjan. Tractability of parameterized completion problems on chordal and interval graphs: Minimum fill-in and physical mapping. In *Proceedings of the 35th Symposium on Foundations of Computer Science*, pages 780–791. IEEE Computer Science Press, Los Alamitos, California, 1994. to appear in *SIAM J. Computing*.
23. T. Kashiwabara and T. Fujisawa. An NP-complete problem on interval graphs. In *IEEE International Symposium on Circuits and Systems (12th)*, pages 82–83, 1979.

24. J. Lewis and M. Yannakakis. The node deletion problem for hereditary properties is NP-complete. *J. Comput. Sys. Sci.*, 20:219–230, 1980.
25. L. Lovás. A characterization of perfect graphs. *J. Combin. Theory*, pages 95–98, 1972.
26. C. Lund and M. Yannakakis. The approximation of maximum subgraph problems. In A. Lingas, R. Karlsson, and S. Carlsson, editors, *Proceedings of International Conference on Automata, Languages and Programming (ICALP '91)*, pages 40–51, Berlin, Germany, 1993. Springer. LNCS 700.
27. F. Margot. Some complexity results about threshold graphs. *DAMATH: Discrete Applied Mathematics and Combinatorial Operations Research and Computer Science*, 49, 1994.
28. A. Natanzon, R. Shamir, and R. Sharan. A polynomial approximation algorithm for the minimum fill-in problem. In *Proceedings of the 30th Annual ACM Symposium on Theory of Computing (STOC'98)*, pages 41–47, New York, May 23–26 1998. ACM Press.
29. R. Ravi, A. Agrawal, and P. Klein. Ordering problems approximated: single processor scheduling and interval graph completion. In *Proc. ICALP 1991*, pages 751–762. Springer, 1991. LNCS 510.
30. J. D. Rose. A graph-theoretic study of the numerical solution of sparse positive definite systems of linear equations. In R. C. Reed, editor, *Graph Theory and Computing*, pages 183–217. Academic Press, N.Y., 1972.
31. R. E. Tarjan and M. Yannakakis. Simple linear-time algorithms to test chordality of graphs, text acyclicity of hypergraphs, and selectively reduce acyclic hypergraphs. *SIAM J. Computing*, 13:566–579, 1984.
32. L. Trevisan, G. Sorkin, M. Sudan, and D. Williamson. Gadgets, approximation, and linear programming. In *Proc. IEEE Symposium on Foundations of Computer Science (FOCS'96)*, pages 617–626, 1996.
33. J. Xue. Edge-maximal triangulated subgraph and heuristics for the maximum clique problem. Technical report, Graduate School of Management, Clark University, Worcester, MA, July 1993.
34. M. Yannakakis. Computing the minimum fill-in is NP-complete. *SIAM J. Alg. Disc. Meth.*, 2, 1981.
35. M. Yannakakis. Edge deletion problems. *SIAM J. Computing*, 10(2):297–309, 1981.

On Minimum Diameter Spanning Trees under Reload Costs*

Hans-Christoph Wirth and Jan Steffan

University of Würzburg
Department of Computer Science,
Am Hubland, 97074 Würzburg, Germany
{wirth,steffan,}@informatik.uni-wuerzburg.de

Abstract We examine a network design problem under the *reload cost* model. Given an undirected edge colored graph, reload costs arise at the nodes of the graph and are depending on the colors of the pair of edges used by a walk through the node.

In this paper we consider the problem of finding a spanning tree of minimum diameter with respect to the underlying reload costs. We present hardness results and lower bounds for the approximability even on graphs with maximum degree 5. On the other hand we provide an exact algorithm for graphs of maximum degree 3.

Keywords: Transportation problems, Network Design, Diameter, Spanning Tree, Node weighted graphs

1   Introduction and Related Work

Network design problems are graph theoretic optimization problems which have a wide area of applications. We consider a scenario where a bunch of different providers runs a subnetwork each. The transportation costs inside of each subnetwork are negligible. Costs arise only at points where the underlying provider changes.

This models situations such as cargo transportation by different means of transportation where usually changing the carrier involves cost and time expensive unloading and reloading of the goods. Another application is modeling the costs arising in data networks. The cost and time needed for data conversion at interchange points between incompatible subnetworks usually dominate the costs arising by routing the information packets within each of the subnetworks.

We model the subnetworks by sets of edges of different colors in an edge colored undirected graph. In our reload cost model, costs arise at each node, and they depend on the pair of colors of the edges used by the walk through that node. For practical applications it is useful to assume that the reload costs satisfy the triangle inequality, since otherwise one could save costs by performing more than one reload job at the same node.

---

* Supported by the Deutsche Forschungsgemeinschaft (DFG), Grant NO 88/15–1

The problem investigated in this paper, DIAMETER-TREE, is to search for a spanning tree of minimum diameter with respect to the underlying reload costs. This is motivated by the fact that bounding the diameter is equivalent to bounding the maximum cost which can arise for one single transportation order.

For general reload cost functions, we can show that DIAMETER-TREE is not approximable within any constant factor even if restricted to graphs of maximum node degree 5. If the reload cost function satisfies the triangle inequality, we give a lower bound of 5 on the approximation factor on general graphs and a bound of 3 on graphs with maximum degree 5. On the other hand, DIAMETER-TREE can be solved exactly when restricted to graphs of maximum degree 3.

This problem is related to the minimum label spanning tree problem, which was introduced by Chang and Leu [CL97]. Here the goal is to find a spanning tree which uses as least as possible many different colors. In [KW98], the authors give both a logarithmic approximation algorithm and a logarithmic lower bound for the minimum label spanning tree problem.

The minimum diameter spanning tree problem on graphs with nonnegative edge lengths has been examined by Hassin and Tamir [HT95]. The authors show the equivalence to the absolute 1-center problem and use the work of Kariv and Hakimi [KH79] to provide an algorithm with running time $O(|E||V| + |V|^2 \log |V|)$.

2   Preliminaries and Problem Formulations

Definition 1 (Reload cost function). *Let X be a finite set of colors. A function $c: X^2 \to \mathbb{N}_0$ is called a* reload cost function, *if for all $x_1, x_2 \in X$*

1. $c(x_1, x_2) = c(x_2, x_1)$,
2. $x_1 = x_2 \Longrightarrow c(x_1, x_2) = 0$,
3. $x_1 \neq x_2 \Longrightarrow c(x_1, x_2) > 0$.

If additionally for all $x_1, x_2, x_3 \in X$,

4. $c(x_1, x_3) \leq c(x_1, x_2) + c(x_2, x_3)$,

the reload cost function is said to satisfy the triangle inequality.

A *graph with reload costs* is given by an undirected graph $G = (V, E)$ (with parallel edges allowed), a function $\chi: E \to X$ coloring the edge set, and a reload cost function c on the set X of colors. Notice that χ is not an edge coloring in the classical sense, i.e., it does not imply that adjacent edges must have different colors.

Reload costs on a path arise at nodes, where the colors of the edges used by the path may change. A path of one edge only has reload costs zero. For a path $p = (e_1, e_2, \dots, e_k)$ consisting of $k > 1$ edges, we have reload costs

$$c(p) := \sum_{i=1}^{k-1} c(\chi(e_i), \chi(e_{i+1}))$$

If there are no ambiguities, we use the shorter notation $c(e, f) := c(\chi(e), \chi(f))$ throughout the paper.

Definition 2. *Let $G = (V, E, \chi, c)$ be a graph with reload costs. Then, the induced reload cost distance function is given by*

$$dist_G^c(v, w) := \min\{\, c(p) \mid p \text{ is a path from } v \text{ to } w \text{ in } G \,\}.$$

In contrast, for an edge length function $l : E \to \mathbb{N}_0$, the *length* of a path p is given by $l(p) := \sum_{i=1}^k l(e_i)$, and the *induced length distance function* is consequently $dist_G^l(v, w) := \min\{\, l(p) \mid p \text{ is a path from } v \text{ to } w \text{ in } G \,\}$.

We are now ready to define the problem under study.

Definition 3 (Problem   Diameter-Tree). *An   instance   of   problem* DIAMETER-TREE *is given by a graph $G = (V, E, \chi, c)$ with reload costs. The goal is to find a spanning tree $T \subseteq E$ of the graph, such that the diameter with respect to the reload costs*

$$diam^c(T) := \max_{v, w \in V} dist_T^c(v, w)$$

is minimized among all spanning trees.

By Δ-DIAMETER-TREE we denote the problem where the reload cost function satisfies the triangle inequality.

A straightforward idea to deal with reload cost problems would be to transform the graph to its *line graph* and map the reload costs to edge lengths in the line graph. (The line graph of a graph $G = (V, E)$ has node set E and an edge (e_1, e_2) if and only if e_1 and e_2 are adjacent edges in G.) Unfortunately, this approach does not work for trees because there is no relationship between a spanning tree of a graph and a spanning tree of its line graph.

Let Π be a minimization problem. A polynomial running time algorithm A is called an *α-approximation algorithm* for Π, if for each instance π of Π with optimal solution $\mathrm{OPT}(\pi)$, the solution $A(\pi)$ produced by the algorithm satisfies $A(\pi) \leq \alpha \cdot \mathrm{OPT}(\pi)$. A problem Π is called *not approximable within α*, if there is no α-approximation algorithm unless $\mathrm{P} = \mathrm{NP}$.

3   Hardness Results

In this section we show that DIAMETER-TREE and even Δ-DIAMETER-TREE is NP-hard and we provide inapproximability results. At the end we will show that the hardness results extend to graphs where the node degree is bounded by 5. In contrast, in section 4 we will give an exact algorithm for graphs with maximum degree 3.

Theorem 1. *Unless $\mathrm{P} = \mathrm{NP}$,* DIAMETER-TREE *is not approximable within any constant factor.*

Proof. We perform a reduction from 3-SAT [GJ79, Problem LO2]. An instance π of 3-SAT is given by a set $A = \{a_1, a_2, \ldots, a_n\}$ of variables and a set $C = \{C_1, \ldots, C_k\}$ of clauses over X. Each clause is of the form $C_i = \{l_{i1}, l_{i2}, l_{i3}\}$, where l_{ij} is a literal, i.e., a variable a or its negation $\bar{a}$. The goal is to find a truth assignment satifying all clauses.

We now construct an instance π' of DIAMETER-TREE. The construction of the graph $G = (V, E)$ is shown in Figure 1. We set $V := \{s\} \cup A \cup C$. For each variable $a \in A$, we introduce two colors $x_a, \bar{x}_a$ representing the positive and negative literal.

For each variable $a \in A$, insert two parallel edges between a and s of color x_a and $\bar{x}_a$, respectively. For each literal l of a clause $C' \in C$, where l is the positive or negative variable a, add an edge between nodes C' and a of color l to the graph. Let the reload cost function c be given as

$$
c(l_1, l_2) := \begin{cases} K, & \text{if } l_1 = \bar{l}_2, \\ 0, & \text{if } l_1 = l_2, \\ 1, & \text{otherwise}, \end{cases}
$$

where $K > 1$ is some large constant. Informally, the reload costs are expensive if two incident edges represent a variable and its negation, while they are low if the edges represent different variables.

Assume that there is a spanning tree T of G such that each clause node C' is connected by a path of cost $< K$ to the root s. Then in T each variable node is incident to edges of one color only. If there is a variable node a which is not connected directly by one single edge to the root, choose the unique edge $e = (a, C_j)$ on the path from a to the root and replace e by the edge (a, s) of the same color. This does not increase the path lengths to the root. At the end, each clause node is connected to the root by a path of two edges of cost zero. Since the tree still spans all clause nodes, the colors of the edges adjacent to the root induce a valid assignment for π. Conversely, it is easy to see that a valid solution for π can be used to construct a spanning tree with the property that each clause node is connected by a path of cost zero to the root.

As a consequence, if π has a valid solution, then an optimum spanning tree has diameter 1. On the other hand, the diameter is $\geq K + 1$ if π admits no valid assignment.

Assume that there is an approximation algorithm for DIAMETER-TREE with performance α. Choose $K > \alpha$. Then the algorithm must solve the instance π' exactly which is equivalent to solving 3-SAT by our observations.   □

To show a non-approximability result for Δ-DIAMETER-TREE, we modify the construction from above. We construct two identical copies G_1, G_2 of the graph given above which are connected by identifying the root nodes. Now, if there is a spanning tree of diameter $< 2K + 1$, then there must be one partial graph G_i with the property that each clause node is connected to the root by a path of cost $< K$. Conversely, if the underlying instance of 3-SAT has a valid assignment, then we can construct a spanning tree in the graph of diameter 1. By choosing

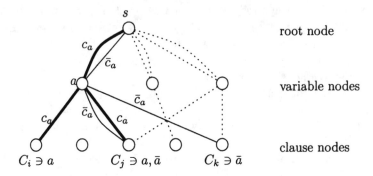

Figure1. Reduction from 3-SAT to Δ-DIAMETER-TREE used in the proof of Theorem 1.

$K = 2$ the triangle inequality is satisfied. As a consequence, for $\alpha < 2K + 1 = 5$, any α-approximation algorithm for Δ-DIAMETER-TREE must in fact solve the underlying instance π' exactly. This is summarized in the following corollary.

Corollary 1. *Unless* P = NP, Δ-DIAMETER-TREE *is not approximable with any factor* $\alpha < 5$. $\qquad\square$

The hardness results from above remain to hold in a slightly weaker form when restricted to graphs of bounded degree.

We modify the construction given in the proof of Theorem 1. The construction is illustrated in Figure 2. Consider variable a. Instead of connecting the clause nodes directly to variable node a, we introduce auxiliary nodes which can be connected to node a by a subgraph of maximum degree 5. To avoid a root node of high degree, we choose a new color and replace the root by a suitable (with respect to the degree bound) tree of edges of the new color. Notice that the construction still can be done in polynomial time.

The new color yields an increase in the diameter of the optimal solutions. If there is a valid solution to the instance of 3-SAT, then there is a spanning tree of diameter 2. Conversely, if the 3-SAT instance admits no valid assignment, then any spanning tree has diameter at least $K + 2$. This yields the following result:

Corollary 2. *Unless* P = NP, DIAMETER-TREE *is not approximable within any constant factor, even when restricted to graphs of maximum degree* 5. $\quad\square$

By setting $K = 2$, the triangle inequality is satisfied. Using a similar technique as in the proof of Corollary 1, we get the following result:

Corollary 3. *Unless* P = NP, Δ-DIAMETER-TREE *is not approximable with any factor* $\alpha < 3$, *even when restricted to graphs of maximum degree* 5. $\quad\square$

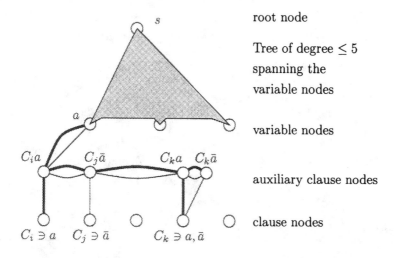

root node

Tree of degree ≤ 5
spanning the
variable nodes

variable nodes

auxiliary clause nodes

clause nodes

Figure2. Reduction for degree 5 bounded graphs.

4   Exact Solution for Graphs with Maximum Degree 3

In this section we provide an exact algorithm for graphs with maximum degree 3. We assume that the reload costs satisfy the triangle inequality which is not a serious restriction in practical applications as stated in the introduction. The main idea is to map the graph G with reload costs to an equivalent graph H with edge lengths and then use known algorithms for finding a minimum diameter spanning tree on an edge weighted graph for solving the problem.

To perform this, we start with graph $H := G$. Then we divide each edge $e = (v, w)$ of H into two edges (v, e^ϕ) and (e^ϕ, w) by placing a new node e^ϕ in the middle of the edge. The length of the edges adjacent to a node is adjusted such that the sum of the edge lengths on a walk through the node equals the reload costs of the related walk in the original graph, i.e.,

$$l(e_1^\phi, v) + l(v, e_2^\phi) = c(e_1, e_2) \qquad \text{for all } v \in V, e_i \text{ incident to } v. \tag{1}$$

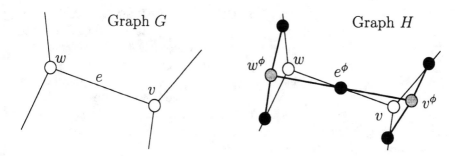

Figure3. The construction performed by CostToLength

Input: Graph $G = (V, E)$ with reload costs c, maximum degree 3

1 $V' \leftarrow V \cup V^\phi \cup E^\phi$

2 $E' \leftarrow \{(v, e^\phi), (e^\phi, w), (v^\phi, e^\phi), (e^\phi, w^\phi) \mid e = (v, w) \in E\}$

3 $\Omega \leftarrow \max\{c(x, x') \mid x, x' \in X\} + 1$

4 **for all** $v \in V$, $e_1 \in E$ incident on v **do**

5   let $d \in \{1, 2, 3\}$ be the degree of v and let $\{e_1, \dots, e_d\}$ be the set of incident edges

6 $l(v, e_1^\phi) \leftarrow \begin{cases} 0, & \text{if } d = 1, \\ 1/2 \cdot c(e_1, e_2), & \text{if } d = 2, \\ 1/2 \cdot \big(c(e_1, e_2) + c(e_1, e_3) - c(e_2, e_3)\big), & \text{if } d = 3. \end{cases}$

7 $l(v^\phi, e_1^\phi) \leftarrow \Omega$

8 **end for**

Output: Graph $H = (V', E')$ and edge lengths l

Algorithm 1: Algorithm CostToLength.

For a node v of degree d, the set of equations (1) consists of $\binom{d}{2}$ equations with d variables which in general does not have a solution if $d > 3$. Therefore we restrict ourselves to graphs of maximum degree 3.

Now for each node $v \in V$, we create a copy v^ϕ which is connected to all neighbors of v by edges of huge length Ω. The complete construction of graph H is performed by Algorithm CostToLength and illustrated in Figure 3.

By this construction we are enabled to express the cost of paths in G in terms of the length of corresponding paths in H as stated in the following lemma.

Lemma 1. *Let $G = (V, E)$ be a graph with reload cost function c. Let H be the graph constructed by Algorithm* CostToLength, *and l be the computed edge lengths. Then for each pair $v, w \in V$,*

$$dist_G^c(v, w) = dist_H^l(v^\phi, w^\phi) - 2\Omega.$$

Moreover, if c satisfies the triangle inequality, then $l \geq 0$.   □

We omit the proof since it is a straightforward application of Equation (1) and of technical nature only.

At this point we call Algorithm MinDiameterSpanningTree as described by Hassin and Tamir [HT95] on the graph H to compute a minimum diameter spanning tree T_1 of graph H with respect to edge lengths l.

We can assume without loss of generality that nodes from V^ϕ do not appear as interior nodes of T_1. This is due to the fact that the path through a node v^ϕ would consist of two Ω-edges and could be replaced by a corresponding path through node v without increasing the diameter of T_1. As a consequence, by removing all edges of weight Ω from T_1, the resulting subgraph would still be a tree spanning the node set $V \cup E^\phi$.

An Ω-length edge (v^ϕ, e^ϕ) is called a *dangling edge*, if its *projection*, i.e., the edge (v, e^ϕ), is not part of the tree. Notice that if T_1 has no dangling edges, then its projection is still a tree which spans V.

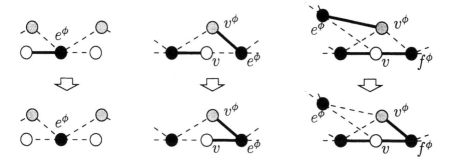

Figure4. Edge swaps performed by Algorithm REMOVE.
Left: line 2; center: line 4; right: line 7

Input: A tree T_1 spanning the node set $V \cup V^\phi \cup E^\phi$, such that no node
        from V^ϕ is an interior node
1 **repeat**
2   if there is a leaf $e^\phi \in E^\phi$ in the tree **then**
3   remove e^ϕ from the tree
4   **else if** there is a dangling edge (v^ϕ, e^ϕ) such that v is a leaf **then**
5   remove v from the tree
6   add (v, e^ϕ) to the tree
7   **else if** there is a dangling edge (v^ϕ, e^ϕ) such that v is an interior node **then**
8   remove v^ϕ from the tree
9   add (v^ϕ, f^ϕ) to the tree, where f^ϕ is a neighbor of v in the tree
10   **end if**
11 **until** no more changes have been made
Output: A tree T_2 spanning $V \cup V^\phi$ with no dangling edges

---

Algorithm 2: Algorithm REMOVE.

Algorithm REMOVE removes dangling edges from T_1 without increasing the diameter of the tree. As a consequence, after applying the algorithm we can use the projection of the modified tree as a solution of the problem on the original graph.

Lemma 2. *Algorithm* REMOVE *performs at most* $|E| + 2|V|$ *iterations.* □

Lemma 3. *The subgraph T_2 produced by Algorithm* REMOVE *is a tree spanning the node set $V \cup V^\phi$. No node from E^ϕ appears as a leaf in T_2. There is no dangling edge in T_2.*

Proof. The first claim clearly holds for the initial graph T_1. Each edge swap operation is performed at leaves and therefore does not affect the connectivity. A node is removed if and only if it is a leaf from set E^ϕ. Dangling edges are removed by the algorithm. □

Input: Graph $G = (V, E)$ with reload costs c, maximum degree 3
1 $(H, l) \leftarrow$ COSTTOLENGTH(G, c)
2 $T_1 \leftarrow$ MINDIAMETERSPANNINGTREE(H, l)
3 $T_2 \leftarrow$ REMOVE(T_1)
4 $E' \leftarrow \{ e = (v, w) \in E \mid (v, e^\phi) \in T_2 \wedge (e^\phi, w) \in T_2 \}$
Output: $T = (V, E')$

Algorithm 3: Algorithm for problem Δ-DIAMETER-TREE on graphs with maximum degree 3.

Lemma 4. *If $T_2 :=$ REMOVE(T_1) is the tree computed by Algorithm REMOVE, then*

$$diam^l(T_2) \leq diam^l(T_1).$$

Proof. We show that in each iteration the diameter does not increase.

Line 2 Removal of a leaf can not increase the diameter.

Line 4 Assume that the diameter of the tree increases by the operation. Then the new diameter is attained by a path p starting with the inserted edge (v, e^ϕ). Since $|V^\phi| = |V| \geq 2$, we can assume that p does not use edge (v^ϕ, e^ϕ). If we replace in p edge (v, e^ϕ) of length $< \Omega$ by edge (v^ϕ, e^ϕ) of length Ω, this yields a path strictly longer than p contradicting the fact that p was a path of maximum length.

Line 7 Consider f^ϕ as the root of the current tree before the operation. Since the condition in line 2 was not satisfied, f^ϕ is not a leaf, hence there are at least two subtrees hanging from f^ϕ. Each of the subtrees must contain at least one Ω-length edge and therefore have height at least Ω, since otherwise one of the conditions in line 2 or line 4 would have been satisfied. Therefore adding the edge (v^ϕ, f^ϕ) of length Ω to the root does not increase the diameter.   □

Algorithm 3 assembles the presented algorithms and constructs the final solution for the original problem on graph G.

Theorem 2. *Let T be the tree returned by Algorithm 3. Then for all spanning trees T' of G,*

$$diam^c(T) \leq diam^c(T'),$$

i.e., the tree T is optimal with respect to $diam^c$.

Proof. (i) Let $p = (e_1, \ldots, e_k)$ be the path between arbitrary nodes v and w in T, $c(p) = dist^c_T(v, w)$ the cost of the path. By construction of T, for each edge $e = (v', w') \in T$, tree T_2 contains the two edges (v', e^ϕ) and (e^ϕ, w'). Therefore, the path q in T_2 from e_1^ϕ to e_k^ϕ uses the nodes $e_2^\phi, \ldots, e_{k-1}^\phi$, and by (1) follows that q is of length $dist_{T_2}^l(e_1^\phi, e_k^\phi) = c(p)$.

We claim that there is in fact a path of length at least $c(p) + 2\Omega$ in T_2. To see this, we show that q can be augmented by a sub-path of length at least Ω at both endpoints. Consider endpoint e_1^ϕ, the other case is similar. By Lemma 3, v^ϕ is spanned by T_2; let (v^ϕ, f^ϕ) be the connecting edge, which is of length Ω. If $f^\phi = e_1^\phi$, we are done. Otherwise, since the edge is not dangling, we have $(f^\phi, v) \in T_2$, and by construction of T, also $(v, e_1^\phi) \in T_2$.

If we choose p as a maximal path, i.e., $c(p) = diam^c(T)$, it follows

$$diam^l(T_2) \geq diam^c(T) + 2\Omega. \tag{2}$$

(ii) Let $T' = (V, E')$ be an arbitrary spanning tree of G. Then we construct a tree T'_H in graph H by choosing the edge set $\cup_{e=(v,w)\in E'}\{(v, e^\phi), (e^\phi, w)\}$. Note that T'_H spans the node set $V \cup E'^\phi$.

For any two nodes $e_1^\phi, e_2^\phi \in E'^\phi$, by construction of the tree and using (1), we have $dist_{T'_H}^l(e_1^\phi, e_2^\phi) \leq diam^c(T')$.

We claim that the remaining nodes can be connected to the tree T'_H such that for each node the distance to the nearest node in E'^ϕ is bounded by Ω. If this claim holds, we have

$$diam^l(T'_H) \leq diam^c(T') + 2\Omega. \tag{3}$$

The claim is true for each node in V. Connect nodes from V^ϕ to T'_H by one Ω-length edge each. Then, the claim also holds for the nodes in V^ϕ. Connect the remaining nodes from $E^\phi \setminus E'^\phi$ to T'_H by an arbitrary edge with endpoint in V each. Since the sum of the lengths of two edges adjacent to a node from V is bounded by Ω, even for the remaining nodes the claim holds.

(iii) Note that the resulting tree T'_H is a spanning tree in H. By Lemma 4 and optimality of T_1, we have

$$diam^l(T_2) \leq diam^l(T_1) \leq diam^l(T'_H). \tag{4}$$

Putting (2), (3), and (4) together, the claim follows. □

We now summarize our results.

Corollary 4. *Algorithm 3 solves problem Δ-DIAMETER-TREE on graphs with degree bound 3. The running time is in $O(|E|^2 \log |E|)$.*

Proof. It remains to show the claim on the running time. The graph H constructed by Algorithm COSTTOLENGTH has $2|V| + |E|$ nodes and $4|E|$ edges. From [HT95] it follows that MINDIAMETERSPANNINGTREE can be implemented to run in time $O(4|E|(2|V|+|E|)+(2|V|+|E|)^2 \log(2|V|+|E|)) \in O(|E|^2 \log |E|)$. By Lemma 2, REMOVE performs at most $O(|E|)$ iterations, each of which needs time $O(|E|)$. Hence, MINDIAMETERSPANNINGTREE dominates the running time and the claim follows. □

5   Conclusion and Open Problems

To our knowledge, reload costs have not been considered in the literature so far. This cost structure is related to node weighted graphs, but the new aspect is that the cost at a node depend on the edges used by the walk through that node.

The following table shows a summary of results presented in this paper. The notion "not approximable" means, assuming $P \neq NP$, there is no polynomial time algorithm which guarantees the specified performance.

Problem	Complexity
DIAMETER-TREE	not approximable within any constant
Δ-DIAMETER-TREE	not approximable within any $\alpha < 5$
DIAMETER-TREE, degree 5	not approximable within any constant
Δ-DIAMETER-TREE, degree 5	not approximable within any $\alpha < 3$
Δ-DIAMETER-TREE, degree 3	polynomially solvable

A natural open question is the complexity of the problem for graphs with maximum degree 4. On the other hand, a major goal is to search for a nontrivial approximation algorithm on general graphs.

It is possible to formulate other well known network design problems under the reload cost model, e.g. the problem of finding a spanning tree of minimum total cost. Here, the notion "total cost" is to be defined more precisely, since it makes sense either to count the number of reloading nodes (i.e., nodes with more than one color incident) or to sum up the costs of all possible paths in the resulting tree.

References

CL97.   R.-S. Chang and S.-J. Leu, *The minimum labeling spanning trees*, Information Processing Letters **63** (1997), 277–282.

GJ79.   M. R. Garey and D. S. Johnson, *Computers and intractability (a guide to the theory of NP-completeness)*, W.H. Freeman and Company, New York, 1979.

HT95.   R. Hassin and A. Tamir, *On the minimum diameter spanning tree problem*, Information processing letters **53** (1995), 109–111.

KH79.   O. Kariv and S. L. Hakimi, *An algorithmic approach to network location problems, part I: The p-center*, SIAM Journal of Applied Mathematics **37** (1979), 513–537.

KW98.   S. O. Krumke and H.-C. Wirth, *On the minimum label spanning tree problem*, Information Processing Letters **66** (1998), no. 2, 81–85.

Induced Matchings in Regular Graphs and Trees

Michele Zito[*]

Department of Computer Science, University of Liverpool,
Liverpool L69 7ZF, UK

Abstract. This paper studies the complexity of the Maximum Induced Matching problem (MIM) in regular graphs and trees. We show that the largest induced matchings in a regular graph of degree d can be approximated with a performance ratio less than d. However MIM is NP-hard to approximate within some constant $c > 1$ even if the input is restricted to various classes of bounded degree and regular graphs. Finally we describe a simple algorithm providing a linear time optimal solution to MIM if the input graph is a tree.

1 Introduction

If $G = (V, E)$ is a graph, a set $M \subseteq E$ is a *matching* in G if for all $e_1, e_2 \in M$ it is $e_1 \cap e_2 = \emptyset$. Let $V(M)$ be the set of vertices belonging to edges in the matching. A matching M is *maximal* if for every $e \in E \setminus M$, $M \cup e$ is not a matching; M is *induced* if for every edge $e = \{u, v\}$, $e \in M$ if and only if $u, v \in V(M)$ and $e \in E$. Let $\nu_I(G)$ denote the maximum cardinality of an induced matching in G. The maximum induced matching problem (MIM) is that of finding an induced matching in G with $\nu_I(G)$ edges.

The problem was introduced in [14] as a variation of the maximum matching problem and motivated as the "risk-free" marriage problem: find the maximum number of pairs such that each married person is compatible with no married person other than the one he (or she) is married to. Induced matchings have stimulated a lot of interest in discrete mathematics because finding large induced matchings is a subtask of finding a *strong edge-colouring* in a graph (see [5,6] and [15,11] for more recent results), a proper colouring of the edges such that no edge is adjacent to two edges of the same colour.

MIM is NP-complete even for bipartite graphs of maximum degree four [14]. One way of coping with the NP-completeness of an optimization problem is to relax the optimality requirement and look for the existence of polynomial time algorithms which guarantee solutions whose size is close to the size of the optimum. In what follows we say that a maximization problem P is approximable with *(performance) ratio* ρ if there is a polynomial time algorithm returning a solution whose size is at least ρ^{-1} times the size of an optimal solution. Not much is known about the approximability of MIM. In Section 2, fairly simple combinatorial arguments allow us to prove the existence of approximation algorithms

---

[*] Supported by EPSRC grant GR/L/77089.

giving a ratio smaller than d, for regular graphs of degree d. In Section 3, we establish a number of non-approximability results. We provide explicit bounds on the performance ratio such that MIM is NP-hard to approximate with ratio less than these bounds in several classes of bounded degree graphs including regular graphs of degree four.

A graph $G = (V, E)$ is a *tree* if it is connected and it has no cycle, it is *chordal* if any cycle of at least four vertices contains an edge connecting two non-consecutive vertices. MIM in chordal graphs can be reduced [3] to finding the largest independent set in a chordal graph and the latter problem admits an optimal polynomial time solution [7]. Since trees are chordal graphs, this argument (and an efficient implementation of Gavril's algorithm) implies the existence of a $O(|V|^2)$ algorithm for finding the maximum induced matching in a tree. In Section 4 we present an alternative algorithm which again solves MIM optimally if the input graph is a tree but it runs in $O(|V|)$ time.

2   Combinatorial Bounds

If $G = (V, E)$ is a graph let $\deg_G v$, the *degree of* v, be the number of vertices that are adjacent to v. Let $V_i(G) = \{v \in V : \deg_G v = i\}$ for all $i = 0, \ldots, |V| - 1$. Notations $V(G)$ and $E(G)$ will be used instead of V and E when necessary to prevent ambiguities. A (δ, Δ)-graph is a graph with minimum degree δ and maximum degree Δ. A (d, d)-graph is a regular graph of degree d. Let (δ, Δ)-MIM (resp. d-MIM) identify MIM when the input is restricted to (δ, Δ)-graphs (resp. regular graphs of degree d). In this section we look at positive approximation results for MIM in regular graphs. We describe two results that "come for free" in the sense that they do not require any involved algorithmic idea and their validity is implied by the combinatorial structure of the matching problem under consideration. The negative results in Section 3 show that there is not much scope for better results.

Definition 1. [9] *An* independent system *is a pair $(E, \mathcal{F})$ where E is a finite set and $\mathcal{F}$ a collection of subsets of E with the property that whenever $F \subset H \in \mathcal{F}$ then $F \in \mathcal{F}$. The elements of $\mathcal{F}$ are called* independent sets. *A* maximal independent set *is an element of $\mathcal{F}$ that is not subset of any other element of $\mathcal{F}$.*

Korte and Hausmann [9] analysed the independent system formed by all matchings in G and proved an upper bound of 2 on the ratio between the sizes of any two maximal matchings. In the next result a similar argument is applied to estimate the maximum ratio between two maximal induced matchings.

Let $\mathcal{M}_I(G)$ be the set of all induced matchings in a graph G. The pair $(E(G), \mathcal{M}_I(G))$ is an independent system. For every $S \subseteq E$ the *lower* (resp. *upper*) *rank* of S, $\underline{\rho}(S)$ (resp. $\bar{\rho}(S)$) is the size of the smallest (resp. largest) maximal induced matching included in S. By [9, Theorem 1.1], if M is a maximal induced matching, then

$$\frac{\nu_I(G)}{|M|} \leq \max_{S \subseteq E} \frac{\bar{\rho}(S)}{\underline{\rho}(S)}$$

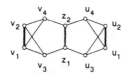

Fig. 1. A cubic graph with large induced matching.

Theorem 1. *Let G be a (δ, Δ)-graph and $(E(G), \mathcal{M}_I(G))$ be given. Then*

$$\max_{S \subseteq E(G)} \frac{\overline{\rho}(S)}{\rho(S)} \leq 2(\Delta - 1).$$

Proof. Let M_1 and M_2 be two maximal induced matchings in G and let $e \in M_2 \setminus M_1$. Clearly $M_1 \cup \{e\} \subseteq E$ and, by the maximality condition, this set is not independent (i.e. it is not an induced matching anymore). Hence there exists $\phi(e) \in M_1$ at distance less than two from e and again since M_2 is maximal and independent $\phi(e) \in M_1 \setminus M_2$. Indeed ϕ defines a function from $M_2 \setminus M_1$ to $M_1 \setminus M_2$. Let f be one of the edges in the range of ϕ. A bound on the number of edges $e \in M_2 \setminus M_1$ that can be the pre-image of $f \in M_1 \setminus M_2$ is needed. There can be at most $2(\Delta - 1)$ such e. The result follows.   □

The last result gives a bound on the ratio of any algorithm that construct a maximal induced matching in a given graph.

Theorem 2. $(1, \Delta)$-MIM *can be approximated with ratio $2(\Delta - 1)$.*

This result can be slightly improved on regular graphs.

Theorem 3. *If G is a (δ, Δ)-graph then $\nu_I(G) \leq \frac{\Delta|V(G)|}{2(\Delta + \delta - 1)}$. Moreover for every $d \geq 3$, with d odd, there exists a regular graph of degree d, with $2(2d - 1)$ vertices and a maximum induced matching of size d.*

Proof. Let G be a (δ, Δ)-graph and M be a maximal induced matching in G. Let $R = V \setminus V(M)$. Each $v \in V(M)$ is adjacent to at least $\delta - 1$ vertices in R. Each $v \in R$ can be adjacent to at most Δ vertices in $V(M)$. Hence $(|V(G)| - 2|M|)\Delta \geq 2|M|(\delta - 1)$ and the result follows.

The second part can be proved by giving a recursive description of a family of graphs $\{G_i\}_{i \in \mathbb{N}^+}$. It is convenient to draw G_i so that all its vertices are on five different layers, called *far-left, mid-left, central, mid-right* and *far-right* layer. Figure 1 shows G_1. Vertices v_1 and v_2 (respectively u_1 and u_2) are in the far-left (respectively far-right) layer. Vertices v_3 and v_4 (respectively u_3 and u_4) are in the mid-left (respectively mid-right) layer. Vertices z_1 and z_2 are in the central layer. Moreover an horizontal axis separates odd-indexed vertices (which are below it) from even-indexed ones (which are above it), with smaller indexes below higher ones.

Let G_{i-1}, for $i \geq 2$, be given. The graph G_i is obtained by adding four central vertices, two mid-left and two mid-right vertices. Since G_1 has two

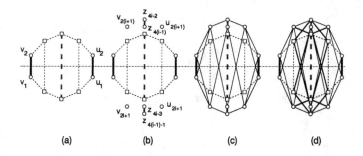

Fig. 2. A regular graph of degree d with large induced matching.

central and two pairs of mid vertices and easy inductions proves that G_{i-1} has $2[2(i-1)-1]$ central vertices and $2(i-1)$ mid-left and mid-right ones. Let $z_{4(i-1)-1}, z_{4(i-1)}, z_{4i-3}, z_{4i-2}$ be the four "new" central vertices, v_{2i+1} and $v_{2(i+1)}$, u_{2i+1} and $u_{2(i+1)}$ the mid-left and mid-right ones. G_i has all edges of G_{i-1} plus the following groups:

1. two edges connecting each of v_{2i+1}, $v_{2(i+1)}$, (respectively u_{2i+1} and $u_{2(i+1)}$) to v_1 and v_2 (respectively u_1 and u_2), plus edges

$$\{v_{2i+1}, z_{4(i-1)-1}\}, \{v_{2i+1}, z_{4(i-1)}\}, \qquad \{v_{2(i+1)}, z_{4i-3}\}, \{v_{2(i+1)}, z_{4i-2}\},$$

$$\{u_{2i+1}, z_{4(i-1)-1}\}, \{u_{2i+1}, z_{4(i-1)}\}, \qquad \{u_{2(i+1)}, z_{4i-3}\}, \{u_{2(i+1)}, z_{4i-2}\}$$

   All these edges are the continuous black lines in Figure 2.(c).
2. A final set of edges connects each of the even index mid vertices with the central vertices of G_{i-1} with indices $4j-2$ and $4j-3$ for $j=0,1,\ldots,i-1$. Each of the odd index mid vertices are connected with the central vertices of G_{i-1} with indices $4(j-1)$ and $4(j-1)-1$ for $j=1,\ldots,i$. The squares in Figure 2 represent all mid vertices in G_{i-1}. The bold solid lines in Figure 2.(d) represent this kind of edges.

Graph G_1 has an induced matching of size three. For each $i \geq 2$ the matching in G_i is obtained by adding the two edges $\{z_{4i-2}, z_{4(i-1)}\}$ and $\{z_{4i-3}, z_{4(i-1)-1}\}$ to the matching in G_{i-1}. □

Theorem 3 is complemented by the following result, giving a lower bound on the size of a particular family of induced matchings.

Theorem 4. *Let* $f(\delta, \Delta) = (4\Delta^2 - 4\Delta + 2)/\delta$. *If* G *is a* (δ, Δ)-*graph then* $\nu_I(G) \geq |V(G)|/f(\delta, \Delta)$. *Moreover for every* $d \geq 2$ *there exists a regular graph of degree* d *with* $d \cdot f(d, d)$ *vertices and a maximal induced matching of size* d.

Proof. Let G be a (δ, Δ)-graph and M a maximal induced matching in G. Each edge in G must be covered by at least an edge in M. Conversely every edge in M can cover at most $2(\Delta - 1)^2 + 2\Delta - 1$ edges. Thus

$$|M| \geq \frac{|E(G)|}{2(\Delta - 1)^2 + 2\Delta - 1} \geq \frac{\delta|V(G)|}{2(2\Delta^2 - 2\Delta + 1)}.$$

A *d-ary depth two tree* T_d is formed by connecting with an edge the roots of two identical copies of a complete d-ary tree on $d^2 - d + 1$ vertices. The graph obtained by taking d copies of T_d all sharing the same set of $(d-1)^2$ leaves is regular of degree d, it has $d \cdot f(d, d)$ vertices and a maximal induced matching of size d. □

Corollary 1. *d-MIM can be approximated with ratio $d - (d-1)/(2d-1)$.*

Proof. Let G be a regular graph of degree d. The proof follows from Theorem 3 and Theorem 4 and the use of any greedy algorithm that returns a maximal induced matching in G. □

3 Hardness of Approximation

In this section we investigate the non-approximability of MIM for various classes of bounded degree graphs. Although several notions of approximation preserving reductions have been proposed (see for example [4]) the L-reduction defined in [13] is perhaps the easiest one to use. Let P be an optimization problem. For every instance x of P, and every solution y of x, let $c_P(x, y)$ be the *cost* of the solution y. Let $opt_P(x)$ be the cost of an optimal solution.

Definition 2. *Let P and Q be two optimization problems. An L-reduction from P to Q is a four-tuple $(t_1, t_2, \alpha, \beta)$ where t_1 and t_2 are polynomial time computable functions and α and β are positive constants with the following properties:*

(1) t_1 maps instances of P to instances of Q and for every instance x of P, $opt_Q(t_1(x)) \leq \alpha \cdot opt_P(x)$.

(2) for every instance x of P, t_2 maps pairs $(t_1(x), y')$ (where y' is a solution of $t_1(x)$) to a solution y of x so that

$$|opt_P(x) - c_P(x, t_2(t_1(x), y'))| \leq \beta |opt_Q(t_1(x)) - c_Q(t_1(x), y')|.$$

Theorem 5. *If P and Q are two maximization problems, there is an L-reduction from P to Q with parameters α and β, and it is NP-hard to approximate P with ratio c then it is NP-hard to approximate Q with ratio $\frac{\alpha\beta c}{(\alpha\beta - 1)c + 1}$.*

Proof. The result is derived from [12, Proposition 13.2]. Suppose by contradiction that there is an algorithm which approximates Q with ratio $\frac{\alpha\beta c}{(\alpha\beta - 1)c + 1}$. For every instance x of P let y' be the result of applying this algorithm to $t_1(x)$. Then, by definition of L-reduction,

$$\frac{opt_P(x) - c_P(x, t_2(t_1(x), y'))}{opt_P(x)} \leq \alpha\beta \frac{opt_Q(t_1(x)) - c_Q(t_1(x), y')}{opt_Q(t_1(x))}$$

By definition of performance ratio it is $\frac{opt_Q(t_1(x))}{c_Q(t_1(x), y')} \leq \frac{\alpha\beta c}{(\alpha\beta - 1)c + 1}$, therefore

$$\frac{opt_Q(t_1(x)) - c_Q(t_1(x), y')}{opt_Q(t_1(x))} \leq 1 - \frac{(\alpha\beta - 1)c + 1}{\alpha\beta c} = \frac{1}{\alpha\beta}\left(1 - \frac{1}{c}\right)$$

and the result follows. □

Let MIS denote the problem of finding a largest independent set in a graph (problem GT20 in [8]). Appellations (δ, Δ)-MIS and d-MIS are defined in the obvious way. There is [10] a very simple L-reduction from MIS to MIM with parameters $\alpha = \beta = 1$. Given a graph $G = (V, E)$, define $t_1(G) = (V', E')$ as follows:

$$V' = V \cup \{v' : v \in V\}, \quad E' = E \cup \{\{v, v'\} : v \in V\}.$$

If U is an independent set in G then $F = \{\{v, v'\} : v \in U\}$ is an induced matching in $t_1(G)$. Conversely if F is an induced matching in $t_1(G)$ the set $t_2(t_1(G), F)$ obtained by picking one endpoint from every edge in F is an independent set in G. Therefore the size of the largest independent set in G is $\nu_I(t_1(G))$.

The s-*padding of a* (δ, Δ)-graph G, G_s, is obtained by replacing every vertex v by a distinct set of vertices $v_1, \ldots, v_s$ with $\{v_i, u_j\} \in E(G_s)$ if and only if $\{u, v\} \in E(G)$. The following result is a consequence of the definition.

Lemma 1. *For any $s \geq 2$, if G is a (δ, Δ)-graph then G_s is a $(s \cdot \delta, s \cdot \Delta)$-graph.*

The key property of the s-padding of a graph is that it preserves the distance between two vertices. If $G = (V, E)$ is a graph then for every $u, v \in V(G)$, $dst_G(u, v)$ is the *distance* between u and v, defined as the number of edges in a shortest path between u and v.

Lemma 2. *For all graphs G and for every $s \geq 2$, $dst_G(u, v) = dst_{G_s}(u_i, v_j)$ for all $u, v \in V(G)$ and all $i, j \in \{1, 2, \ldots, s\}$.*

Lemma 3. *For all graphs G and for every $s \geq 2$, $\nu_I(G) = \nu_I(G_s)$.*

Proof. Let M be an induced matching in G. Define $M_s = \{\{u_1, v_1\} \in E(G_s) : \{u, v\} \in M\}$. By Lemma 2 all edges in M_s are at distance at least two. Conversely if M_s is an induced matching in G_s define $M = \{\{u, v\} \in E(G) : \{u_i, v_j\} \in M_s$ for some $i, j \in \{1, \ldots, s\}\}$. M is an induced matching in G. $\quad\square$

The following Lemmas show how to remove vertices of degree one, two and three from a $(1, \Delta)$-graph.

Lemma 4. *Any $(1, \Delta)$-graph G can be transformed in polynomial time into a $(2, \Delta)$-graph G' such that $|V_1(G)| = \nu_I(G') - \nu_I(G)$.*

Proof. Given a $(1, \Delta)$-graph G, the graph G' is obtained by replacing each vertex v of degree one in G by the gadget G_v shown in Figure 3. The edge $\{v, w\}$ incident to v is attached to v_0. The resulting graph has minimum degree two and maximum degree Δ. If M is an induced matching in G it is easy to build an

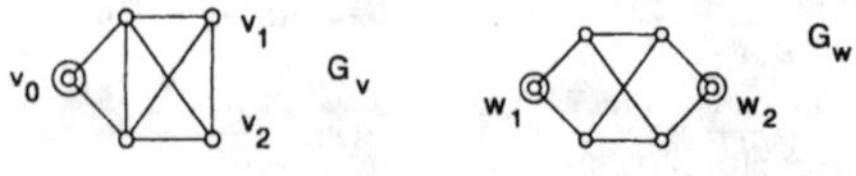

Fig. 3. Gadgets replacing vertices of degree one and two.

induced matching in G' of size $|M|+|V_1(G)|$. Conversely every induced matching M' in G' will contain exactly one edge from every gadget G_v. Replacing (if necessary) each of these edges by the edge $\{v_1, v_2\}$ could only result in a larger matching. The matching obtained by forgetting the gadget-edges is an induced matching in G and its size is (at least) $|M'| - |V_1(G)|$. $\square$

Lemma 5. *Any $(2, \Delta)$-graph G can be transformed in polynomial time into a $(3, \Delta)$-graph G' such that $|V_2(G)| = \nu_I(G') - \nu_I(G)$.*

Proof. Let G be a $(2, \Delta)$-graph. Every vertex w of degree two is replaced by the graph G_w in Figure 3. The two edges $\{u, w\}$ and $\{v, w\}$ adjacent to w are replaced by edges $\{u, w_1\}$ and $\{v, w_2\}$. Let G' be the resulting $(3, \Delta)$-graph. If M is a maximal induced matching in G, a matching M' in G' is obtained by taking all edges in M and adding one edge from each of the graphs G_w. Figure 4 shows all the relevant cases. If $w \in V(M)$ then without loss of generality we can assume that $w_1 \in V(M')$ and one of the two edges adjacent to w_2 can be added to M'. If $w \notin V(M)$ then any of the four central edges in G_w can be added to M'. After these replacements no vertex in the original graph gets any closer to an edge in the matching. Inequality $\nu_I(G') \geq \nu_I(G) + |V_2(G)|$ follows from the argument above applied to a maximum induced matching in G.

Conversely for any induced matching M' in G' at most one edge from each copy of G_w belongs to M'. The copies of G_w with $M' \cap E(G_w) = \emptyset$ are called *empty*, all others are called *full*. Inequality $\nu_I(G) \geq \nu_I(G') - |V_2(G)|$ is proved by the following claims applied to a maximum induced matching in G'.

Claim 1 *Any maximal induced matching M' in G' can be transformed into another induced matching M'' in G' with $|M'| \leq |M''|$ and such that all gadgets in M'' are full.*

Claim 2 $M =_{df} M'' \cap E(G)$ *is an induced matching in G.*

To prove the first claim, an algorithm is described which, given an induced matching $M' \subseteq E(G')$, fills all empty gadgets in M'. The algorithm visits in turn all gadgets in G' that have been created by the reduction and performs the following steps:

(1) If the gadget G_w under consideration is empty some local replacements are performed that fill G_w.

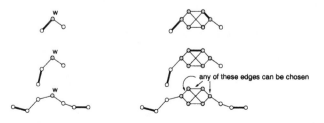

Fig. 4. Possible ways to define the matching in G' given the one in G.

Fig. 5. Filling an empty gadget, normal cases.

(2) The gadget G_w is then marked as "checked".

(3) A *maximality restoration* phase is performed in which, as a consequence of the local replacements in Step (1), some edges might be added to the induced matching.

Initially all gadgets are "unchecked". Let G_w be an unchecked gadget. If G_w is full the algorithm simply marks it as checked and carries on to the next gadget. Otherwise, since M' is maximal, at least one of the two edges adjacent to vertices w_1 and w_2 must be in M' for otherwise it would be possible to extend M' by picking any of the four central edges in G_w. Without loss of generality let $\{u, w_1\} \in M'$. Figure 5 shows all possible cases. If vertex v does not belong to another gadget then either of the configurations on the left of Figure 5 is replaced by the one shown on the right. If v is part of another gadget few subcases need to be considered. Figure 6 shows all possible cases and the replacement rule. In all cases after the replacement the neighbouring gadget is marked as checked. Notice that all replacement rules do not decrease the size of the induced matching. Also as the process goes by, new edges in $E(G)$ can only be added to the current matching during the maximality restoration phase. To prove the second claim, assume by contradiction that two edges $e = \{u, v\}$ and $f = \{w, y\}$ in M are at distance one. Notice that $dst_{G'}(e, f) = dst_G(e, f)$ unless all the shortest paths beween them contain a vertex of degree two. The existence of e and f is contradicted by the fact that M' and M'' are induced matchings in G' and all gadgets in G' are filled by M''.   □

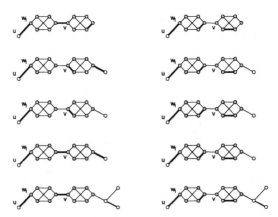

Fig. 6. Filling an empty gadget, special cases.

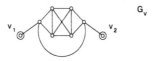

Fig. 7. Gadget connecting pairs of vertices of degree three.

Lemma 6. *Any $(3, \Delta)$-graph G can be transformed in polynomial time into a $(4, \Delta)$-graph G' such that $|V_3(G)| = \nu_I(G') - 2 \cdot \nu_I(G)$.*

Proof. Let G be a $(3, \Delta)$-graph. The graph G' is defined by taking two copies of G and connecting pairs of corresponding vertices of degree three with the gadget shown in Figure 7. The result follows by an argument similar to that in the proof of Lemma 4. □

The non-approximability of $(ks, (\Delta + 1)s)$-MIM (for $k = 1, 2, 3, 4$) and $4s$-MIM follows from Theorem 5 and known results on independent set [1,2].

Theorem 6. *Let $h(\delta, \Delta, c) = \frac{\lfloor f(\delta, \Delta) + 1 + \lfloor \delta/3 \rfloor \rfloor c}{f(\delta, \Delta)c + 1}$. Define $g(0, \Delta, c) = c$ and for $i > 0$, $g(i, \Delta, c) = h(i, \Delta, g(i - 1, \Delta, c))$. For every $\Delta \geq 3$, let c_Δ be a constant such that it is NP-hard to approximate $(1, \Delta)$-MIS with ratio c_Δ. Then for $k = 1, 2, 3, 4$ and every integer $s > 0$ it is NP-hard to approximate $(ks, (\Delta+1)s)$-MIM with ratio $g(k - 1, \Delta, c_\Delta)$.*

Proof. The result for $k = 1$ follows from the L-reduction at the beginning of the section for $s = 1$ and a further L-reduction based on s-paddings for $s \geq 2$. For $k \in \{2, 3, 4\}$ If G has minimum degree $k - 1$, Theorem 4 implies $\nu_I(G) \geq |V_{k-1}(G)|/f(k - 1, \Delta)$. The result follows using these bounds along with the reductions in Lemma 4, 5, and 6. □

Theorem 7. *Let c_0 be a constant such that 3-MIS is NP-hard to approximate with ratio c_0. Then for every integer $s > 0$ it is NP-hard to approximate $4s$-MIM with ratio $\frac{3913c_0}{3858c_0 + 55}$.*

Proof. The reduction at the beginning of this section and Theorem 5 imply that it is NP-hard to approximate $(1, 4)$-MIM with ratio c_0. If the original cubic graph G has n vertices, then $t_1(G)$ has $|V_1(t_1(G))| = |V_4(t_1(G))| = n$, no vertex of degree two or three, $5n/2$ edges and the maximum number of edges at distance at most one from a given edge is 19. We call one such graph a *special* $(1, 4)$-graph.

Claim 3 *There is an L-reduction from $(1, 4)$-MIM restricted to special $(1, 4)$-graphs to $(3, 4)$-MIM with parameters $\alpha = \frac{43}{5}$ and $\beta = 1$.*

If G is a $(1, 4)$-graph with $|V_2(G)| = |V_3(G)| = 0$, then replacing each vertex v of degree one with the gadget in Figure 3, gives a $(3, 4)$-graph G'. The properties of special $(1, 4)$-graphs and the same argument used to prove Theorem 4 imply $\nu_I(G) \geq \frac{5}{38}|V_1(G)|$. Therefore

$$\nu_I(G') = \nu_I(G) + |V_1(G)| \leq \nu_I(G) + \frac{38}{5}\nu_I(G) = \frac{43}{5}\nu_I(G)$$

Also, for every matching M' in G', define $t_2(G', M')$ as described in Lemma 4. It is $\nu_I(G) - |t_2(G', M')| \leq \nu_I(G') - |M'|$ and the claim is proved.

Therefore, by Theorem 5, $(3, 4)$-MIM is hard to approximate with ratio $c_1 = \frac{43c_0}{38c_0+5}$. The *special* $(3, 4)$-graphs H generated by the last reduction have again a lot of structure. It is $|V_3(H)| = |V_4(H)|$, $|E(H)| = 7|V_3(H)|/2$ and again the maximum number of edges at distance at most one from a given edge is 23.

Claim 4 *There is an L-reduction from $(3, 4)$-MIM restricted to special $(3, 4)$-graphs to 4-MIM with parameters $\alpha = \frac{91}{11}$ and $\beta = 1$.*

The reduction was described in Lemma 6. Theorem 4 and the properties of special $(3, 4)$-graphs imply $\nu_I(H) \geq \frac{11}{69}|V_3(H)|$. Therefore $\nu_I(H') \leq \frac{91}{11}\nu_I(H)$ and thus, by Theorem 5, 4-MIM is hard to approximate with ratio $\frac{91c_1}{80c_1+11}$. Finally by Lemma 3 there is an L-reduction from 4-MIM to $4s$-MIM (for $s \geq 2$) with parameters $\alpha = \beta = 1$. $\qquad\qquad\square$

4   Polynomial Time Solution on Trees

Although NP-complete for several classes of graphs including planar or bipartite graphs of maximum degree four and regular graphs of degree four, the problem of finding the largest induced matching admits a polynomial time solution on trees [3]. The algorithmic approach suggested by Cameron reduces the problem to that of finding the largest independent set in a graph H that can be defined starting from the given tree. If $G = (V, E)$ is a tree, the graph $H = (W, F)$ has $|V|-1$ vertices, one for each edge in G and there is an edge between two members of W if and only if the two original edges in G are either incident or connected by a single edge. Notice that $|F| = O(|V|^2)$. Moreover each induced matching in G is an independent set of the same size in H. Gavril's algorithm [7] finds the largest independent set in a chordal graph with n vertices and m edges in $O(n+m)$ time. Since the graph H is chordal, the largest induced matching in the tree can be found in $O(|V|^2)$ time. In this section we describe a simpler and more efficient way of finding a maximum induced matching in a tree. If $G = (V, E)$ is a tree we choose a particular vertex $r \in V$ to be the *root* of the tree (and we say that G is *rooted* at r). If $v \in V \setminus \{r\}$ then parent(v) is the unique neighbour of v in the path from v to r; if parent$(v) \neq r$ then grandparent$(v) = $ parent(parent(v)). In all other cases parent and grandparent are not defined. If $u = $ parent(v) then v is u's *child*. All children of the same node are *siblings* of each other. Let $c(v)$ be the number of children of node v. The *upper neighbourhood* of v (in symbols UN(v)) is empty if $v = r$, it includes r and all v's siblings if v is a child of r and it includes v's siblings, v's parent and v's grandparent otherwise. $E($UN$(v))$ is the set of edges in G connecting the vertices in UN(v).

Claim 5 *If $G = (V, E)$ is a tree and M is an induced matching in G then $|M \cap E($UN$(v))| \leq 1$, for every $v \in V$.*

Note that if M is an induced matching in G, any node v in the tree belongs to one of the following types with respect to the set of edges $E($UN$(v))$:

Type 1. the node $\{v, \text{parent}(v)\}$ is part of the matching,
Type 2. either $\{\text{parent}(v), \text{grandparent}(v)\}$ or $\{\text{parent}(v), w\}$ (where w is some siblings of v) belongs to the matching,
Type 3. Neither **Type 1.** nor **Type 2.** applies.

The algorithm for finding the largest induced matching in a tree G with n vertices handles an $n \times 3$ matrix Value such that $\text{Value}[i, t]$ is the size of the matching in the subtree rooted at i if vertex i is of type t.

Lemma 7. *If G is a tree with n vertices, $\text{Value}[i, t]$ can be computed in $O(n)$ time for every $i \in \{1, \ldots, n\}$ and $t = 1, 2, 3$.*

Proof. Let G be a tree with n vertices. We assume G is in adjacency list representation. If h is the height of the tree, some linear preprocessing is needed to define an array level[i] (for $i = 0, \ldots, h$) such that level[i] contains all vertices at distance i from the root.

The matrix Value can be filled in a bottom-up fashion starting from the deepest vertices of G. If i is a leaf of G then $\text{Value}[i, t] = 0$ for $t = 1, 2, 3$. In filling the entry corresponding to node $i \in V$ of type t we only need to consider the entries for all children of i.

(1) $\text{Value}[i, 1] = \sum_{k=1}^{c(i)} \text{Value}[j_k, 2]$. Since $\{i, \text{parent}(i)\}$ will be part of the matching, we cannot pick any edge from i to one of its children. The matching for the tree rooted at i is just the union of the matchings of the subtrees rooted at each of i's children.

(2) $\text{Value}[i, 2] = \sum_{k=1}^{c(i)} \text{Value}[j_k, 3]$. We cannot pick any edge from i to one of its children here either.

(3) If node i has $c(i)$ children then $\text{Value}[i, 3]$ is the maximum between $\sum_{k=1}^{c(i)} \text{Value}[j_k, 3]$ and a number of terms

$$s_{j_k} = 1 + \text{Value}[j_k, 1] + \sum_{l \neq k} \text{Value}[j_l, 2]$$

If the upper neighbourhood of i is unmatched we can either combine the matchings in the subtrees rooted at each of i's children (assuming these children are of type 3) or add to the matching an edge from i to one of its children j_k (the one that maximises s_{j_k}) and complete the matching for the subtree rooted at i with the matching for the subtree rooted at j_k (assuming j_k is of type 1) and that of the subtrees rooted at each of i's other children (assuming these children are of type 3).

Option (3) above is the most expensive involving the maximum over a number of sums equal to the degree of the vertex under consideration. Since the sum of the degrees in a tree is linear in the number of vertices the whole table can be computed in linear time. $\square$

Theorem 8. *MIM can be solved optimally in polynomial time if G is a tree.*

Proof. The largest between $\text{Value}[r, 1]$, $\text{Value}[r, 2]$ and $\text{Value}[r, 3]$ is the size of the largest matching in G. By using appropriate date structures it is also possible to store the actual matching. The complexity of the whole process is $O(n)$. $\square$

5   Conclusions

In this paper we investigated the complexity of finding a largest induced matching in a graph. We suggested a couple of simple heuristics for solving the problem with $O(\Delta)$-bounded ratio on graphs with maximum degree Δ. The definition of algorithms achieving better performance ratios is an open problem. In Section 3 we complemented these positive results with a number of non-approximability results. In particular there is a constant c such that it is NP-hard to find induced matchings whose size approximates $\nu_I(G)$ with ratio c even if G is regular of degree $4s$ for any integer $s > 0$. We believe that similar results hold for cubic graphs and, using the s-padding technique described in Section 3, for regular graphs of degree $3s$ for any integer $s > 0$. Proving hardness on regular graphs of degree d for every integer $d > 2$ maybe harder. Finally we presented an algorithm which solves MIM optimally if the input graph is a tree. Our algorithm is simple in that it does not reduce the original problem to another one, its complexity improves the one of the best algorithm known and it is clearly optimal in the sense that to define an induced matching in a tree $\Omega(n)$ operations are needed.

References

1. N. Alon, U. Feige, A. Wigderson, and D. Zuckerman. Derandomized Graph Products. *Computational Complexity*, 5:60–75, 1995.
2. P. Berman and M. Karpinski. On Some Tighter Inapproximability Results. In *Proc. 26th I.C.A.L.P.*, Springer - Verlag, 1999.
3. K. Cameron. Induced Matchings. *Discr. Applied Math.*, 24(1-3):97–102, 1989.
4. P. Crescenzi. A Short Guide to Approximation Preserving Reductions. In *Proc. 12th Conf. on Comput. Complexity*, pages 262–273, Ulm, 1997.
5. P. Erdős. Problems and Results in Combinatorial Analysis and Graph Theory. *Discr. Math.*, 72:81–92, 1988.
6. R. J. Faudree, A. Gyárfas, R. H. Schelp, and Z. Tuza. Induced Matchings in Bipartite Graphs. *Discr. Math.*, 78(1-2):83–87, 1989.
7. F. Gavril. Algorithms for Minimum Coloring, Maximum Clique, Minimum Covering by Cliques and Maximum Independent Set of a Chordal Graph. *SIAM J. Comp.*, 1(2):180–187, June 1972.
8. M. R. Garey and D. S. Johnson. *Computer and Intractability, a Guide to the Theory of NP-Completeness*. Freeman and Company, 1979.
9. B. Korte and D. Hausmann. An Analysis of the Greedy Heuristic for Independence Systems. *Ann. Discr. Math.*, 2:65–74, 1978.
10. S. Khanna and S. Muthukrishnan. Personal communication.
11. J. Liu and H. Zhou. Maximum Induced Matchings in Graphs. *Discr. Math.*, 170:277–281, 1997.
12. C. Papadimitriou. *Computational Complexity*. Addison-Wesley, 1994.
13. C. H. Papadimitriou and M. Yannakakis. Optimization, Approximation and Complexity Classes. *J. Comp. Sys. Sciences*, 43:425–440, 1991.
14. L. J. Stockmeyer and V. V. Vazirani. NP-Completeness of Some Generalizations of the Maximum Matching Problem. *I.P.L.*, 15(1):14–19, August 1982.
15. A. Steger and M. Yu. On Induced Matchings. *Discr. Math.*, 120:291–295, 1993.

Mod-2 Independence and Domination in Graphs

Magnús M. Halldórsson[1,3], Jan Kratochvíl[2*], and Jan Arne Telle[3]

[1] University of Iceland, Reykjavik, Iceland
mmh@hi.is
[2] Charles University, Prague, Czech Republic
honza@kam.ms.mff.cuni.cz
[3] University of Bergen, Bergen, Norway
telle@ii.uib.no

Abstract. We develop an $O(n^3)$ algorithm for deciding if an n-vertex digraph has a subset of vertices with the property that each vertex of the graph has an even number of arcs into the subset. This algorithm allows us to give a combinatorial interpretation of Gauss-Jordan and Gauss elimination on square boolean matrices. In addition to solving this independence-mod-2 (even) set existence problem we also give efficient algorithms for related domination-mod-2 (odd) set existence problems on digraphs. However, for each of the four combinations of these two properties we show that even though the existence problem on digraphs is tractable, the problems of deciding the existence of a set of size exactly k, larger than k, or smaller than k, for a given k, are all NP-complete for undirected graphs.

1 Introduction

A large class of well-studied independence and domination properties in graphs can be characterized by two sets of nonnegative integers σ and ρ. A (σ, ρ)-set S in a graph has the property that the number of neighbors every vertex $u \in S$ (or $u \notin S$) has in S, is an element of σ (of ρ, respectively) [7]. In a recent paper [3] it is shown that deciding if a given graph has a $(\{0\}, \rho)$-set, *i.e.* vertices in S forming an independent set with further ρ-imposed domination constraints, is NP-complete whenever there is a non-negative integer $x \notin \rho$ with $x + 1 \in \rho$, unless ρ are exactly the positive numbers, and polynomial in all other cases. For the cases $0 \notin \rho$ the vertices in S form a dominating set. In the present paper we consider cases of independence and domination modulo 2, where σ and ρ are either the set of all even numbers or the set of all odd numbers. We denote these sets EVEN and ODD, respectively, and consider both decision and optimization versions of the four cases of (σ, ρ) equal to (EVEN, EVEN), (EVEN, ODD), (ODD, ODD), (ODD, EVEN).

In the next section, we develop an $O(n^3)$ algorithm to decide if a given graph G has an (EVEN, EVEN)-set, *i.e.* a set S such that each vertex of G has an even number of neighbors in S. This disproves a 1994 conjecture stating that no non-trivial (σ, ρ)-set existence problems were solvable in polynomial time [2]. In

* Research support in part by Czech research grants GAUK 194/1996 and 158/1999, and GAČR 201/1996/0194.

Widmayer et al. (Eds.): WG'99, LNCS 1665, pp. 101–109, 1999.

section 3 we show that quite trivially this problem is equivalent to determining if the adjacency matrix of the given graph is singular, and that our algorithm is a combinatorial interpretation of Gauss-Jordan elimination on square boolean matrices. A recent paper giving combinatorial interpretations of various matrix algorithms left such a view of Gaussian elimination, for general matrices, as an open problem [5]. We provide a partial answer to this question by using the connection with (EVEN, EVEN)-sets to give a combinatorial interpretation of Gaussian elimination for square boolean matrices.

In section 4 we give polynomial algorithms also for the three remaining existence problems. In section 5 we consider the complexity of deciding if a given graph has a desired set of size at least k, at most k, or exactly k, for a given integer k. By reductions from NP-complete coding problems asking for codewords of given length, we show that these maximization, minimization and exact versions of all four problems are NP-complete.

2   Existence of Independence Mod-2 Sets

We will be viewing an undirected graph as a directed graph with arcs uv and vu for each edge $\{u,v\}$. We first generalize the (σ, ρ) problems to directed graphs with loops, and denote the set of out-neighbors of vertex v by $v_{out} = \{u : vu \in E\}$ and its in-neighbors by $v_{in} = \{u : uv \in E\}$. If the graph G is not clear from context we write $v_{out}(G), v_{in}(G)$.

Definition 1. *A nonempty subset of vertices S of a directed graph $G = (V, E)$ is a (σ, ρ)-set if $|v_{out} \cap S| \in \sigma$ for any $v \in S$ and $|v_{out} \cap S| \in \rho$ for any $v \in V \setminus S$.*

Note that we could also have chosen to count in-neighbors, since in the graph with all arcs reversed this would define the exact same vertex subsets as (σ, ρ)-sets. This simple transformation implies that decision problems over (σ, ρ)-sets will have the same time complexity regardless of whether we count in-neighbors or out-neighbors.

Our algorithm for deciding if a graph G has an (EVEN, EVEN)-set will consist of repeatedly applying a graph operation that will maintain the property of interest. This will give a series of graphs $G^{(0)}, G^{(1)}, ..., G^{(i)}$, starting with the input graph and ending with a graph for which it will be trivial to decide if it has an (EVEN, EVEN)-set. Let $\oplus$ be the symmetric difference operator, *i.e.* $A \oplus B = \{x \in A \cup B : x \notin A \cap B\}$. The main observation is that for any two vertices u and r, (EVEN, EVEN)-sets are invariant under the operation:

$$u_{out} := u_{out} \oplus r_{out}$$

Lemma 1. *Let G' be the graph G altered by $u_{out}(G') := u_{out}(G) \oplus r_{out}(G)$ for two vertices u, r. Then S is an (EVEN, EVEN)-set of G if and only if S is an (EVEN, EVEN)-set of G'.*

Proof. Outgoing neighbors for any vertex $x \neq u$ are identical in G and G'. For the forward direction of the proof it therefore suffices to show that $|u_{out}(G') \cap S| =$

$|u_{out}(G) \cap S| + |r_{out}(G) \cap S| - 2|u_{out}(G) \cap r_{out}(G) \cap S|$ is even. But since S is an (EVEN, EVEN)-set of G, all 3 terms in the right-hand side of the above equality are even and thus so is their sum. Conversely, we have $|u_{out}(G) \cap S| = |u_{out}(G') \cap S| - |r_{out}(G) \cap S| + 2|u_{out}(G) \cap r_{out}(G) \cap S|$ also even for similar reasons.

Our algorithm will for each of the n vertices in G maintain an *in-flag* and an *out-flag*, initially all lowered. In the ith stage of the algorithm we choose a vertex c with lowered in-flag that has at least one incoming neighbor r with lowered out-flag. In addition to raising the in-flag of c and the out-flag of r the code for this stage consists of the loop:

$$\text{for each } u \in c_{in} \setminus \{r\} \text{ do } u_{out} := u_{out} \oplus r_{out}$$

In the resulting graph $G^{(i)}$ the vertex c will have only the single incoming neighbor r. Once flags are raised they are never lowered, thus after n successful stages each vertex would have exactly one incoming and one outgoing neighbor. Clearly, such a graph can have no (EVEN, EVEN)-set. However, if there is a vertex in $G^{(i)}$ with lowered in-flag which has no incoming neighbor with lowered out-flag, then an (EVEN, EVEN)-set exists and we halt. Before proving this fact we give the algorithm formally below. Sets C and R represent the subsets of vertices having raised in-flags and out-flags, respectively.

∃ (EVEN, EVEN)-SET ALGORITHM
***
input: digraph $G = (V, E)$
$C := R := \emptyset$
$i := 0$
while $(i < n)$ and $(\not\exists x \in V \setminus C : x_{in} \subseteq R)$ do
$\quad \{ \ i := i + 1$
$\qquad$ pick $c \in V \setminus C$ and set $C := C \cup \{c\}$
$\qquad$ pick $r \in c_{in} \setminus R$ and set $R := R \cup \{r\}$
$\qquad$ for each $u \in c_{in} \setminus \{r\}$ do $u_{out} := u_{out} \oplus r_{out} \ \}$
if $(i = n)$ then $\not\exists$ (EVEN, EVEN)-set
else $\{$ let $x \in V \setminus C : x_{in} \subseteq R$
$\qquad S := \{x\} \cup \{v \in C : v \in y_{out} \wedge y \in x_{in}\}$ is an (EVEN, EVEN)-set$\}$

Lemma 2. *If $i < n$ upon completion of the algorithm then S is an (EVEN, EVEN)-set of the current graph $G^{(i)}$.*

Proof. A vertex $v \notin R$ has no outgoing edges to C so $|v_{out} \cap S| = 0$. Note that $|x_{in}| = |\{v \in C : v \in y_{out} \wedge y \in x_{in}\}|$ since $x_{in} \subseteq R$ and each vertex of R has exactly one, distinct, outgoing neighbor in C. Each vertex $v \in x_{in}$ has therefore 2 outgoing neighbors in S, namely x and $v_{out} \cap C$, while any vertex $w \in R$ with $w \notin x_{in}$ has no outgoing neighbors in S.

By applying Lemma 1 inductively it follows that the algorithm for existence of (EVEN, EVEN)-sets is correct. Its time complexity is $O(n^3)$ since in each of the at most n stages the chosen vertex c has at most n incoming neighbors that each have their at most n outgoing neighbors updated.

Theorem 1. *The algorithm decides, in time $O(n^3)$, if the input graph has an (EVEN,EVEN)-set or not.*

3   Gaussian Elimination on Boolean Matrices

Consider what the existence of an (EVEN, EVEN)-set S in a graph G implies for the boolean adjacency matrix A_G of G. Clearly, the columns corresponding to vertices in S sum to the all-zero vector (over GF(2)). Conversely, any non-empty set of columns summing to the all-zero vector is linearly dependent and the corresponding vertices form an (EVEN, EVEN)-set. Thus the matrix A_G has less than full rank, *i.e.* is singular, *i.e.* has determinant zero, if and only if G has an (EVEN, EVEN)-set.

Theorem 2. *A square boolean matrix is singular if and only if its associated directed graph has an (EVEN, EVEN)-set.*

Note that the algorithm given for the existence of (EVEN, EVEN)-sets works for any digraph, even one with self-loops. In fact, viewing it as a matrix algorithm over GF(2) it is equivalent to Gauss-Jordan elimination, as follows: In the main loop of the algorithm a new column ($c \notin C$) is processed, a non-zero pivot (entry rc) is chosen from the remaining pivot rows ($r \notin R$), and row operations are performed to make all other entries in column c equal to zero. If the algorithm completes all n stages then we are left with a permutation matrix, and otherwise we find a set of columns that are linearly dependent.

Even if it has the same asymptotic time complexity, Gaussian elimination is usually preferred over Gauss-Jordan in practice, as the constant term is smaller. Let us consider Gaussian elimination as an algorithm for determining existence of (EVEN, EVEN)-sets. The changes from the previous algorithm are in: (i) labelling of chosen vertices for ease, (ii) all in-neighbors of c^i in R (previously only r^i) are left untouched in the main loop, and (iii) definition of (EVEN, EVEN)-set S.

GAUSS ∃ (EVEN, EVEN)-SET ALGORITHM
input: digraph $G = (V, E)$
$C := R := \emptyset$
$i := 0$
while $(i < n)$ and $(\nexists x \in V \setminus C : x_{in} \subseteq R)$ do
    $\{\ i := i + 1$
       pick $c^i \in V \setminus C$ and set $C := C \cup \{c^i\}$
       pick $r^i \in c^i_{in} \setminus R$ and set $R := R \cup \{r^i\}$
       for each $u \in c^i_{in} \setminus R$ do $u_{out} := u_{out} \oplus r^i_{out}\ \}$
if $(i = n)$ then $\nexists$ (EVEN, EVEN)-set
else $\{\ S := \{x\}$
       for $k := i$ downto 1   if $|r^k_{out} \cap S|$ is odd then $S := S \cup \{c^k\}$
       S is an (EVEN, EVEN)-set$\}$

Lemma 3. *The GAUSS ∃ (EVEN, EVEN)-set algorithm is correct.*

Proof. Assume the algorithm completes with $i < n$. Then each $v \in V \setminus R$ has zero out-neighbors to x by the halting condition of the main loop, and zero outgoing neighbors to $\{c^1, c^2, ..., c^i\}$ as the only arcs to c^k left after iteration k of the main loop are from $\{r^1, ..., r^{k-1}\} \subseteq R$. Since r^k has an arc to c^k, but none to $\{c^1, ..., c^{k-1}\}$ the reverse ordering of the final loop in the definition of S implies that each $r^k \in R$ will have an even number of out-neighbors to S. Hence, S is an (EVEN, EVEN)-set.

On the other hand, if $i = n$, we show by reverse induction on k that c^k cannot belong to an (EVEN, EVEN)-set S. Assume $c^n, ..., c^{k+1} \notin S$, for $k \leq n$. We cannot have $c^k \in S$ as the only out-neighbor of r^k among $\{c^1, ..., c^k\}$ is c^k, so that r^k would then have had exactly one out-neighbor in S.

We thus have a combinatorial interpretation of Gaussian elimination for square boolean matrices.

4   Existence of Domination Mod-2 Sets

In this section we prove the following result.

Theorem 3. *The existence of (σ, ρ)-sets of type (ODD, ODD), (ODD, EVEN) and (EVEN, ODD) in directed graphs can be decided in polynomial time.*

Proof. Let G have n vertices and let A_G be its adjacency matrix. We denote by **1** and **0** the all-one and all-zero vectors of dimension n and by **I** the $n \times n$ identity matrix. We have observed that G has an (EVEN, EVEN)-set if and only if there is a non-zero vector $\mathbf{x}$ such that $A_G\mathbf{x} = \mathbf{0}$. Similarly, a vector $\mathbf{x}$ is the characteristic vector of an (ODD, ODD)-set if and only if $A_G\mathbf{x} = \mathbf{1}$. Similarly, for an (ODD, EVEN)-set we have $(A_G + \mathbf{I})\mathbf{x} = \mathbf{0}$ and for an (EVEN, ODD)-set we have $(A_G + \mathbf{I})\mathbf{x} = \mathbf{1}$. Thus, deciding the existence of these kinds of sets can be done in polynomial time by solving linear equations.

5   Existence of Sets of a Given Size

In this section we show that deciding the existence of independence and domination mod-2 sets of a given size k, whether exactly k, at least k or at most k, is NP-complete even for undirected graphs. Note that the properties studied are not hereditary, so that a graph may for example have an (EVEN, EVEN)-set of size k, but none of size larger or smaller than k. Our reductions will be from NP-complete problems in coding theory, that for our purposes can be described as follows:

Codeword of given weight: Given a binary $r \times c$ matrix H and an integer w, is there a vector $\mathbf{x}$ with w ones s.t. $H\mathbf{x} = \mathbf{0}$?

This problem on binary linear codes was shown NP-complete in [1]. The problem **Codeword of maximal weight**, asking for a vector of weight **at least** w is also NP-complete for binary codes [6]. Finally, the problem **Codeword of minimal weight** for binary linear codes, asking for a non-zero vector of weight **at most** w was conjectured NP-complete in [1], and finally proven to be so in a recent paper [9]. These problems are equivalent to asking if the orthogonal complement of the linear space generated by the columns of H contains a non-zero vector of weight w, at least w, or at most w (in other words, if there are exactly w, at least w, or nonempty set of at most w columns of H that sum up to the all-zero vector). They are thus very close to (EVEN, EVEN)-set problems. However, inputs to the (EVEN, EVEN)-set problems are square matrices, and for undirected graphs also symmetric matrices with zeros on the diagonal. We first show NP-completeness for the maximum, minimum and exact versions of the (EVEN, EVEN)-set undirected graph problems, and then use these results to give reductions for the other three properties.

Theorem 4. *The problems Codeword of maximal weight and Codeword of minimal weight remain NP-complete for symmetric matrices with all-zero diagonals.*

Proof. The problems are clearly in NP. We first resolve the maximal weight version by giving a polynomial-time reduction from the NP-complete problem Codeword of maximal weight. Given a boolean $r \times c$ matrix H and an integer w we construct a symmetric matrix with all-zero diagonals G such that G has a codeword of size at least $k = 2r + w$ iff H has a codeword of weight at least w. G will have the following form:

$$\begin{pmatrix} \mathbf{0} & \mathbf{0} & H \\ \mathbf{0} & \mathbf{0} & H \\ H^t & H^t & \mathbf{0} \end{pmatrix}$$

where H^t is the transpose of H, the lower-right $\mathbf{0}$ is the $c \times c$ all-zero matrix, and the other $\mathbf{0}$s are $r \times r$ all-zero matrices. This is a square $(2r + c) \times (2r + c)$ symmetric matrix with zeros on the diagonal. Since the leftmost $2r$ columns sum to the all-zero vector, we conclude that this matrix has a set of at least $2r + w$ columns summing to the all-zero vector iff H has a set of at least w columns summing to the all-zero vector.

We next resolve the minimal weight version by reduction from Codeword of minimal weight. Given a boolean $r \times c$ matrix H and an integer w, we construct a symmetric all-zero diagonal matrix G such that G has a codeword of size at most $k = w$ iff H has a codeword of weight at most w.

We may assume wlog that r is even, since we could add an all-zero row to H otherwise. G will have $(w + 1) \times (w + 1)$ blocks $W_{ij}, i, j = 1, 2, \ldots, w + 1$ where $W_{1,w+1} = H$ and $W_{w+1,1} = H^t$. The blocks $W_{1,w} = W_{i,w+1-i} = W_{i,w+2-i}$ for $i = 2, \ldots, w$ will contain the symmetric permutation matrix P of size r by r with the unique 1-entry in each row and column in position $(r+1-i, i), i = 1..r$ (since

r is even P has zeroes on the diagonal.) All other blocks are all-zero matrices of appropriate size. For the case $w = 3$ the matrix G thus becomes:

$$\begin{pmatrix} 0 & 0 & P & H \\ 0 & P & P & 0 \\ P & P & 0 & 0 \\ H^t & 0 & 0 & 0 \end{pmatrix}$$

This is a $3r + c$ by $3r + c$ ($wr + c$ by $wr + c$) symmetric matrix with zeros on the diagonal consisting of 4 by 4 ($w + 1$ by $w + 1$) blocks. The placement of the permutation matrices ensures that choosing a column from any but the rightmost column of blocks will force a choice of a column from all the columns of blocks, *i.e.* forcing a choice of at least $w + 1 = 4$ columns. Hence the matrix has a set of at most $w = 3$ columns summing to the all-zero vector iff all columns come from the rightmost block, *i.e.* from H. A similar argument applies to the general case.

Corollary 1. *Given an undirected graph G and an integer k, deciding if G has a non-empty (EVEN, EVEN)-set of size at least k, at most k, or exactly k is NP-complete.*

The corollary is immediate since the minimum and maximum versions are equivalent to the analogous codeword problems and the exact version follows by a Cook reduction from either of the other two. We turn to the other problems.

Theorem 5. *The maximum, minimum and exact versions of the (ODD, ODD), (ODD, EVEN) and (EVEN, ODD) problems are all NP-complete, even for undirected graphs.*

Proof. Since the reductions are quite straightforward we only give a sketch of the constructions involved and let the reader fill in details. Given graph G as input to a known NP-complete problem (as specified below), Figure 1 shows the constructed graphs $G_1, ..., G_5$ for five separate NP-completeness reductions for maximization and minimization versions. NP-completeness of the exact versions will follow from this. For each graph in Figure 1 is shown two vertices of the graph G, and the corresponding subgraph that is attached to every vertex of G to form G_i. The two unique possibilities for membership in a (σ, ρ)-set S' of G_i are shown, with black vertices belonging to S' and white not.

G_1 has an (ODD,EVEN)-set of size $2k$ iff G has an (EVEN,EVEN)-set of size k. G_1 is used to reduce from min and max (EVEN,EVEN) to min and max (ODD,EVEN), respectively.

Remaining reductions are all from max (EVEN,EVEN) and assume that G has n vertices.

G_2 has an (EVEN,ODD)-set of size $2k + n$ iff G has an (EVEN,EVEN)-set of size k, and is used to show that max (EVEN,ODD) is NP-complete.

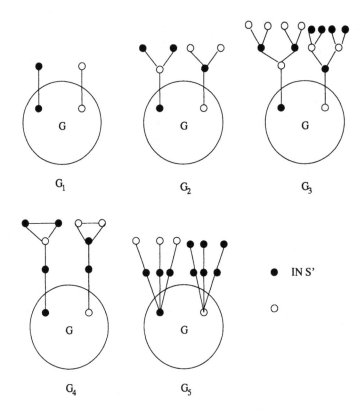

Fig. 1. The constructed graphs $G_1, ..., G_5$ for five separate reductions. For each graph is shown two vertices of the given graph G, and the corresponding subgraph that is attached to every vertex of G to form G_i. The two unique possibilities for membership in a (σ, ρ)-set S' of G_i are shown, with black vertices belonging to S' and white not.

G_3 has an (EVEN,ODD)-set of size $5n - 2k$ iff G has an (EVEN,EVEN)-set of size k, and is used to show that min (EVEN,ODD) is NP-complete.

G_4 has an (ODD,ODD)-set of size $2k + 2n$ iff G has an (EVEN,EVEN)-set of size k, and is used to show that max (ODD,ODD) is NP-complete.

G_5 has an (ODD,ODD)-set of size $6n - 2k$ iff G has an (EVEN,EVEN)-set of size k, and is used to show that min (ODD,ODD) is NP-complete.

6   Conclusion

We have resolved the complexity of (σ, ρ)-set existence, maximization, minimization and exact size problems for the cases where $\sigma, \rho \in \{EVEN, ODD\}$. The only other cases of polynomial-time solvable (σ, ρ)-set existence problems we know are either the trivial cases, for example $\sigma = \{0\}, \rho = \{1, 2, ...\}$ where the answer is always positive since every graph has an independent dominating set, those solvable by a simple greedy algorithm, see [8], or by exhaustive search, see [3]. We believe these are the only easy cases.

Conjecture 1. The only (σ, ρ)-set existence problems solvable in polynomial time, apart from the trivial cases and those resolved by exhaustive search or a simple greedy algorithm, are when $\sigma, \rho \in \{EVEN, ODD\}$.

To decide if a graph had an (EVEN,EVEN)-set we essentially did Gaussian elimination on its boolean adjacency matrix. The more general graph property resolved by Gaussian elimination on square matrices over the finite field Z_p for a prime p is: Given an edge-weighted digraph, can we assign vertex weights (not all zero) in such a way that after multiplying each edge weight by the weight of its sink vertex, the weights of edges leaving each vertex sum to zero mod p?

References

1. E. Berlekamp, R.J. McEliece and H.C.A. van Tilborg, On the inherent intractability of certain coding problems, *IEEE Trans. Inform. Theory. Vol.29, No.3, 1978, 384-386.*
2. M. Halldórsson, unpublished.
3. M. Halldórsson, J. Kratochvíl and J.A. Telle, Independent sets with domination constraints, *Proceedings ICALP'98 - 25th International Colloquium on Automata, Languages and Programming, Aalborg, Denmark, July 1998, LNCS vol. 1443, 176-187*
4. M. Mahajan and V. Vinay, Determinant: combinatorics, algorithms, complexity, *Chicago Journal of Theoretical Computer Science, 1997:5, 1997.* Preliminary version SODA'97.
5. M. Mahajan and V. Vinay, Determinant: Old Algorithms, New Insights, in *Proceedings SWAT'98 - 6th Scandinavian Workshop on Algorithm Theory, Stockholm, Sweden, July 1998, LNCS Vol. 1432, 276-287.*
6. S.C. Ntafos and S.L. Hakimi, On the complexity of some coding problems, *IEEE Trans. Inform. Theory, Vol. 27, 1981, 794-796.*
7. J.A. Telle, Characterization of domination-type parameters in graphs, *Proceedings of 24th Southeastern International Conference on Combinatorics, Graph Theory and Computing -Congressus Numerantium Vol.94 1993, 9-16.*
8. J.A. Telle, Complexity of domination-type problems in graphs, *Nordic Journal of Computing 1(1994), 157-171.*
9. A. Vardy, The intractability of computing the minimum distance of a code, *IEEE Trans. Inform. Theory. Vol.43 No. 6, 1997, 1757-1766.* Preliminary version STOC'97.

NLC$_2$-Decomposition in Polynomial Time

Extended Abstract

Öjvind Johansson

Department of Numerical Analysis and Computing Science,
Royal Institute of Technology, SE-100 44 Stockholm, Sweden
ojvind@nada.kth.se
http://www.nada.kth.se/~ojvind

Abstract. NLC$_k$ is a family of algebras on vertex-labeled graphs introduced by Wanke. An NLC-decomposition of a graph is a derivation of this graph from single vertices using the operations in question. The width of the decomposition is the number of labels used, and the NLC-width of the graph is the smallest width among its NLC-decompositions. Many difficult graph problems can be solved efficiently with dynamic programming if an NLC-decomposition of low width is given for the input graph. It is unknown though whether arbitrary graphs of NLC-width at most k can be decomposed with k labels in polynomial time. So far this has been possible only for $k = 1$, which corresponds to cographs. In this paper, an algorithm is presented that works for $k = 2$. It runs in $O(n^4 \log n)$ time and uses $O(n^2)$ space. Related concepts: clique-decomposition, clique-width.

1 Introduction

In [15] Wanke introduces an algebra for a class of vertex-labeled graphs called NLC$_k$. This class consists of all graphs that can be obtained from single vertices with labels in $[k] = \{1, \ldots, k\}$ using the two operations of the algebra, union and relabeling, defined as follows: *Union* requires two disjoint graphs, and permits edges to be drawn between these. More precisely, all such edges will be added that match a set of ordered label pairs accompanying the union operator. *Relabeling* requires just one graph, and changes its labels according to a likewise specified mapping from $[k]$ to $[k]$. With $k = 1$, one obtains the class already known as *cographs*, described in [2] for example. (In such a comparison, we are referring to edge-structure only, since cographs are unlabeled.)

A similar algebra has been defined in [6]. The main difference is that in the latter, no edges can be added when two graphs are united; edge-drawing is a separate unary operator.

By a *decomposition* of a graph with respect to one of these algebras, we mean a derivation of this graph from single vertices using the operations in question. Specifically, the terms *NLC-decomposition* and *clique-decomposition* refer to the two algebras discussed above. The *width* of a decomposition is the number of distinct labels actually used, and the *NLC-width* and *clique-width* of

Widmayer et al. (Eds.): WG'99, LNCS 1665, pp. 110–121, 1999.

a graph are the smallest widths among all its NLC-decompositions and clique-decompositions respectively.

The relationship between these algebras has been studied in [10]. It was found that a clique-decomposition can be transformed into an NLC-decomposition of the same graph and vice versa. In the first direction, no additional labels are needed. In the second direction, at most a doubling of the label set may be necessary. Thus, the NLC-width of a graph is bounded by the clique-width, which in turn is bounded by two times the NLC-width. NLC-width 1 (cographs) corresponds exactly to clique-width 1 and 2. Less clear is the relationship between the classes with NLC-width 2 and clique-width 3 for example. They may intersect properly. An indication for this is given by the fact that neither of the two bounds above can be improved by a constant multiplicative factor.

NLC-decompositions and clique-decompositions have a binary tree structure. What makes them important is that decomposing a graph can be an excellent first step in solving more particular problems on it. Many problems which are hard for arbitrary graphs can be solved with dynamic programming in polynomial or even linear time on graphs which can be decomposed using a bounded number of labels, assuming that the graph is given in such a decomposed form. For example, decision, optimization, and enumeration problems expressible in MS1 logic, such as 3-Colorability, MaxClique and #MaxClique, can be solved in linear time on graphs given as clique-decompositions of width at most k [4,5]. And *P-recognizable* problems, such as Hamiltonian Circuit (which is not MS1-expressible [4]), can be solved in polynomial time on graphs given as NLC-decompositions of width at most k [15]. Note here that in theory it does not really matter which decomposition we have. For the transformations between NLC-decompositions and clique-decompositions mentioned above can in fact be carried out in linear time.

However, these transformations do not necessarily preserve minimality of width. Since the time complexities of the dynamic programming algorithms in question grow quickly with increasing k, it is of practical interest to use "first-hand" decompositions for that algebra which best captures a particular graph problem. It is by no means clear though that the problems of finding NLC-decompositions and clique-decompositions of minimal width are equally hard. In fact, it is unknown whether arbitrary graphs of NLC-width (clique-width) at most k can be NLC-decomposed (clique-decomposed) with k labels in polynomial time. For cographs, algorithms follow easily from [3], for example. A more recent result concerns certain families of graphs with restrictions on the number of induced P_4s [11]. (A P_4 is shown in Fig. 3.)

With the algorithm in this paper, it is now possible to NLC-decompose, using a minimum number of labels, all graphs of NLC-width at most 2 in polynomial time. Concerning these, one can note that although cographs are equivalent with P_4-free graphs [2], a graph with NLC-width 2 — as well as one with clique-width 3 — can have an exponential number of induced P_4s. (Consider for example those fourpartite graphs whose edges we can define by letting each vertex part correspond to one of the vertices in a P_4.)

It was pointed out in [6] that clique-decomposition can refine the *modular decomposition* of a graph. This refinement idea works equally well for NLC-decomposition. In either case, a minimum-width decomposition of a graph G can be obtained from minimum-width decompositions of the *quotient* graphs in the modular decomposition of G. Accordingly, the algorithm presented in this paper uses modular decomposition as a first step. Thus, in Sect. 4 we discuss modular decomposition of labeled graphs, and we summarize some properties of the resulting quotient graphs. In Sect. 5 we then show how to NLC-decompose these quotient graphs, as long as their NLC-width is at most 2. We indeed exploit some observations particular to NLC-width 2, and no generalization to higher width seems readily obtainable.

2   Preliminaries

Unless stated otherwise, a *graph* G is assumed to be undirected, but it may be either labeled (see below) or unlabeled. $V(G)$ and $E(G)$ denote the vertex and edge sets of G.

With a *labeled graph* G we mean an unlabeled graph, also denoted unlab(G), together with a labeling function, lab_G, mapping each vertex in $V(G)$ to a positive integer. In this context, we often use the set $[k] = \{1, \ldots, k\}$. If all vertices in a set $V \subseteq V(G)$ have (that is, are mapped to) the same label (by lab_G), we say that V is *uniformly labeled* (in G). If this holds for $V = V(G)$, G is uniformly labeled.

Two graphs G_1 and G_2 are disjoint when $V(G_1) \cap V(G_2) = \emptyset$. Then, if G_1 and G_2 are both unlabeled or both labeled, their *disjoint union* G is defined as follows: In either case, $V(G) = V(G_1) \cup V(G_2)$ and $E(G) = E(G_1) \cup E(G_2)$. In case G_1 and G_2 are both unlabeled, so is G. In case G_1 and G_2 are both labeled, G is labeled too, with $lab_G(u) = lab_{G_1}(u)$ for all $u \in V(G_1)$, and $lab_G(u) = lab_{G_2}(u)$ for all $u \in V(G_2)$.

For a set of vertices V in a graph G, $G|V$ denotes the subgraph of G induced by V. The usual definition for unlabeled graphs is extended to labeled graphs in the obvious way.

3   NLC-Decomposition

In this section we give basic definitions related to NLC-decomposition, as well as a first lemma. We begin with the fundamental operations and expressions.

Definition 1 (Union [15]). *Let G_1 and G_2 be disjoint graphs labeled with numbers in $[k]$, and let $S \subseteq [k]^2$ (that is, S is a set of ordered label pairs). Then $\times_S(G_1, G_2)$ is defined as the graph obtained by forming the disjoint union of G_1 and G_2, and adding to that all edges $\{u, v\}$ satisfying $u \in V(G_1)$, $v \in V(G_2)$, and $(lab_{G_1}(u), lab_{G_2}(v)) \in S$. See Fig. 1.*

Definition 2 (Relabeling [15]). *Let G be a graph labeled with numbers in $[k]$, and let R be a mapping from $[k]$ to $[k]$. Then $\circ_R(G)$ is the labeled graph G' defined by $\mathrm{V}(G') = \mathrm{V}(G)$, $\mathrm{E}(G') = \mathrm{E}(G)$, and $lab_{G'}(u) = R(lab_G(u))$ for all $u \in \mathrm{V}(G')$. See Fig. 1.*

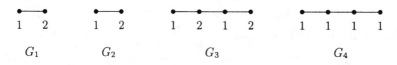

$$G_1 \qquad G_2 \qquad G_3 \qquad G_4$$

Fig. 1. Union and relabeling. $G_3 = \times_{\{(2,1)\}}(G_1, G_2)$ and $G_4 = \circ_{\{(1,1),(2,1)\}}(G_3)$.

Definition 3 (NLC$_k$-term). *An NLC$_k$-term D is a graph-producing expression in the above operators $\times_S$ and $\circ_R$, as well as the unary operator λ_i, where $i \in [k]$. The latter produces a graph consisting of a single vertex labeled with i. (The operand may be the name of the vertex, or a bullet symbol, $\bullet$, making the vertex nameless but unique.) $\mathrm{G}(D)$ is the graph produced by D, and $\mathrm{L}(D)$ is the set of all labels in graphs produced by subexpressions of D, including D itself.*

Example 1. Let $D = \times_{\{(2,1)\}}\left(\times_{\{(1,2)\}}(\lambda_1(\bullet), \lambda_2(\bullet)), \times_{\{(1,2)\}}(\lambda_1(\bullet), \lambda_2(\bullet))\right)$. Then D is an NLC$_2$-term, and $\mathrm{L}(D) = \{1, 2\}$. D produces the graph G_3 in Fig. 1.

It is often convenient to view NLC$_k$-terms as rooted ordered binary trees. See Fig. 2.

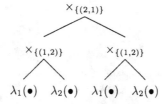

Fig. 2. The NLC$_2$-term in Example 1 expressed with a binary tree.

We are now prepared to define the class NLC$_k$, as well as what will be our most frequently used concepts.

Definition 4 (NLC$_k$ [15]). NLC$_k$ *is the class of all (labeled) graphs produced by NLC$_k$-terms.*

Definition 5 (NLC$_k$-decomposition, NLC-decomposition [10]). *An NLC$_k$-decomposition of a graph G is an NLC$_k$-term D such that $G = G(D)$ if G is labeled, and such that $G = \text{unlab}(G(D))$ if G is unlabeled. If D exists, G is said to be NLC$_k$-decomposable. An NLC-decomposition of G is an NLC$_k$-decomposition of G for some unspecified value of k. Note that G always is "NLC-decomposable".*

Definition 6 (Width). *The width of an NLC$_k$-term D is $|L(D)|$.*

Definition 7 (NLC-width [10]). *The NLC-width of a graph G, width$_{\text{NLC}}(G)$, is the smallest width among all NLC-decompositions of G.*

The NLC$_k$-decomposition problem. A graph G is given, unlabeled or labeled with numbers in $[k]$. The task is to find an NLC$_k$-decomposition of G, if that exists.

Example 2. The NLC$_2$-term D in Example 1 is an NLC$_2$-decomposition of G_3 in Fig. 1, as well as of its unlabeled variant called P_4, shown in Fig. 3. D has width 2. The reader is invited to show that there is no NLC-decomposition of P_4 with width 1. Thus P_4 has NLC-width 2.

Fig. 3. The graph P_4.

Definition 8 (Restriction). *Let D be an NLC$_k$-term, and let $V \subseteq V(G(D))$. Then $D|V$ denotes the restriction of D to V, the expression obtained by deleting the terms for vertices not in V, and removing superfluous operations in the obvious way. Evidently, $G(D|V) = G(D)|V$.*

Since restricting a decomposition does not increase its width, we immediately have:

Lemma 1. *If H is an induced subgraph of G, then width$_{\text{NLC}}(H) \leq$ width$_{\text{NLC}}(G)$, and H is NLC$_k$-decomposable if G is.*

4   Modular Decomposition

In general, the modular decomposition of a structure S is a derivation of S with the implicit or explicit help of a substitution operation. Each substitution step is described by a *quotient* structure. In the full paper, we extend ordinary graph substitution so that it can produce labeled graphs as well. We then use the results in [13] to define and characterize the modular decomposition, $D_M(G)$, of any labeled or unlabeled graph G. In particular, when G is labeled, a set $M \subseteq V(G)$ is a *module* of G if and only if

    (i) M is nonempty;

   (ii) for each vertex $v \in V(G) - M$, v has edges, either to all vertices in M, or to none of them; and

 (iii) either M is uniformly labeled, or $M = V(G)$;

whereas if G is unlabeled, $M \subseteq V(G)$ is a module of G if and only if (i) and (ii) hold. As usual, a graph is *prime* if its only modules are $V(G)$ and the singleton subsets thereof.

Moreover, whether G is labeled or not, each quotient graph Q of $D_M(G)$ satisfies one of the following, (where "properly" means "having three or more vertices"):

- Q has two vertices.
- Q is properly degenerate, implying that it is complete or discrete, and either unlabeled or uniformly labeled.
- Q is properly prime.

It is shown in the full paper that we can compute $D_M(G)$ in $O(n^2)$ time by first expressing G as a two-structure [7,8], and then using the algorithm in [12].

5   NLC$_2$-Decomposition

In this section we solve the NLC$_k$-decomposition problem for $k = 2$. But before we restrict our choice of k, let us state the following from the full paper:

Proposition 1. *Let G be a graph with more than one vertex. Then the NLC-width of G equals the highest NLC-width among the quotient graphs in the modular decomposition of G, and an NLC-decomposition of G with this width can be obtained from NLC-decompositions of these quotient graphs.*

So let Q be a quotient graph in the modular decomposition of a graph G. Q may be labeled or unlabeled. If Q has two vertices, its NLC-width is at most 2, and if Q is properly degenerate, its NLC-width is 1. In each of these cases, it is trivial to find an NLC-decomposition of Q with minimum width.

From here on, we study the remaining case — Q is properly prime — and we restrict our discussion to NLC$_2$-decomposition. Thus, Q will be unlabeled or labeled with numbers in $\{1, 2\}$. A key observation then is that in NLC$_2$, any nontrivial use of the relabeling operator will produce a graph where all vertices have the same label. This means that these vertices will be treated equally by subsequent operations. Thereby, they will make up a module of the final graph, which will not be prime.

Therefore, to NLC$_2$-decompose, if possible, a properly prime graph Q, we shall do as follows:

- If Q is non-uniformly labeled, then we use Algorithm 1, which searches for a relabeling-free NLC$_2$-decomposition of Q.

- If Q is unlabeled, then we use Algorithm 2. It searches for a non-uniform labeling of Q that permits a relabeling-free NLC_2-decomposition, as determined by Algorithm 1.
- If Q is uniformly labeled, then we use Algorithm 2 to search for an NLC_2-decomposition D' of $unlab(Q)$. If we find D', we can easily construct an NLC_2-decomposition $\circ_R(D')$ of Q.

5.1   Algorithm 1

The input to this algorithm is a graph G labeled with numbers in $\{1, 2\}$. The output is a relabeling-free NLC_2-decomposition D of G, if such a decomposition exists.

The algorithm is of divide-and-conquer type. If the decomposition D above exists, it has the form $\times_S (D_1, D_2)$, where S is one of the 16 subsets of $\{(1, 1), (1, 2), (2, 1), (2, 2)\}$. More interesting, as soon as we find a partition $\{V_1, V_2\}$ of the vertices of G such that $G = \times_S (G_1, G_2)$ for $G_1 = G|V_1$, $G_2 = G|V_2$, and S among the 16 possibilities above, we know that G has a relabeling-free NLC_2-decomposition if and only if G_1 and G_2 do. For if D is a relabeling-free NLC_2-decomposition of G, then the restrictions $D|V_1$ and $D|V_2$ are relabeling-free NLC_2-decompositions of G_1 and G_2. And conversely, if D_1 and D_2 are relabeling-free NLC_2-decompositions of G_1 and G_2, then $\times_S (D_1, D_2)$ is a relabeling-free NLC_2-decomposition of G.

To find, if possible, a partition $\{V_1, V_2\}$ such that $G = \times_S (G_1, G_2)$, where $G_1 = G|V_1$ and $G_2 = G|V_2$, we are going to try each S, if needed. But first, we select any vertex u in G and specify that u shall belong to V_2. This will be no restriction, since $\times_S (G_1, G_2) = \times_{S'} (G_2, G_1)$, where S' is obtained by reversing each pair in S. We now try each relation S as follows: First, we let $V_2 = \{u\}$. For each vertex v in $V_1 = V(G) \setminus u$, we check if u has an edge to v if and only if it should, according to S. If not, we move v from V_1 to V_2. Each time we move a vertex to V_2, we check this vertex with respect to those left in V_1; we compare with S, and move more vertices if needed. Continuing like this, we end up either with a valid partition, or with an empty V_1. If a partition is found, we continue to partition V_1 and V_2, and so on, until all obtained sets have size one. All in all, this can be done in $O(n^3)$ time (where $n = |V(G)|$), and only $O(n)$ space is needed besides the input.

5.2   Algorithm 2

The input to this algorithm is a properly prime unlabeled graph G. The output is a relabeling-free NLC_2-decomposition D of G, if such a decomposition exists.

As before, if D exists, it has the form $\times_S (D_1, D_2)$, where S is a subset of $\{(1, 1), (1, 2), (2, 1), (2, 2)\}$. However, G is now unlabeled, and the number of interesting possibilities for S is then smaller. Firstly, D produces a labeled graph such that $G = unlab(G(D))$. Our freedom in choosing the labeling makes many possibilities for S equivalent. Secondly, many values of S would make G contain modules. These values can be excluded.

It so turns out that we only have to try three possibilities for S: $\{(1,1)\}$, $\{(1,1),(2,2)\}$, and $\{(1,2),(2,1),(2,2)\}$. (A detailed argument can be found in the full paper.) For the rest of this paper, we devote ourselves to the first possibility. To try the second possibility is similar but a little simpler. (See the full paper.) The third possibility is easily reduced to the first by edge complementation of G.

Thus, we now let $S = \{(1,1)\}$. To find a decomposition of G on the form $\times_S(D_1, D_2)$, we may go through all edges $\{v_1, v_2\}$ in G, and determine for each the satisfiability of $G = \mathrm{unlab}(\times_S(G_1, G_2))$, where G_1 and G_2 are required to be NLC$_2$-decomposable and to contain v_1 and v_2 respectively, both of which must then be labeled with 1. We will later on develop this idea a little further, in order to reduce the number of edges $\{v_1, v_2\}$ that we have to go through.

We assume from now on that v_1 and v_2 have been fixed like this. The fact that $S = \{(1,1)\}$ then implies that as soon as we place any other vertex v in G_1 or G_2, we know what its label must be. For example, if v is placed in G_1, its label must be 1 if it has an edge to v_2, and 2 otherwise. Therefore, given a subset V of $\mathrm{V}(G)$ containing v_1 possibly, but not v_2, let $\mathrm{G}_{left}(V)$ denote the graph on V whose edges are induced by G, and in which a vertex is labeled with 1 if it has an edge to v_2, and 2 otherwise. In the same way, but with the roles of v_1 and v_2 swapped, we define $\mathrm{G}_{right}(V)$.

The fixation of v_1 and v_2 not only helps us to label the vertices in $V^* = \mathrm{V}(G) \setminus \{v_1, v_2\}$ once they have been placed in G_1 or G_2, but it also creates a useful dependency among these vertices with respect to their placement. For $i, j \in \{1, 2\}$, let an i–j-vertex — a vertex of *type* i–j — be a vertex in V^* which will be labeled with i if placed in G_1, and with j if placed in G_2. Notice that each vertex in V^* is an i–j-vertex for some i and j. As an example of the dependency, let us look at two 1–1-vertices u and v. If there is no edge between u and v, then they must be placed together, either in G_1 or in G_2, since $\times_S(G_1, G_2)$ produces edges between 1-labeled vertices in G_1 and 1-labeled vertices in G_2.

We use a directed graph, G_{dep}, to reflect this dependency. G_{dep} is unlabeled, has vertex set V^*, and there is an edge from u to v in G_{dep}, also written $u \to v$, whenever the existence or not of an edge between u and v does not match S when u is placed in G_2 and v is placed in G_1. So if $u \to v$, then u cannot be placed in G_2 without v being placed there too. We let $u \leftrightarrow v$ mean that both $u \to v$ and $v \to u$ hold, and we let $u \mid v$ mean that neither $u \to v$ nor $v \to u$ holds. Finally, we define $\lesssim$ to be the reflexive and transitive closure of the relation $\to$.

A partition $\{V_1^*, V_2^*\}$ of V^* is said to *respect* $\lesssim$ if $u \lesssim v$ does not hold for any vertices $v \in V_1^*$ and $u \in V_2^*$. Notice that given a partition $\{V_1^*, V_2^*\}$ of V^* (where we allow one of V_1^* and V_2^* to be empty), $G = \mathrm{unlab}(\times_S(G_1, G_2))$ is true for $G_1 = \mathrm{G}_{left}(v_1 \cup V_1^*)$ and $G_2 = \mathrm{G}_{right}(v_2 \cup V_2^*)$ if and only if $\{V_1^*, V_2^*\}$ respects $\lesssim$. As soon as this is the case, we can use Algorithm 1 to search for NLC$_2$-decompositions D_1 and D_2 of G_1 and G_2. If they exist, $D = \times_S(D_1, D_2)$ is an NLC$_2$-decomposition of G, and $\{V_1^*, V_2^*\}$ is said to be a *successful* partition. If D_1 and D_2 do not both exist, we can try another partition of V^*. Below we show that if we choose these partitions carefully, we only need to try $O(\log(|\mathrm{V}(G)|))$

of them. If we have not found D after that, we can conclude that we have to continue with a new fixation of v_1 and v_2.

To bound the number of partitions we have to consider, we first collect vertices into *clusters*. If C is a strongly connected component in G_{dep}, then all vertices of C must be placed together, either in G_1 or in G_2. We then say that C is a cluster of V^*. For clusters C_1 and C_2, we may write $C_1 \lesssim C_2$ if $u \lesssim v$ for some $u \in C_1$ and $v \in C_2$. However, unless stated otherwise, clusters will be assumed distinct, and we will write $C_1 < C_2$ instead of $C_1 \lesssim C_2$. (To have both $C_1 \lesssim C_2$ and $C_2 \lesssim C_1$ is then not possible.) If neither $C_1 < C_2$ nor $C_2 < C_1$ holds, we write $C_1 \parallel C_2$.

In agreement with previous notation, we also write $C_1 \rightarrow C_2$ if $u \rightarrow v$ for some $u \in C_1$ and $v \in C_2$, and we write $C_1 \mid C_2$ if neither $C_1 \rightarrow C_2$ nor $C_2 \rightarrow C_1$ holds. Of course, we never have "$C_1 \leftrightarrow C_2$". Note that $C_1 \rightarrow C_2$ implies $C_1 < C_2$. Conversely, $C_1 \parallel C_2$ implies $C_1 \mid C_2$.

We can get a deeper understanding of clusters by looking at $\rightarrow$ for specific pairs of vertex types. The reader is referred to the full paper for these details. To summarize our findings, we call the vertex types 1–1, 1–2, and 2–1, *determining*. We have:

- Each cluster contains one or more vertices of at least one determining type. (Here we have used the fact that G is connected.)
- If t is a determining type in a cluster C_1, and a cluster C_2 contains a vertex of some other determining type, then $C_1 \rightarrow C_2$ or $C_2 \rightarrow C_1$.
- If two clusters, C_1 and C_2, contain exactly one and the same determining type, then $C_1 \mid C_2$.

Using these statements only, it is not hard show that if C_1 and C_2 are clusters satisfying $C_1 \parallel C_2$, and if C is a third cluster, then $C < C_1$ implies $C < C_2$, and $C_1 < C$ implies $C_2 < C$.

It is now easy to see that we can group (in a unique way) clusters into *boxes*, so that we satisfy the following *box structure properties*:

- There is a linear order, $<$, on the boxes.
- Each box contains at least one cluster.
- If B_1 and B_2 are boxes with $B_1 < B_2$, then $C_1 < C_2$ for any clusters $C_1 \in B_1$ and $C_2 \in B_2$.
- If C_1 and C_2 are clusters in the same box, then $C_1 \parallel C_2$.

We define boxes like this, and for simplicity, we let each box denote the union of its clusters. We can observe that a partition $\{V_1^*, V_2^*\}$ of V^* respects $\lesssim$ if and only if the following *monotonicity conditions* are satisfied:

- When V_1^* contains a box B_1, it also contains each box $B < B_1$.
- When V_2^* contains a box B_2, it also contains each box B such that $B_2 < B$.
- At most one box is *split* by the partition — that is, has some clusters in V_1^* and some in V_2^*.

Thereby we are ready to discuss the partitioning procedure. We will use a somewhat informal language — the boxes are assumed to be ordered from left to right, so that if $B_1 < B_2$, we can formulate this as "B_1 is to the left of B_2".

We first try to partition in between boxes. We describe this by extending the total order to include *separator* elements between the boxes, and at the ends. Given a separator s, we partition V^* as $\{V_1^*, V_2^*\}$, where V_1^* is the union of all boxes to the left of s, and V_2^* is the union of all boxes to the right of s. As described previously, we then define $G_1 = G_{left}(v_1 \cup V_1^*)$ and $G_2 = G_{right}(v_2 \cup V_2^*)$. From what we already know about partitions respecting $\lesssim$, we note, with the help of Lemma 1:

- If G_1 is not NLC$_2$-decomposable, any successful partition $\{V_1', V_2'\}$, must satisfy $V_1' \subset V_1^*$.
- If G_2 is not NLC$_2$-decomposable, any successful partition $\{V_1', V_2'\}$, must satisfy $V_2' \subset V_2^*$.

We can therefore use binary search among separators with one of the following results:

- We find a successful partition.
- We find a partition such that neither G_1 nor G_2 is NLC$_2$-decomposable. We can conclude that there is no successful partition for the current fixation of v_1 and v_2.
- We find separators s_l and s_r immediately to the left and right of some box, B, such that when s_l is used, G_1 is NLC$_2$-decomposable but G_2 is not, and such that when s_r is used, G_2 is NLC$_2$-decomposable but G_1 is not. We can conclude that if there exists a successful partition, it must split B.

In the last case, we must examine B more closely. As we shall see, we only need to try one more partition, and we can find it as follows: First, for each cluster C in B, we use Algorithm 1 to search for NLC$_2$-decompositions of $G_{left}(C)$ and $G_{right}(C)$. If only one of these is decomposable, there is no doubt about in what part of a successful partition that C must be placed. (If neither $G_{left}(C)$ nor $G_{right}(C)$ is decomposable, the conclusion is of course simple.) We may now be left with a number of clusters for whose placement we have not yet seen any restrictions. Let us call them *remaining* clusters. Fortunately, all of them can safely be placed together. It is the vertex types in B that matter: When B contains 1–2-vertices, the remaining clusters can be placed in V_1^*. When B contains 2–1-vertices, the remaining clusters can be placed in V_2^*. And when B contains 1–1-vertices, the remaining clusters can be placed anywhere. The detailed arguments can be found in the full paper.

Let us now summarize: To determine the satisfiability of $G = \text{unlab}(\times_S(G_1, G_2))$, where $S = \{(1,1)\}$, and where G_1 and G_2 are required to be NLC$_2$-decomposable and to contain v_1 and v_2 respectively, we first group the vertices in $V^* = V(G) \setminus \{v_1, v_2\}$ into clusters by computing the strongly connected components of G_{dep} — the dependency graph with respect to v_1 and v_2. This can be done with two depth-first searches, as described in [1]. The time needed is linear in the size of G_{dep}, which is $O(n^2)$, where $n = |V(G)|$. We assume here that G_{dep} is stored explicitly.

We thereafter compute the box structure. This we do by inserting one cluster C at a time. Either C fits in an existing box, or it must be placed in a new one. This new box will appear either between two unaffected old boxes (or at an end), or between the divided contents of an old box. The arrangement of all clusters can easily be computed in $O(n^2)$ time.

We are now set to search for a successful partition of V^*. The binary search phase involves $O(\log n)$ partitions, each of which takes $O(n^3)$ time to check with Algorithm 1. If needed, we continue with the "box-splitting" phase. We then call Algorithm 1 twice for each cluster in the box in question. The total time for this sums to $O(n^3)$. The final partition can then be checked, again in $O(n^3)$ time. All in all, we use $O(n^3 \log n)$ time and $O(n^2)$ temporary space for each fixation of v_1 and v_2.

To find out if G has an NLC$_2$-decomposition on the form $\times_S (D_1, D_2)$, we can now repeat the above procedure for each edge $\{v_1, v_2\}$. However, without making things more than marginally more complicated, we can get by with only $n - 1$ such repetitions. By the symmetry of S, we can take any vertex $u \in V(G)$ and require that it shall belong to $G_2 = G(D_2)$. First, we let $v_2 = u$, and we let each neighbor of u play the role of v_1. If this does not lead us to a successful partition, we know that u must be labeled with 2. This in turn brings all neighbors of u to G_2. Next, we let one of these neighbors, u', play the role of v_2, and we let each neighbor of u' that is not already in G_2 play the role of v_1, and so on.

The new thing here is that not only v_1 and v_2 are fixed, but other vertices are fixed too — some to G_2, and some of these even to the label 2. It is argued in the full paper that this does not cause any problems. Thus, we can find a possible NLC$_2$-decomposition of G on the form $\times_S (D_1, D_2)$ in $O(n^4 \log n)$ time and $O(n^2)$ space.

5.3   Concluding Analysis

Given a graph G that is unlabeled or labeled with numbers in $\{1, 2\}$, we can now find an NLC$_2$-decomposition D_Q (if existing) of each quotient graph Q in the modular decomposition of G in $O(n_Q{}^4 \log n_Q)$ time, where $n_Q = |V(Q)|$. It is shown in the full paper that the sum of n_Q for all Q is bounded by $2n$, (where $n = |V(G)|$). Thus, the total time to find D_Q for all Q is $O(n^4 \log n)$. These decompositions can then easily be pieced together in linear time into an NLC$_2$-decomposition of G.

Acknowledgments

I am grateful to Stefan Arnborg for his advice and comments.

References

1. Thomas H. Cormen, Charles E. Leiserson, and Ronald L. Rivest. *Introduction to Algorithms*. The MIT Press, 1990.
2. D. G. Corneil, H. Lerchs, and L. Stewart Burlingham. Complement reducible graphs. *Discrete Applied Mathematics*, 3:163–174, 1981.
3. D. G. Corneil, Y. Perl, and L. K. Stewart. A linear recognition algorithm for cographs. *SIAM Journal on Computing*, 14:926–934, 1985.
4. B. Courcelle, J. A. Makowsky, and U. Rotics. On the fixed parameter complexity of graph enumeration problems definable in monadic second order logic. To appear in Discrete Applied Mathematics.
5. B. Courcelle, J. A. Makowsky, and U. Rotics. Linear time solvable optimization problems on graphs of bounded clique width. In *Graph-Theoretic Concepts in Computer Science, 24th International Workshop, WG'98*, volume 1517 of *Lecture Notes in Computer Science*, pages 1–16. Springer, 1998.
6. Bruno Courcelle and Stephan Olariu. Clique-width: A graph complexity measure— preliminary results and open problems. In *Proceedings of the Fifth International Workshop on Graph Grammars and Their Application to Computer Science*, 1994.
7. A. Ehrenfeucht and G. Rozenberg. Theory of 2-structures, part i: Clans, basic subclasses, and morphisms. *Theoretical Computer Science*, 70:277–303, 1990.
8. A. Ehrenfeucht and G. Rozenberg. Theory of 2-structures, part ii: Representation through labeled tree families. *Theoretical Computer Science*, 70:305–342, 1990.
9. Andrzej Ehrenfeucht, Harold N. Gabow, Ross M. McConnell, and Stephen J. Sullivan. An $O(n^2)$ divide-and-conquer algorithm for the prime tree decomposition of two-structures and modular decomposition of graphs. *Journal of Algorithms*, 16:283–294, 1994.
10. Öjvind Johansson. Clique-decomposition, NLC-decomposition, and modular decomposition — relationships and results for random graphs. *Congressus Numerantium*, 132:39–60, 1998.
11. J. A. Makowsky and U. Rotics. On the clique-width of graphs with few P_4's. To appear in IJFCS.
12. R. M. McConnell. An $O(n^2)$ incremental algorithm for modular decomposition of graphs and 2-structures. *Algorithmica*, 14:229–248, 1995.
13. R. H. Möhring. Algorithmic aspects of the substitution decomposition in optimization over relations, set systems and boolean functions. *Annals of Operations Research*, 4:195–225, 1985/6.
14. R. H. Möhring and F. J. Radermacher. Substitution decomposition for discrete structures and connections with combinatorial optimization. *Annals of Discrete Mathematics*, 19:257–356, 1984.
15. Egon Wanke. k-NLC graphs and polynomial algorithms. *Discrete Applied Mathematics*, 54:251–266, 1994.

On the Nature of Structure and Its Identification

Benno Stein and Oliver Niggemann

Dept. of Mathematics and Computer Science—Knowledge-based Systems,
University of Paderborn, D–33095 Paderborn, Germany
{stein,murray}@uni-paderborn.de

Abstract. When working on systems of the real world, abstractions in the form of graphs have proven a superior modeling and representation approach. This paper is on the analysis of such graphs. Based on the paradigm that a graph of a system contains information about the system's structure, the paper contributes within the following respects:

1. It introduces a new and lucid structure measure, the so-called weighted partial connectivity, Λ, whose maximization defines a graph's structure (Section 2).
2. It presents a fast algorithm that approximates a graph's optimum Λ-value (Section 3).

Moreover, the proposed structure definition is compared to existing clustering approaches (Section 4), resulting in a new splitting theorem concerning the well-known minimum cut splitting measure. A key concept of the proposed structure definition is its implicit determination of an optimum number of clusters.

Different applications, which illustrate the usability of the measure and the algorithm, round off the paper (Section 5).

1 What Is Structure?

"Structure defines the organization of parts as dominated by the general character of the whole."

This informal definition reflects the common sense understanding of the notion "structure". Structure information is some kind of meta information and may take different shapes. However, the nature of structure can often be captured by a graph. Figure 1, for example, shows a gantry crane, its graph representation in the form of the component graph G, and two abstractions, say contractions of G, that can be interpreted as the crane's structure. The paper in hand is on the automatic detection of such structure information.

To allow of a more formal definition of the term structure, the following abstraction is useful.

1. The system, the "whole", is mapped onto a graph, $G = \langle V, E \rangle$. The system's elements form the set of nodes, V; the relations between the elements are represented by the set of (weighted) edges, E.

Widmayer et al. (Eds.): WG'99, LNCS 1665, pp. 122–134, 1999.

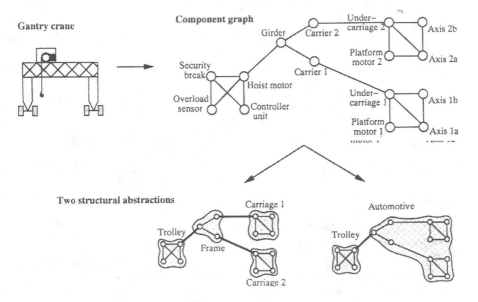

Fig. 1. Gantry crane with component graph and two structural abstractions.

2. The system's structure, its "general character", is reflected by the *distribution* of G's edges.

This understanding of a system's structure relies on the following paradigms.

1. *Modular Character.* The system (say, the graph $G = \langle V, E \rangle$) can be decomposed into several modules or functions such that each element of the system (say, each node $v \in V$) belongs to exactly one module.
2. *Connectivity.* Modules are defined implicitly, merely exploiting the graph-theoretical concept of connectivity: The connectivity between nodes assigned to the same module is assumed to be higher than the connectivity between any two nodes from two different modules.
3. *Contraction.* The system's structure is the contraction of G where a single node is substituted for all nodes belonging to the same module.

Remarks. Point 1 reflects hierarchy or decentralization aspects of a system or an organization. Point 2 is based on the observation, that the elements within a module are closely related; the modules themselves, however, are coupled by narrow interfaces only. A similar observation can be made respecting organizational or biological structures. Point 3 states that structure information can be derived by a simple abstraction.

These structuring paradigms may not apply to all kinds of systems—but, for a broad class of (technical) systems they form a useful set of assumptions.

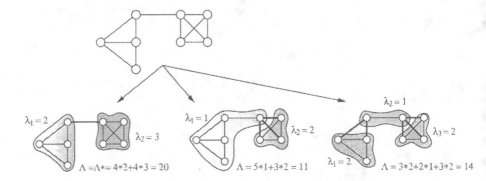

Fig. 2. Graph decompositions and related Λ values.

2   Quantifying a Graph's Structure

The structure of a system G has been introduced as some contraction of G. This descriptive definition can be quantified by means of a new measure called "weighted partial connectivity", Λ, which is introduced now. The weighted partial connectivity is defined for a decomposition of a graph G, and it is based on the graph-theoretical concept of *edge connectivity*.

Let $G = \langle V, E \rangle$ be the graph abstraction of the interesting system.[1]

1. $\mathcal{C}(G) = (C_1, \ldots, C_n)$ is a *decomposition* of G into n subgraphs induced on the C_i, if $\bigcup_{C_i \in \mathcal{C}} = V$ and $C_i \cap C_{j, j \neq i} = \emptyset$. The induced subgraphs $G(C_i)$ are called *cluster*. $E_{\mathcal{C}} \subseteq E$ consists of the set of edges between the clusters.
2. The *edge connectivity* $\lambda(G)$ of a graph G denotes the minimum number of edges that must be removed to make G a not-connected graph: $\lambda(G) = \min\{|E'| : E' \subset E \text{ and } G' = \langle V, E \setminus E' \rangle \text{ is not connected}\}$.

Definition 2.1 (Λ). Let G be a graph, and let $\mathcal{C} = (C_1, \ldots, C_n)$ be a decomposition of G. The *weighted partial connectivity* of $\mathcal{C}$, $\Lambda(\mathcal{C})$, is defined as

$$\Lambda(\mathcal{C}) := \sum_{i=1}^{n} |C_i| \cdot \lambda_i, \quad \text{where}$$

$\lambda(C_i) \equiv \lambda_i$ designates the edge connectivity of $G(C_i)$.
Figure 2 illustrates the weighted partial connectivity Λ.

Definition 2.2 (Connectivity Structure). Let G be a graph, and let $\mathcal{C}^*$ be a decomposition of G that maximizes Λ:

$$\Lambda(\mathcal{C}^*) \equiv \Lambda^* := \max\{\Lambda(\mathcal{C}) \mid \mathcal{C} \text{ is a decomposition of } G\}$$

Then the contraction $H = \langle \mathcal{C}^*(G), E_{\mathcal{C}^*} \rangle$ is called *connectivity structure* (or simply: *structure*) of the system represented by G.

Figure 3 shows that Λ-maximization means structure identification.

---

[1] Concepts and definitions of graph theory are used in their standard way; they are adopted from [10,7].

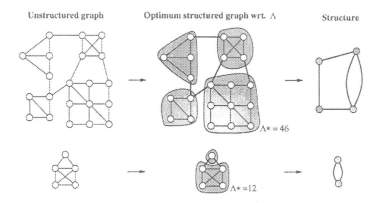

Fig. 3. Examples for decomposing a graph according to our structure definition.

Remarks. A key feature of the above structure measure is its *implicit definition of a structure's number of clusters.*

Two rules of decomposition, which are implied in our structure definition, are worth to be noted.

(i) If for a (sub)graph $G = \langle V, E \rangle$ and a decomposition $(C_1, \ldots, C_n)$ the *strong splitting condition*

$$\lambda(G) < \min\{\lambda_1, \ldots, \lambda_n\}$$

is fulfilled, G will be decomposed. Note that the strong splitting condition is commensurate for decomposition, and its application lessens the mean value of the standard deviations of the clusters' connectivity values λ_i. Obviously this splitting rule follows the human sense when identifying clusters in a graph, and there is a relation to the Min-Cut-splitting approach, which is derived in Section 4.

(ii) If for no decomposition $\mathcal{C}$ the strong splitting condition holds, G will be decomposed only, if for some $\mathcal{C}$ the condition $|V| \cdot \lambda(G) < \Lambda(\mathcal{C})$ is fulfilled. This inequality forms a necessary condition for decomposition—it is equivalent to the following special case of the structure definition: $\max\{\Lambda(\{V\}), \Lambda(\mathcal{C})\} = \Lambda(\mathcal{C})$, because $\Lambda(\{V\}) \equiv |V| \cdot \lambda(G)$.

The weighted partial connectivity, Λ, can be made independent of the graph size by dividing it by the graph's node number $|V|$. The resulting normalized Λ value is designated by $\bar{\Lambda} \equiv \frac{1}{|V|} \cdot \Lambda$.

3   Operationalizing Structure Identification

In this section a fast clustering algorithm optimizing the weighted partial connectivity Λ is presented. This algorithm implements a local heuristic and is suboptimal.

Initially, the algorithm assigns each node of a graph its own cluster. Within the following re-clustering steps, a node adopts the same cluster as the majority

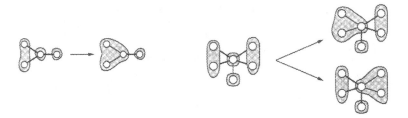

Fig. 4. A definite majority clustering situation (left) and an undecided majority clustering situation (right).

3   Operationalizing Structure Identification

In this section a fast clustering algorithm optimizing the weighted partial connectivity Λ is presented. This algorithm implements a local heuristic and is suboptimal.

Initially, the algorithm assigns each node of a graph its own cluster. Within the following re-clustering steps, a node adopts the same cluster as the majority of its neighbors belong to. If there exist several such clusters, one of them is chosen randomly. If re-clustering comes to an end, the algorithm terminates.

The left hand side of Figure 4 shows the definite case: most of the neighbors of the central node belong to the left cluster, and the central node becomes a member of that cluster. In the situation depicted on the right hand side, the central node has the choice between the left and the right cluster.

We now write down this algorithm formally.

MAJORCLUST.
Input. A graph $G = \langle V, E \rangle$.
Output. A function $c : V \mapsto \mathbf{N}$, which assigns a cluster number to each node.

(1) $n = 0$, $t = \textit{false}$
(2) $\forall v \in V$ **do** $n = n + 1$, $c(v) = n$ **end**
(3) **while** $t = \textit{false}$ **do**
(4) $t = \textit{true}$
(5) $\forall v \in V$ **do**
(6) $c^* = i$ **if** $\left| \{u : \{u, v\} \in E \wedge c(u) = i\} \right|$ is max.
(7)   **if** $c(v) \neq c^*$ **then** $c(v) = c^*, t = \textit{false}$
(8)   **end**
(9) **end**

The runtime complexity of MAJORCLUST is $\Theta(|E| \cdot |C_{max}|)$, where $C_{max} \subseteq V$ designates a maximum cluster. In the While-loop (line 3 to 8) each edge of G is investigated twice; within each pass, a growing cluster is enlarged by at least one node; if no node changes its cluster MAJORCLUST terminates. Note that this evaluation neglects "pathological" cases, where the algorithm oscillates between two (or more) decompositions. However, such a situation constitutes neither a clustering nor a runtime problem: It can be detected easily since all

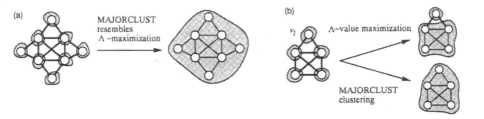

Fig. 5. The local behavior of MAJORCLUST may lead to sub-optimum Λ values.

nodes are either stable or in an undecided constellation. This advisements and experimental results (see Section 5) show the usability of the algorithm for large graphs with several thousand nodes.

The algorithm's greatest strength, its restriction to local decisions, is bound up with its sub-optimality. In every step only a node's neighbors are considered, resulting in an excellent runtime behavior. On the other hand, by disregarding global criteria like the connectivity, MAJORCLUST cannot always find the optimum solution. Figure 5 illustrates this.

The optimum solution for graph (a) is one cluster, which is also the solution as found by MAJORCLUST. For graph (b), a splitting into the two clusters $\{v_1\}$ and $V \setminus \{v_1\}$ is optimum. MAJORCLUST cannot find this decomposition—working strictly locally, it behaves exactly as on graph (a) and creates only one cluster.

3.1 Extension for Weighted Graphs

It is both useful and obvious to extend our structure identification approach by introducing edge weights. The amount of the weight $w(e)$ models the importance of an edge e. A prerequisite for this is a generalization of $\Lambda(\mathcal{C})$ by introducing the *weighted edge connectivity* $\bar{\lambda}$ of a graph as follows:
$\bar{\lambda}(G) = \min\{\sum_{e \in E'} w(e) : E' \subset E \text{ and } G' = \langle V, E \setminus E' \rangle \text{ is not connected}\}$. Using this definition all results from Section 2 can be directly extended to graphs with edge weights.

In the same way the algorithm MAJORCLUST is altered: Every node v now adapts the same cluster as the *weighted* majority of its neighbors, i.e. every neighbor counts according to the weight (i.e. importance) of the edge connecting it to v.

4   Existing Clustering Approaches

Clustering data given as graphs has been a focus of research for years. The existing approaches can be classified as follows.

Hierarchical versus Non-hierarchical Algorithms. Hierarchical algorithms create a tree of node subsets by successively subdividing or merging the graph's nodes.

In order to obtain a unique clustering, a second step is necessary that prunes this tree at adequate places.

Hierarchical algorithms can be further classified into divisive and agglomerative approaches. Divisive algorithms start with each vertex being its own cluster and union clusters iteratively. For agglomerative algorithms on the other hand, the entire graph initially forms one single cluster which is successively subdivided. Examples for divisive algorithms are Min-cut-clustering [10,20] or dissimilarity-based algorithms e. g. [11]. Typical agglomerative algorithms are k-nearest-neighbor or linkage methods [4,16,6].

Non-Hierarchical algorithms subdivide the graph into clusters within one step. Examples are clustering techniques based on Minimal-Spanning-Trees [22], self-organizing Kohonen networks [9] or approaches which optimize a given goal criterion [1,13,14,13].

Exclusive versus Non-exclusive Algorithms. Exclusive clustering algorithms assign every node to exactly one cluster, while non-exclusive algorithms assign to a node a membership value respecting each cluster. The algorithms mentioned above are of exclusive type; an example for a non-exclusive algorithm is Fuzzy clustering [21].

Clustering versus Partitioning. The clustering algorithms described above do not impose any constraint on cluster sizes. Partitioning algorithms as used in the fields of parallel computing or VLSI design typically demand homogeneous cluster sizes. Examples for partitioning algorithms can be found in [10,8].

Λ-maximization and MAJORCLUST can be classified as non-hierarchical and exclusive. MAJORCLUST finds a fast, but possibly suboptimal solution for the problem of Λ-maximization. I. e., unlike most optimization approaches, Λ-maximization as performed by MAJORCLUST does not rely on slow optimization techniques and can be used for large graphs.

The clustering quality of the Λ criterion and the MAJORCLUST algorithm will be illustrated by the following two comparisons with well-known clustering techniques as well as by different applications in Section 5.

4.1   Clustering Based on the Minimum Cut

MAJORCLUST is a divisive approach, that recursively subdivides a graph at its smallest cut. The following theorem relates Min-cut-clustering to clustering by means of Λ-maximization.

Theorem 4.1 (Strong Splitting Condition). Applying the strong splitting condition (see Section 2) results in a decomposition at minimum cuts.

To proof this theorem we first show that $\lambda(G)$ equals the cardinality of the minimum cut of G.

Proof of Lemma. Let $\mu(G)$ denote the minimum cut of G. $\lambda(G) \leq |\mu(G)|$ because the removal of all edges belonging to the cut splits G into two components. $\lambda(G) \geq |\mu(G)|$ because in G there exists $v_1, v_2 \in V$ so that exactly $\lambda(G)$ edge

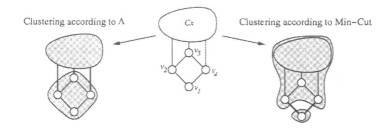

Fig. 6. Weighted partial connectivity (Λ-) maximization versus Min-Cut-clustering.

disjoint paths connect them. By removing one edge from each path, v_1 will not be connected to v_2 anymore, therefore exists a cut with $\lambda(G)$ edges.

Proof of Theorem. Let $cut(V_i, V_j)$ denote the edges between $G(V_i)$ and $G(V_j)$. From $\lambda(G) \leq \min\{\lambda_1, \ldots, \lambda_r\}$ follows $|\mu(G)| \leq \min\{|\mu(G(V_1))|, \ldots, |\mu(G(V_r))|\}$, i.e. no cut in $G(V_i), i = 1, \ldots, r$ is smaller than $\mu(G)$. Since every cut $\mu'(G)$ except of $cut(V_1, \ldots V_r)$ decomposes at least one $G(V_i)$, $\mu'(G)$ must consist of more than $|\mu(G)|$ edges. It follows that $cut(V_1, \ldots V_r)$ must be minimum.

When the strong splitting condition does not hold, an optimum decomposition according to the structuring value need not be the same decomposition as found using the minimum cut. This is because of the latter's disregard for cluster sizes. Figure 6 is such an example. Here C_x refers to a clique with $x \geq 3$ nodes. An optimum solution according to the weighted partial connectivity Λ (which is also closer to human sense of esthetics) consists of one cluster $\{v_1, v_2, v_3, v_4\}$ and a second cluster C_x. An algorithm using the minimum cut would only separate v_1.

The reader may also notice that, as mentioned before, maximizing the weighted partial connectivity implies an optimum number of clusters, while the minimum cut approach lacks any criterion for the number of necessary division steps.

4.2   Clustering Based on Nearest-Neighbor Strategies

Nearest-Neighbor clustering is an agglomerative approach that iteratively merges the two closest clusters. Its widespread use results in several variations [3,4,16,6]. The following qualitative comparison to MAJORCLUST does not take all existing variations into consideration, but we claim that Λ-maximization and MAJOR-CLUST respectively can indeed overcome typical problems inherent to Nearest-Neighbor clustering concepts.

1. Nearest-Neighbor clustering, like all hierarchical algorithms, does not define the (optimal) number of clusters. Λ-maximization (as well as MAJORCLUST) implicitly defines both number and sizes of the clusters.

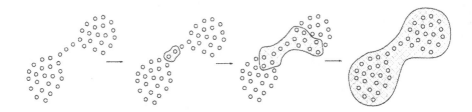

Fig. 7. The (undesired) chaining behavior of Nearest-Neighbor methods.

2. The greedy nature of Nearest-Neighbor methods (unlike as in MAJORCLUST, nodes are never reassigned to another cluster) leads to the so-called chaining effect [3], as illustrated in Figure 7.
3. The step of transforming the tree as created by a Nearest-Neighbor algorithm into a unique clustering often depends on extra parameters such as the minimum cluster or vertex distance. This results in difficulties if clusters have strongly varying point densities or inter-cluster distances. Λ-maximization and MAJORCLUST do not depend on additional parameters and behave more sensible in such clustering situations.
4. Nearest-Neighbor methods rely on distance information only. They thus disregard connectivity information. For weighted graphs this may lead to clusters which lack the human sense of esthetics, for unweighted graphs this behavior may result in a failure to find any clusters.

5   Application

This section outlines three applications for structure identifications.[2]

5.1   Monitoring Computer Networks

Monitoring traffic, i. e. recording inter-computer communications, is substantial for administrating and analyzing computer networks. The amount of traffic between all pairs of computers in the network is recorded in the so-called traffic matrix. By interpreting the traffic matrix as the adjacency matrix of a weighted graph, cluster identification techniques can be applied to the problem of traffic analysis. Figure 8 illustrates the procedure.

Being faced with rather large traffic matrices (> 400 nodes, > 800 edges), human experts need to fall back upon computer support for cluster identification. However, this clustering problem is difficult to solve since no features about communication structures are known beforehand, which makes this problem an ideal testbed for MAJORCLUST.

---

[2] Aside from the presented applications, structure identification has been investigated for the preprocessing of configuration knowledge bases and the topological analysis of fluidic systems [5,18].

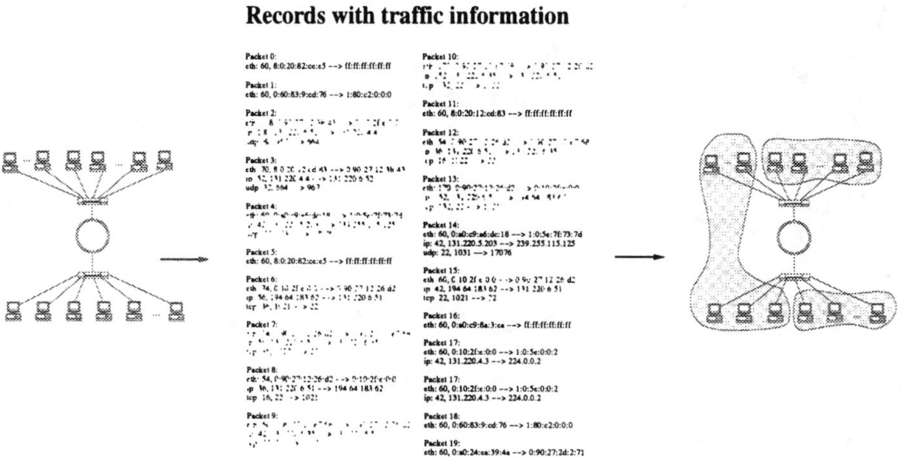

Records with traffic information

Fig. 8. Communication clusters can be identified in the traffic matrix of a network.

Cooperations with network experts allowed the application of MAJORCLUST under realistic conditions. MAJORCLUST revealed several interesting structures in traffic matrixes: (*i*) subnets were subdivided according to main applications, (*ii*) project member in different subnets were identified, and (*iii*) computer serving similar purposes were clustered together.

These insights are helpful in several ways. Firstly, they render a general understanding of traffic structure possible, e. g. clusters combining computers in different subnets mean high traffic on the backbone. Secondly, they provide additional information for planning tasks in the form of modification hints for the network architecture.

5.2   Visualizing Knowledge Bases

Automatic graph visualization is a key problem when supporting human understanding of complex data structures. To reduce the complexity of the visualization problem, one strategy is to apply a Divide-and-Conquer approach [2,14].

The role of clustering in this connection is to tackle the divide task, i. e., to break down a large graph into useful subgraphs. By a second step the resulting clusters are arranged on a grid, and by a third step the nodes within each cluster are positioned.

We have operationalized and applied this concept for the analysis and visualization of resource-based configuration knowledge bases. A resource-based knowledge base textually describes the configuration objects by tuples comprising an object's supplied and demanded functions. From such a description a global overview can be created that envisions the closely connected modules and their functional interplay. Figure 9 shows a part of a visualized knowledge-base

Configuration knowledge base

Fig. 9. Part of a configuration knowledge base with analyzed and visualized structure.

for the resource-based configuration of telecommunication system. Details and related information can be found in [12,17,19,15,2].

5.3 Clustering Metric Data

Data for clustering is often given as positions in a metric space, which can canonically be transfered into a graph. Based on this graph, the structure measure Λ, operationalized in the form of MAJORCLUST, can be applied for clustering. Figure 10 shows a set of points and the identified clusters. Recall that MAJORCLUST did not need meta information about the number and the size of the clusters. Input for MAJORCLUST was the totally connected graph of points. However, instead of connecting all pairs of vertices, connecting a vertex solely to its nth closest neighbors improves the performance. Both the quality of the clusters found and the runtime needed by MAJORCLUST have been examined using our VisioDat tool.

6 Summary

The paper presented a new approach to quantify the structure of graphs. Following this approach, a domain, a problem, or a system can syntactically be

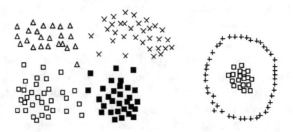

Fig. 10. Clusters in a set of points in a metric space.

analyzed regarding its structure—provided that a graph constitutes the adequate modeling paradigm.

The proposed structure measure, the weighted partial connectivity Λ, relies on subgraph connectivity, which is weighted with the subgraphs' sizes. The subgraphs in turn are determined by that decomposition of a graph that maximizes Λ. Hence, cluster number as well as cluster size of the structure are defined implicitly by the optimization—a characteristic which makes this approach superior to other clustering concepts. Λ-maximization resembles the human sense when trying to identify a graph's structure: Rather than searching for a given number of clusters, the density distribution of a graph's edges is analyzed.

Aside from the mathematical definition, the fast algorithm MAJORCLUST operationalizing Λ-maximization has been developed. Applications from the fields of monitoring, visualization, and configuration revealed both usability (the detected structures are reasonable) and applicability (efficient runtime behavior). Structure processing as proposed here thus provides a powerful knowledge preprocessing concept.

References

1. T. Bailey and J. Cowles. Cluster definition by the optimization of simple measures. *IEEE Transactions on Pattern Analysis and Machine Intelligence*, September 1983.
2. L. A. R. Eli, B. Messinger and R. R. Henry. A divide-and-conquer algorithm for the automatic layout of large directed graphs. *IEEE Transactions on Systems, Man, and Cybernetics*, January/February 1991.
3. B. S. Everitt. Cluster analysis. *Edward Arnold, a division of Hodder & Stoughton*, 1992.
4. K. Florek, J. Lukaszewiez, J. Perkal, H. Steinhaus and S. Zubrzchi. Sur la liason et la division des points d'un ensemble fini. *Colloqium Mathematicum*, 1951.
5. T. Hesse and B. Stein. Hybrid Diagnosis in the Fluidic Domain. *Proc. EIS 98, International ICSC Symposium on Engineering of Intelligent Systems, University of La Laguna, Tenerife, Spain*, Feb. 1998.
6. S. C. Johnson. Hierarchical clustering schemes. *Psychometrika 32*, 1967.
7. D. Jungnickel. *Graphen, Netzwerke und Algorithmen*. BI Wissenschaftsverlag, 1990.
8. B. Kernighan and S. Lin. Partitioning graphs. *Bell Laboratories Record*, January 1970.
9. T. Kohonen. Self Organizing and Associate Memory. *Springer-Verlag*, 1990.
10. T. Lengauer. *Combinatorical algorithms for integrated circuit layout*. Applicable Theory in Computer Science. Teubner-Wiley, 1990.
11. P. MacNaughton-Smith, W.T. Williams, M.B. Dale and L.G. Mockett. Dissimilarity analysis. *Nature 202*, 1964.
12. O. Niggemann, B. Stein, and M. Suermann. On Resource-based Configuration— Rendering Component-Property Graphs. In J. Sauer and B. Stein, editors, *12. Workshop "Planen und Konfigurieren"*, tr-ri-98-193, Paderborn, Apr. 1998. University of Paderborn, Department of Mathematics and Computer Science.
13. T. Roxborough and Arunabha. Graph Clustering using Multiway Ratio Cut. In S. North, editor, *Graph Drawing*, Lecture Notes in Computer Science, Springer Verlag, 1996.
14. R. Sablowski and A. Frick. Automatic Graph Clustering. In S. North, editor,

15. G. Sander. Graph Layout through the VCG Tool. Technical Report A/03/94, 1994.
16. P. H. A. Sneath. The application of computers to taxonomy. *J. Gen. Microbiol.* *17*, 1957.
17. B. Stein. *Functional Models in Configuration Systems.* Dissertation, University of Paderborn, Department of Mathematics and Computer Science, 1995.
18. B. Stein and E. Vier. Computer-aided Control Systems Design for Hydraulic Drives. *Proc. CACSD 97, Gent,* Apr. 1997.
19. K. Sugiyama, S. Tagawa, and M. Toda. Methods for Visual Understandig of Hierarchical System Structures. *IEEE Transactions on Systems, Man, and Cybernectics, Vol. SMC-11, No. 2,* 1981.
20. Z. Wu and R. Leahy. An optimal graph theoretic approach to data clustering: Theory and its application to image segmentation. *IEEE Transactions on Pattern Analysis and Machine Intelligence,* November 1993.
21. J.-T. Yan and P.-Y. Hsiao. A fuzzy clustering algorithm for graph bisection. *Information Processing Letters 52,* 1994.
22. C. T. Zahn. Graph-Theoretical Methods for Detecting and Describing Gestalt Clusters. *IEEE Transactions on computers Vol. C-20, No. 1,* 1971.

On the Clique–Width of Perfect Graph Classes
Extended Abstract

Martin Charles Golumbic and Udi Rotics*

Department of Mathematics and Computer Science
Bar-Ilan University
Ramat-Gan, Israel
{golumbic,rotics}@macs.biu.ac.il

Abstract. Graphs of clique–width at most k were introduced by Cour-celle, Engelfriet and Rozenberg (1993) as graphs which can be defined by k-expressions based on graph operations which use k vertex labels. In this paper we study the clique–width of perfect graph classes.

On one hand, we show that every distance–hereditary graph, has clique–width at most 3, and a 3–expression defining it can be obtained in linear time. On the other hand, we show that the classes of unit interval and permutation graphs are not of bounded clique–width. More precisely, we show that for every $n \in \mathcal{N}$ there is a unit interval graph I_n and a permutation graph H_n having n^2 vertices, each of whose clique–width is exactly $n+1$. These results allow us to see the border within the hierarchy of perfect graphs between classes whose clique–width is bounded and classes whose clique–width is unbounded.

Finally we show that every $n \times n$ square grid, $n \in \mathcal{N}$, $n \geq 3$, has clique–width exactly $n + 1$.

1 Introduction

The notion of clique–width of graphs was first introduced by Courcelle, Engelfriet and Rozenberg in [CER93], as graphs which can be defined by k-expressions based on graph operations which use k vertex labels. The clique–width of a graph G, denoted by $cwd(G)$, is defined as the minimum number of labels needed to construct G, using the 3 graph operation: disjoint union ($\oplus$), connecting vertices with specified labels (η) and renaming labels (ρ). More details, are given in section 2.

A detailed study of clique–width is [CO98]. Clique–width has analogous prop-erties to tree–width: If the clique–width of a class of graphs $\mathcal{C}$ is bounded by k (and the k–expression can be computed from its corresponding graph in time $T(|V| + |E|)$, then every decision, optimization, enumeration or evaluation prob-lem on $\mathcal{C}$ which can be defined by a Monadic Second Order formula ψ can be solved in time $c_k \cdot O(|V| + |E|) + T(|V| + |E|)$ where c_k is a constant which

---

* Supported in part by postdoctoral fellowships at Bar-Ilan University and the Uni-versity of Toronto.

depends only on ψ and k, where $|V|$ and $|E|$ denote the number of vertices and edges of the input graph, respectively. For details, cf. [CMRa,CMRb].

In this paper we study the clique–width of perfect graph classes. We first show that:

Theorem 1. *For every distance–hereditary graph G, $cwd(G) \leq 3$, and a 3–expression defining it can be constructed in time $O(|V| + |E|)$.*

Let $d_G(x,y)$ denote the length of the shortest path connecting vertices x and y in the graph G. A graph G is called *distance hereditary* if for every connected induced subgraph H of G, $d_G(x,y) = d_H(x,y)$ holds for every pair of vertices from H. These graphs were introduced by E. Howorka [How77] and have been studied intensively in recent years, cf. [DM88,HM90,Dra94,DNB97,BD98]. Linear time $O(|E| + |V|)$ algorithms were presented for all the following problems on distance hereditary graphs: dominating set [BD98], Steiner tree [BD98], maximum weighted clique [HM90], maximum weighted stable set [HM90], diameter [Dra94], and diametral pair [DNB97].

Since all these problems are in the class of Monadic Second Order Logic optimization problems presented in [CMRa], it follows from Theorem 1 above and from Theorem 4 of [CMRa] that all these problems, and many others, have linear time solutions on the class of distance–hereditary graphs. For example:

Corollary 1. *All the following problems have linear time $O(|V| + |E|)$ solution on the class of distance–hereditary graphs: minimum dominating set, minimum connected dominating set, minimum Steiner tree, maximum weighted clique, maximum weighted stable set, diameter, domatic number for fixed k, vertex cover, and k–colorability for fixed k.*

Other problems which are known to have linear time solutions on the class of distance hereditary graphs are: central vertex [Dra94], radius [Dra94], minimum r–dominating clique [Dra94],and the connected r–domination problem [BD98]. These problems cannot be added to the list of problems mentioned in Corollary 1 above, since they are not included in the class of Monadic Second Order Logic optimization problems presented in [CMRa].

Clearly Theorem 1 above also holds for any subclass of the class of distance hereditary graphs. For example:

Corollary 2. *Let C be any of the following graph classes, (defined in [PW99]): block graphs, block duplicate graphs, restricted block duplicate graphs, restricted unimodular chordal graphs, (6,2)–chordal bipartite graphs and Ptolematic graphs. For every graph $G \in C$, $cwd(G) \leq 3$, and a 3–expression defining it can be constructed in time $O(|V| + |E|)$.*

We say that a class of graphs C is *not of bounded clique–width* if there is no fixed integer k, such that for every graph $G \in C$, $cwd(G) \leq k$. We continue by showing that:

Theorem 2. *The class of unit interval graphs is not of bounded clique–width.*

Theorem 3. *The class of permutation graphs is not of bounded clique–width.*

Since many graph classes contain the classes of unit interval or permutation graphs, it follows that many perfect graph classes are not of bounded clique–width. For example:

Corollary 3. *All the following graph classes (defined in [Gol80]) and their complements are not of bounded clique–width: interval graphs, circle graphs, circular arc graphs, unit circular arc graphs, proper circular arc graphs, directed path graphs, undirected path graphs, comparability graphs, chordal graphs, and strongly chordal graphs.*

The reason the complements of all graph classes mentioned in Corollary 3 are not of bounded clique–width, is that for every graph G, $cwd(\overline{G}) \leq 2 * cwd(G)$, (cf. [CO98]).

Finally, we show that:

Theorem 4. *For every $n \times n$ square grid G, $n \in \mathcal{N}$, $n \geq 3$, $cwd(G) = n + 1$.*

Corollary 4. *For every $n \times m$ rectangular grid G, $n, m \in \mathcal{N}$, $n, m \geq 3$, $min\{n, m\} + 1 \leq cwd(G) \leq min\{n, m\} + 2$.*

Theorem 4 above improves the result of Makowsky and Rotics (cf. [MR99]), who showed that for every $n \times n$ square grid G, $cwd(G) \geq n/3$. The clique–width of the 2×2 grid is easily seen to equal 2.

In this extended abstract we just sketch the proofs of the theorems mentioned above. The detailed proofs will be presented in the full paper.

2   Background

In this section we define the notions of graph operations and clique–width, as presented in [CO98].

Definition 1 ((k–graph)). *A k-graph is a labeled graph with (vertex) labels in $\{1, 2, \ldots, k\}$. A k-graph G, is represented as a structure $\langle V, E, V_1, \ldots, V_k \rangle$, where V and E are the sets of vertices and edges respectively, and $V_1, \ldots, V_k$ form a partition of V, such that V_i is the set of vertices labeled i in G. Note that some V_i's may be empty. A non-labeled graph $G = \langle V, E \rangle$, will be considered as a 1-graph with all vertices labeled by 1.*

Definition 2 (($G \oplus H$)). *For k-graphs G, H such that $G = \langle V, E, V_1, \ldots, V_k \rangle$ and $H = \langle V', E', V_1', \ldots, V_k' \rangle$ and $V \cap V' = \emptyset$ (if this is not the case then replace H with a disjoint copy of H), we denote by $G \oplus H$, the disjoint union of G and H such that:*

$$G \oplus H = \langle V \cup V', E \cup E', V_1 \cup V_1', \ldots, V_k \cup V_k' \rangle$$

Note that $G \oplus G \neq G$.

Definition 3 $((\eta_{i,j}(G)))$. *For a k-graph G as above we denote by $\eta_{i,j}(G)$, where $i \neq j$, the k-graph obtained by connecting all the vertices labeled i to all the vertices labeled j in G. Formally:*

$$\eta_{i,j}(G) = \langle V, E', V_1, \ldots, V_k \rangle \ , \ where$$

$$E' = E \cup \{(u,v) : u \in V_i, \ v \in V_j\}$$

Definition 4 $((\rho_{i \to j}(G)))$. *For a k-graph G as above we denote by $\rho_{i \to j}(G)$ the k-graph obtained by the renaming of i into j in G such that:*

$$\rho_{i \to j}(G) = \langle V, E, V_1', \ldots, V_k' \rangle, \ where$$

$V_i' = \emptyset$, $V_j' = V_j \cup V_i$, and $V_p' = V_p$ for $p \neq i, j$.

These graph operations have been introduced in [CER93] for characterizing graph grammars. For every vertex v of a graph G and $i \in \{1, \ldots, k\}$, we denote by $i(v)$ the k-graph consisting of one vertex v labeled by i.

Example 1. A clique with four vertices u, v, w, x can be expressed as:

$$\rho_{2 \to 1}(\eta_{1,2}(2(u) \oplus \rho_{2 \to 1}(\eta_{1,2}(2(v) \oplus \rho_{2 \to 1}(\eta_{1,2}(1(w) \oplus 2(x)))))))$$

Definition 5 $((k\text{--expression}))$. *With every graph G one can associate an algebraic expression which defines G built using the 3 types of operations mentioned above. We call such an expression a k–expression defining G, if all the labels in the expression are in $\{1, \ldots, k\}$. Trivially, for every graph G, there is an n–expression which defines G, where n is the number of vertices of G.*

Definition 6 ((**The clique–width of a graph** G, $cwd(G)$)). *Let $C(k)$ be the class of graphs which can be defined by k–expressions. The clique–width of a graph G, denoted $cwd(G)$, is defined by: $cwd(G) = Min\{k : G \in C(k)\}$.*

$C(1)$ is the class of edge-less graphs, cographs are exactly the graphs of clique–width at most 2, and trees have clique–width at most 3 (cf. [CO98]).

In the following sections when considering a k–expression t which defines a graph G, it will often be useful to consider the tree structure, denoted as $tree(t)$, corresponding to the k–expression t. For that we shall need the following definitions.

Definition 7 $((tree(t)))$. *Let t be any k–expression, and let G be the graph defined by t. We denote by $tree(t)$ the parse tree constructed from t in the usual way. The leaves of this tree are the vertices of G, and the internal nodes correspond to the operations of t, and can be either binary corresponding to $\oplus$ or unary corresponding to η or ρ.*

Definition 8 (($tree(a,t)$)). *Let t be any k–expression, a be any node in t, we denote by $tree(a,t)$ the subtree of $tree(t)$ rooted at a.*

Definition 9 ((t_1 is a sub–expression of t_2)). *Let t_1 be a k–expression and let t_2 be an l–expression, $k \leq l$. We say that t_1 is a sub–expression of t_2 if there exists a node a such that $tree(t_1)$ is the sub–tree of $tree(t_2)$ rooted at a. In other words $tree(t_1)$ is equal to $tree(a, t_2)$.*

Definition 10 ((The label of a vertex v at an internal node a)). *Let t be any k–expression, and let G be the graph defined by t. Let a be any internal node of $tree(t)$ and let v be any vertex of G occurring in $tree(a, t)$, i.e. v is a leaf of $tree(a, t)$. The labels of v may change by the ρ operations in t. However, whenever an operation is applied on a sub–expression t_1 of t which contains v, the label of v (like the labels of all the other vertices occurring in t_1) is well defined. The label of v at a is defined as the label that v has immediately before the operation a is applied on the subtree of $tree(t)$ rooted at a.*

3   Distance Hereditary Graphs

Let $d_G(x,y)$ denote the length of the shortest path connecting vertices x and y in the graph G. Recall that a graph G is called *distance hereditary* if for every connected induced subgraph H of G, $d_G(x,y) = d_H(x,y)$ holds for every pair of vertices from H. For every vertex x, we denote by $N(x)$ the set of all neighbors of x (not including x). A *leaf* is a vertex having exactly one neighbor. We say that x and y are *twins* if they have the same neighborhood outside x and y, i.e. $N(x) - \{y\} = N(y) - \{x\}$. The vertices x and y are called *true twins* (resp. *false twins*) if x and y are twins and x is adjacent (resp. not adjacent) to y.

Definition 11 ((Pruning sequence, cf. [HM90])). *Let G be a graph with n vertices denoted by $v_1, \ldots, v_n$, and let $S = \{s_2, \ldots, s_n\}$ be a sequence of pairs of the form $\langle (v_i, v_j), type \rangle$, where $j < i$ and type is either leaf, true or false. We say that S is a pruning sequence for G, if for $2 \leq i \leq n$, if $s_i = \langle (v_i, v_j), leaf \rangle$ (resp. if $s_i = \langle (v_i, v_j), false \rangle$, or $s_i = \langle (v_i, v_j), true \rangle$) then the subgraph of G induced by $\{v_1, \ldots, v_i\}$ is obtained from the subgraph induced by $\{v_1, \ldots, v_{i-1}\}$ by adding the vertex v_i and making it a leaf (resp. a false twin, or a true twin) of the vertex v_j.*

Theorem 5 (Hammer and Maffray [HM90]). *For every connected graph G, G is distance hereditary if and only if there exists a pruning sequence for G. Moreover, there is a linear time algorithm which constructs a pruning sequence for a given graph G, if it exists, or claims that there is no pruning sequence for G.*

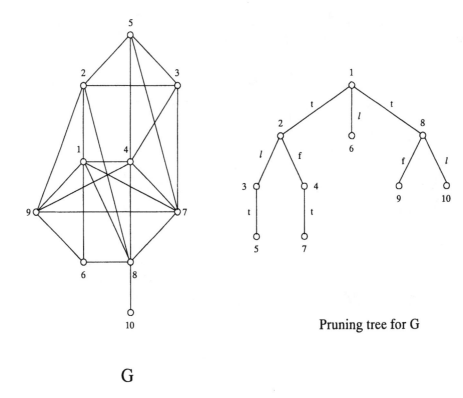

Pruning tree for G

G

Fig. 1. A connected distance hereditary graph G and a pruning tree for G.

Definition 12 ((Pruning–tree)). *Let G be a graph having n vertices denoted by $v_1, \ldots, v_n$, and let $S = \{s_2, \ldots, s_n\}$ be a pruning sequence for G. The pruning–tree corresponding to the pruning sequence S of G, is the labeled ordered tree T constructed as follows:*

    (i) Set T_1 as the tree consisting of a single root vertex v_1, and set $i := 1$.
    (ii) Set $i := i + 1$. If $i > n$ then set $T := T_n$ and stop.
    (iii) Let $s_i = \langle (v_i, v_j), leaf \rangle$ (resp. $s_i = \langle (v_i, v_j), false \rangle$, or $s_i = \langle (v_i, v_j), true \rangle$), then set T_i as the tree obtained from T_{i-1} by adding the new vertex v_i and making it a rightmost son of the vertex v_j, and labeling the edge connecting v_i to v_j by leaf (resp. by false or true).
    (iv) Go back to step (ii) above.

Example 2. Figure 1 illustrates a connected distance hereditary graph G. The vertices of G are denoted by $\{1, 2, \ldots, 10\}$. Figure 1 also illustrates the pruning–tree corresponding to the pruning sequence S of G defined by

$$S = \{ \langle (2,1), true \rangle, \langle (3,2), leaf \rangle, \langle (4,2), false \rangle, \langle (5,3), true \rangle, \langle (6,1), leaf \rangle, \\ \langle (7,4), true \rangle, \langle (8,1), true \rangle, \langle (9,8), false \rangle, \langle (10,8), leaf \rangle \}.$$

In the figure we denoted true (resp. false or leaf) shortly by t (resp. f or l).

Definition 13 ((T_a)). *Let T be any rooted tree, and let a be any node occurring in T. We denote by T_a the sub–tree of T rooted at a.*

Definition 14 ((True/false twin son, leaf son, twin descendant)). *Let G be a graph having a pruning sequence S, T be the pruning–tree corresponding to S and let v and u be any two vertices of G. We say that v is a true twin son (resp. false twin son, leaf son) of u, if v is a son of u in T and the edge connecting v to u in T is labeled with true (resp. false, leaf). We say that v is a twin descendant of u, if v is either the same vertex as u, or v is a descendant of u in T such that all the edges of the path connecting v to u in T are labeled with true or false.*

Lemma 1. *Let G be a graph having a pruning sequence S with corresponding pruning–tree T, and let a be any internal node in T whose sons in T are denoted by $a_1, \ldots, a_l$ ordered from left to right. For all $1 \leq i < j \leq l$, and for every two vertices v and u occurring in T_{a_i} and T_{a_j} respectively, v is adjacent to u in G if and only if a_i is either a leaf son or a true twin son of a, a_j is either a true or false twin son of a, and v and u are twin descendants of a_i and a_j respectively.*

Let $A \subseteq V$ be a subset of the vertices of $G = \langle V, E \rangle$. We denote by $G[A]$ the subgraph of G induced by A. Furthermore, if $T_{a_1}, \ldots, T_{a_k}$ are disjoint sub–trees of a pruning–tree, then $G[T_{a_1} \cup \ldots \cup T_{a_k}]$ is the subgraph of G induced by the vertices of $T_{a_1} \cup \ldots \cup T_{a_k}$.

The following lemma follows immediately from Lemma 1 above.

Lemma 2. *Let G be a graph having a pruning sequence S, with corresponding pruning–tree T, and let a be any internal node in T whose sons ordered from left to right are $a_1, \ldots, a_l$. For $1 \leq i \leq l$, we have the following:*

(i) *If a_i is a false twin son of a, then $G[\{a\} \cup T_{a_i} \cup T_{a_{i+1}} \cup \ldots \cup T_{a_l}]$ is equal to the disjoint union of $G[\{a\} \cup T_{a_{i+1}} \cup \ldots \cup T_{a_l}]$ and $G[T_{a_i}]$.*

(ii) *If a_i is either a leaf or a true twin son of a, then $G[\{a\} \cup T_{a_i} \cup T_{a_{i+1}} \cup \ldots \cup T_{a_l}]$ can be constructed by taking the disjoint union of $G[\{a\} \cup T_{a_{i+1}} \cup \ldots \cup T_{a_l}]$ and $G[T_{a_i}]$, and connecting all the twin descendants of a_i to a and to all the twin descendants of $a_{i+1}, \ldots, a_l$.*

Theorem 1 *For every distance hereditary graph G, $cwd(G) \leq 3$, and a 3–expression defining it can be constructed in time $O(|V| + |E|)$.*
Proof:
Let G be a distance hereditary graph. We assume that G is connected, since if G is not connected we can construct a 3–expression for G by applying the disjoint union operation (i.e. the $\oplus$ operation) on the 3–expressions obtained for the connected components of G. By Theorem 5 above there is a pruning sequence S for G, which can be obtained in linear time. Let T be the pruning–tree corresponding to the pruning sequence S.

Claim. For each internal node a of the pruning tree T, there is a 3–expression t_a which defines the labeled graph G', such that $G' = G[T_a]$, all the twin descendants of a are labeled with 2 in G', and all the other vertices of G' are labeled with 1.

proof of claim: We shall prove the claim by induction on the height of sub–trees of T. The claim trivially holds for all the sub–trees of T of height 1. Suppose the claim holds for all the sub–trees of T of height $n-1$. Let a be any internal node of T such that T_a is of height n and let $a_1, \ldots, a_l$ be the sons of a ordered from left to right. By the induction hypothesis there are 3–expressions $t_{a_1}, \ldots, t_{a_l}$ which defines the disjoint labeled graphs $G[T_{a_1}], \ldots, G[T_{a_l}]$, respectively, such that all the vertices which are twin descendants of $a_1, \ldots, a_l$ are labeled with 2 and all other vertices in these graphs are labeled with 1. We construct the expression t_a as follows:

Procedure A

(i) Set $e_{l+1} := 2(a)$ and set $i := l + 1$.
(ii) Set $i := i - 1$. If $i = 0$ then set $t_a := e_1$ and stop.
(iii) If a_i is either a leaf son or a true twin son of a then set

$$e_i := \rho_{3\rightarrow2}(\eta_{2,3}(t_{a_i} \oplus \rho_{2\rightarrow3}(e_{i+1})))$$

(iv) If a_i is a false twin son of a then set $e_i := t_{a_i} \oplus e_{i+1}$.
(v) Go back to step (ii) above

From Lemma 2 above it follows that for $1 \leq i \leq l$, the 3–expression e_i constructed by the above procedure, defines the graph $G[\{a\} \cup T_{a_i} \cup \ldots \cup T_{a_l}]$. Hence, the 3–expression t_a constructed by the above procedure (which is equal to e_1), defines the graph $G[T_a]$. Since in the graph defined by t_a, all the twin descendants of a are labeled with 2 and all the other vertices are labeled with 1, this completes the proof of Claim 3.

Let x be the root of the pruning–tree T. By the above claim there is a 3–expression t_x which defines the graph G. Moreover, using Procedure A above, it is easy to see that the 3–expression t_x which defines G can be constructed in linear time, and by that the proof of Theorem 1 is completed.   □

4   Unit Interval Graphs and Permutation Graphs Are Not of Bounded Clique–Width

In this section we show that the classes of unit interval graphs and permutation graphs are not of bounded clique–width. Below (cf. definition 15) we define the graph I_n which is a unit interval graph with n^2 vertices (cf. Fact 1). Informally, the vertices of the graph I_n can be thought as being arranged in an $n \times n$ square array, such that all the vertices occurring in the same column form a clique, vertices in non-consecutive columns are not connected, and a vertex $v_{i,j}$ occurring in row i and column j is adjacent to all the vertices occurring in column $j + 1$ and in rows $1, \ldots, i - 1$. Figure 2 illustrates the graph I_4, and Figure 3 shows its representation as intersecting intervals.

Definition 15 ((The graph I_n)). *We denote by I_n the graph $\langle V, E \rangle$, where the set of vertices V is defined by:*

$$V = \{ v_{i,j} : 1 \le i \le n, \ 1 \le j \le n \}$$

and the set of edges E is defined by: $E = E' \cup E''$, where

$$E' = \{ (v_{i_1,j}, v_{i_2,j}) : 1 \le i_1 \le n,$$
$$1 \le i_2 \le n, 1 \le j \le n, i_1 \ne i_2 \}$$

$$E'' = \{ (v_{i_1,j}, v_{i_2,j+1}) : 1 \le j \le n - 1,$$
$$2 \le i_1 \le n, 1 \le i_2 \le i_1 - 1 \}$$

Fact 1 *For every $n \in \mathcal{N}$, the graph I_n is a unit interval graph.*

Fact 1 above can be verified by constructing for every $n \in \mathcal{N}$, a unit interval graph presentation for the graph I_n , similar to the one illustrated in Figure 3 for the graph I_4. For example, let $\varepsilon = 1/2n$ and define the (closed) interval corresponding to $v_{i,j}$ to be $J_{i,j} = [j + i\varepsilon, j + 1 + (i - 1)\varepsilon]$, $(1 \le i, j \le n)$.

Lemma 3. *For every $n \in \mathcal{N}$, $n \ge 2$, $cwd(I_n) = n + 1$.*

Theorem 2 follows immediately from Lemma 3.

A graph $G = \langle V, E \rangle$ is a *permutation graph* if and only if there are two linear ordering of its vertices R_1 and R_2, such that for every two vertices v and u in G, v is adjacent to u if and only if v occurs before u in the linear order R_1 and v occurs after u in the linear order R_2, cf. [Gol80]. Below (cf. definition 16) we define the graph H_n which is a permutation graph (cf. Fact 2). Informally, the vertices of the graph H_n can be put in an $n \times n$ square array, such that all the vertices occurring in the same column form a clique, vertices in non-consecutive columns are not connected, a vertex v occurring in row i and an *odd* column j is adjacent to all the vertices occurring in column $j + 1$ and in rows $1, \ldots, i - 1$, and a vertex v occurring in row i and *even* column j is adjacent to all the vertices occurring in column $j + 1$ in rows $i + 1, i + 2, \ldots, n$.

Definition 16 ((The graph H_n)). *We denote by H_n the graph $\langle V, E \rangle$, where the set of vertices V is defined by:*

$$V = \{ v_{i,j} : 1 \le i \le n, \ 1 \le j \le n \}$$

and the set of edges E is defined by: $E = E' \cup E'' \cup E'''$, where

$$E' = \{ (v_{i_1,j}, v_{i_2,j}) : 1 \le i_1 \le n,$$
$$1 \le i_2 \le n, 1 \le j \le n, i_1 \ne i_2 \}$$

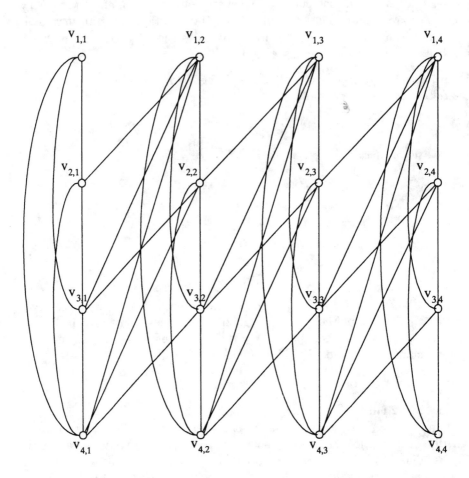

Fig. 2. The graph I_4

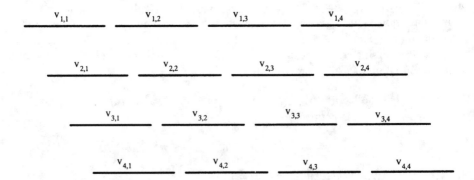

Fig. 3. The unit interval representation of the graph I_4

$$E'' = \{ (v_{i_1,j}, v_{i_2,j+1}) : 1 \le j \le n - 1, j \text{ odd},$$
$$2 \le i_1 \le n, 1 \le i_2 \le i_1 - 1 \}$$

$$E''' = \{ (v_{i_1,j}, v_{i_2,j+1}) : 2 \le j \le n - 1, j \text{ even},$$
$$1 \le i_1 \le n - 1, i_1 + 1 \le i_2 \le n \}$$

Fact 2 *For every $n \in \mathcal{N}$, the graph H_n is a permutation graph.*

Lemma 4. *For every $n \in \mathcal{N}$, $n \ge 2$, $cwd(H_n) = n + 1$.*

Theorem 3 follows immediately from Lemma 4.

5 Square Grids

In this section we show that every $n \times n$ square grid, $n \ge 3$, has clique–width exactly $n + 1$. Throughout this section we denote by $v_{i,j}$ the vertex of the grid occurring in row i and column j.

Lemma 5. *For every $n \times n$ square grid G, $cwd(G) \le n + 1$.*

Proof:
[Sketch] Let G be an $n \times n$ square grid. We shall prove the lemma by constructing an $n + 1$–expression f which defines G. For that we first construct an $n + 1$–expression c which defines the subgraph G_L of the grid G induced by the vertices occurring in the lower triangle of G, such that all the vertices of the diagonal of G_L are labeled with labels from 1 to n, and all the other vertices of G_L are labeled with $n + 1$. Similarly we construct an $n + 1$–expression d which defines the subgraph of the grid G induced by the vertices occurring in the subgraph G_R of G induced by the upper triangle of G. Finally, we construct the $n + 1$–expression f, by adding all the vertices of the main diagonal of the grid G, and connecting them to the vertices in the diagonals of the graphs G_L and G_R. □

We now show that $n + 1$ is also the lower bound for $cwd(G)$. Recall (cf. definition 8 above) that for a k–expression t and for every internal node a of $tree(t)$, we denote by $tree(a, t)$ the sub–tree of $tree(t)$ rooted at a.

Lemma 6. *Let G be an $n \times n$ square grid, $n \in \mathcal{N}$, $n \ge 3$, let t be a k–expression which defines G, let a be the highest $\oplus$ node in $tree(t)$, let b and c be the highest $\oplus$ nodes in the sub–trees rooted at the left and right sons of a, respectively. If neither graph defined by $tree(b, t)$ and $tree(c, t)$ contains a full row of the grid G, then $k \ge n + 1$. Similarly, if neither graph defined by $tree(b, t)$ and $tree(c, t)$ contains a full column of the grid G, then $k \ge n + 1$.*

Lemma 7. *Let G be an $n \times n$ square grid, let t be a k–expression which defines G, let d be an internal $\oplus$ node in $tree(t)$. If the graph defined by $tree(d, t)$ contains a full row of the grid and does not contain a full column of the grid, then $k \ge n + 1$. Similarly, if the graph defined by $tree(d, t)$ contains a full column of the grid and does not contain a full row of the grid, then $k \ge n + 1$.*

Lemma 8. *Let G be an $n \times n$ square grid, let t be a k-expression which defines G, let d be an internal $\oplus$ node in $tree(t)$ and let e and f be the highest $\oplus$ nodes in the sub-trees rooted at the left and right sons of d, respectively. If the graph defined by $tree(d, t)$ contains a full row of the grid and a full column of the grid, and neither graph defined by $tree(e, t)$ or $tree(f, t)$ contains a full row or a full column of the grid, then $k \geq n + 1$.*

Theorem 4 *For every $n \times n$ square grid G, $n \in \mathcal{N}$, $n \geq 3$, $cwd(G) = n + 1$.*
Proof:
Let G be an $n \times n$ square grid $n \in \mathcal{N}$, $n \geq 3$. By Lemma 5 above $cwd(G) \leq n+1$. We shall show that $cwd(G) > n$, which implies that $cwd(G) = n + 1$. Suppose that there is a k-expression t which defines G, and $k \leq n$. Let a be the highest $\oplus$ node in $tree(t)$, let b and c be the highest $\oplus$ nodes in the sub-trees rooted at the left and right sons of a, respectively. If neither graph defined by $tree(b, t)$ and $tree(c, t)$ contains a full column of the grid, or neither contains a full row of the grid, then by Lemma 6 above $k \geq n + 1$, a contradiction. Hence, we assume without loss of generality, that the graph defined by $tree(c, t)$ contains a full row of the grid. Suppose there exist a node d in $tree(c, t)$, such that either $tree(d, t)$ contains a full row of the grid and does not contain a full column of the grid, or $tree(d, t)$ contains a full column of the grid and does not contain a full row of the grid. In either case, by Lemma 7 above $k \geq n + 1$, a contradiction. Hence, there exist a node d in $tree(c, t)$ such that $tree(d, t)$ contains a full row and a full column of the grid, and both $tree(e, t)$ and $tree(f, t)$ do not contain a full column or row of the grid, where e and f are the highest $\oplus$ nodes in the sub-trees rooted at the left and right sons of d, respectively. In this case, by Lemma 8 above $k \geq n + 1$, a contradiction.

Since we have considered all possible cases, we conclude that the assumption that $k \leq n$, was not correct, which implies that $k \geq n + 1$. $\qquad\square$

Corollary 4 *For every $n \times m$ rectangular grid G, $n, m \in \mathcal{N}$, $n, m \geq 3$, $min\{n, m\} + 1 \leq cwd(G) \leq min\{n, m\} + 2$.*

Acknowledgments

We are indebted to Bruno Courcelle, Johann Makowsky, Derek Corneil and Michel Habib for their helpful comments.

References

BD98.   A. Brandstädt and F.F. Dragan. A linear-time algorithm for connected r-domination and steiner tree on distance-hereditary graphs. *Networks*, 31:177–182, 1998.

CER93.   B. Courcelle, J. Engelfriet, and G. Rozenberg. Handle-rewriting hypergraph grammars. *J. Comput. System Sci.*, 46:218–270, 1993.

CMRa. B. Courcelle, J.A. Makowsky, and U. Rotics. Linear time solvable optimization problems on certain structured graph families, extended abstract. Graph Theoretic Concepts in Computer Science, 24th International Workshop, WG'98, volume 1517 of Lecture Notes in Computer Science, pages 1-16. Springer Verlang, 1998.

CMRb. B. Courcelle, J.A. Makowsky, and U. Rotics. On the fixed parameter complexity of graph enumeration problems definable in monadic second order logic. To appear in Disc. Appl. Math.

CO98. B. Courcelle and S. Olariu. Upper bounds to the clique-width of graphs. to appear in Disc. Appl. Math.
(http://dept-info.labri.u-bordeaux.fr/~courcell/ActSci.html), 1998.

DM88. A. D'Atri and M. Moscarini. Distance–hereditary graphs Steiner trees and connected domination. SIAM J. Comput., 17:521-538, 1988.

DNB97. F.F. Dragan, F. Nicolai, and A. Brandstädt. LexBFS-orderings and powers of graphs. Graph Theoretic Concepts in Computer Science, 22th International Workshop, WG'96, volume 1197 of Lecture Notes in Computer Science, pages 166-180, 1997.

Dra94. F.F. Dragan. Dominating cliques in distance-hereditary graphs. Algorithm theory—SWAT'94, volume 824 of Lecture Notes in Computer Science, pages 370-381, 1994.

Gol80. M. C. Golumbic. Algorithmic Graph Theory and Perfect Graphs. Academic Press, New York, 1980.

HM90. P. L. Hammer and F. Maffray. Completely separable graphs. Disc. Appl. Math., 27:85-99, 1990.

How77. E. Howorka. A characterization of distance-hereditary graphs. Q. J. Math. Oxford Ser. (2), 28:417-420, 1977.

MR99. J.A. Makowsky and U. Rotics. On the classes of graphs with few P_4's. To appear in the International Journal of Foundations of Computer Science (IJFCS), 1999.

PW99. U. N. Peled and J. Wu. Restricted unimodular chordal graphs. To appear in Journal of Graph Theory, 1999.

An Improved Algorithm for Finding Tree Decompositions of Small Width

Ljubomir Perković[1] and Bruce Reed[2]

[1] Department of Mathematics and Computer Science,
Drexel University, Philadelphia PA 19104, USA
`lperkovi@mcs.drexel.edu`
[2] Equipe Combinatoire, CNRS,
Universite Pierre et Marie Curie, Paris, France
`reed@ecp6.jussieu.fr`

Abstract. We present a modification of Bodlaender's linear time algorithm that, for constant k, determines whether an input graph $G = (V, E)$ has treewidth k and, if so, constructs a tree decomposition of G of width at most k. Our algorithm has the following additional feature: if G has treewidth greater than k then a subgraph G' of G of treewidth greater than k is returned along with a tree decomposition of G' of width at most $2k$. A consequence is that the fundamental disjoint rooted paths problem can now be solved in $O(n^2)$ time. This is the primary motivation for this paper.

1 Introduction

The notions of tree decompositions and treewidth of a graph were introduced in the 1980's by Robertson and Seymour [10] and they have since found a large number of applications (see Bodlaender [5] for a tutorial on graph treewidth and its applications). We shall define tree decompositions and treewidth precisely in the next section. For now it will suffice to think of a tree decomposition of a graph G as a decomposition of the vertex set of G into pieces that fit together in a "tree-like way". The width of a tree decomposition is the maximum size of a piece in the tree decomposition and the treewidth of the graph G is the minimum treewidth over all possible tree decompositions of G.

It turns out that if G has small treewidth (i.e. bounded by a constant) then dynamic programming can be used to solve many optimization problems on G, just as dynamic programming can be used to solve optimization problems on trees. Examples of problems include edge coloring, Hamiltonian cycle, clique, vertex coloring and various routing problems, all of which can be solved in linear time. In fact, given a tree decomposition of G of constant width, one can solve in linear time any optimization problem on G that can be expressed in monadic second order logic. Furthermore, other problems such as graph isomorphism can be solved in polynomial time (for references see [5]). This is what spurred an interest in finding tree decompositions of small (constant) width in

Widmayer et al. (Eds.): WG'99, LNCS 1665, pp. 148–154, 1999.

graphs which have them. The first algorithm for finding an $O(k)$ tree decomposition (for some constant k) was due to Robertson and Seymour [10] and had an $O(n^{f(k)})$ running time on graphs with n vertices (Robertson and Seymour never bothered to compute $f(k)$). This was first improved by Arnold, Corneil and Proskurowski [3] who gave an $O(n^{k+2})$ algorithm, then again by Robertson and Seymour [12] who gave an $O(n^2)$ algorithm, and then by Lagergren [7] who gave an $O(n \log^2 n)$ algorithm and by Reed [8] who gave an $O(n \log n)$ algorithm. Finally, Bodlaender [4] gave a linear time algorithm for computing a tree decomposition of width k of a graph G of treewidth at most k.

We describe our modification of Bodlaender's algorithm which has the following additional feature: if G has treewidth greater than k, our algorithm returns a subgraph G' of G and a tree decomposition of G' of width at most $2k$. We illustrate the importance of this extra information by considering a fundamental application of treewidth due to Robertson and Seymour [12]:

Problem 1 (The l-disjoint rooted path problem).

Let $S = \{s_1, s_2, ..., s_l\}$ and $T = \{t_1, t_2, ..., t_l\}$ be disjoint sets of vertices of a graph G. Is there a set of vertex disjoint paths $\{P_1, P_2, ..., P_l\}$ such that each P_i contains vertices s_i and t_i?

In general, the l-disjoint rooted path problem is NP-complete. In a seminal paper, Robertson and Seymour [12] developed an $O(n^3)$ algorithm which solves the l-disjoint rooted path problem, for a fixed l. One step of their recursive algorithm consists of attempting to find a tree decomposition of bounded width for G. If such a tree decomposition is found, they apply dynamic programming to solve the problem (i.e. to find the disjoint paths). Otherwise, they find an unnecessary vertex in $O(n^2)$ time and recursively solve the problem on $G - v$. This gives an $O(n^3)$ time algorithm for the l-disjoint rooted path problem. One of the various ways in which they find an unnecessary vertex is to construct a tree decomposition of width at most $2k$ of a subgraph of G of width greater than k and then use dynamic programming to find the vertex to be deleted.

Reed [8], [9], by using his $O(n \log n)$ treewidth decomposition algorithm and modifying Robertson and Seymour's algorithm somewhat, developed an $O(n^2 \log n)$ algorithm for the l-disjoint rooted paths problem. Now, we can parachute our modification of Bodlaender's algorithm into Reed's algorithm to improve its performance so that it runs in $O(n^2)$ time. This is the primary motivation for our result. We also note that the above discussion applies also to the more general k-realizations problem (see [10,11,12,13]), implying that the k-realization problem can be solved in $O(n^2)$ time.

2   Definitions and Preliminary Lemmas

A tree decomposition $(T, \mathcal{X})$ of a graph $G = (V, E)$ consists of a tree T and of a family $\mathcal{X}$ of subsets X_t of V, one for each node t of T, such that:

(i) $\bigcup_{t \in T} X_t = V$,

(ii) for all $(v, w) \in E$ there exists $t \in T$ such that $v, w \in X_t$,

(iii) for all t_0, t_1 and t_2 of T, if t_1 is on the unique path from t_0 to t_2 then $X_{t_0} \cap X_{t_2} \subseteq X_{t_1}$.

The width of a tree decomposition is $\max_{t \in T}\{|X_t| - 1\}$. The treewidth of G (denoted by $TW(G)$) is the minimum of the widths of its tree decompositions.

We will call a tree decomposition $(T, \mathcal{X})$ of $G = (V, E)$ *standard* if $|T| \leq |V|$. Note that any tree decomposition $(T'\mathcal{X}')$ of G can be transformed into a standard tree decomposition $(T, \mathcal{X})$ of same width k. This can be done in $O(|T'|k)$ time by initially setting $T = T'$ and $\mathcal{X} = \mathcal{X}'$ and then recursively identifying nodes $s, t \in T$ such that $(s, t) \in T$ and $|X_s \cup X_t| \leq k + 1$, and assigning $X_s \cup X_t$ to the identified new vertex.

We now state a few definitions and well known lemmas that we will need (see [4] for proofs of the lemmas).

Lemma 1. *If $TW(G) \leq k$ then $|E| < k|V|$.*

Let $S = \{v \in V : d(v) \leq 4k\}$.

Corollary 1. $|S| \geq \frac{|V|}{2}$.

Lemma 2. *Let $(T, \mathcal{X})$ be a tree decomposition of G.*

(i) *If $X \subseteq V$ induces a clique in G then there exists $t \in T$ such that $X \subseteq X_t$;*

(ii) *For $X_1, X_2 \subseteq V$, if (X_1, X_2) induces a complete bipartite subgraph in G then there is $t \in T$ such that $X_1 \in X_t$ or $X_2 \in X_t$.*

The following easily follows.

Corollary 2. *If $v, w \in V$ have $k + 1$ common neighbors then any tree decomposition of G of width at most k is a tree decomposition of $G + (u, v)$.*

Let $G^* = (V^*, E^*)$ be obtained by identifying the vertices of some matching M of G (and removing resulting multiple edges) and let $(T^*, \mathcal{X}^*)$ be a standard tree decomposition of G^* of width k.

Lemma 3. *Let $T = T^*$ and let $\mathcal{X}$ be the collection of subsets X_t obtained from X_t^* by replacing every identified vertex $v \in X_t^*$ with the original two endpoints, for all $t \in T$. Then $(T, \mathcal{X})$ is a standard tree decomposition of G of width at most $2k + 1$ and thus $TW(G) \leq 2TW(G^*) + 1$. Furthermore, for fixed k, the tree decomposition $(T, \mathcal{X})$ can be computed in linear $(O(|V|))$ time from $(T^*, \mathcal{X}^*)$.*

Lemma 4. $TW(G^*) \leq TW(G)$.

3   The Tree Decomposition Algorithm

Given a constant k, the specifications for our algorithm are:

Input: $G = (V, E)$
Output: Either
  I a standard tree decomposition (T, X) of G of width no greater than k,
    or
  II a subgraph G' of G of treewidth greater than k and a standard tree
    decomposition (T', X') of G' of width at most $2k$.
Running time: $O(|V|)$.

If $|V| \leq 12k^3$, the base case, we simply apply a brute force algorithm. The main step of the recursive part of our algorithm is to find a new graph $G^* = (V^*, E^*)$ with $|V^*| \leq (1 - \frac{1}{16k^2})|V|$ such that either

(a) G^* is a subgraph of G and, given a standard tree decomposition of G^* of
    width k, we can find in linear time a standard tree decomposition of G of
    width k, or
(b) G^* is obtained by identifying the edges of some matching M in G.

We then recursively apply our algorithm on G^*. Then, depending on whether we are in case (a) or (b) and on the type of output the recursive call returns, we do one of several things.

In case (a), if the output of the recursive call is of type I, we find in linear time a standard tree decomposition of G of width k. In case (a), if the output is of type II, we are clearly done. In case (b), if the output is of type I, we apply lemma 3 to find a standard tree decomposition (T, X) of G of width at most $2k + 1$. In case (b), if the output is of type II, we apply lemma 3 to find a subgraph G' of G of treewidth greater than k and a standard tree decomposition (T', X') of G' of width at most $4k + 1$. In both cases, we apply the Bodlaender and Kloks [6] algorithm, specified as follows for constants k and l:

Input: A graph G and a standard tree decomposition (T, X) of G of width at
    most l.
Output: A standard tree decomposition of G of width at most k, if G has one.
Running time: $O(|V|)$.

In fact, the Bodlaender and Kloks algorithm can be easily modified, without affecting the running time, so that if $TW(G) > k$ then the algorithm returns a subgraph G' of G of treewidth greater than k and a standard tree decomposition of G' of width at most $2k$ ($k + 1$ is possible too). We use this modified version of the Bodlaender and Kloks algorithm, to obtain outputs of type I or II for the original input graph G.

In the remainder of this section, we describe the steps involved in finding G^* and modifying the output from the recursive calls to obtain output for the original problem. We show they can be performed in linear ($O(|V|)$) time. It follows that the total running time $T(|V|)$ of our algorithm is bounded above by $T(1 - \frac{1}{16k^2})|V| + O(|V|)$ and thus $T(|V|) = O(|V|)$.

Our algorithm modifies and improves Bodlaender's original algorithm in two ways. First, while Bodlaender's algorithm just stops when it obtains a certificate that $TW(G) > k$, we do additional work. Second, we have streamlined the construction of G^* (the graph the algorithm recurses on) so the algorithm is simpler and the graph G^* is smaller. A payoff is that the recursion is applied to a smaller graph. In fact, our algorithm iterates no more than $O(k^2)$ times while Bodlaender's algorithm may require $O(k^8)$ recursive iterations.

We now turn to the details of our algorithm. We can assume that $|E| \leq k|V|$ (note that this implies $|E| = O(|V|)$); otherwise $TW(G) > k$, by lemma 1, and we run our algorithm on a subgraph of G obtained by removing all but $k|V|$ edges from G. We also assume that $k \geq 2$. The case $k = 0$ is just testing whether G has no edge and the case $k = 1$ is testing whether G has no cycle. Both tests are easy to do in linear time, as are the constructions of the standard tree decompositions. Let $V = \{v_1, v_2, ..., v_{|V|}\}$.

3.1   Matching Contraction (Case (b))

Let S be the vertices of V of degree no greater than $4k$. We construct a matching M that is maximal with respect to vertices in S by first ordering the vertices in S as $v_{i_1}, ..., v_{i_{|S|}}$, and then, for each v_{i_j} in turn, if there is a neighbor $v \in V$ of v_{i_j} which is in no matching edge picked so far, we add (v_{i_j}, v) to the matching.

If $|M| \geq \frac{|V|}{16k^2}$, we identify all endpoints of M, and remove multiple edges, to obtain the graph G^*. To remove multiple edges in linear time, we apply bucket sort on the edges of G^* twice, once for each coordinate; since all multiple edges will show up in consecutive positions, it is easy to remove all copies in one traversal of the edges.

3.2   Subgraph Construction (Case (a))

If $|M| < \frac{|V|}{16k^2}$, let A be the subset of vertices in S that are endpoints of edges in M, let B be the set of the remaining endpoints (i.e. not in S) and let C be the set of neighbors of vertices in A that are not in $A \cup B$. Note that $|A| < \frac{|V|}{8k^2}$, $|B| < \frac{|V|}{16k^2}$ and $|C| < \frac{|V|}{2k}$. Finally, we set $D = S - A - C$ and $F = V - S - B - C$ (see figure 1). There is no edge between a vertex in D and a vertex in $F \cup D \cup C \cup A$. Thus all edges incident to vertices in D are also incident to vertices in B and $|D| > \frac{|V|}{2} - \frac{|V|}{8k^2} - \frac{|V|}{2k} > \frac{|V|}{5}$ (by corollary 1 and since $k \geq 2$).

We now attempt to assign the vertices in D to non-adjacent *pairs* of neighbors of v in B so that no pair is assigned more than $k + 1$ vertices. To do this, we construct a data structure that allows us to perform the assignments in linear time. Let Q be a list containing triples of the form $((v_i, v_j), u)$ for all vertices $u \in D$ and all $v_i, v_j \in N(v) (\subseteq B)$ with $i < j$ and $(v_i, v_j) \notin E$. We can construct Q in linear time by visiting every vertex v of D and considering all pairs of vertices in $N(v)$. We sort Q twice using Bucket sort, once for each coordinate of the pair. Note that there are only $O(|V|)$ items in Q (since $d(u) \leq 4k$ for all $u \in D$) and all triples of the form $((v_i, v_j), ...)$ are in consecutive positions after

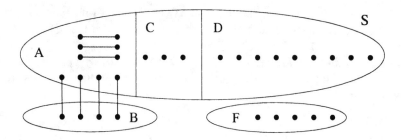

Fig. 1. Matching M and subsets A, B, C, D and F of V

the sort. We can thus easily find all common neighbors in D of pairs of vertices $v_i, v_j \in B$.

The following lemma explains our assignments:

Lemma 5. *If U is the set of unassigned vertices in D and if $(T, \mathcal{X})$ is a tree decomposition of $G - U$ of width at most k then for every $u \in U$ there is $t \in T$ such that $N(u) \subset X_t$.*

Proof. If v, w are a pair of non-adjacent neighbors in B of $u \in U$ then they have at least $k + 1$ common neighbors in $G - U$. By corollary 2, it follows $(T, \mathcal{X})$ is a tree decomposition of $G - U + (v, w)$. But then $(T, \mathcal{X})$ is also a tree decomposition of $G - U + \{(v, w) : v, w \in N(u)\}$. Thus, by lemma 2, there must be some $t \in T$ such that $N(u) \in X_t$. For all $u \in U$ with only one neighbor in B, the lemma is trivially true.   □

Now we consider three cases.

There exists $u \in U$ of degree greater than k. We set $G^* = (V^*, E(V^*))$ where

$$V^* = u + N(u) + \{v \in D : v \text{ is assigned to a pair in } N(u)\}$$

By applying the same argument as in the proof of lemma 5, we obtain that in any tree decomposition $(T^*, \mathcal{X}^*)$ of G^* there is $t \in T^*$ such that $N(u) + u \in X_t$, implying that $TW(G^*) > k$. Because $|V^*| \leq 1 + 4k + \binom{4k}{2}(k+1) \leq 12k^3$ (since $k \geq 2$), we just apply the base case of our algorithm on G^*.

Every $u \in U$ has degree at most k and $|U| \geq \frac{|V|}{8k^2}$. We recursively apply our algorithm on $G^* = (V - U, E(V - U))$.

From a standard tree decomposition $(T^*, \mathcal{X}^*)$ of G^* of width at most k, we construct a tree decomposition $(T, \mathcal{X})$ of G of width at most k by initially setting $T = T^*$ and $\mathcal{X} = \mathcal{X}^*$. Then, for every $u \in U$, we find $t \in T$ such that $N_G(u) \subseteq X_t$

and we add a new node t_v to T with $X_{t_v} = N_G(v) + v$ and we connect t_v to t in T. By lemma 5, t must exist and $(T, \mathcal{X})$ is clearly a standard tree decomposition of G.

In order to find, for all $v \in U$, some $t \in T$ such that $N_G(v) \subseteq X_t$ in linear time, we need a special data structure, the same that Bodlaender used in his algorithm. See [4] for details.

Every $u \in U$ has degree at most k and $|U| < \frac{|V|}{8k^2}$. If $|U| < \frac{|V|}{8k^2}$, let D' be the set of all but $\lceil \frac{|V|}{8k^2} \rceil$ vertices in $D - U$. Then $|D'| \geq |D| - |U| - \lceil \frac{|V|}{8k^2} \rceil > \frac{|V|}{5} - \frac{|V|}{4k^2} - 1 > \frac{|V|}{8}$ (for $k \geq 2$). The number of vertex pairs in B assigned to a vertex in D' is then greater than $\frac{|V|}{8(k+1)}$. We recursively apply our algorithm on the subgraph G^* induced by $B \cup D'$. Note that $|V(G^*)| < (1 - \frac{1}{8k^2})|V|$. Furthermore, $TW(G^*) > k$ as seen from the following argument. Suppose that $TW(G^*) \leq k$. Every vertex $u \in D'$ assigned to a pair $(v, w) \in B$ can be contracted with v or w to form a minor of G with an additional edge in B. By lemma 4 and lemma 1, it follows that $|E(B) + \{$assigned pairs in $B\}| < |B|k < \frac{|V|}{16k^2}k = \frac{|V|}{16k}$, a contradiction for every positive k.

References

1. Arnborg, S., Lagergren, J., Seese, D.: Easy problems for tree-decomposable graphs. J. of Algorithms **12** (1991) 308–340
2. Arnborg, S., Courcelle, B., Proskurowski, A., Seese, D.: An algebraic theory of graph reduction. J. Assoc. Comput. Mach. **40** (1993) 1134–1164
3. Arnborg, S., Corneil, D. G., Proskurowski, A.: Complexity of finding embeddings in a k-tree. SIAM J. Algebra Discrete Methods **8** (1987) 277–284
4. Bodlaender, H. L.: A linear-time algorithm for finding tree-decompositions of small treewidth. SIAM J. Comput. **25:6** (1996) 1305–1317
5. Bodlaender, H. L.: A partial k-arboretum of graphs with bounded treewidth. Theoret. Comput. Sci. **209:1-2** (1998) 1–45
6. Bodlaender, H. L., Kloks, T.: Efficient and constructive algorithms for the pathwidth and treewidth of graphs. J. Algorithms **21:2** (1996) 358–402
7. Lagergren, J.: Efficient parallel algorithms for graphs of bounded tree-width. J. Algorithms **20** (1996) 20-44
8. Reed, B.: Finding approximate separators and computing treewidth quickly. Proc. 24th STOC (1992) 221–228
9. Reed, B.: Manuscript (1992)
10. Robertson, N., Seymour P. D.: Graph Minors II. Algorithmic aspects of tree-width. J. Algorithms **7** (1986) 309–322
11. Robertson, N., Seymour P. D.: Graph Minors VI. Disjoint paths across a disk. J. Combin. Theory Ser. B **45** (1986) 115–138
12. Robertson, N., Seymour, P.D.: Graph Minors XIII. The disjoint path problem. J. Combin. Theory Ser. B **63** (1995) 65-110
13. Robertson, N., Seymour P. D.: Graph Minors VII. Disjoint paths on a surface. J. Combin. Theory Ser. B **??** (1988) 212–254

Efficient Analysis of Graphs with Small Minimal Separators

K. Skodinis *

University of Passau,
94030 Passau, Germany,
skodinis@fmi.uni-passau.de

Abstract We consider the class $\mathcal{C}^*$ of graphs whose minimal separators have a fixed bounded size. We give an $O(nm)$-time algorithm computing an optimal tree-decomposition of every graph in $\mathcal{C}^*$ with n vertices and m edges. Furthermore we make evident that many NP-complete problems are solvable in polynomial time when restricted to this class. Both claims hold although $\mathcal{C}^*$ contains graphs of arbitrarily large tree-width.

1 Introduction

One of the most important techniques for designing efficient graph algorithms is "Divide and Conquer". The application of this technique supplies a decomposition of the input graph. Such a graph decomposition is the *tree-decomposition* introduced by Robertson and Seymour [20]. Here, for convenience, we deal with simple connected graphs. A tree-decomposition of a graph $G = (V, E)$ is a tree T with nodes representing subsets X_i of V, such that (1) for every edge $\{u, v\} \in E$ there is an X_i containing both u and v and (2) for every vertex $u \in V$ all X_i containing u form a subtree in T. The width of T is the maximal cardinality of the subsets X_i minus 1. The tree-width of G is the minimal width of all possible tree-decompositions of G.

Tree-width is an important complexity parameter. It is well known that many NP-hard problems are solvable in polynomial (or linear) time when restricted to graphs with small tree-width [3,2,8]. In fact all graph properties definable in monadic second order logic with quantification over vertex and edge sets can be decided in polynomial time for graphs of fixed bounded tree-width. Such problems are for example CLIQUE, MINIMUM FILL-IN, VERTEX COVER, MAXIMUM INDEPENDENT SET, GRAPH COLORING, and HAMILTONIAN CIRCUIT.

Unfortunately given a graph G and an integer k the question whether or not graph G has tree-width k is NP-complete [1]. Moreover this problem remains NP-complete even if the input graph is bipartite [13]. Tree-width is computable in polynomial time for cographs [7], circular arc graphs [21], chordal bipartite graphs [16], and permutation graphs [6]. Bodlaender has shown [5] that for a

---

* The work of the author was supported by the German Research Association (DFG) grant BR 835/7-1

fixed k determining whether or not the tree-width of a graph G is at most k and finding a corresponding tree-decomposition can be done in linear time (but exponential in k).

Another notion related to tree-width is the notion of *minimal separators*. A vertex subset S of a graph is a separator for nonadjacent vertices u and v if the removal of S from the graph separates u and v into different connected components. As usual S is a minimal separator for u and v (minimal u,v-separator) if no proper subset of S separates u and v in this way.
It is well known that a graph is chordal if and only if every minimal separator induces a complete subgraph, see [11]. The relationship between tree-width and minimal separators is summed up by the following result shown in [1]: A graph G has tree-width at most k if and only if there exists a minimal separator S with at most k vertices such that every connected component of $G - S$ augmented by the completely connected separator vertices has tree-width at most k. For further relationships see [19].

In order to motivate our paper consider a class $\mathcal{C}$ of graphs $G = (V, E)$ having at most $p(|V|)$ different minimal separators, where $p(n)$ is a fixed polynomial. For many examples of such classes as permutation graphs, circle graphs, circular arc graphs, and chordal bipartite graphs, the tree-width is computable in polynomial time. It is of interest to know whether or not the polynomial time complexity is a consequence of the fact that these classes have polynomial many minimal separators. If so, it is also an important generalization to find a polynomial time algorithm computing the tree-width and the optimal tree-decompositions for all classes $\mathcal{C}$. This leads to the following open question stated in [14,15]: Is the tree-width (and the corresponding tree-decomposition) of every graph in $\mathcal{C}$ computable in polynomial time?

Here we give an answer to this question for the nontrivial subclasses $\mathcal{C}^*$ containing all graphs whose minimal separators have size bounded by a fixed integer. Note that every class $\mathcal{C}^*$ can be recognized in polynomial time since every graph of $\mathcal{C}^*$ has at most polynomial many separators and all minimal separators of a graph can be listed in polynomial time per separator [17]. In Figure 1 we show some examples of such graph classes.

Before we present our result let us make clear that there is no direct relationship between *tree-width* and the *maximum size* of the minimal separators of a graph.
Consider for some large n the complete bipartite graph $K_{2,n}$ with color vertex sets $\{u_1, u_2\}$ and $\{v_1, v_2, \ldots, v_n\}$. The tree-width of $K_{2,n}$ is small (2) but the maximum size of the minimal separators of $K_{2,n}$ is large (n). In fact $\{v_1, v_2, \ldots, v_n\}$ is a minimal u_1, u_2-separator of size n. Conversely consider the split graph $K_n \cup \{v\}$ for some large n consisting of a clique K_n with n vertices and a new vertex v only connected with a vertex u from K_n. The tree-width of $K_n \cup \{v\}$ is large ($n - 1$) but the maximum size of the minimal separators is small (1). In fact $\{u\}$ is the unique minimal separator of $K_n \cup \{v\}$. Hence there is no relationship between the tree-width and the maximum size of the minimal

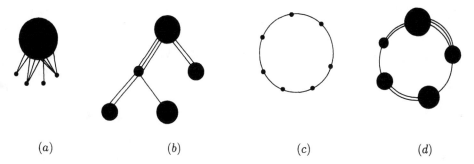

(a) (b) (c) (d)

Figure 1. (a) split graphs whose isolated vertices have bounded degree, (b) tree-like graphs whose nodes are arbitrary cliques and two neighbor cliques are connected with a bounded number of edges, (c) cycles, and (d) circle-like graphs whose nodes are arbitrary cliques and two neighbor cliques are connected with a bounded number of edges.

separators of a graph. In particular C^* contains all split graphs $K_n \cup \{v\}$, $n \geq 1$, and therefore it contains graphs with arbitrarily large tree-width.

In this paper using the Tarjan algorithm for clique separable graphs [22] we show that the tree-width problem is solvable in $O(nm)$ time for every graph in C^* with n vertices and m edges. The optimal tree-decomposition is also computable in $O(nm)$ time. Moreover many NP-complete problems become solvable in polynomial time for these graphs although they can have arbitrarily large tree-width. Such problems are for example CLIQUE, MINIMUM FILL-IN, VERTEX COVER, MAXIMUM INDEPENDENT SET, and GRAPH COLORING. The HAMILTONIAN CIRCUIT problem remains among others NP-complete (it is solvable in polynomial time for graphs with fixed bounded tree-width). That means not all problems which are solvable in polynomial time for graphs with fixed bounded tree-width are also solvable in polynomial time for graphs having minimal separators of fixed bounded size.

The question remains open for the class C if the size of the minimal separators is not bounded. However we can show that in this case even the CLIQUE problem stays NP-complete (it is solvable in polynomial time if the minimal separators of the input graph have a fixed bounded size).

2   Preliminaries

In this section we review some basic notions. A graph $G = (V, E)$ is a pair of vertex set V and edge set E. For a subset V' of V, $G[V']$ denotes the subgraph induced by V' in G. We deal with finite, undirected graphs without loops and multiple edges. Furthermore we assume that a graph is connected; otherwise we first restrict our attention to its connected components and then we extend our considerations to the whole graph. An important graph class are the split graphs, see [9].

Definition 1. *Let $G = (V, E)$ be a graph. G is a split graph if there exists a partition $V = K \cup I$ of its vertex set into a clique K and an independent set I. There is no restriction on edges between vertices of K and vertices of I.*

For our considerations the notion of vertex separator is of importance.

Definition 2. *Let $G = (V, E)$ be a graph. A subset S of vertices is a (vertex) separator if there are two nonadjacent vertices u and v of G such that the removal of S from G separates u and v into distinct connected components. S is also called u, v-separator of G. S is minimal if no proper subset of S is an u, v-separator of G.*

The next lemma is easy to verify.

Lemma 1. *Let G be a graph. A vertex set S is an u, v-separator of G if and only if every path connecting the vertices u and v contains at least one vertex of S.*

For a graph $G = (V, E)$ and an u, v-separator S the connected component of $G[V - S]$ containing u will be denoted by C_u. The following lemma is an exercise in [11].

Lemma 2. *Let S be a minimal u, v-separator of a graph G. Then every vertex of S has a neighbor in C_u and a neighbor in C_v.*

Now we define tree-decompositions and the tree-width of a graph.

Definition 3. *Let $G = (V, E)$ be a graph. A tree-decomposition of G is a pair (X, T), where $X = \{X_i \mid i \in I\}$ is a collection of subsets of V and $T = (I, F)$ is a tree with one vertex for each X_i such that*

1. *for each edge $\{u, v\}$ of G there is an $X_i \in X$ which contains both u and v,*
2. *for each vertex u of G the set of vertices $\{i \mid u \in X_i\}$ forms a subtree of T.*

The width of a tree-decomposition (X, T) is $\max_{i \in I}(|X_i| - 1)$. The tree-width $twd(G)$ of G is the minimum width over all its tree-decompositions.

We say that a tree-decomposition of a graph G is optimal if its width is equal to $twd(G)$.

Given a graph G and an integer k the tree-width problem is the question whether or not $twd(G) \leq k$. This problem is NP-complete in general, see [1]. It remains NP-complete even for bipartite graphs [13]. Bodlaender [5] gave for any fixed k a linear time algorithm deciding whether or not an input graph has tree-width at most k and if so computing an optimal tree-decomposition. The time complexity of this algorithm is exponential in k and therefore inpracticable for large k.

There are some restricted classes of graphs for which the tree-width problem is solvable in polynomial time. Such classes are for example cographs [7], circular arc graphs [21], chordal bipartite graphs [16] and permutation graphs [6].

For computations of tree-decompositions the following well known lemma is very useful, see [4].

Lemma 3. *Let (X, T) be a tree-decomposition of a graph G. For every clique K in G there is an element X_i of X such that $K \subseteq X_i$.*

The next lemma shows that the intersection of two adjacent sets X_i and X_j in every tree-decomposition of a graph G is a separator in G, see [20].

Lemma 4. *Let (X, T) be a tree-decomposition of a graph G and $X_i, X_j \in X$ be two sets connected by an edge in T. If neither $X_i \subseteq X_j$ nor $X_j \subseteq X_i$ then $X_i \cap X_j$ is a separator in G.*

The last notions of this section concern decompositions of graphs based on clique separators introduced by Gavril [10] and examined among others by Tarjan [22]. Let us denote these decompositions by *clique-decompositions* in order to distinguish them from *tree-decompositions* of Definition 3. A clique-decomposition can be represented by a binary tree whose leaves are atoms (graphs having no clique separators) and whose internal nodes are clique separators. The subtrees rooted at an internal node S correspond to the connected components of the graph separated by clique separator S. Notice that a clique-decomposition is neither unique nor necessarily a tree-decomposition (see Figures 1 and 2 in [22] for a demonstration). Moreover the separator cliques used by the clique-decompositions need not be minimal.

Tarjan [22] gave an algorithm computing a clique-decomposition in $O(nm)$ time for every input graph with n vertices and m edges. In order to recall the algorithm we need the following notions.

An *elimination ordering* π of a graph $G(V, E)$ is a numbering of its vertices from 1 to $|V|$. The *fill-in* F_π of G caused by ordering π is a set of "new" edges $\{u, v\}$, where $\{u, v\} \notin E$ and there is a path $u = u_1 u_2 \cdots u_k = v$ in G such that $\pi(u_i) < min\{\pi(u), \pi(v)\}$ for every $i = 2, \ldots, k - 1$. An elimination ordering π is *minimum* if $|F_\pi|$ is minimum over all possible orderings and *minimal* if there is no ordering π' such that $F_{\pi'} \subset F_\pi$.

Now we recall the Tarjan algorithm computing a clique-decomposition of every n-vertex and m-edge graph $G = (V, E)$ in $O(nm)$ time.

Step 1: Find a minimal ordering π and compute $S(v) = \{w \mid \pi(w) > \pi(v), \{v, w\} \in E \cup F(\pi)\}$ for every vertex v. Then repeat the following step for every vertex v in increasing order with respect to π.

Step 2: Let C_1 be the connected component of $G[V - S(v)]$ containing v and let $C_2 = V - (S(v) \cup C)$. If $S(v)$ is a clique in G and $C' \neq \emptyset$ decompose G into $G_1 = G[C_1 \cup S(v)]$ and $G_2 = G[C_2 \cup S(v)]$ separated by $S(v)$. Replace G by G_2.

The most important fact for our considerations is that the actual graph G_1 is an *atom*, i.e. G_1 is either a clique or a graph having no clique separator.

3   Tree-Width and Optimal Tree-Decomposition of Graphs with Small Minimal Separators

In this section we consider the class $\mathcal{C}^*$ of graphs whose minimal separators have a size bounded by some fixed integer k. As mentioned in the introduction $\mathcal{C}^*$ contains all split graphs $K_n \cup \{v\}$, $n \geq 1$, where vertex v is connected with only one vertex u of K_n. Since graph $K_n \cup \{v\}$ has tree-width $n - 1$ and only one minimal separator $\{u\}$ of size 1 the considered class contains graphs of arbitrarily large tree-width.

We give a polynomial algorithm which solves the tree-width problem and computes the optimal tree-decomposition of any input graph in $\mathcal{C}^*$.

We begin with a notation. Let $G = (V, E)$ be a graph, S a minimal separator of G, and C a component of $G[V - S]$. We denote by $\widehat{C}$ the induced subgraph $G[C \cup S]$ completed by all edges connecting every vertex pair of S.

The next lemma shows that every minimal u, v-separator of $\widehat{C}$ is a minimal u, v-separator of G.

Lemma 5. *Let $G = (V, E)$ be a graph, S a minimal separator of G, and C a component of $G[V - S]$. Then every minimal separator of $\widehat{C}$ is a minimal separator of G.*

Proof. Let S' be a minimal u, v-separator of $\widehat{C}$. First we show that S' is also an u, v-separator of G. Assume S' is not an u, v-separator of G. By Lemma 1 there exists a path $p = u w_1 \cdots w_r v$ connecting vertices u, v in G and containing no vertices of S'. If p contains no vertex of S then all vertices $w_1, \ldots, w_r$ have to be in C because C is a component of $G[V - S]$. But then at least one vertex of $\{w_1, \ldots, w_r\}$ is in S' which contradicts to the assumption. If p contains vertices of S then we eliminate all vertices on the path which are not in $\widehat{C}$. Let p' be the obtained part of p. Since S is a clique in $\widehat{C}$, p' is also a path in $\widehat{C}$. But then at least one vertex of p' is in S' which again is a contradiction.

Finally we show that S' is also a minimal u, v-separator of G. Assume there is an u, v-separator S'' of G such that S'' is a proper subset of S'. We show that there is a path p in G connecting u and v in G such that p contains no elements of S''. This obviously contradicts to Lemma 1.

Let x be an element of $S' - S''$. Since S' is a minimal u, v-separator of $\widehat{C}$ by Lemma 2 there is a path $u w_1 \cdots w_r w_{r+1} \cdots w_{r+t} v$ in $\widehat{C}$ with $w_r = x$ such that all vertices on this path except x are not in S' and therefore not in S''. If every edge $\{w_i, w_{i+1}\}$, $1 \leq i \leq r + t - 1$, is in G then $u w_1 \cdots w_r w_{r+1} \cdots w_{r+t} v$ is the desired path p.

If an edge $\{w_i, w_{i+1}\}$ is not in G then both vertices $\{w_i, w_{i+1}\}$ have to be in S. Consider a component C' of $G[V - S]$ different to C such that every vertex of S has a neighbor in C'. C' has to exist since Lemma 2. We replace every undesired edge $\{w_i, w_{i+1}\}$ on the path by a path containing only edges of C' and edges between vertices of C' and vertices of S. Since all these edges are in G the obtained path is the desired path p.

Now we examine the size of the maximal clique of $\widehat{C}$.

Lemma 6. *Let $G = (V, E)$ be a graph whose minimal separators have size at most k. Let S be a minimal separator of G and C a component of $G[V - S]$. Then every clique K of $\widehat{C}$ with size larger than $2k$ is a clique of G.*

Proof. Consider a clique K of $\widehat{C}$ with $|K| > 2k$. If $K \cap S$ is either empty or a clique of G then K is also a clique of G. If not then K has to contain at least two nonadjacent vertices s and s' of S both connected to all vertices of $K - S$. That means $|K - S| \leq k$ or else the vertices s and s' would have at least $k + 1$ common neighbors and therefore every minimal s, s'-separator of G would have a size at least $k + 1$ which is impossible. But then $|K| \leq |K - S| + |S| \leq k + k = 2k$ which is a contradiction.

The following theorem shows that G has a tree-decomposition with a simple structure.

Theorem 1. *Let $G = (V, E)$ be a graph whose minimal separators have size at most k. Then there is a (not necessarily optimal) tree-decomposition (X, T) of G such that each set of X with more than $2k$ vertices induces a clique in G.*

Proof. It is an induction on the number of vertices of G. If G is a clique the claim is obvious. If not then G has a minimal u, v-separator, say S. Let $C_1, \ldots, C_l$ be the components of $G[V - S]$ and $\widehat{C}_i$, $1 \leq i \leq l$, the induced subgraph $G[C_i \cup S]$ completed by all edges connecting every vertex pair of S. By Lemma 5 the size of all minimal separators of $\widehat{C}_i$ is bounded by k. By the inductive hypothesis $\widehat{C}_i$ has a tree-decomposition (X^i, T^i) such that each set of X^i with more than $2k$ elements induces a clique in $\widehat{C}_i$ and by Lemma 6 also a clique in G. By Lemma 3 we can assume that every tree-decomposition (X^i, T^i) is rooted by a set of X^i containing all vertices of S because S induces a clique in every $\widehat{C}_i$. Now the tree with root S and with the above subtrees (X^i, T^i), $1 \leq i \leq l$, is a tree-decomposition of G meeting the requirements of the claim.

By Theorem 1 we obtain:

Corollary 1. *Let G be a graph and K a maximum clique of G. If the size of all minimal separators of G is bounded by an integer k then $twd(G) \leq \max\{|K| - 1, 2k - 1\}$.*

Proof. Consider the tree-decomposition (X, T) constructed by Theorem 1. Let $X_i \in X$ be the vertex set having the most vertices of G. By Theorem 1 we obtain:
If $|X_i| > 2k$ then X_i is a clique in G with $|X_i| = |K|$ and by the definition of tree-width we have $twd(G) \leq |X_i| - 1 = \max\{|K| - 1, 2k - 1\}$.
If $|X_i| \leq 2k$ then $|X_i| \geq |K|$ and by Lemma 3 and the definition of tree-width we have $twd(G) \leq |X_i| - 1 \leq \max\{|K| - 1, 2k - 1\}$.

Theorem 2. *Let $G = (V, E)$ be a graph but not a clique whose minimal separators have size at most k. If $twd(G) \geq 2k$ then there exists a clique separator in G.*

Proof. By Theorem 1 graph G has a tree-decomposition (X,T) such that each set of X with more than $2k$ vertices induces a clique in G. Let $X_i \in X$ be a set with more than $2k$ vertices and e be an edge in T connecting X_i with some set $X_j \in X$ (X_i exists since $twd(G) \geq 2k$).

We can assume that neither $X_i \subseteq X_j$ nor $X_j \subseteq X_i$ otherwise we contract edge e in T and we consider the obtained smaller tree-decomposition of G. By Lemma 4, $X_i \cap X_j$ is a clique separator in G.

Now we are ready to state our algorithm computing an optimal tree-decomposition of every graph $G = (V, E)$ whose minimal separators have a size bounded by a fixed integer k.

At first we apply the algorithm of Tarjan to compute a clique-decomposition of G. We obtain a binary tree B whose leaves are atoms and whose internal nodes are clique separators, see Figure 2. It is easy to see that B has at most $n - 1$ atoms.

Then we compute the optimal tree-decomposition of every atom $G_1, \ldots, G_{r+1}$. There are two cases: If an atom $G_i = (V_i, E_i)$ is a clique then $(X_i, T_i) = (V_i, \{t\})$ is the trivial optimal tree-decomposition of G_i. If not then G_i has no clique separator and by Theorem 2 every atom G_i has tree-width at most $2k$. Hence its optimal tree-decomposition can be found in $O(|V_i|)$ time using the algorithm from Bodlaender [5].

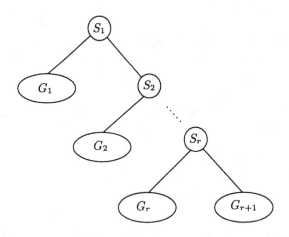

Figure 2. Binary tree representation of a clique-decomposition

function *atom_treedecomp* $(G : atom) : optimal\ tree\text{-}decomposition$
begin
    if (G is a clique) **then** $(X,T) = (\{V\}, \{t\})$;
    else Find optimal (X,T) of G using the algorithm from Bodlaender;
    atom_treedecomp $(G) := (X,T)$
end

Finally using the clique-decomposition of G and the tree-decompositions of the atoms we compute an optimal tree-decomposition in bottom up fashion. We assume that the clique-decomposition of G has r internal nodes and $r+1$ leaves.

function $find\_treedecomp$: *optimal tree-decomposition*
begin
    $(X,T) := atom\_treedecomp\ (G_{r+1})$
    for $i := r$ **downto** 1 **do**
      begin
        Reroot T by t corresponding to a set containing S_i;
        $(X',T') := atom\_treedecomp\ (G_i)$;
        Reroot T' by t' corresponding to a set containing S_i;
        $X^* := X \cup X'$; $V_{T^*} := V_T \cup V_{T'}$; $E_{T^*} := E_T \cup E_{T'} \cup \{tt'\}$;
        $(X,T) := (X^*,T^*)$
      end;
    $find\_treedecomp := (X,T)$
end

The correctness of our algorithm can be shown easily by an induction on the number of the vertices of the input graph and by Theorems 1 and 2. We examine the complexity of the algorithm.

Lemma 7. *Let $G = (V,E)$ be a graph with n vertices and m-edges. If the size of all minimal separators of G is bounded by a fixed integer k then the computation of an optimal tree-decomposition of G takes at most $O(nm)$ time.*

Proof. The construction of the binary tree representation B of the clique-decomposition of graph G takes at most $O(nm)$ time [22]. Finding optimal tree-decompositions of all atoms in B takes $O(n^2)$ time since there are at most n atoms each of size at most n and an optimal tree-decomposition of an atom can be computed in $O(n)$ [5]. Finally the construction of an optimal tree-decomposition of G by using the optimal tree-decompositions of the atoms takes at most $O(n)$ time.
Thus the total time is $O(nm)$ (but exponential in k).

We obtain the following Theorem:

Theorem 3. *Let G be a graph with n vertices and m edges whose minimal separators have size bounded by a fixed integer k. Then an optimal tree-decomposition (X,T) of G can be found in $O(nm)$ time. Moreover every set of X with more than $2k$ elements induces a clique in G.*

The obtained optimal tree-decomposition has a simple structure. We can use it to show that many in general NP-complete problems become polynomial in this case. Such problems are for example CLIQUE, MINIMUM FILL-IN, GRAPH COLORING, and MAXIMUM INDEPENDENT SET, see [22].
To the contrary many other problems remain NP-complete for this class although

they are solvable in polynomial time for the class of graphs with bounded tree-width. An example is the HAMILTONIAN CIRCUIT problem. The proof is based on the fact that the HAMILTONIAN CIRCUIT problem remains NP-complete even for planar bipartite graphs with maximum degree 3 [12]. Consider such a graph $G = (V, V', E)$. First observe that it may be assumed that $|V| = |V'|$, otherwise there is no hamiltonian circuit in G. Then construct the split graph H from G by completing one of its color sets, say V. Now all minimal u, v-separators of H have size ≤ 3 and H has a hamiltonian circuit if and only if G has one. Thus the HAMILTONIAN CIRCUIT problem is NP-complete even for graphs whose minimal u, v-separators have size ≤ 3.

Now we consider graphs having polynomial many minimal separators. In this case the size of the separators is not necessarily bounded by some fixed integer k. We don't know whether or not there exists a polynomial algorithm computing the optimal tree-decomposition of G. However we can show that in this case even the CLIQUE problem remains NP-complete.

Lemma 8. *Let C be a class of graphs $G = (V, E)$ having at most $2^8 |V|^2$ minimal separators. The CLIQUE problem remains NP-complete even when it is restricted to C.*

Proof. It is a reduction from a special version of 3-SAT in which every variable is restricted to appear at most three times and every literal at most twice. It is well known that 3-SAT remains NP-complete even for expressions meeting these requirements, see [18] (pp 183).

Consider the usual reduction from 3-SAT to CLIQUE: Given an instance ϕ with n clauses $F_i = (x_i \vee y_i \vee z_i)$ we create $3n$ vertices, 3 for every clause. We identify these 3 vertices with the literals of the corresponding clause. Then we add an edge between two vertices if and only if their associated literals belong to different clauses and are not opposite. Let $G(\phi) = (V, E)$ be the obtained graph. $G(\phi)$ contains a clique with size $\geq n$ if and only if ϕ is satisfiable.

Since $|V| = 3n$ it is sufficient to show that the number of all minimal separators of $G(\phi)$ is $\leq 2^8 3n(3n - 1)$. There are two cases to consider: Let u and v be nonadjacent vertices of G.

Case 1: u and v correspond to literals x_i and z_i of the same clause F_i, see Figure 3(a).

In this case all vertices of $G(\phi)$ except at most 5 are connected to both u and v. These 5 vertices are the vertex associated with the literal y_i of F_i and the (at most 4) vertices associated with the literals $\neg x_i$ and $\neg z_i$ occuring in the clauses of ϕ except F_i. Thus every minimal u, v-separator has to contain at least $n - 5$ vertices of $G(\phi)$. So there exist at most $2^5 < 2^8$ different u, v-separators since a set with 5 elements has at most 2^5 different subsets.

Case 2: u and v correspond to literals x_i of F_i and z_j of F_j, see Figure 3(b).

Again all vertices of $G(\phi)$ except at most 8 are connected to both u and v. These 8 vertices are the 4 ones associated with the literals y_i, z_i, x_j, and y_j of F_i and F_j and the (at most 4) ones associated with the literals $\neg x_i$ and $\neg z_j$ occuring in the clauses of ϕ except F_i and F_j. The same argument as before shows that in this case there exist at most 2^8 different minimal u, v-separators.

Since there exist $3n(3n-1)$ pairs (u,v) in $G(\phi)$ the proof is complete.

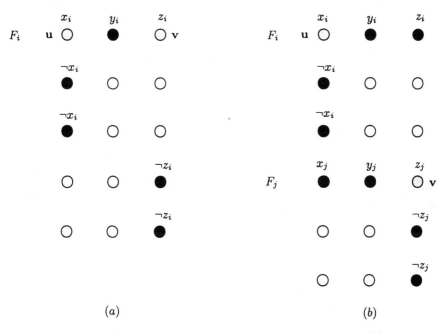

(a)   (b)

Figure 3. All vertices except the black ones are connected to both u and v.

References

1. S. Arnborg, D. Corneil, and A. Proskurowski. Complexity of finding embeddings in a k-tree. *SIAM J. Alg. Disc. Meth.*, 8:277–284, 1987.
2. S. Arnborg, J. Lagergren, and D. Seese. Easy problems for tree-decomposable graphs. *J. Algorithms*, 12:308–340, 1991.
3. S. Arnborg and A. Proskurowski. Linear time algorithms for NP-hard problems restricted to partial k-trees. *Disc. Appl. Math.*, 23:11–24, 1989.
4. H. Bodlaender. Dynamic programming algorithms on graphs with bounded treewidth. In *Proceedings of the International Colloquium on Automata, Languages and Programming*, volume 317 of *Lect. Notes in Comput. Sci.*, pages 105–119. Springer-Verlag, New York/Berlin, 1988.
5. H. Bodlaender. A linear time algorithm for finding tree-decompositions of small treewidth. *SIAM J. Comput.*, 25(6):1305–1317, 1996.
6. H. Bodlaender, T. Kloks, and D. Kratsch. Treewidth and pathwidth of permutation graphs. *SIAM J. Disc. Math.*, 8:606–616, 1995.
7. H. Bodlaender and R. Möhring. The pathwidth and treewidth of cographs. *SIAM J. Disc. Math.*, 6:181–188, 1993.

8. R. Borie, R. Parker, and C. Tovey. Automatic generation of liner-time algorithms from predicate calculus describtions of problems on recursively constructed graph families. *Algorithmica*, 7:555–581, 1992.
9. A. Brandstädt. *Graphen und Algorithmen*. B. G. Teubner, Stuttgard, 1994.
10. F. Gavril. Algorithms on clique separable graphs. *Disc. Math.*, 19:159–165, 1977.
11. M. Golumbic. *Algorithmic Graph Theory and Perfect Graphs*. Academic Press, New York, 1980.
12. A. Itai, C. Papadimitriou, and J. Szwarcfiter. Hamilton paths in grid graphs. *SIAM J. Comput.*, 11(4):676–686, 1982.
13. T. Kloks. *Treewidth. Computations and Approximations*, volume 842 of *Lect. Notes in Comput. Sci.* Springer-Verlag, New York/Berlin, 1994.
14. T. Kloks, H. Bodlaender, H. Müller, ahd D. Kratsch. Computing treewidth and minimum fill-in: All you need are the minimal separators. In *Proceedings of the European Symposium on Algorithms*, volume 726 of *Lect. Notes in Comput. Sci.*, pages 260–271. Springer-Verlag, New York/Berlin, 1993.
15. T. Kloks, H. Bodlaender, H. Müller, and D. Kratsch. Erratum to the ESA'93 proceedings. In *Proceedings of the European Symposium on Algorithms*, volume 855 of *Lect. Notes in Comput. Sci.*, page 508. Springer-Verlag, New York/Berlin, 1994.
16. T. Kloks and D. Kratsch. Treewidth of chordal bipartite graphs. *J. Algorithms*, 19:266–281, 1995.
17. T. Kloks and D. Kratsch. Listing all minimal separators of a graph. *SIAM J. Comput.*, 27(3):605–613, 1998.
18. C. Papadimitriou. *Computational Complexity*. Addison-Wesley Publishing Company, 1994.
19. A. Parra and P. Scheffler. How to use the minimal separators of a graph for its chordal triangulation. In *Proceedings of the International Colloquium on Automata, Languages and Programming*, volume 944 of *Lect. Notes in Comput. Sci.*, pages 123–134. Springer-Verlag, New York/Berlin, 1995.
20. N. Robertson and P. Seymour. Graph minors X. Obstructions to tree-decompositions. *J. Comb. Theory Series B*, 52:153–190, 1991.
21. R. Sundaram, K. Sher Singh, and C. Pandu Rangan. Treewidth of circular-arc graphs. *SIAM J. Disc. Math.*, 7:647–655, 1994.
22. R. Tarjan. Decomposition by clique separators. *Disc. Math.*, 55:221–232, 1985.

Generating All the Minimal Separators of a Graph

Anne Berry, Jean-Paul Bordat, and Olivier Cogis

LIRMM
161, Rue Ada, 34 392, Montpellier, France
{aberry,bordat,cogis}@lirmm.fr

Abstract. We present an efficient algorithm which computes the set of minimal separators of a graph in $O(n^3)$ time per separator, thus gaining a factor of n^2 on the current best-time algorithms for this problem.
Our process is based on a new structural result, derived from the work of Kloks and Kratsch on listing all the minimal separators of a graph.

Keywords: Graph. Minimal Separator. Enumeration Algorithm.

1 Introduction

In this paper, we address the issue of computing the global set of minimal separators of a given graph G.

The use of minimal separators as a structural tool has become a modern research topic in graph theory, with many algorithmic applications (see for example [6], [15], [13], [3], [1], [2]).

Although a graph may have an exponential number of minimal separators, it is important for many hard problems to be able to enumerate the set of minimal separators efficiently.

As an example, the NP-Complete graph problems TREEWIDTH and MINIMUM FILL-IN have been extensively studied (see [11], [5], [7], [6], etc.). Bodlaender and al. in [6] conjecture that when the number of minimal separators of a graph is polynomially bounded, these problems can be solved in polynomial time. Bouchitté and Todinca (see [7]) define the concept of Potential Maximal Clique, which uses minimal separators to define the maximal cliques obtained by some minimal triangulation process. They show that when the number of Potential Maximal Cliques is polynomially bounded, these problems can be solved in polynomial time. The question remains open as to whether any graph class with a polynomially bounded number of minimal separators has a polynomially bounded number of Potential Maximal Cliques, a conjecture also related to the structural aspects of the global set of minimal separators of a graph.

Recent research has been done to compute the set of minimal separators of a graph ([9], [10], [14]). [9] proposed a process with a global complexity of $O(n^6)$ per separator, which was later streamlined to $O(n^5)$ in ([10]), a complexity also independently obtained by [14].

Widmayer et al. (Eds.): WG'99, LNCS 1665, pp. 167–172, 1999.

All three papers use the following technique:

1. Choose an arbitrary pair $\{a, b\}$ of non-adjacent vertices, and compute the set of minimal ab-separators.
2. Repeat this process on every possible pair $\{x, y\}$ of non-adjacent vertices, which requires $O(n^2)$ passes.

In this paper, we present a process which computes all the minimal separators in a single pass, thus gaining a factor of $O(n^2)$, since the second step need not be executed.

To do this, we extend a very interesting idea from [9] which uses a minimal ab-separator S to generate another minimal ab-separator S^* which is "close to S".

We present a more general structural result which, given a minimal separator S, enables us to generate a family of minimal separators which are close to S, working in all directions at the same time, instead of restricting ourselves to the two components containing a and b as in [9].

Our new insight derives from our work on a general Galois lattice representing all the minimal separators of a graph (see [4]), of which every minimal ab-separator lattice as defined by Escalante (see [8]) is a sub-order. In this general lattice, as all directions are equivalent, we see no reason to favour a pair $\{a, b\}$ of vertices. Moreover, we consider minimal separators as *neighborhoods* of connected components, a technique used in [1] and [7].

2   Preliminaries

A graph is denoted $G = (V, E)$, with $\mid V \mid = n$, $\mid E \mid = m$. For $x \in V$, $N(x) = \{y \neq x | xy \in E\}$. For $C \subseteq V$, $N(C) = (\cup_{x \in C} N(x)) \setminus C$. $G(C)$ denotes the subgraph induced by the set of vertices C.

For $X \subseteq V$, $\mathscr{C}(X)$ denotes the set of connected components of $G(V \setminus X)$ (connected components are vertex sets).

A vertex set S is called a *separator* if $\mid \mathscr{C}(S) \mid \geq 2$, an *ab-separator* if a and b are in different connected components of $\mathscr{C}(S)$, a *minimal ab-separator* if no proper subset of S is an ab-separator.

S is called a *minimal separator* if there is some pair $\{a, b\}$ of vertices such that S is a minimal ab-separator. We express this by the following equivalent property: S is a *minimal separator* iff there are two components C_1, $C_2 \in \mathscr{C}(S)$ such that $N(C_1) = N(C_2) = S$; C_1 and C_2 are called *full components* of $\mathscr{C}(S)$. S is a minimal ab-separator iff a and b belong to different full components of $\mathscr{C}(S)$.

$\mathscr{S}(G)$ denotes the set of minimal separators of G.

3   Global Listing of the Minimal Separators

Theorem 1. *Given a graph* $G = (V, E)$, $\mathscr{S}(G)$ *is the set* $\mathscr{F}$ *of subsets of V inductively defined by:*

- _Basis_ $\mathscr{B}$ is the set of all minimal separators included in some neighborhood $N(v)$ for some $v \in V$.
- _Rule_ $\mathscr{R}$: $X \in \mathscr{F}, x \in X, C \in \mathscr{C}(X \cup N(x))$ implies $N(C) \in \mathscr{F}$.

$\mathscr{F}$ is defined as the closure of $\mathscr{B}$ under production rule $\mathscr{R}$.

Proof. $\mathscr{F} \subseteq \mathscr{S}(G)$

By definition, $\mathscr{B} \subseteq \mathscr{S}(G)$.

Assume that $S \in \mathscr{S}(G), x \in S$ and $C \in \mathscr{C}(S \cup N(x))$. Let $S' = N(C)$. As $S \subseteq S \cup N(x)$, there must be some component C_1 of $\mathscr{C}(S)$ which contains C; clearly, $N(C) \subseteq C_1 \cup S$. Let us define $A = N(C) \cap C_1$ and $B = N(C) \cap S$, and note that $A \subseteq N(x) \cap C_1$, $B \subseteq S \setminus \{x\}$, and $S' = N(C) = A \cup B$, with $A \cap B = \emptyset$.

Since S is minimal, there must be a full component $C_2 \neq C_1$ of $\mathscr{C}(S)$.

C_2 obviously induces a connected subgraph of $G(V \setminus S')$ for $S' = A \cup B$, with $A \subseteq C_1$ and $B \subseteq S$.

Let C' be the component of $\mathscr{C}(S')$ which contains C_2. Because $x \in S, N(x) \cap C_2 \neq \emptyset$, so $x \in C'$. Now let $t \in S'$. If $t \in A$, $N(t) \cap C' \supseteq \{x\}$, and if $t \in B$, $N(t) \cap C' \supseteq N(t) \cap C_2 \neq \emptyset$. Thus every vertex in S' has a neighbor in C', as well as in C, by definition of S'. C and C' are thus full components of S', and therefore the vertex set S' which has been generated is a minimal separator.

$\mathscr{S}(G) \subseteq \mathscr{F}$

Conversely, assume that there is some minimal ab-separator S, for some pair of non-adjacent vertices $\{a, b\}$ in V, which does not belong to $\mathscr{F}$. Let C_a and C_b be the components of $\mathscr{C}(S)$ that contain a and b resp., and let S_a be the unique minimal ab-separator contained in the neighborhood of a, C_a^a and C_b^a being the components of $\mathscr{C}(S_a)$ that contain a and b resp.

As $C_b \cap N(a) = \emptyset$, any path in $G(C_b)$ must be a path in $G(C_b^a)$, so $C_b \subseteq C_b^a$. As $S_a \in \mathscr{B} \subseteq \mathscr{F}$, there is some minimal ab-separator $S' \in \mathscr{F}$, C_a' and C_b' being the components of $\mathscr{C}(S')$ which contain a and b resp., such that $C_b \subset C_b'$ and C_b' is minimal w.r.t. inclusion for this property.

Because S is a minimal ab-separator, $S = N(C_b)$. Let $B = N(C_b) \cap S' = S \cap S'$. Since S and S' are different minimal ab-separators, one must have $B \neq S'$. Let x be in $S' \setminus B$. Note that $b \notin N(x)$. Let C_b'' be the component of $\mathscr{C}(S' \cup N(x))$ that contains b. $C_b'' \subset C_b'$, as $N(x) \cap C_b' \neq \emptyset$.

On the other hand $C_b'' \supseteq C_b$, for, because $x \notin N(C_b)$, by definition of x, any path in $G(C_b)$ is also a path of $G(C_b'')$.

As $S' \in \mathscr{F}$, $x \in S'$, and $C_b'' \in \mathscr{C}(S' \cup N(x))$, one has $N(C_b'') \in \mathscr{F}$.

Therefore, $C_b \subseteq C_b'' \subset C_b'$ contradicts the minimality of C_b'.

Remark 1. In [9], given a minimal ab-separator S, a new minimal ab-separator S^* is defined as follows: let C_a be the component of $G(V \setminus S)$ that contains a, let x be in S, let Δ be the minimal ax-separator included in $N(x)$ in graph $G(C_a \cup \{x\})$, and let C_a' be the induced component containing a. Let N be the set of vertices of S which have no neighbor in C_a'. Then $S^* = (S \cup \Delta) \setminus N$. It is easy to see that $S^* = N(C_a')$ and that C_a' is in $\mathscr{C}(S \cup N(x))$. Our construction rule therefore extends the work based on C_a' to every other component in $\mathscr{C}(S \cup N(x))$.

4 Generation Algorithm

The inductive definition of the set of minimal separators of a graph expressed in Theorem 1 yields the following algorithm.

Algorithm AllMinSep

input : A graph $G = (V, E)$, with $|V| = n$ and $|E| = m$.

output : The set $\mathscr{S}(G)$ of minimal separators of G.

begin

  INITIALIZATION:

  $\mathscr{S} \leftarrow \emptyset$;

  foreach $v \in V$ **do**

    foreach *component* $C \in \mathscr{C}(\{v\} \cup N(v))$ **do**

      $\mathscr{S} \leftarrow \mathscr{S} \cup \{N(C)\}$;

  GENERATION:

  $\mathscr{T} \leftarrow \emptyset$;

  while $\mathscr{S} \setminus \mathscr{T} \neq \emptyset$ **do**

    let $S \in \mathscr{S} \setminus \mathscr{T}$;

    foreach $x \in S$ **do**

      $\mathscr{S} \leftarrow \mathscr{S} \cup \{N(C) | C \in \mathscr{C}(S \cup N(x))\}$;

    $\mathscr{T} \leftarrow \mathscr{T} \cup \{S\}$;

  $\mathscr{S}(G) \leftarrow \mathscr{S}$.

end

Theorem 2. *Given graph G, an execution of Algorithm AllMinSep computes the set $\mathscr{S}(G)$ of minimal separators of G.*

Proof. Algorithm AllMinSep is a straightforward application of rule $\mathscr{R}$, starting from basis $\mathscr{B}$ until saturation is obtained. Therefore, $\mathscr{T} \subseteq \mathscr{S} \subseteq \mathscr{S}(G)$ is an invariant for the while–loop. Because $|\mathscr{T}|$ increases by one each time the body of the while–loop is performed, the algorithm terminates.

Theorem 3. *Algorithm AllMinSep can be implemented to compute the minimal separators of G in time $O(n^3 |\mathscr{S}(G)|)$.*

Proof. The components of $G(V \setminus X)$ for some $X \subseteq V$, together with their neighborhood in X, can be computed in $O(m)$ time : a global graph search using X as a limit will scan and yield each component, as well as its neighborhood. Adding a new separator $N(C)$ to $\mathscr{S}$ first requires testing whether $N(C)$ already appears in $\mathscr{S}$. This can be implemented in $O(n)$ time with a SEARCH-INSERT procedure using a suitable data structure based on a lexicographical tree (see [12]). Sets $\mathscr{S}$ and $\mathscr{T}$ can be implemented using a traditional FIFO queue structure.

Initialization: for each $v \in C$, an $O(m)$ search can produce $O(n)$ separators, each of them processed by SEARCH-INSERT, which yields an $O(n^3)$ overall complexity.

While–loop: in the same fashion, the internal for–loop requires an $O(m)$ search, plus $O(n^2)$ time for adjunctions to $\mathscr{S}$, thus the processing of one separator $S \in \mathscr{S}$ requires $O(n^3)$ operations.

In conclusion, Algorithm AllMinSep has a global complexity of $O(n^3|\mathscr{S}(G)|)$.

5 Conclusion

Though we have gained a factor of n^2 on the previous algorithms for solving the problem of generating the minimal separators of a graph, it is obvious that a given minimal separator may be recalculated many times by our algorithm; we feel that it could be improved.

We have defined in this paper a separator $S \cup N(x)$, which is not, in general, a minimal separator, but which has the interesting property that all of the connected components it defines have minimal separators as neighborhoods. This is true of any minimal separator, but also for any $x \cup N(x)$, as well as for any Potential Maximal Clique (see [7]). This new type of object may be interesting to investigate in a systematic fashion.

We leave open the question which arises as to whether the subproblem of computing the set of minimal ab-separators for a given pair $\{a, b\}$ has the same complexity as the general problem of computing the whole set of minimal separators.

Acknowledgements

A. Berry thanks D. Kratsch and H. Müller for many interesting and fruitful discussions.

References

1. A. Berry: A Wide-Range Efficient Algorithm for Minimal Triangulation. In Proceedings of the Tenth Annual ACM-SIAM Symposium on Discrete Algorithms (SODA'99), Baltimore, Jan. 1999.
2. A. Berry: Désarticulation d'un graphe. PhD Dissertation LIRMM, 1998.
3. A. Berry, J.–B. Bordat: Separability Generalizes Dirac's Theorem. Discrete Applied Mathematics 84 (1998) 43–53.
4. A. Berry, J.–P. Bordat: Orthotreillis et séparabilité dans un graphe non-orienté. Mathématiques, Informatique et Sciences Humaines 146 (1999) 5–17.
5. H. Bodlaender, T. Kloks, D. Kratsch: Treewidth and Pathwidth of Permutation Graphs. SIAM Journal on Discrete Mathematics 8 (1995) 606–616.
6. H. Bodlaender, T. Kloks, D. Kratsch, H. Müller: Computing Treewidth and Minimum Fill-in: All You Need are the Minimal Separators. In Proceedings of the First Annual European Symposium on Algorithms (ESA'93), Vol. 726 of Lecture Notes

in Computer Science, 260–271, Springer-Verlag, 1993, and erratum in Proceedings of the Second Annual European Symposium on Algorithms (ESA'94), Vol. 855 of Lecture Notes in Computer Science, 508, Springer-Verlag, 1994.

7. V. Bouchitté, I. Todinca: Minimal Triangulations for Graphs with "Few" Minimal Separators. In Proceedings of the Sixth Annual European Symposium on Algorithms (ESA'98), Vol. 1461 of Lecture Notes in Computer Science, 344–355, Springer-Verlag, 1998.

8. F. Escalante: Schnittverbände in Graphen. Abhandlungen aus dem Mathematischen Seminar des Universität Hamburg 38 (1972) 199–220.

9. T. Kloks, D. Kratsch: Finding All Minimal Separators of a Graph. In Proceedings of the Eleventh Symposium on Theoretical Aspects of Computer Science (STACS'94), Vol. 775 of Lecture Notes in Computer Science, 759–768, Springer-Verlag, 1994.

10. T. Kloks, D. Kratsch: Listing All Minimal Separators of a Graph. SIAM Journal on Computing 27 (1998) 605–613.

11. T. Kloks, D. Kratsch, J. Spinrad: On Treewidth and Minimum Fill-in of Asteroidal Triple-free Graphs. Theoretical Computer Science 175 (1997) 309–335.

12. L. Nourine, O. Raynaud: A Fast Algorithm for Building Lattices, Res. Rep. LIRMM 1998.

13. A. Parra: Structural and Algorithmic Aspects of Chordal Graph Embeddings. PhD Dissertation, Technische Universität Berlin 1996.

14. H. Sheng, W. Liang: Efficient Enumeration of All Minimal Separators in a Graph. Theoretical Computer Science 180 (1997) 169–180.

15. I. Todinca, Aspects algorithmiques des triangulations minimales des graphes, PhD Dissertation, LIP, ENS Lyon, 1999.

Two Broadcasting Problems in Faulty Hypercubes

Stefan Dobrev* and Imrich Vrťo**

Institute of Mathematics, Slovak Academy of Sciences
Department of Informatics, P.O.Box 56, 840 00 Bratislava, Slovak Republic

Abstract. We consider two broadcasting problems in the n-dimensional hypercube under the shouting communication mode, i.e. any node of a network can inform all its neighbours in one time step. In addition, during any time step a number of links of the network can be faulty. Moreover the faults are dynamic. The first problem is to find an upper bound on the number of time steps necessary to complete the broadcasting if at most $n - 1$ links are faulty in any step. Fraigniaud and Peyrat [10] proved that $n + O(\log n)$ time steps are sufficient. De Marco and Vaccaro [8] decreased the upper bound to $n + 7$ and showed a worst case lower bound $n + 2$ for $n \geq 3$. We prove that $n + 2$ time steps are sufficient. The second problem from [8] is to find the maximal number k such that the broadcasting time remains n if at most k faults are allowed in any step. We prove that k equals either $n - 2$ or $n - 3$. Our method is related to the isoperimetric problem in graphs and can be applied to other networks.
Keywords: broadcasting, fault tolerance, distributed computing, hypercube, isoperimetric problem

1 Introduction

Broadcasting is the standard communication problem in interconnection networks when a node has to send a message to all other nodes. There are many applications of the broadcasting problem in parallel and distributed computing [9,12,13]. Recently, a lot of attention has been paid to fault-tolerant dissemination of information in networks [15]. In this paper we consider the shouting communication mode in which any node can inform all its neighbours in one time step. In addition, we assume that during any time step a number of links less than the edge-connectivity of the network can be faulty. The problem is to find an upper bound on the number of time steps necessary to complete broadcasting under this additional assumption. This model was introduced by Santoro and Widmayer [17]. We will concentrate ourselves on the n-dimensional hypercube. In the first problem, we assume that at most $n - 1$ links are faulty in any step. Fraigniaud and Peyrat [10] proved that $n + O(\log n)$ time steps are sufficient. An $O(n)$ upper bound also follows from a more general result of Chlebus, Diks

* Supported by the VEGA grant No. 1/4315/97.
** Supported by the VEGA grant No. 95/5305/277.

Widmayer et al. (Eds.): WG'99, LNCS 1665, pp. 173–178, 1999.

and Pelc [5]. De Marco and Vaccaro [8] decreased the upper bound to $n+7$ and proved a worst case lower bound $n+2$ for $n \geq 3$. We show that $n+2$ time steps are sufficient.

The second problem, mentioned in [8], is to find the maximal number k such that the broadcasting time remains n if at most k faults are allowed in any step. We prove that k equals either $n-2$ or $n-3$.

Our method is related to the isoperimetric problem in graphs and can be applied to other networks.

2 The Model

Let Q_n be a network of processors connected as an n-dimensional hypercube with 2^n nodes and $n2^{n-1}$ links, that is, the graph whose vertices are all binary strings of lengths n and two vertices are joined by an edge if the corresponding strings differ precisely in one position. Note that the n-dimensional hypercube is a vertex transitive graph and its edge connectivity is n. Initially, a node of Q_n knows a message. This message needs to be sent to all other nodes. The links are bidirectional. The computation is synchronous. In one time step a node is able to send its message to all its neighbours. This is called the shouting or the all-port mode. In each time step a number links less than the edge connectivity are faulty, i.e. the message transmitted along the faulty link is not delivered. The faults are dynamic in the sense that the set of faulty links can change during the execution of the broadcast. Our problem is to determine $T(n)$, the minimum time to broadcast the message in the hypercube Q_n.

3 Vertex Isoperimetric Inequality for Hypercube

Let H be a subgraph of a graph G. The vertex boundary of H is the set of vertices not in H joined to some vertex in H. The vertex isoperimetric problem can be stated as follows: For a fixed m, of all subgraphs of G of cardinality m, which one has the smallest vertex boundary? This is a discrete anologue of a well known isoperimetric problems studied in mathematics for more than 2000 years [6,16]. There are several applications of the vertex isoperimetric problem in graphs, the most frequent is the bandwidth problem and the vertex separator (bisection) [7,11], but only for a few classes of graphs a satisfiable solution is known [1,2]. We introduce the result for the n-dimensional hypercube Q_n from [14].

Let A be a subset of vertices of Q_n. Let $S(A)$ denote the set of all vertices in distance at most 1 from A in Q_n, i.e. $S(A)$ is the union of the vertex boundary of A with A.

Lemma 1. Let $1 \leq |A| \leq 2^n - 1$. The number $|A|$ has a unique representation in the form

$$|A| = \sum_{i=r+1}^{n} \binom{n}{i} + \sum_{i=s}^{r} \binom{n_i}{i},$$

where $1 \leq s \leq n_s < n_{s+1} < ... < n_r < n$. Then

$$|S(A)| \geq \sum_{i=r}^{n} \binom{n}{i} + \sum_{i=s}^{r} \binom{n_i}{i-1}.$$

Remark. The inequality above is the best possible and can be achieved taking $|A|$ vertices in the simplicial order, i.e. $x = (x_{n-1}, ..., x_1, x_0)$ preceds $y = (y_{n-1}, ..., y_1, y_0)$ if either $\sum_{i=0}^{n-1} x_i < \sum_{i=0}^{n-1} y_i$ or $\sum_{i=0}^{n-1} x_i = \sum_{i=0}^{n-1} y_i$ and y preceds x lexicographically. A special case $|A| = \sum_{i=0}^{r} \binom{n}{i}$ is more instructive. Then $|S(A)| \geq \sum_{i=0}^{r+1} \binom{n}{i}$. This means that the set with the minimal vertex boundary is the set of all vertices in a distance at most r from $(0,0,...,0)$.

4   Time Bounds

First we use Lemma 1 to determine the number of time steps which are sufficient to complete the broadcasting in the presence of $n - 1$ dynamic faults.

Theorem 1. *The minimum broadcasting time $T(n)$ in the n-dimensional hypercube with $n - 1$ dynamic link faults satisfies*

$$T(n) \leq \begin{cases} 3, \text{if } n = 2 \\ n + 2, \text{if } n > 2 . \end{cases}$$

Proof. The broadcasting scheme is simple. In each time step each node sends the message to all its neighbours. The analysis follows. Let $n \geq 10$. By A_k we will denote suitable sets of nodes which know the message after the k-th time step. Observe that the number of nodes that know the message after the k-th call is at least $|S(A_{k-1})| - n + 1$. Clearly, there exist sets A_0, A_1 and A_2 s.t. $|A_0| = 1, |A_1| = 2$ and $|A_2| = 1 + n$. Consider the set A_2. Its cardinality is

$$|A_2| = \binom{n}{n} + \binom{n}{n-1}.$$

By Lemma 1,

$$|S(A_2)| \geq \binom{n}{n} + \binom{n}{n-1} + \binom{n}{n-2}.$$

Because of the $n - 1$ faulty links, we have

$$|S(A_2)| - n + 1 \geq \binom{n}{n} + \binom{n}{n-1} + \binom{n}{n-2} - (n-1)$$

$$\geq \binom{n}{n} + \binom{n}{n-1} + \binom{n-1}{n-2} + \binom{n-2}{n-3}.$$

Define A_3 to be a subset of nodes that know the message after the 3-rd step and satisfies

$$|A_3| = \binom{n}{n} + \binom{n}{n-1} + \binom{n-1}{n-2} + \binom{n-2}{n-3}.$$

We claim that for $3 \leq k \leq \lfloor (n+3)/2 \rfloor$, there exists a set A_k, s.t.

$$|A_k| = \sum_{i=n-k+2}^{n} \binom{n}{i} + \binom{n-1}{n-k+1} + \binom{n-k+1}{n-k}.$$

We proceed by induction on k. The claim is true for $k = 3$. Assume the claim holds for some $k - 1$, where $3 \leq k - 1 < \lfloor (n+3)/2 \rfloor$. Hence we have a set A_{k-1} with

$$|A_{k-1}| = \sum_{i=n-k+3}^{n} \binom{n}{i} + \binom{n-1}{n-k+2} + \binom{n-k+2}{n-k+1}.$$

By Lemma 1, setting $r = n - k + 2, s = n - k + 1$ and $n_r = n - 1, n_s = n - k + 2$ we get

$$|S(A_{k-1})| \geq \sum_{i=n-k+2}^{n} \binom{n}{i} + \binom{n-1}{n-k+1} + \binom{n-k+2}{n-k}.$$

Because of the $n - 1$ faulty links, we have for $k - 1 < \lfloor (n+3)/2 \rfloor$ and $n \geq 10$

$$|S(A_{k-1})| - n + 1 \geq \sum_{i=n-k+2}^{n} \binom{n}{i} + \binom{n-1}{n-k+1} + \binom{n-k+2}{n-k} - (n-1)$$

$$\geq \sum_{i=n-k+2}^{n} \binom{n}{i} + \binom{n-1}{n-k+1} + \binom{n-k+1}{n-k}.$$

Define A_k to be a subset of nodes which know the message after the k-th step and satisfies

$$|A_k| = \sum_{i=n-k+2}^{n} \binom{n}{i} + \binom{n-1}{n-k+1} + \binom{n-k+1}{n-k},$$

which proves the claim.

Now we use a dual argument. By B_k we will denote suitable subsets of nodes, which do not know the message after the k-th step. Assume that after the $(n + 2)$-nd step there exists at least one node which does not know the message. Observe that the number of nodes that do not know the message after the $(k - 1)$-st step is at least $|S(B_k)| - n + 1$. Clearly there exist sets B_{n+2}, B_{n+1} and B_n s.t. $|B_{n+2}| = 1, |B_{n+1}| = 2$ and $|B_n| = 1 + n$. This results in a similar backward recurrent computation as for $|A_k|$, starting with the set B_n which satisfies

$$|B_n| = \binom{n}{n} + \binom{n}{n-1} + \binom{n-1}{n-2} + \binom{n-2}{n-3}.$$

Thus for $k \geq \lfloor (n+3)/2 \rfloor$ we find a set B_k with

$$|B_k| = \sum_{i=k}^{n} \binom{n}{i} + \binom{n-1}{k-1} + \binom{k-1}{k-2}.$$

Finally set $k = \lfloor (n+3)/2 \rfloor$. Then

$$2^n \geq |A_k| + |B_k|$$

$$\geq \sum_{i=n-k+2}^{n} \binom{n}{i} + \binom{n-1}{n-k+1} + \binom{n-k+1}{n-k}$$

$$+ \sum_{i=k}^{n} \binom{n}{i} + \binom{n-1}{k-1} + \binom{k-1}{k-2}$$

$$= \sum_{i=n-k+2}^{n} \binom{n}{i} + \sum_{i=0}^{n-k} \binom{n}{i}$$

$$+ \binom{n-1}{k-2} + \binom{n-1}{k-1} + \binom{n-k+1}{n-k} + \binom{k-1}{k-2}$$

$$= \sum_{i=0}^{n} \binom{n}{i} + \binom{n-k+1}{n-k} + \binom{k-1}{k-2} = 2^n + \binom{n-k+1}{n-k} + \binom{k-1}{k-2},$$

a contradiction.

Now assume that $n < 10$. The cases for $n \leq 4$ are mentioned in [10]. For $5 \leq n \leq 9$ the claim can be verified manually in a similar way as above, but computing the cardinalities of the sets A_k, B_k precisely from Lemma 1. $\qquad\square$

Using essentially the same method as above we are able to prove:

Theorem 2. For $n \geq 3$, the minimum broadcasting time $T(n)$ in Q_n with $n-3$ dynamic link faults satisfies

$$T(n) = n.$$

Remark. The above approach also yields that if we allow $n-2$ dynamic faults in any step then $T(n) \leq n+1$. We conjecture that $T(n) = n$ in this case too.

References

1. Bezrukov, S., Isoperimetric problems in discrete spaces, in: Extremal Problems for Finite Sets, (P. Frankl, Z. Füredi, G. Katona, D. Miklos, eds.), J. Bolyai Soc. Math. Studies, Akadémia Kiadó, Budapest, 1994, 59-91.
2. Bollobás, B., Combinatorics, Chapter 16.: Isoperimetric Problems, Cambridge University Press, Cambridge, 1986, 123-130.
3. Bollobás, B., Leader, I., Compressions and isoperimetric inequalities, *J. Combinatorial Theory A* **56** (1991), 46-62.
4. Bollobás, B., Leader, I., An isoperimetric inequality on the discrete torus, *SIAM J. on Discrete Mathematics* **3** (1990), 32-37.
5. Chlebus, B., Diks, K., Pelc, A., Broadcasting in synchronous networks with dynamic faults, *Networks* **27** (1996), 309-318.
6. Chung, F. R. K., Spectral Graph Theory, Chapter 2.: Isoperimetric Problems, Regional Conference Series in Mathematics Number 92, American Mathematical Society, Providence, RI, 1997.

7. F. R. K. Chung, Labelings of graphs, in: *Graph Theory 3*, (W. Beineke, R. Wilson, eds.), Academic Press, 1988, 152-167.
8. De Marco, G., Vaccaro, U., Broadcasting in hypercubes and star graphs with dynamic faults, *Information Processing Letters* **66** (1998), 321-326.
9. Fraigniaud, P., Lazard, E., Methods and problems of communication in usual networks, *Discrete Applied Mathematics* **53** (1994), 79-133.
10. Fraigniaud, P., Peyrat, C., Broadcasting in a hypercube when some calls fail, *Information Processing Letters* **39** (1991), 115-119.
11. Heath, L., Rosenberg, A., Graphs Separators, with Applications, 1999.
12. Hedetniemi, S.M., Hedetniemi, S.T., and Liestman, A., A survey of gossiping and broadcasting in communication networks, *Networks* **18** (1986), 319-349.
13. Hromkovic, J., Klasing, R., Monien, B., Paine, R., Dissemination of information in interconnection networks (broadcasting and gossiping), in: Combinatorial Network Theory, (Ding-Zhu Du, D. F. Hsu, eds.), Kluwer Academic Publishers, 1995, 125-212.
14. Katona, G.O.H., The Hamming-sphere has minimum boundary, *Studia Scientarum Mathematicarum Hungarica* **10** (1975), 131-140.
15. Pelc, A., Fault tolerant broadcasting an gossiping in communication networks, *Networks* **26** (1996), 143-156.
16. Pólya, G., Szegö, Isoperimetric Inequalities in Mathematical Physics, Princeton University Press, Princeton, 1951.
17. Santoro, N., Widmayer, P., Distributed function evaluation in the presence of transmission faults, in: *Proc. Intl. Symposium on Algorithms, SIGAL'90*, Lecture Notes in Computer Science 450, Springer Verlag, Berlin, 1990, 358-369.

Routing Permutations in the Hypercube[*]

Olivier Baudon[1], Guillaume Fertin[1], and Ivan Havel[2][**]

[1] LaBRI U.M.R. C.N.R.S. 5800, Université Bordeaux I
351 Cours de la Libération, F33405 Talence Cedex
{fertin,baudon}@labri.u-bordeaux.fr
[2] Faculty of Mathematics and Physics, Charles University
Malostranské nám. 25, 118 00 Praha, Czech Republic
haveli@math.cas.cz

Abstract. We study an n-dimensional directed symmetric hypercube H_n, in which every pair of adjacent vertices is connected by two arcs of opposite directions. Using the computer, we show that for H_4 and for any permutation on its vertices, there exists a system of pairwise arc-disjoint directed paths from each vertex to its target in the permutation. This gives the answer to Szymanski's conjecture [Szy89] for dimension 4.

In addition to this study, we consider in H_n the so-called *2-1 routing requests*, that is routing requests where any vertex of H_n can be used twice as a source, but only once as a target. We give two such routing requests which cannot be routed in H_3. Moreover, we show that for any dimension $n \geq 3$, it is possible to find a 2-1 routing request g_n such that g_n cannot be routed in H_n : in other words, for any $n \geq 3$, H_n is not (2-1)-rearrangeable.

Keywords: Hypercubes, routing permutations, Szymanski's conjecture, 2-1 routing requests.

1 Introduction

The directed symmetric hypercube H_n of dimension $n \geq 1$ has the set of vertices V_n with $|V_n| = 2^n$ and the set of arcs A_n with $|A_n| = n2^n$. From several possible equivalent definitions we are choosing the following : V_n consists of all the integers i such that $0 \leq i \leq 2^n - 1$ and for $i, j \in V_n$, (i, j) is an arc of H_n iff the binary representation of i and j differ in only one coordinate. If this coordinate is in ν-th position in the binary string, we will say that (i, j) is in dimension ν. For every ν, there are 2^n arcs in dimension ν in H_n. Observe that H_n is symmetric, i.e. for each $i, j \in V_n$, $(i, j) \in A_n$ iff $(j, i) \in A_n$.

We also define below a $h - k$ *routing request*, definition that can be found in [GT97].

---

[*] This work has been supported by the Barrande Cooperative Research Grant #97137
[**] Supported by Grant #201/98/1451 of GACR

Definition 1.1 (Routing Request). *A routing request on a directed graph G is a multi-set R of ordered pairs of vertices of G. For each pair $\{s, t\}$ in a routing request, s is called the* source *and t is called the* target *of the pair. A routing request R is said to be h − k if each vertex appears in R at most h times as a source and at most k times as a target. A routing request on G is called a* partial permutation *if it is 1-1, and a* permutation *if it is 1-1 and has exactly $|V(G)|$ pairs.*

Szymanski [Szy89] considered the following problem : given a hypercube H_n and a permutation R on the vertices of H_n, is it possible to realize these source-target pairs by arc-disjoint paths ? That is, is it possible to find for each $\{s_i, t_i\} \in R$ a directed path P_i from s_i to t_i such that the paths P_i are pairwise arc-disjoint ? Or, in the terminology of interconnection networks : is the hypercube *rearrangeable* ? Szymanski [Szy89] conjectured that the answer is yes for any $n \geq 1$; he proved it for any $n \leq 3$, with the stronger property that all the requests are satisfied with shortest paths. In the following, we will refer to Szymanski's conjecture with the shortest paths property as the *strong Szymanski's conjecture*.

Lubiw [Lub90] gave a counterexample of the strong Szymanski's conjecture, concerning the shortest paths assumption. She gave an example of a permutation in H_5 which cannot be realized by arc-disjoint and shortest paths. Moreover, Darmet [Dar92] also gave a counterexample of the strong Szymanski's conjecture concerning the shortest paths, but in dimension 4. It is presented in Fig. 1, which gives a 1-1 routing request π on the vertices of H_4. Note that π is a partial permutation, but, still, π cannot be routed by arc-disjoint and shortest paths. Any permutation realizing at least the requests from π would fail to be routed by arc-disjoint and shortest paths as well.

x	0	1	2	3	4	5	6	7	8	9	10	11	12	13	14	15
$\pi(x)$	10	2	14		8	3	15	11					9	1	13	

Fig. 1. A 1-1 routing request π in H_4 which cannot be routed with shortest paths

Note that all the source-target pairs which are given in the table are such that $dist(x, \pi(x)) = 2$ in H_4. This is the base of the proof : suppose we want to find a shortest path for the source-target pair $\{6, 15\}$. Then we can either route via 14, or via 7. Suppose we route via 14, that is $6 \rightarrow 15 : 6, 14, 15$. Hence the only solution to route the pair $\{2, 14\}$ is $2 \rightarrow 14 : 2, 10, 14$, which means that $0 \rightarrow 10 : 0, 8, 10$, hence $4 \rightarrow 8 : 4, 12, 8$, and consequently $12 \rightarrow 9 : 12, 13, 9$. In that case, the arcs $(14, 15)$ and $(12, 13)$ have both been used. Hence it is impossible to route from 14 to 13 with a shortest path in an arc-disjoint fashion. Similarly, if we decide to route from 6 to 15 via 7, we end up with a contradiction of the same sort.

Note that we will see in Theorem 2.1 that all the permutations in H_4 that realize at least π can be routed with non shortest paths.

Since we have managed to find a counterexample in dimension 4 concerning the strong Szymanski's conjecture, it is easy to see that for any dimension $n \geq 4$, we can find a permutation for which it is not possible to route with shortest paths, since in any hypercube H_n with $n \geq 4$, H_4 is a subgraph of H_n. For this, we take any subgraph isomorphic to H_4 in H_n and we apply π on this subgraph ; then we "complete" π to get a permutation in H_4.

However, until now, it has not been proved whether Szymanski's conjecture (i.e. the rearrangeability of H_n, with $n \geq 4$, without the shortest paths condition) holds. Note that many authors [GT97,CS93,ST94,BR90] have solved part of the problem, whether by proving that some families of permutations could be routed in any H_n, or by proving that by doubling some of the arcs in one or several dimensions in the hypercube, any permutation could be routed in H_n. For a good survey of the results concerning the latter, we refer to [GT97].

In this paper, we first show in Section 2 that Szymanski's conjecture holds for $n = 4$; for this, we use a computer program. We will see that this proof implies the use of 2-1 routing requests. Hence, in Section 3, we will focus on the (2-1)-rearrangeability of H_n, that is the rearrangeability of H_n with respect to the 2-1 routing requests. In H_3, we show that at least two such (2-1) routing requests cannot be routed by arc-disjoint paths ; this can be generalized, and we show in Section 3.2 that for any $n \geq 3$, H_n is not (2-1)-rearrangeable.

2   On the Rearrangeability of H_4

Here, we prove Szymanski's conjecture for $n = 4$. This is done using a computer program, whose steps are detailed below. First, we observe that deleting all the arcs in dimension i $(0 \leq i \leq n-1)$ in H_n results in a disconnected graph, each of the two connected components being a copy of H_{n-1}. Hence, we will call a *cut in dimension* i $(0 \leq i \leq n-1)$ the deletion of all the directed egdes of H_n in dimension i. The two copies of H_{n-1} that we get that way are called *subcubes*. For example, the *cut in dimension 0* of H_4 gives us two subcubes of dimension 3, say $H_{3,0}$ and $H_{3,1}$, where $H_{3,0}$ (resp. $H_{3,1}$) is the subgraph of H_4 induced by the vertices p with $0 \leq p \leq 7$ (resp. with $8 \leq q \leq 15$).

H_4 has 16 vertices, and therefore there are 16! permutations on the vertices of H_4. Hence, thanks to a brute force method, we could have tried to route each and every permutation on the vertices of H_4. However, a deeper study of H_3 will save us many unnecessary computations.

2.1   Converting the Problem from H_4 to H_3

The main idea here is to wonder whether any 2-1 routing request can be routed with arc-disjoint paths in H_3. Indeed, if we manage to prove this, then it is not difficult to see that any permutation can be routed by arc-disjoint paths in H_4 : suppose we have a permutation π in H_4, and let us cut H_4 in dimension 0. Let $V_{3,0} = \{v \in V_4 \mid 0 \leq v \leq 7\}$ and $V_{3,1} = \{v \in V_4 \mid 8 \leq v \leq 15\}$. Now let us route π in two steps. The first step is the following : for each $v \in V_{3,0}$ (resp. $v \in V_{3,1}$), if $\pi(v) \in V_{3,1}$ (resp. $\pi(v) \in V_{3,0}$) then route v to $(v+8)$ (resp. to $(v-8)$) using the arc $(v, v+8)$ (resp. $(v, v-8)$) first. For all the pairs $\{v, \pi(v)\}$ such that v and $\pi(v)$ are in the same $V_{3,i}$ $(0 \le i \le 1)$, do nothing. This first routing step clearly is pairwise arc-disjoint. Now, look at the subgraphs of H_4 induced respectively by $V_{3,0}$ and $V_{3,1}$: each of them is a hypercube of dimension 3, and π induces on each of them a 2-1 routing request by our method. Hence, if any 2-1 routing request can be routed by arc-disjoint paths in H_3, then H_4 is rearrangeable.

To know whether any 2-1 routing request on H_3 can be routed by arc-disjoint paths, we use the computer. For a better understanding, we give an overview of the algorithm used : first, for each request $\{s_i, t_i\}$, we are allowed certain paths depending on the distance from s_i to t_i in H_3. These paths are the following.

- if $dist(s_i, t_i)=1$, we are only allowed the shortest path, that is the arc (s_i, t_i) ;
- if $dist(s_i, t_i)=2$, we can use the 2 shortest paths or the 6 paths of length 4 ;
- if $dist(s_i, t_i)=3$, we can use the 6 shortest paths or the 6 paths of length 5.

Note that, for each request, we order the possible paths by priority (in that case, the shortest paths will be placed first, then the non shortest ones).

The algorithm is the following : for each request, take the allowed path with higher priority. If at least one arc of this path is already used by a previous request, then try the second path, etc., till one path is such that no arc has been used before. If it is not possible, then backtrack to the previous request, and do the same thing recursively till we can find a path P with no arc already used. In that case, use the path P, and try to route the next request. If no path P is found, the routing request is said to be failing. If a path is found for each of the requests, then the given 2-1 routing request can be routed in H_3 by arc-disjoint paths.

Thanks to the computer, we are able to show that most of the 2-1 routings on H_3 can be routed by arc-disjoint paths. In fact, only 72 of them did not get through our algorithm. Let us call them the *72 failing routing requests*. Thanks to the numerous automorphisms of H_3, we can show that only two of them are non equivalent : these are the 2-1 routing requests f_3 and g_3 defined in Section 3.1.

Our algorithm does not try each and every possible path for a given request. However, we show in Section 3.1 that f_3 and g_3 *cannot* be routed in H_3.

2.2   Getting Back to H_4

Now let us consider one of those 72 2-1 failing routing requests, say ρ. The aim is to consider ρ as the "projection" on H_3 of a permutation π of the vertices of H_4 and to retrieve all the possible permutations π corresponding. Depending on which of the 4 dimensions we decide to cut H_4, and, having done so, on which of the 2 subcubes of dimension 3 we consider, there are $4 \cdot 2 = 8$ possibilities. Once we have decided this, we have to "rebuild" π from the informations given by ρ. In each ρ among the 72 failing routing requests, three of the eight vertices are used twice as a source ; hence, only five distinct vertices are used as sources. Consider the following example (Figure 2), where we consider ρ in the subgraph $H_{3,0}$ induced by a cut in dimension 0.

x	0	1	2	3	4	5	6	7
$\rho(x)$			5	4	3	6	1	
			7		2		0	

Fig. 2. A 2-1 routing request ρ in H_3

In that case, we see that 2 is taken twice as a source. Hence the corresponding permutations in H_4 will either have $\{2, 5\}$ and $\{10, 7\}$, or $\{2, 7\}$ and $\{10, 5\}$, as source-target pairs. The same goes for a vertex which is only taken once as a source. Take, for instance, vertex 3. The corresponding permutations in H_4 either could have $\{3, 4\}$ as a source-target pair, or $\{11, 4\}$. As we have five vertices which are sources at least once, this gives us $2^5 = 32$ possible different sets of 8 requests in H_4. This fixes only 8 requests ; hence, there are $8! = 40320$ possibilities for the 8 remaining requests in H_4.

Consequently, for each of the 8 considered subcubes of dimension 3, and for each 2-1 routing request ρ in this subcube, we need to test $32 \cdot 40320 = 1290240$ permutations π in H_4. Thanks to the computer, it is very easy and fast to verify that those permutations in H_4 can be routed by arc-disjoint paths. Indeed, suppose we have cut H_4 in dimension 0, and that we are looking at $H_{3,0}$, i.e. the hypercube of dimension 3 induced by the vertices $0 \leq p \leq 7$. In that case, for each of the 72 2-1 failing routing requests ρ, we have to test the rearrangeability of H_4 on the permutations $\pi_{i,\rho}$ ($1 \leq i \leq 32 \cdot 40320$). For a given $\pi_{i,\rho}$, let us cut H_4 in a different dimension (say 1) and see, in each of the two subcubes induced by the cutting, if the 2-1 routing requests induced by this cut is among the 72 failing ones. If this is the case, let us try by cutting in another dimension (say 2), etc.

It appears that, for each of the 72 2-1 routing requests ρ, no $\pi_{i,\rho}$ is such that, by cutting H_4 in one of the three other dimensions, the new 2-1 routing requests given in each of the two subcubes are among the 72 failing ones. Consequently, if a permutation π is such that a cut in dimension $0 \leq d \leq 3$ induces one of the 72 failing routing requests in at least one of its two subcubes of dimension 3, then there exists $0 \leq d' \neq d \leq 3$ such that a cut in dimension d' does not imply this situation. Hence the following theorem.

Theorem 2.1. *Any permutation π on the vertices of H_4 can be routed by arc-disjoint paths, that is H_4 is rearrangeable.*

3   2-1 Routing Requests in H_n

In Section 3.1, we study two examples of 2-1 routing requests, f_3 and g_3 (cf. Section 2.1), and prove that they cannot be routed in H_3 with arc-disjoint paths. Starting from g_3, we show in Section 3.2 a recursively constructed 2-1 routing request on H_n, for which no arc-disjoint routing can be found ; this proves Theorem 3.1.

3.1 H_3 is not (2-1)-Rearrangeable

We have seen in Section 2.1 that among the 72 failing 2-1 routing requests, only two of them are non equivalent by automorphism of H_3. We denote denote those two routing requests f_3 and g_3, which are as follows :

x	0	1	2	3	4	5	6	7	x	0	1	2	3	4	5	6	7
$f_3(x)$	3		5	4		0	1		$g_3(x)$				5	4	7	2	0
	6		7			2							6	3			1

Fig. 3. f_3 (left) and g_3 (right)

We are going to show that none of them can be routed in H_3 with arc-disjoint paths. First, we introduce some auxiliary notions and notational conventions. We call an arc (u, v) of H_n a d-arc (downwards going arc) if $u > v$. Since we consider H_n, we have : (u, v) is a d-arc iff $u - v = 2^i$, with $0 \leq i \leq n - 1$. An arc (u, v) is called a u-arc (upwards going arc) if it is not a d-arc. Note that if H_n is drawn using a "level" representation in such a way that 0 is the lowest and $2^n - 1$ the highest vertex in the drawing, then u-arcs are really directed upwards and d-arcs downwards (cf. for instance Fig. 4) ; in that case, we say the we *arrange* H_n with respect to the pair $(2^n - 1, 0)$.

Assume $s, t \in V_n$. Then obviously all shortest directed paths from s to t in H_n use the same number of d-arcs (resp. u-arcs) ; let us denote it $d(s, t)$ (resp. $u(s, t)$). Hence, for a routing request $R = \{s_1, t_1\}, \ldots, \{s_r, t_r\}$ in H_n, we define $d(R)$ and $u(R)$ as follows : $d(R) = \Sigma_{i=1}^r d(s_i, t_i)$ and $u(R) = \Sigma_{i=1}^r u(s_i, t_i)$. Observe that in R, identical ordered pairs are allowed and also that any directed path from s to t uses at least $d(s, t)$ d-arcs and $u(s, t)$ u-arcs. Moreover, for any directed path from s to t, the difference between the number of d-arcs and u-arcs, that is $d(s, t) - u(s, t)$, remains constant. Finally we define, for $v \in V_n$: $v^{in} = \{(u, v); u \in V_n \text{ and } (u, v) \in A_n\}$ and $v^{out} = \{(v, u); u \in V_n \text{ and } (v, u) \in A_n\}$. Now we are ready to prove the following.

Proposition 3.1. *Neither f_3 nor g_3 can be routed by arc-disjoint paths in H_3.*

Proof. We have $V_3 = \{0, 1, \ldots, 7\}$ and $A_3 = \{(0, 1), (1, 0), \ldots, (7, 6)\}$ with $|A_3| = 24$. We also verify easily that $d(f_3) = 9$, $u(f_3) = 11$, $d(g_3) = 12$ and $u(g_3) = 8$.

We first assume that there is a routing by arc-disjoint paths in H_3 for f_3, let us denote it ρ. Analyzing the sources and targets of f_3 we conclude that ρ must use the arcs $(0, 4)$, $(1, 5)$, $(2, 6)$ and $(3, 7)$, i.e. all the arcs leading from the subcube induced in H_3 by the vertices $\{0, 1, 2, 3\}$ to the subcube induced by the vertices $\{4, 5, 6, 7\}$. (The reason is that 4 pairs of f_3 have their sources in the first subcube and targets in the second one.) It follows quite similarly that ρ must also use the arcs $(0, 1)$, $(2, 3)$, $(4, 5)$ and $(6, 7)$. Further, we conclude that ρ uses not more than 2 arcs from each of the following sets: 1^{out}, 4^{out} and 5^{in} (since, e.g. 1 is a target but not a source, hence exactly one path from ρ ends

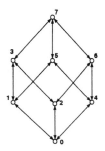

Fig. 4. The hypercube H_3 drawn in "level representation"

in 1 and no path from ρ begins there). Since ρ uses $(1,5)$ and $(4,5)$, it does not use $(7,5)$.

Let us look at the set 1^{out}: at least one of its arcs is not used by ρ; since we showed above that $(1,5)$ must be used, there are 2 cases to be considered:

1. $(1,3)$ is not used by ρ : then all the remaining 11 u–arcs have to be used by ρ and ρ is necessarily a shortest path routing. It follows that $0 \to 3 : 0, 2, 3$, $0 \to 6 : 0, 4, 6$, $2 \to 7 : 2, 6, 7$, $2 \to 5 : 2, 0, 1, 5$. Now consider $6 \to 1$: it can start neither with $(6,7)$ (used already) nor with $(6,2)$ (there is no way out from 2), hence $6 \to 1 : 6, 4, 5, 1$; finally, $5 \to 0 : 5, 4, 0$. This is a contradiction, since all the 3 arcs from 4^{in} are already used and therefore $3 \to 4$ cannot be done.
2. $(1,0)$ is not used by ρ : because of symmetry and the fact that at most 2 arcs from 4^{out} may be used by ρ, we conclude that $(4,0)$ is not used by ρ either. Since $d(f_3) = 9$ and the d–arcs $(1,0)$, $(4,0)$, $(7,5)$ are not used by ρ, ρ has to be a shortest path routing. This is a contradiction, because, obviously, $5 \to 0$ cannot be routed by a shortest path. This contradiciton completes the analysis of the second case and we are done with f_3.

Now we consider function g_3 : arrange H_3 with respect to the pair $(7,0)$. We have seen that $d(g_3) = 12$ and $u(g_3) = 8$. But there are 12 d–arcs altogether, hence all the d–arcs must be used and ρ consists of shortest paths. Consider the subset $S = \{3, 5, 7\}$ of V_3. Observe that, in ρ, there are 5 paths with targets in $V_3 \setminus S$, which start in S. On the other hand, there are 5 arcs going from a vertex in S to a vertex in $V_3 \setminus S$. We conclude that $5 \to 3 : 5, 7, 3$. Further, ρ uses not more than 2 arcs from 6^{out}, since it is a target and not a source in g_3. However, no d-arc can be unused, because $d(g_3) = 12$; hence ρ does not use $(6,7)$, and we conclude that $4 \to 7$ cannot be managed by a shortest path. This contradiciton accomplishes the whole proof. □

3.2 2-1 Routing Requests in H_n

In this Section, we are going to define recursively a 2-1 routing request g_n in the hypercube of dimension n, H_n, and show in Theorem 3.1 that for any $n \geq 3$, g_n cannot be routed in H_n with arc-disjoint paths. This shows that for any $n \geq 3$,

H_n is not (2-1)-rearrangeable. The idea here is to find an "equivalent" of function g_3 of Section 3.1 for any dimension $n \geq 3$, and to generalize the arguments that lead us to show that no routing could achieve g_3 with arc-disjoint paths. First, let us define recursively such a 2-1 routing request, g_n.

Definition 3.1 (2-1 routing request g_n). *Let g_3 be the 2-1 routing request defined for H_3 in Section 3.1. For any $n \geq 3$, let g_{n+1} be defined as follows. For any $i \in [0; 2^n - 1]$, if $g_n(i) \in \emptyset$, then $g_{n+1}(i) \in \emptyset$; otherwise :*

- $g_{n+1}(i) = g_n(i) + 2^n$;
- $g_{n+1}(i + 2^n) = g_n(i)$.

Note that for any $n \geq 3$, any vertex of V_n is used exactly once as a target. In order to show that g_n cannot be routed by arc-disjoint paths in H_n for any $n \geq 3$, we first show that if a routing ρ_n could achieve this, then it must be by shortest paths. For this, we arrange H_n with respect to the pair $(2^n - 1, 0)$, and we show by induction that $d(g_n)$, the minimum number of d–arcs used to route g_n, satisfies : $d(g_n) = n \cdot 2^{n-1}$. We have seen that this is true for $n = 3$, since $d(g_3) = 12$. Now suppose it is true for a fixed n, and let us show that it holds for $n + 1$. For this, we refer to the recursive definition of g_{n+1}, and we discuss the number of d-arcs ; for any $i \in [0; 2^n - 1]$, we have :

- $g_{n+1}(i) = g_n(i) + 2^n$. This means that, for any $i \to g_n(i)$ in g_n, we have $i \to g_n(i) + 2^n$ in g_{n+1}. This means that for any request of this type in g_{n+1}, we will need at least 1 more u–arc, but as many d–arc as in g_n.
- $g_{n+1}(i + 2^n) = g_n(i)$. This means that, for any $i \to g_n(i)$ in g_n, we have $i + 2^n \to g_n(i)$ in g_{n+1}. This means that for any request of this type in g_{n+1}, we will need at least 1 more d–arc, and as many d–arc as in g_n. Since we know that there are 2^n targets in g_n, this means that at least 2^n more d–arcs are needed to route g_{n+1}.

Consequently, we see that $d(g_{n+1}) = d(g_n) + (d(g_n) + 2^n)$, that is $d(g_{n+1}) = (n+1) \cdot 2^n$. This proves , by induction, that for any $n \geq 3$, $d(g_n) = n \cdot 2^{n-1}$, and this shows that any routing ρ_n which satisfies the 2-1 routing request g_n will be of shortest paths, since we have exactly $n \cdot 2^n$ d–arcs.

Now let us define in H_n the set S_n of vertices as follows : S_n is the set of vertices which have 2 targets in g_n. Note that from now on, we will denote the vertices $u \in [0; 2^n - 1]$ of H_n by their binary representation $B(u)$. More precisely, any vertex $u \in [0; 2^n - 1]$ in H_n will be noted as follows $B(u) = mx_2x_1x_0$, where m consists of the $(n-3)$ leading bits of $B(u)$, and x_i is the bit of weight i in $B(u)$. An arc from u to v, (u, v), will then be denoted by $(B(u), B(v))$. Thanks to this representation, we will be able to prove Properties 3.1 and 3.2, which will help us to prove the main result, i.e. Theorem 3.1.

Property 3.1. For any $n \geq 3$ and for any m a $(n-3)$-bits string, every arc of the form $(m101, m111)$ is used in ρ_n to route a particular request $s \to t$, where $s, t \in S_n$.

Proof. We first characterize the vertices of S_n, thanks to their binary representation. Indeed, we can show by induction that $S_n = \{m011, m101, m111 \mid m \in \{0, 1\}^{n-3}\}$. This is true by definition for $n = 3$ (i.e., $S_3 = \{011, 101, 111\} = \{3, 5, 7\}$) ; moreover, since by definition of g_{n+1}, only the vertices $0m011$, $0m101$, $0m111$ and $1m011$, $1m101$ and $1m111$ have two targets by g_{n+1}, we get the result.

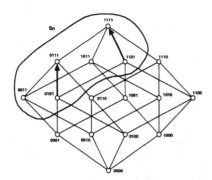

Fig. 5. Property 3.1 illustrated in H_4

The aim now is to show that every routing request of the form $s \to t$ with $s, t \in S_n$ must take place within S_n, that is no directed path P used to route such a request could be such that one of its vertices $u \notin S_n$. For this, let us first count the number of arcs going from a vertex in S_n to a vertex in $V_n \setminus S_n$, and let us then count the number of requests $s \to t$ with $s \in S_n$ and $t \in V_n \setminus S_n$.

We see that there are exactly $5 \cdot 2^{n-3}$ arcs going from a vertex of S_n to a vertex of $V_n \setminus S_n$. This is straightforward since, for a vertex $u \in S_n$, any change of a single bit among the n-3 leading ones will give another vertex v of S_n. Hence, the only way to get an arc from a vertex $u \in S_n$ to a vertex $v \notin S_n$ is to change one of the last 3 bits. This leads to 9 possibilities for any fixed m (3 possibilities for each vertex $m011$, $m101$ and $m111$), but we see that $(m011, m111)$, $(m101, m111)$, $(m111, m011)$ and $(m111, m101)$ are arcs whose both end vertices are in S_n. Hence, for a fixed m, there are 5 arcs going from a vertex $u \in S_n$ to a vertex $v \notin S_n$. Since m is a $(n$-3)-bits string, we have on the whole $5 \cdot 2^{n-3}$ such arcs.

Now, let us count the number of requests of the form $s \to t$ with $s, t \in S_n$. We show the following by induction : every request $s \to t$ with $s, t \in S_n$ is for every $s = m101$ and $t = \overline{m}011$, where m is a $(n$-3)-bits string and $\overline{m}$ is the string obtained from m by changing every 0-bit (resp. 1-bit) into a 1-bit (resp. a 0-bit). This is true for $n = 3$, since the only such request is $5 \to 3$, that is $101 \to 011$. If this is true for a fixed n, then it is easy to see, thanks to Definition 3.1, that this is true for $n + 1$ too (that is $g_{n+1}(0m101) = 1\overline{m}011$ and $g_{n+1}(1m101) = 0\overline{m}011$ for every m a $(n$-3)-bits string). Hence we see that the number of requests of the form $s \to t$, with $s, t \in S_n$ is equal to 2^{n-3}. Since $|S_n| = 3 \cdot 2^{n-3}$ and since each vertex $u \in S_n$ has two targets, this means that we need at least $5 \cdot 2^{n-3}$

directed paths going from vertices of S_n to vertices of $V_n \setminus S_n$. However, we have exactly $5 \cdot 2^{n-3}$ arcs going from a vertex in S_n to a vertex in $V_n \setminus S_n$. From this, we conclude that every path P used to route a request of the form $s \to t$ with $s, t \in S_n$ remains in S_n, that is no intermediate vertex in P can be in $V_n \setminus S_n$.

Starting from this point, let us show the property. For any request of the form $s \to t$ with $s, t \in S_n$ (that is $s = m101 \to t = \overline{m}011$), since we need to route by shortest paths, we only have two options. They are as follows:

$$m101 \to \ldots \to \underline{p101 \to p001} \to \ldots \to q001 \to q011 \to \ldots \to \overline{m}011 \ \text{ or}$$

$$m101 \to \ldots \to p'101 \to p'111 \to \ldots \to q'111 \to q'011 \to \ldots \to \overline{m}011$$

where p, q, p' and q' are some $(n-3)$-bits strings.

However, the first option cannot occur : indeed, the underlined step implies the use of an arc going from vertex $p101 \in S_n$ to vertex $p001 \notin S_n$, which contradicts the above arguments. Consequently, only the second option is valid, which shows that the arc $(p'101, p'111)$ is used to route the request $m101 \to \overline{m}011$ for any fixed m. Since there are 2^{n-3} such requests, since we use arc-disjoint paths and since there are exactly 2^{n-3} different possibilities for p', we conclude that every arc of the form $(m101, m111)$, for any m a $(n-3)$-bits string, is used for a particular routing request $s \to t$, with $s, t \in S_n$. This proves the property.   □

Property 3.2. For any $n \geq 3$, every arc of the form $(m110, m111)$ is necessarily unused in ρ_n, for every m a $(n-3)$-bits string.

Proof. It is possible to prove by induction that, for every $n \geq 3$, every vertex of the form $m110$ (for m a $(n-3)$-bits string) has no target in g_n. This is true by definition for any $n = 3$, and if this is true for a fixed n, we see that it is still true for $n+1$ since $g_{n+1}(0m110)$ and $g_{n+1}(1m110)$ are defined thanks to $g_n(m110)$, which belongs to the empty set.

Since no vertex of the form $m110$ has an image by g_n, we know that there is at least one arc from $m110^{out}$ which is unused by ρ_n. The aim here is to show that this unused arc is necessarily $(m110, m111)$. For this, let us detail all the possible cases for the $(n-3)$-bits string m, with decreasing $|m|_1$, where $|m|_1$ is the number of 1-bits in m.

If $|m|_1 = n-3$, that is $m = 1111\ldots111$, we know we need to use all the d-arcs (because $d(g_n) = n2^{n-1}$ for any $n \geq 3$) ; thus the only possible unused arc is the only existing u-arc going out of $m110$, that is $(m110, m111)$. Now let $|m|_1 = n-4$. Then m has exactly one bit equal to 0. Since the number of bits equal to 0 correspond to the number of u-arcs going out of $m110$, we have two options to choose the unused arc in ρ_n : either it is $(m110, m'110)$ (where $m' = 111\ldots111$), or it is $(m110, m111)$. But if we suppose $(m110, m'110)$ unused, this means that there is one more arc from $m'110^{out}$ which is unused by ρ_n. But we have seen that there could only be one unused arc going out of $m'110$, otherwise a d-arc would be unused, which is not possible. Hence the unused arc must be $(m110, m111)$. This argument is illustrated in Fig. 6, where $n = 4$. The

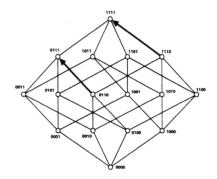

Fig. 6. Property 3.2 illustrated in H_4

same argument goes on for every m a $(n-3)$-bits string, with decreasing $|m|_1$: indeed, step by step, we see that any m with $|m|_1 = p$ has $n - p - 2$ u-arcs. These arcs are either arcs of the form $(m110, m'110)$ (where m' is obtained from m by changing exactly one 0-bit into a 1-bit), or the arc $(m110, m111)$. But since $|m'|_1 = p + 1$, this case has already been considered before, and we have seen that no u-arc other than $(m'110, m'111)$ can be unused (otherwise, this would mean that a new u-arc from $m'110^{out}$ is unused by ρ_n, a contradiction). Hence the only u-arc which can be unused by ρ_n is $(m110, m111)$.

Consequently, for any m a $(n\text{-}3)$-bits string, the arc $(m110, m111)$ is unused by ρ_n. $\qquad\square$

Theorem 3.1. g_n *cannot be routed in* H_n*, for any* $n \geq 3$*, and thus* H_n *is not (2-1)-rearrangeable for any* $n \geq 3$*.*

Proof. It is not difficult to see that, for any $n \geq 3$, the request $4 \to 2^n - 1$ always exists in g_n. Indeed, we see that $4 \to 7$ exists in H_3. Moreover, if we have $4 \to 2^n - 1$ in H_n for a fixed n, then the request $4 \to 2^{n+1} - 1$ exists in H_{n+1}, since by definition $g_{n+1}(4) = g_n(4) + 2^n$, that is $g_{n+1}(4) = 2^n - 1 + 2^n = 2^{n+1} - 1$ for any $n \geq 3$.

Now let us show that the request $4 \to 2^n - 1$ cannot be routed in H_n for any $n \geq 3$. For this, let us consider the binary representation of the source and target in this request. In that case $B(4) = 000\ldots00100$, and $B(2^n - 1) = 11111\ldots11111$. Let $m_0 = 000\ldots0000$ and $\overline{m_0} = 111\ldots1111$ be such that $|\overline{m_0}|_1 = n - 3$. In that case, $4 \to 2^n - 1$ becomes $m_0100 \to \overline{m_0}111$. Since we know we need to use arc-disjoint and shortest paths, there are only two routing schemes for this request. They are as follows:

$$m_0100 \to \ldots \to p_0100 \to p_0110 \to \ldots \to \underline{q_0110 \to q_0111} \to \ldots \to \overline{m_0}111 \text{ or}$$

$$m_0100 \to \ldots \to p_0'100 \to p_0'101 \to \ldots \to \underline{q_0'101 \to q_0'111} \to \ldots \to \overline{m_0}111$$

where p_0, q_0, p_0' and q_0' are some $(n\text{-}3)$-bits strings.

However, in each of those two cases, there is one step which cannot be achieved, which we have underlined. Indeed, in the first case, we know by Property 3.1 that any arc $(m101, m111)$ is already used to route a request of the form

$s \to t$, with $s, t \in S_n$, for every m a $(n\text{-}3)$-bits string. Since $4 \notin S_n$ for all $n \geq 3$, we conclude that we cannot use the first scheme. Similarly, Property 3.2 yields that any arc of the form $(m110, m111)$ cannot be used for any routing request, for every m a $(n\text{-}3)$-bits string. This shows that the request $4 \to 2^n - 1$ of g_n cannot be routed in H_n for any $n \geq 3$, which proves the theorem. $\square$

Remark 3.1. Let f_4 be the 2-1 routing request on H_4 obtained by applying to f_3 the same operation which gave g_4 from g_3. We note that f_4 can be routed in H_4 with arc-disjoint paths.

4   Conclusion

In the first part of this paper, we use the computer to prove the rearrangeability of H_4. This proof relies on the "splitting" of any permutation in H_4 into two 2-1 routing requests (one in each of the H_3 obtained by cutting H_4 in dimension 0), which we try to route by arc-disjoint paths in their respective subcube H_3. We note that this method is the one employed by Szymanski [Szy89] to prove the rearrangeability of H_3.

Starting from what was a study of 1-1 routing requests in H_n and the rearrangeability of H_n, we have mainly studied and proved the non-rearrangeability of H_n with respect to 2-1 routing requests. Though we have answered the question of the (2-1)-rearrangeability of H_n, the question of the (1-1)-rearrangeability of H_n for any $n \geq 5$ remains open.

References

BR90.   R. Boppana and C.S. Raghavendra. Optimal self-routing of linear-complement permutations in hypercubes. In *DISTMEMCC: 5th Distributed Memory Computing Conference*, pages 800–808. IEEE Computer Society Press, 1990.

CS93.   S.B. Choi and A.K. Somani. Rearrangeable circuit-switched hypercube architectures for routing permutations. *Journal of Parallel and Distributed Computing*, 19:125–130, 1993.

Dar92.   A. Darmet. Private communication, 1992.

GT97.   Q-P Gu and H. Tamaki. Routing a permutation in the hypercube by two sets of edge disjoint paths. *Journal of Parallel and Distributed Computing*, 44(2):147–152, 1997.

Lub90.   A. Lubiw. Counterexample to a conjecture of Szymanski on hypercube routing. *Information Processing Letters*, 35:57–61, 1990.

ST94.   A.P. Sprague and H. Tamaki. Routing for involutions of a hypercube. *Discrete Applied Mathematics*, 48:175–186, 1994.

Szy89.   T. Szymanski. On the permutation capability of a circuit-switched hypercube. In *Proc. Internat. Conf. on Parallel Processing*, pages I-103 – I-110, 1989.

ZS90.   K. Zemoudeh and A. Sengupta. Routing frequently used bijections on hypercube. In *DISTMEMCC: 5th Distributed Memory Computing Conference*, pages 824–832. IEEE Computer Society Press, 1990.

An Optimal Fault-Tolerant Routing
for Triconnected Planar Graphs

Koichi Wada[1], Yoriyuki Nagata[1], and Wei Chen[1]

Nagoya Institute of Technology
Gokiso-cho, Syowa-ku, Nagoya 466-8555, JAPAN
{wada,y-nagata,chen}@phaser.elcom.nitech.ac.jp

Abstract. We study the problem of designing fault-tolerant routings
for a communication network which is a triconnected planar network
of processors in the surviving route graph model. The surviving route
graph for a graph G, a routing ρ and a set of faults F is a directed graph
consisting of nonfaulty nodes with a directed edge from a node x to a
node y iff there are no faults on the route from x to y. The diameter of
the surviving route graph could be one of the fault-tolerance measures
for the graph G and the routing ρ. In this paper, we show that we can
construct a routing for any triconnected planar graph with a triangle face
such that the diameter of the surviving route graphs is two(thus optimal)
for any faults $F(|F| \leq 2)$. We also show that the optimal routing can be
computed in linear time.

1 Introduction

Consider a communication network or an undirected graph G in which a limited
number of link and/or node faults F might occur. A *routing* ρ for a graph defines
at most one path called *route* for each ordered pair of nodes. We assume that it
must be chosen without knowing which components might be faulty.

Given a graph G, a routing ρ and a set of faults F, the *surviving route graph*
$R(G, \rho)/F$ is defined to be a directed graph consisting of all nonfaulty nodes in
G, with a directed edge from a node x to a node y iff the route from x to y is
intact. The diameter of the surviving route graph (denoted by $D(R(G, \rho)/F)$)
could be one of the fault-tolerance measures for the graph G and the routing ρ
[1,2]. Many results have been obtained for the diameter of the surviving route
graph [3,6,7,8]. This model can be used to evaluate the fault-tolerance for ATM
and/or optical networks [10].

If the diameter of the surviving route graph is minimal over all routings on a
given graph, the routing is said to be *optimal*. As long as faults are assumed to
occur, the diameter of the surviving route graph becomes more than one. Thus,
if the diameter of the surviving route graph is two, the routing is optimal. It is
shown that an optimal routing can be constructed for any n-node k-connected
graph G such that $n \geq 7k^3 \lceil log_2 n \rceil$ [6]. Although an optimal routing can be
defined for any k by this construction, G must have a lot of nodes in order
to obtain an optimal routing. Even for the cases that $k = 2, 3$ we need nodes

Widmayer et al. (Eds.): WG'99, LNCS 1665, pp. 191–201, 1999.

greater than or equal to $560, 2268$, respectively. It is also shown that an optimal routing can be constructed for every biconnected graph [9,10]. However, it is an open question whether or not we can construct an optimal routing for every k-connected graph for fixed $k \geq 3$.

In this paper, we show a sufficient condition with which an optimal routing can be constructed for triconnected graphs. Using the condition, we show that we can construct an optimal routing for every triconnected planar graph with a triangle face, that is a triangle is located on a face. Our construction is based on a canonical ordering [5] which characterizes triconnected planar graphs. We also show that the optimal routing can be defined in linear time.

The remainder of the paper is organized as follows. In Section 2 some definitions are introduced. In Section 3 we show a sufficient condition with which an optimal routing can be constructed for triconnected graphs. In Section 4, we show that we can construct an optimal routing for triconnected planar graphs with a triangle face. We put concluding remarks in Section 5.

2   Preliminary

In this section, we give definitions and terminology. We refer readers to [4] for basic graph terminology.

Unless otherwise stated, we deal with an undirected graph $G = (V, E)$ that corresponds to a network. For a node v of G, $N_G(v) = \{u|(v, u) \in E\}$ and $deg_G(v) = |N_G(v)|$. A complete graph with three nodes is called a *triangle*. Two paths with common endpoints are said to be *node-disjoint* if their intermediate nodes are disjoint. A set of paths with common endpoints is said to be *node-disjoint* if every pair of paths in the set is node-disjoint. Two paths are said to *cross* if these paths have common nodes or edges except the endpoints. A graph G is *k-connected* if there exist k node-disjoint paths between every pair of distinct nodes in G. Usually 2-connected graphs are called *biconnected graphs* and 3-connected graphs are called *triconnected graphs*. The *distance* between nodes x and y in G is the length of the shortest path between x and y and is denoted by $dis_G(x, y)$. The *diameter* of G is the maximum of $dis_G(x, y)$ over all pairs of nodes in G and is denoted by $D(G)$. Let P_1 and P_2 be a path from u to v and a path from v to w, respectively. A concatenation of two paths P_1 and P_2 is denoted by $P_1 \cdot P_2$.

A graph is *planar* if it can be embedded in the plane so that no two edges intersect except at an endpoint in common. A *plane* graph is a planar graph with a fixed planar embedding.

Let $G = (V, E)$ be a graph and let x and y be nodes of G. Define $P_G(x, y)$ to be the set of all simple paths from the node x to the node y in G, and $P(G)$ to be the set of all simple paths in G. A *routing* is a partial function $\rho : V \times V \rightarrow P(G)$ such that $\rho(x, y) \in P_G(x, y)(x \neq y)$. The path specified to be $\rho(x, y)$ is called the *route from x to y*. The length of the route $\rho(x, y)$ is denoted by $|\rho(x, y)|$. For the routes $\rho(x_{i-1}, x_i)(1 \leq i \leq p)$, define $[\rho(x_0, x_1), \rho(x_1, x_2), \ldots, \rho(x_{p-1}, x_p)]$ to be $\rho(x_0, x_1) \cdot \rho(x_1, x_2) \cdot \ldots \cdot \rho(x_{p-1}, x_p)(p \geq$

1). We call $[\rho(x_0, x_1), \rho(x_1, x_2), \ldots, \rho(x_{p-1}, x_p)]$ a *route sequence of length p from x_0 to x_p.*

For a graph $G = (V, E)$, let $F \subseteq V \cup E$ be a set of nodes and edges called a set of *faults*. We call $F \cap V(= F_V)$ and $F \cap \overset{\cdot}{E}(= F_E)$ the set of *node faults* and the set of *edge faults*, respectively. If an object such as a route or a route sequence does not contain any element of F, the object is said to be *fault free*.

For a graph $G = (V, E)$, a routing ρ on G and a set of faults $F(= F_V \cup F_E)$, the *surviving route graph*, $R(G, \rho)/F$, is a directed graph with node set $V - F_V$ and edge set $E(G, \rho, F) = \{< x, y > | \rho(x, y)$ *is defined and fault free*$\}$.

A routing ρ is *bidirectional* if $\rho(x, y) = \rho(y, x)$ for any node pair (x, y) in the domain of ρ. Note that if the routing ρ is bidirectional, the surviving route graph $R(G, \rho)/F$ can be represented as an undirected graph. In what follows we only consider bidirectional routings and undirected surviving route graphs.

Given a graph G and a routing property P, a routing ρ on G is *optimal* with respect to P if $max_{F s.t. |F| \le k} (D(R(G, \rho)/F))$ is minimal over all routings on G satisfying P. Note that from the definition of the optimality, if $D(R(G, \rho)/F)$ is 2 for any set of faults F such that $|F| \le k$, the routing is obviously optimal with respect to any property. If the property P is known, we simply call the routing optimal.

A routing ρ is said to *be computed* in $O(f)$ time if we can construct a data structure in $O(f)$ time from which any route $\rho(x, y)$ can be obtained in $O(|\rho(x, y)|)$ time.

3 Sufficient Condition for an Optimal Routing

In this section, we show a sufficient condition that an optimal routing can be defined for a triconnected graph G with at most 2 faults.

Let $G = (V, E)$ be a triconnected graph with a triangle $T = (\{t, \ell, r\}, \{(t, \ell), (t, r), (\ell, r)\})$. For $i \in T$ and $x \in V - T$, $P_i(x)$ denotes a path from x to i.

Condition OR

1. For any $x \in V - T$, $P_t(x)$, $P_\ell(x)$ and $P_r(x)$ are node-disjoint.
2. For x and $y(x \ne y) \in V - T$, there is at most one pair $(i, j)(i \ne j$ and $i, j \in T)$ such that $P_i(x)$ and $P_j(y)$ cross.

Note that $P_i(x)$ and $P_i(y)$ may cross for the same i.

We define a routing ρ for a triconnected graph with a triangle satisfying the condition (OR).

routing ρ(Fig. 1)

(a) For $u, v \in T$, $\rho(u, v) = \rho(v, u) := (u, v)$.
(b) For $u \in T$ and $v \in V - T$, $\rho(u, v) = \rho(v, u) := P_u(v)$.
(c) For $u, v \in V - T$,
   if $P_i(u)$ and $P_j(v)(i \ne j)$ cross, $\rho(u, v) = \rho(v, u) := P_j(u) \cdot (j, i) \cdot P_i(v)$.

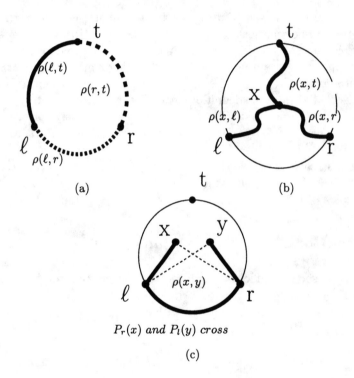

t

(a)

(b)

$P_r(x)$ and $P_l(y)$ cross

(c)

Fig. 1. Routing ρ.

The routing ρ is well defined because there is at most one pair $(i, j)(i, j \in T, i \neq j)$ such that $P_i(u)$ and $P_j(v)$ cross. Note that if there is no pair (i, j) such that $P_i(u)$ and $P_j(v)$ cross, the route $\rho(u, v)$ is not defined.

Theorem 1. *Let $G = (V, E)$ be a triconnected graph with a triangle satisfying the condition (OR).*
Then $D(R(G, \rho)/F) \leq 2$ for any set of faults $F(|F| \leq 2)$ in G.

Proof. Let $R = R(G, \rho)/F$. Let u and v be any pair of distinct nonfaulty nodes in G. We assume that $\rho(u, v)$ contains at least one fault if it is defined, otherwise $dis_R(u, v) = 1$ and the theorem holds.

(1) Suppose that $u, v \in T$. From the assumption F contains (u, v). Let w be the other node in T than u and v and let z be any node in $V - T$. Since $[\rho(u, w), \rho(w, v)] = (u, w) \cdot (w, v)$ and $[\rho(u, z), \rho(z, v)] = P_u(z) \cdot P_v(z)$, these two route sequences are node-disjoint. Therefore, at least one route sequence does not contain another fault in F. Thus, $dis_R(u, v) \leq 2$.

(2) Suppose that $u \in T$ and $v \in V - T$. Let z and w be other nodes in T than u. Since $[\rho(v, z), \rho(z, u)] = P_z(v) \cdot (z, u)$ and $[\rho(v, w), \rho(w, u)] = P_w(v) \cdot (w, u)$ are node-disjoint, at least one route sequence between them is fault free, because F contains at most one element other than an element on $\rho(v, u) = P_u(v)$. Thus, $dis_R(u, v) \leq 2$.

(3) Suppose that $u, v \in V - T$. There are two cases whether the route $\rho(u, v)$ is defined or not.

If the route $\rho(u, v)$ is not defined, $P_i(u)$ and $P_j(v)$ do not cross for any distinct pair $(i, j) \in T$. Therefore, the three route sequences $[\rho(u, i), \rho(i, v)](i \in T)$ are node-disjoint. Since $|F| \leq 2$, at least one of them is fault free. Thus, $dis_R(u, v) \leq 2$.

If the route $\rho(u, v)$ is defined and contains at least one fault in F, we assume that $P_i(u)$ and $P_j(v)$ cross and we notice that $\rho(u, v) = \rho(v, u) = P_j(u) \cdot (j, i) \cdot P_i(v)$. Let k be the other node in T than i and j. If there are 2 faults on the route $\rho(u, v)$, $[\rho(u, k), \rho(k, v)]$ does not contain any fault because $P_k(u)$ and $P_k(v)$ neither cross $P_j(u)$ nor $P_i(v)$. If there is one fault on the route $\rho(u, v)$, either $P_j(u)$ or $P_i(v)$ does not contain any fault because $P_j(u)$ and $P_i(v)$ do not cross. We can assume that $P_i(v)$ does not contain any fault. In this case the node i is fault free. Since $[\rho(u, k), \rho(k, v)] = P_k(u) \cdot P_k(v)$ and $[\rho(u, i), \rho(i, v)] = P_i(u) \cdot P_i(v)$ are node-disjoint and $P_k(u)$, $P_k(v)$, $P_i(u)$ and $P_i(v)$ do not cross $P_j(u)$, at least one of these route sequences is fault free. Thus, $dis_R(u, v) \leq 2$.

4   Optimal Routing for Triconnected Planar Graphs

In this section, we show that every triconnected planar graph G with a triangle face satisfies the condition (OR) and therefore an optimal routing can be constructed for G. Our construction is based on a canonical ordering of triconnected planar graphs which is a generalization of an s-t numbering for biconnected graphs [5].

4.1   Canonical Ordering

Let $G = (V, E)$ be a triconnected plane graph with a vertex v_1 on the outerface. Let $\pi = (V_1, \ldots, V_K)$ be an ordered partition of V, that is, $V_1 \cup \cdots \cup V_K = V$ and $V_i \cap V_j = \phi$ for $i \neq j$. Define G_k to be the subgraph of G induced by $V_1 \cup \cdots \cup V_k$ and denote by C_k the outerface of G_k. π is said to be a *canonical ordering* of G if:

1. V_1 consists of $\{v_1, v_2\}$, where v_2 lies on the outerface and $(v_1, v_2) \in E$.
2. V_K is a singleton $\{v_n\}$, where v_n lies on the outerface, $(v_1, v_n) \in E$ and $v_n \neq v_2$.
3. Each $C_k(k > 1)$ is a cycle containing (v_1, v_2).
4. Each G_k is biconnected and internally triconnected, that is, removing two interior vertices of G_k does not disconnect it.
5. For each k in $2, \ldots, K - 1$, one of the following conditions holds:

    V_k is a singleton, $\{z_1\}$, where z_1 belongs to C_k and has at least one neighbor in $G - G_k$.

    V_k is a chain, $\{z_1, \ldots, z_\ell\}$, where each z_i has one neighbor in $G - G_k$, and where z_1 and z_ℓ each have one neighbor on C_{k-1}, and these are the only two neighbors of V_k in G_{k-1}.

Proposition 1. [5] *Every triconnected planar graph G with predefined vertex v_1 on the outerface has a canonical ordering. And it can be computed in linear time and space.*

For a triconnected plane graph G with the canonical ordering, the nodes are numbered with a bijective function $g : V \to \{1, 2, \ldots, |V| = n\}$ as follows:

1. $g(v_1) = 1$, $g(v_2) = 2$ and $g(v_n) = n$.
2. For $V_k = \{z_1, \ldots, z_\ell\}(\ell \geq 1)(2 \leq k \leq K - 1)$, $g(z_i) = |V_1 \cup \ldots \cup V_{k-1}| + i$.

In what follows, the nodes in G are denoted as integers with the numbering g. A path $(p_1, p_2, \ldots, p_\ell)$ is said to be *decreasing (increasing)* if $p_i > p_{i+1}(p_i < p_{i+1})$ for any $i(1 \leq i \leq \ell - 1)$. Fig. 2 shows an example of a triconnected plane graph and its canonical ordering.

4.2   Optimal Routing for a Triconnected Planar Graph

Let $G = (V = \{1, 2, \ldots, n = |V|\}, E)$ be a triconnected planar graph with a triangle face. We assume that G is embedded such that the triangle is located on the outerface and a canonical ordering can be done for the embedding.

We define three paths $P_t(x)$, $P_\ell(x)$ and $P_r(x)$. We set $\ell = 1$, $r = 2$ and $t = n$ and $T = \{1, 2, n\}$.

We may assume that for the cycle $C_k = (c_{k_1}(= 1), c_{k_2}(= 2), \ldots, c_{k_p}(= 1))$, $c_{k_1}, c_{k_2}, \ldots, c_{k_{p-1}}$ and c_{k_p} appear counter-clockwise in this order. $c_{k_i}(2 \leq i \leq p - 1)$ is called the *left neighbor* of $c_{k_{i-1}}$ and the *right neighbor* of $c_{k_{i+1}}$ in C_k.

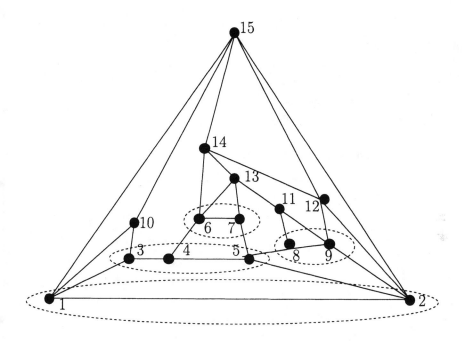

Fig. 2. An example of a triconnected plane graph and its canonical ordering.

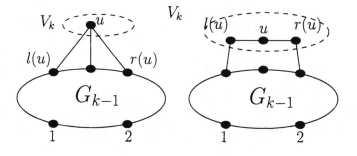

Fig. 3. $\ell(u)$ and $r(u)$.

To each node $u \in V - T$, we define the left-node $\ell(u)$ and the right node $r(u)$ as follows.

If $u \in V_k (2 \le k \le K - 1)$, $\ell(u)$ is defined to be the left neighbor of u in C_k and $r(u)$ is defined to be the right neighbor of u in C_k (Fig. 3).

$P_1(u)\ P_2(u)\ P_n(u)\ (u \in V - T)$

$P_1(u) = (v_1(= u), v_2, \ldots, v_p(= 1))$ where $v_i = \ell(v_{i-1})(2 \le i \le p)$,

$P_2(u) = (v_1(= u), v_2, \ldots, v_p(= 2))$ where $v_i = r(v_{i-1})(2 \le i \le p)$ and

$P_n(u) = (v_1(= u), v_2, \ldots, v_p(= n))$ where $v_i = max\{u | u \in N_G(v_{i-1})\}$.

Fig. 4 shows an example of the three paths.

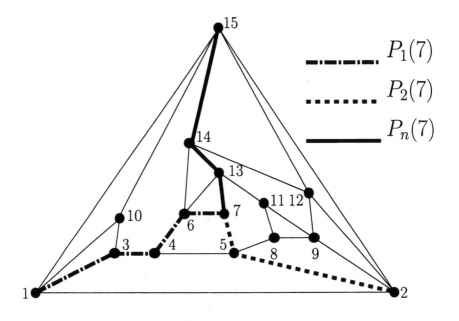

Fig. 4. The three paths $P_1(x)$, $P_2(x)$ and $P_n(x)$.

Lemma 1. *Let $G = (V = \{1, 2, \ldots, n = |V|\}, E)$ be a triconnected plane graph with a triangle and a canonical ordering on G is assumed. The three paths $P_1(x), P_2(x) and P_n(x)$ defined on G satisfy the condition (OR).*

Proof. From the definition, it is easily verified that $P_1(x)$, $P_2(x)$ and $P_n(x)$ are node-disjoint for any $x \in V - T$ and if $x > y$, $P_n(x)$ and $P_1(y)$ do not cross and $P_n(x)$ and $P_2(y)$ do not cross. Therefore, in order to show that these paths satisfy the condition (OR), there is at most one case among the following four cases for any nodes x and y $(x > y)$.

(I) $P_2(x)$ and $P_1(y)$ cross.
(II) $P_1(x)$ and $P_2(y)$ cross.
(III) $P_2(x)$ and $P_n(y)$ cross.
(IV) $P_1(x)$ and $P_n(y)$ cross.

Case:1((I) and (II) do not occur at the same time.)(Fig. 5)

If $P_2(x)$ and $P_1(y)$ cross, since $x > y$ the node x is on or inside the cycle $(1,n) \cdot (n,2) \cdot P_2(y) \cdot P_1(y)$. If $P_1(x)$ and $P_2(y)$ cross, $P_1(x)$ and $P_2(y)$ must have nodes in common except the endpoints. However, since the node y is on or inside the cycle $P_n(x) \cdot (n,2) \cdot P_2(x)$, $P_1(x)$ must got through nodes on $P_2(x)$ or $P_n(x)$. It means that $P_1(x)$ and $P_2(x)$(or $P_n(x)$) must cross. This is a contradiction since these three paths are node-disjoint.

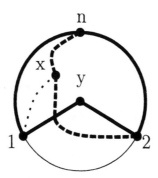

Fig. 5. (I) and (II) do not occur at the same time.

Case:2((I) and (III) do not occur at the same time.)

Suppose that $P_2(x)$ and $P_n(y)$ cross. Let a be the minimum numbered node among nodes $P_2(x)$ and $P_n(y)$ have in common(Fig. 6(a)). Since $P_2(x)$ traverses right neighbors, the path from a to 2 in $P_2(x)(P_2(a))$ is on or inside the cycle $P_n(y) \cdot (n,2) \cdot P_2(y)$. Thus, $P_1(y)$ and $P_2(a)$ do not have nodes in common except the case $y = a$. Therefore, $P_1(y)$ and $P_2(a)$ do not cross.

Next, we show that the path from x to a in $P_2(x)$ and $P_1(y)$ do not cross. Since the node a is located on $P_n(y)$, it holds that $a \geq y$.

If $a > y$, since the path from a to x in $P_2(x)$ is increasing and the path from y to 1 in $P_1(y)$ is decreasing these paths do not have nodes in common. If $a = y$, $P_2(x)$ and $P_n(y)$ have other common nodes except a(Fig. 6(b)). Let b be the least numbered node among them. From the definition of $P_n(y)$, b is the largest numbered node among the neighbors of a and the edge (a, b) is on $P_n(y)$. Thus, $P_1(y)$ does not contain the edge (a, b). Also the path from b to x in $P_2(x)$ is increasing and the path y to 1 in $P_1(y)$ decreasing. Therefore, $P_1(y)$ and the path from x to a in $P_2(x)$ do not cross.

Case:3((I) and (IV) do not occur at the same time.)
Case:4((II) and (III) do not occur at the same time.)

These cases can be shown similar to Case:1.

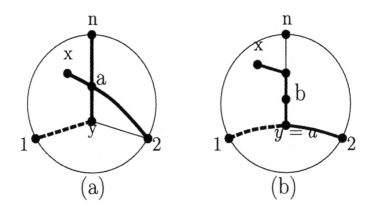

Fig. 6. (I) and (III) do not occur at the same time.

Case:5((II) and (IV) do not occur at the same time.)

This case can be shown similar to Case:2.

Case:6((III) and (IV) do not occur at the same time.)(Fig. 7)

Suppose that $P_1(x)$ and $P_n(y)$ cross. Let a be the minimum numbered node among nodes $P_1(x)$ and $P_n(y)$ have in common. Since the common part between $P_1(x)$ and $P_n(y)$ is either a node or a path from the definition, the path from a to n in $P_n(y)(P_n(a))$ is on or inside the cycle $P_1(x) \cdot (1, n) \cdot P_n(x)$. Note that if $a = x$, $P_1(x)$ and $P_n(y)$ do not cross from the definition. Thus, $P_n(a)$ and $P_2(x)$ do not have nodes in common. If the path from y to a in $P_n(y)$ and $P_2(x)$ have nodes in common, $P_n(y)$ and $P_1(x)$ never cross from the construction of $P_n(y)$. Therefore, $P_2(x)$ and $P_n(y)$ do not cross.

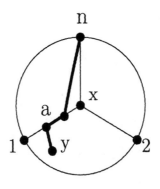

Fig. 7. (III) and (IV) do not occur at the same time.

We obtain the following theorem from Lemma 1 and Proposition 1.

Theorem 2. *Let $G = (V, E)$ be a triconnected planar graph with a triangle face. Then $D(R(G, \rho)/F) \leq 2$ for any set of faults $F(|F| \leq 2)$ in G. Moreover, ρ is constructed in $O(|E|)$ time.*

Since we can make a triconnected planar graph with a triangle face from every triconnected planar graph adding at most one edge, we have the following theorem.

Theorem 3. *Let G be a triconnected planar graph. We can construct an optimal routing for G adding at most one edge.*

5 Concluding Remarks

In this paper, we show that we can construct an optimal routing for every triconnected planar graph with a triangle. It is remained to find an optimal routing for every triconnected graph which is always planar and every k-connected graph $(k \geq 4)$.

Acknowledgement

This research is supported in part by a Scientific Researchof the Ministry of Education, Science and Culture, Japan under Grant: 10680352 and 10205209.

References

1. A.Broder, D.Dolev, M.Fischer and B.Simons: "Efficient Fault Tolerant Routing in Network," *Information and Computation*75,52–64(1987).
2. D.Dolev, J.Halpern , B.Simons and H.Strong: "A New Look at Fault Tolerant Routing," *Information and Computation*72,180–196(1987).
3. P.Feldman: "Fault Tolerance of Minimal Path Routing in a Network," *Proc. 17th ACM STOC*,pp.327–334(1985).
4. F.Harary, Graph Theory, *Addison-Wesley*, Reading, MA(1969).
5. G. Kant: "Drawing Planar Graphs Using the Canonical Ordering," *Algorithmica*, 16, 1, 4-32 (1996).
6. K.Kawaguchi and K.Wada: "New Results in Graph Routing," *Information and Computation*, 106, 2, 203–233 (1993).
7. D.Peleg and B.Simons: "On Fault Tolerant Routing in General Graph," *Information and Computation*74,33–49(1987).
8. K.Wada and K.Kawaguchi: "Efficient Fault-Tolerant Fixed Routings on $(k + 1)$-connected Digraphs," *Discrete Applied Mathematics*, 37/38, 539–552 (1992).
9. K.Wada, Y.Luo and K.Kawaguchi: "Optimal Fault-Tolerant Routings for Connected Graphs," *Information Processing Letters*, 41, 3, 169–174 (1992).
10. K.Wada, W.Chen, Y.Luo and K.Kawaguchi: "Optimal Fault-Tolerant ATM-routings for Biconnected Graphs," *The 23rd International Workshop on Graph-Theoretic Concepts in Computer Science*, Lecture Notes in Computer Science 1335, 354–367 (1997).

Optimal Irreversible Dynamos in Chordal Rings

Paola Flocchini[1], Frédéric Geurts[2], and Nicola Santoro[3]

[1] University of Ottawa, Canada
flocchin@site.uottawa.ca
[2] Université Libre de Bruxelles, Belgium
fgeurts@ulb.ac.be
[3] Carleton University Ottawa,Canada
santoro@scs.carleton.ca

1 Introduction

Let G be a simple connected graph where every node is colored either *black* or *white*. Consider now the following repetitive process on G: each node recolors itself, at each local time step, with the color held by the majority of its neighbors. Since the clocks are not necessarily synchronized, the process can be asynchronous.

Depending on the initial assignment of colors to the nodes and on the definition of majority, different *dynamics* can occur. These dynamics have been extensively studied in the context of *synchronous* systems (mostly cellular automata); researchers mostly concentrated their efforts on determining *periodic* behaviors, transients, number and form of fixed points for different graph structures (rings, infinite lines, general finite graphs, general infinite graphs) with different majority functions (simple majority, strong majority, weighted threshold functions, convex functions) and different coloring sets (e.g., see [1,11,16,17,22]).

In the context of *distributed computing*, this repetitive process is particularly important in that it describes the impact that a set of initial faults can have in majority-based systems (where black nodes correspond to faulty elements and white to non-faulty ones). Consider for example systems where majority voting among various copies of crucial data are performed between neighbours at each step [20]: if the majority of its neighbors is faulty (e.g., has corrupted data), a non-faulty element will exhibit a faulty behavior (e.g., its data will become corrupted) and will therefore be indistinguishable from a faulty one. In this paper, we are interested in the patterns of initial faults which may lead the entire system to a faulty behaviour (e.g., every entity has corrupted data); in terms of system dynamics, these are the patterns for which the system converges to a *monochromatic* fixed point. These patterns are called *dynamos* (short for "dynamic monopolies") and their study has been introduced by Peleg [21]. If the initial faults are permanent, the dynamo is said to be *irreversible*; if instead the initial faults can be mended by the majority rule, the dynamo will be called *reversible*.

Surprisingly, very little is known about dynamos. Most of the results are known for the *static* version of this process; that is, considering only a single step in the evolution [3,5,13,20].

Widmayer et al. (Eds.): WG'99, LNCS 1665, pp. 202–214, 1999.

Recently, researchers have started to focus directly on dynamos. In particular, reversible monotone dynamos have been studied in general graphs [21] and tori [9]; irreversible dynamos have been investigated in tori [8], butterfly and similar interconnection networks [14].

In this paper we study irreversible dynamos for infinite and sufficiently large finite *chordal rings*, i.e. ring networks where each node is also directly connected to the nodes at a fixed set of distances, under the *simple majority* rule: a node becomes black if a simple majority of its neighbours are black [1] . More specifically, we concentrate on two large clasess of chordal rings, namely weakly and strongly chorded rings. Given a dynamo for these graphs, its *weight* is the number of black nodes it contains, and its *length* is the size of the (smallest) segment of the ring containing all the back elements. We are interested in characterizing *optimal dynamos*, that is dynamos of *minimal length* with the *minimal weight*.

Chordal rings are a particular case of *circulant graphs*, and are also known in the literature as *distributed loop networks*. Weakly and strongly chorded rings include many of the most studied chordal rings, ranging from double loops to fan graphs to complete graphs. Special classes of chordal rings (in particular the so called double loops and triple loops) have been widely studied to analyze their fault tolerant properties (for a survey see [4]). Subclasses of chordal rings have been intensively studied under a variety of scenarios and for a large number of problems and applications ranging from routing (e.g., [12]) to election (e.g., [2,15,19]) to broadcast (e.g., [15]). Of particular relevance are also the existing studies on *catastrophic fault patterns* for redundant linear arrays (e.g., see [6,18,23]); in fact, these patterns are exactly irreversible dynamos for infinite chordal rings under a *directional unanimity* rule: a node becomes black if all its neighbours in the same direction are black.

In the following sections of this paper, we first establish some basic bounds on the weight and length of dynamos for general chordal rings (§2). We then consider weakly chorded rings (§3). We establish a *lower bound* on the weight of dynamos of minimal length, and prove that the bound is *tight*; the upper bound proof is *constructive*. We also provide a *complete characterization* of the optimal dynamos for the well known class of double loop networks and the subclass of triple loops. In Section §4, we study strongly chorded rings and their optimal dynamos. Also for this class of graphs we establish *tight bounds*, and show how to construct optimal dynamos.

In the following, due to space constrains, theorems and lemmas are stated without proofs, or their proofs are only sketched; the interested reader is referred to [7] for further technical details.

2   Definitions and Basic Bounds

A *circulant graph* of size n and link structure $\langle d_1, d_2, \cdots, d_h \rangle$ ($d_i \leq d_{i+1}$, $d_h \leq \lfloor \frac{n}{2} \rfloor$), is a graph on n nodes $x_0, x_1, \cdots, x_{n-1}$ where each node x_i is connected

---

[1] In the terminology of [21], simple majority is the "self-not-included, prefer-black" model.

to the nodes x_{i+d_j} and x_{i-d_j} $(1 \leq j \leq h)$. We will often use "x_i sees x_j" to mean that node x_i and node x_j are connected to each other and all operations on the indices are modulo n. We shall denote such a graph by $\mathcal{C}\{d_1, d_2, \cdots, d_h\}$. A *chordal ring* (or *loop network*) is any circulant graph with $d_1 = 1$.

A (p, k)-*chorded ring*, $p \leq k$, is the circulant graph $\mathcal{C}\{1, 2, \cdots, p, k\}$. We shall denote such a graph by $\mathcal{C}\langle p, k \rangle$. Examples of (p, k)-chorded rings are rings $(p = k = 1)$, double loop networks $(p = 1, k > 1)$, fan networks $(p = k - 1)$, and complete graphs $(p = k = \lfloor \frac{n}{2} \rfloor)$. Depending on the relationship between p and k a chorded ring will be said to be *weakly* $(p < \lfloor \frac{k}{2} \rfloor)$ or *strongly* $(p \geq \lfloor \frac{k}{2} \rfloor)$.

Let G be the chordal ring $\mathcal{C}\{d_1, d_2, \cdots, d_h\}$. In the following, and unless otherwise stated, we will assume that all d_i are distinct

Consider the following repetitive process on G. Initially, each node is in one of two states: *black* or *white* (denoted in the following also by "(0)" and "(1)", respectively). An initially black node does not change its value (*irreversibility*). Any other node recolors itself at each local time step with the color held by the majority of its neighbors; in case of tie, it becomes black. Notice that the assumption of irreversibility of the initial black nodes implies that the evolution of the graph is monotonic, i.e., once a node becomes black, it remains black forever.

Since the clocks are not necessarily synchronized, the process can be asynchronous. However, since the evolution of the system is monotonic, we can assume synchronicity w.l.g. [10,21].

A *pattern* P is a sequence of values of consecutive nodes in the ring at a given global time; its *weight* is the number of black values it contains; its *length* is its total number of nodes. A pattern is *bounded* if its first and last values are both black; it is *converging* if, after a finite number of steps, all its nodes become black.

A *configuration* Y is a global state representing the colors of all the nodes at a global time. The *window* $W(Y)$ of Y is the smallest pattern containing all the black nodes (or one of the smallest if the smallest is not unique); we denote by $weight(Y)$ and $length(Y)$ respectively the weight and length of $W(Y)$.

An *irreversible dynamo* (or simply *dynamo*) is a configuration from which an all black configuration is reached. A dynamo is *optimal* if it has *minimum weight for the minimum length*.

Let us first consider the minimum *weight*; i.e., the minimum number of initial black nodes needed to reach an all-black configuration. An *upper bound* is easily established, observing that any configuration Y containing d_h consecutive blacks is obviously a dynamo for $\mathcal{C}\{1, d_2, \cdots, d_h\}$.

Property 1 *Let Y be a configuration in $\mathcal{C}\{1, d_2, \cdots, d_h\}$ which contains d_h consecutive blacks (i.e., the pattern $(1)^{d_h}$); then Y is a dynamo.*

An obvious *lower bound* on the weight of any dynamo X in G is: $weight(X) \geq h$, since otherwise no node will ever become black. It is interesting to note that there exist infinite families of chordal rings having dynamos achieving this lower bound, for example, as we will show later, this holds for strongly chorded rings. Summarizing,

Property 2 Let X be a dynamo for $C\{1, d_2, \cdots, d_h\}$; if X has minimum weight, then we have $h \leq weight(X) \leq d_h$.

In the case of chorded rings this becomes:

Corollary 1 Let X be an optimal dynamo for $C\langle p, k \rangle$, $p < k$; then: $p + 1 \leq weight(X) \leq k$

Consider now $length(X)$; i.e., the size of the window containing the all the black nodes of a dynamo X. An obvious lower bound follows from the definition: $weight(X) \leq length(X)$ since all black nodes must be contained in the window. A stronger lower bound is stated in the next property.

Property 3 Every dynamo X for $C\{1, d_2, \cdots, d_h\}$, is such that $d_h \leq length(X)$.

In the case of chorded rings this becomes:

Corollary 2 Let X be an optimal dynamo for $C\langle p, k \rangle$, $p < k$; then: $k \leq length(X)$

3   Weakly Chorded Rings

A chorded ring $C\langle p, k \rangle$ is said to be *weakly* if $p < \lfloor \frac{k}{2} \rfloor$. In this Section we consider only weakly chorded rings and establish tight bounds on their optimal dynamos. We then study in details two particular cases: the (simple) well known class of double loop networks and the (more complex) subclass of triple loops $C\langle 2, k \rangle$; for both we provide a complete characterization of the optimal dynamos.

3.1   Lower Bound

Any dynamo of length k is length-minimal (see Corollary 2). In this section we establish a lower bound on the weight of such a dynamo.

We first characterize a *forbidden* pattern for dynamos of length k. Let V be the set of nodes of the chordal ring $C\langle p, k \rangle$, a *white block* $B \subseteq V$ is a subset of V composed of all white vertices, each of which has at least $p + 2$ neighbours in B; hence nodes of a white block will never become black. Thus,

Lemma 1 No dynamo for $C\langle p, k \rangle$ can contain a white block.

We now characterize several types of white blocks. The first one concerns infinite graphs.

Lemma 2 Let $C\langle p, k \rangle$ be infinite, R_i be the set of consecutive nodes $x_i \ldots x_{i+p}$, and $R[i]$ be the set of nodes $R[i] = \cup_{j=0}^{\infty} R_{i \pm jk}$. If all nodes in $R[i]$ are white, then $R[i]$ is a white block.

Proof. Each $R_{i \pm jk}$ forms a white clique of degree p. Moreover, each node x_l in $R[i]$ is linked to the nodes $x_{l \pm k}$ which both belong to $R[i]$. Thus $R[i]$ is a white block.

The following lemma gives a type of white blocks for sufficiently large chorded rings.

Lemma 3 *Let $A_i = a_0 \ldots a_p$ and $B_i = b_0 \ldots b_{n-2k+p}$ be the sets of consecutive nodes $x_i \ldots x_{i+p}$ and $x_{i+k} \ldots x_{i+n-k+p}$ of a configuration of a length-n weakly chorded ring $C\langle p, k \rangle$, and let all nodes in $A_i \cup B_i$ be white. Then, $A_i \cup B_i$ is a white block if one of the following conditions holds:*

1. *$p = 1$ and $n \geq 4k - 2$ (i.e., $|B_i| \geq 2k$);*
2. *$p > 1$ and $n \geq 3k - p$ (i.e., $|B_i| \geq k + 1$).*

Based on Lemmas 2 and 3, we define the notion of *large enough* weakly chorded ring as an infinite ring or as a ring which fulfills one of the two conditions of Lemma 3. From this point on, we will always assume that the weakly chorded ring is large enough.

¿From Lemmas 1–3 it follows that in a large enough chorded ring, a dynamo of length k cannot contain in its window, $p + 1$ consecutive white nodes:

Theorem 1 *In a large enough $C\langle p, k \rangle$, a dynamo of length k cannot contain $p + 1$ consecutive white nodes in its window.*

We now characterize a pattern whose presence is *prescribed* inside the window of a dynamo of length k.

Theorem 2 *Let $C\langle p, k \rangle$ be a weakly chorded ring. In $C\langle p, k \rangle$, a dynamo of length k contains at least a bounded pattern P of $length(P) \leq 2p + 1$ and $weight(P) \geq p + 1$.*

Let $W(X) = w_0, w_1, \ldots w_{k-1}$ be the window of a dynamo X of length k. From Theorem 2 we have that any dynamo X of length k must contain a bounded pattern with length at most $2p + 1$ and weight at least $p + 1$. Let $I(X) = w_{i(X)}, \ldots, w_{i(X)+q(X)-1}$ denote a such a pattern, where $q(X) \leq 2p + 1$ denotes its length.

Lemma 4 *In $C\langle p, k \rangle$, for every dynamo X of length k:*

$$weight(X) \geq p + 1 + \lceil \frac{i(X)}{p+1} \rceil + \lceil \frac{k - i(X) - q(X)}{p+1} \rceil$$

Proof. By Theorem 1, a pattern of $p + 1$ consecutive white nodes is forbidden in a dynamo of length k; it follows that the portion of the window $w_0, \ldots, w_{i(X)-1}$ preceding the pattern $I(X)$ must contain at least $\lceil \frac{i(X)}{p+1} \rceil$ blacks and the portion $w_{i(X)+q(X)}, \ldots, w_{k-1}$ following $I(X)$ must contain at least $\lceil \frac{k-i(X)-q(X)}{p+1} \rceil$ blacks. Since $I(X)$ contains at least $p + 1$ blacks, it follows that $weight(X) \geq p + 1 + \lceil \frac{i(X)}{p+1} \rceil + \lceil \frac{k-i(X)-q(X)}{p+1} \rceil$.

We can now establish a lower bound on the weight of a dynamo of length k.

Theorem 3 *In* $C\langle p, k \rangle$, *for every dynamo* X *of length* k:

$$weight(X) \geq p + 1 + \lceil \frac{k - 2p - 1}{p + 1} \rceil$$

Proof. From Lemma 4, it suffices to notice that, for all $i(X)$ $(0 \leq i(X) \leq k - q(X) - 1)$, $p + 1 + \lceil \frac{i(X)}{p+1} \rceil + \lceil \frac{k-i(X)-q(X)}{p+1} \rceil \geq p + 1 + \lceil \frac{k-q(X)}{p+1} \rceil$. Moreover, for all $q(X)$ $(p + 1 \leq q(X) \leq 2p + 1)$, $p + 1 + \lceil \frac{k-q(X)}{p+1} \rceil \geq p + 1 + \lceil \frac{k-2p-1}{p+1} \rceil$.

3.2 Optimal Dynamos

We now show that there are dynamos of length k and furthermore their weight matches the lower bound established in the previous section. We first introduce the notion of initiating and filling pattern.

We call *initiating pattern* for $C\langle p, k \rangle$ a converging bounded pattern of length at most $2p+1$ whose weight is at least $p+1$. An initiating pattern P is *maximum* if $length(P) = 2p + 1$ and $weight(P) = p + 1$.

Lemma 5 *There exist maximum initiating patterns for any* $C\langle p, k \rangle$.

Proof. Consider, for example, $(1)(0)^p(1)^p$.

We call *right (left) filling pattern* for $C\langle p, k \rangle$ a pattern which, when located to the right (left) of a pattern of length and weight p (i.e., $(1)^p$), is always converging. Examples of fillings patterns are given in the following lemma.

Lemma 6 *The patterns* $(0)^i(1)$ *and* $(1)(0)^i$, $i \leq p$, *are right and left filling, respectively.*

By definition of initiating and filling patterns, dynamos can be constructed using the following method:

General Construction

1. Choose an arbitrary initiating pattern, place it anywhere in the length-k window.
2. Complete the right (left) portion of the window with right (left) filling patterns.

Theorem 4 *The construction described above for* $C\langle p, k \rangle$ *forms a dynamo of length* k.

Proof. By definition, the initiating pattern converges to a sequence of at least $p + 1$ consecutive 1s; this enables the sequential convergence of the attached filling patterns. Hence the entire window becomes black; by Property 1, such configuration is converging.

We now establish an upper bound on the weight of configuration which matches the lower bound previously established.

Theorem 5 *There exist dynamos of length k for $C\langle p, k \rangle$ whose weight is $p+1+ \lceil \frac{k-2p-1}{p+1} \rceil$.*

Proof. It is sufficient to place any maximum initiating pattern at the beginning of the window and complete the window with $\lfloor \frac{k-(2p+1)}{p+1} \rfloor$ filling patterns of the form $(0)^p$ (1) and one filling pattern of the form $(0)^{k-(2p+1)-(p+1)\lfloor \frac{k-(2p+1)}{p+1} \rfloor - 1}(1)$.

From Theorems 3 and 5 we have that:

Theorem 6 *A dynamo X is optimal for weakly chorded $C\langle p, k \rangle$ iff:* $length(X) = k$ *and* $weight(X) = p + 1 + \lceil \frac{k-2p-1}{p+1} \rceil$.

3.3   Case Study: Double Loop Networks

In this Section we give a complete characterization of the optimal dynamos for the class of *double loop networks*. These graphs are exactly the chorded rings $C\langle 1, k \rangle$, $k > 1$, and coincide with the class of chordal rings of degree four.

The following is a corollary of Theorem 6:

Corollary 3 *A dynamo X is optimal in a double loop network iff $length(X) = k$ and $weight(X) = 2 + \lceil \frac{k-3}{2} \rceil = \lceil \frac{k+1}{2} \rceil$.*

The construction described in the proof of Theorem 5 allows to construct some optimal dynamos; in the case of double loop networks, we can actually characterize all of them.

Theorem 7 *In any $C\langle 1, k \rangle$: if k is odd the only optimal dynamo is $(10)^{\lfloor \frac{k}{2} \rfloor}(1)$; if k is even there are $\lceil \frac{k}{2} \rceil$ optimal dynamos of the form $(10)^\alpha$ (1) $(10)^\beta$ (1) with $\alpha + \beta = \frac{k-2}{2}$.*

Notice that the double loop networks coincide with the class of tori with snake-like connections on the rows (torus cordalis). In [8,9], diffent types of tori (among which the torus cordalis) have been investigated and lower and upper bounds on the size of irreversible and monotone dynamos are given.

3.4   Case Study: Triple Loop $C\langle 2, k \rangle$

We now study the triple loop network $C\langle 2, k \rangle$. The following is a corollary of Theorem 6:

Corollary 4 *A dynamo X is optimal in a Triple Loop $C\langle 2, k \rangle$ iff:* $length(X) = k$ *and* $weight(X) = 3 + \lceil \frac{k-5}{3} \rceil = \lceil \frac{k+4}{3} \rceil$.

The construction described in the proof of Theorem 5 allows to construct some optimal dynamos; in the following we will characterize them all.

Optimal Dynamos

Theorem 8 *Any dynamo of length k in $C\langle 2, k\rangle$ contains a pattern with two consecutive black nodes and does not contain any pattern of three consecutive whites.*

Theorem 9 *Any configuration in a window of length k which contains two consecutive black nodes and which does not contain three consecutive white nodes is a dynamo for $C\langle 2, k\rangle$.*

Proof. Let $W(X) = x_1 \ldots x_k$ be such a configuration and let, w.l.g $x_i = 1$ and $x_{i+1} = 1$; we shall consider two cases depending on i.
Case 1, $i \leq k - 4$: Since $x_{i+2}x_{i+3}x_{i+4}$ cannot be a block of three consecutive whites, the pattern $x_i x_{i+1} x_{i+2} x_{i+3} x_{i+4}$ has length 5 and weight ≥ 3. Moreover, it is easy to verify that no matter which of the nodes $x_{i+2}x_{i+3}x_{i+4}$ is black, the pattern is converging; thus, it is an initiating pattern. So, X contains an initiating pattern. Since a block of three consecutive whites is forbidden, the portion of the window at the left of the pattern $x_i x_{i+1} x_{i+2} x_{i+3} x_{i+4}$ can contain only blocks of the form **(1)**, **(10)**, and **(100)** which are all left filling patterns (Lemma 6) and the portion of the window at the right of X can contain only blocks of the form **(1)**, **(01)**, and **(001)** which are all right filling patterns. Thus X is a dynamo.
Case 2, $i > k - 4$: In this case, the proof consists in showing that $x_{i-4}x_{i-3}x_{i-2}x_{i-1}x_i$ is an initiating pattern.

From Theorems 8 and 9 it follows that:

Theorem 10 *A configuration of length k is a dynamo iff it contains two consecutive black nodes and does not contain three consecutive white nodes.*

We now introduce a set of configurations that we will prove to be optimal dynamos.

- *Case $k = 3i + 2$:* Dynamos whose windows have the form:
  (1) $(100)^\alpha (11)(001)^\beta$
  with $\alpha + \beta = i$. The number of such configurations is $\frac{i+1!}{i!} = i + 1$.
- *Case $k = 3i + 1$:* Dynamos whose windows have one of the following two forms:
  (2) $(100)^\alpha\ (10)\ (100)^\beta\ (11)\ (001)^\gamma$
  (3) $(100)^\alpha\ (11)\ (001)^\beta\ (01)\ (001)^\gamma$
  with $\alpha + \beta + \gamma = i - 1$. The number of such configurations is $\frac{i+1!}{i-1!} = i(i + 1)$.
- *Case $k = 3i$:* Dynamos whose windows have one of the following four forms:
  (4) $(100)^\alpha\ (11)\ (001)^\beta\ (1)(001)^\gamma$
  (5) $(100)^a(10)\ (100)^b\ (10)(100)^c\ (11)\ (001)^d$
  (6) $(100)^a(10)\ (100)^b\ (11)(001)^c\ (01)\ (001)^d$
  (7) $(100)^a(11)\ (001)^b\ (01)(001)^c\ (01)\ (001)^d$

with $a + b + c + d = i - 2$ and $\alpha + \beta + \gamma = i - 1$. The number of such configurations is $\frac{1}{2}i(i + 1) + \frac{1}{2}(i - 1)i(i + 1)$.

Theorem 11 *In $C\langle 2, k \rangle$ chordal rings, the configurations listed above are optimal dynamos.*

Proof. Each configuration contains a block **(11)** and no block **(000)**, thus, by Theorem 9, they are dynamos. Moreover, it is easy to verify that their weight is $3 + \lceil \frac{k-5}{3} \rceil$ which, by Corollary 4, is the weight of an optimal dynamo.

Completeness To prove that the optimal dynamos above are the only ones, we introduce a different problem which will be shown to be equivalent to the problem of construction all optimal dynamos for chordal rings $C\langle 2, k \rangle$.

The Ring Problem: Given a ring of size m where every node is white, colour black the minimum number of nodes in such a way that no three consecutive nodes are white.
    We will need the following result on the number of blacks in such strings.

Corollary 5 *Any solution to the ring problem on a ring of size m contains $\lceil \frac{m}{3} \rceil$ blacks.*

Property 4 *The obvious solutions to the ring problem are the following circular strings:*

1. *if $m = 3j$: $(\mathbf{001})^j$;*
2. *if $m = 3j + 1$: $(\mathbf{1})(\mathbf{001})^j$, and $(\mathbf{01})(\mathbf{001})^a(\mathbf{01})(\mathbf{001})^b$, with $a + b = j - 1$;*
3. *if $m = 3j + 2$: $(\mathbf{001})^a(\mathbf{01})(\mathbf{001})^b$, with $a + b = j$.*

We now show that there is a correspondence between the class of optimal dynamos of length k and the solutions to the ring problem, where $m = k - 2$.

Theorem 12 *There is a correspondence between the class of optimal dynamos for $C\langle 2, k \rangle$ and the solutions to the ring problem with $m = k - 2$.*

Proof. Let $W(X) = x_1, \ldots, x_k$ be the window of an optimal dynamo. By Theorem 8 we know that X contains at least two consecutive blacks, say x_i and x_{i+1}. Let us construct a string Y as follows: $Y = x_1, \ldots, x_i, x_{i+2}, \ldots, x_{k-1}$. By Theorem 8, Y does not contain three consecutive whites. Moreover, since x_1 is black, it follows that Y does not contain three consecutive whites even considered as a circular string. Since X is an optimal dynamo, the number of blacks in X is $\lceil \frac{k+4}{3} \rceil$; it follows that $weight(Y) = \lceil \frac{k+4}{3} \rceil - 2 = \lceil \frac{k-2}{3} \rceil$ which means that Y is a solution to the ring problem.
    Viceversa, let $X = x_1, \ldots, x_{k-2}$ be a circular string corresponding to a solution of the ring problem. Let x_i and x_j be two arbitrary blacks (i.e., $x_i = 1$ and $x_j = 1$); we construct the corresponding optimal dynamo Y as follows:

$Y = x_i, x_{i+1}, \ldots, x_{k-2}, x_1, \ldots x_j, x_j, x_{j+1} \ldots x_{i-1}, x_i.$ Such a configuration starts and ends with blacks by construction; moreover it does not contain any block of three consecutive whites and it contains at least a block of two consecutive blacks (x_j, x_j), thus it is a dynamo (Theorem 9). Furthermore, by Corollary 5, we know that the number of blacks in X is $\lceil \frac{m}{3} \rceil$, it follows that $weight(Y) = 2 + \lceil \frac{m}{3} \rceil = \lceil \frac{k+4}{3} \rceil$ which means that Y is optimal.

We now prove that the transformation used in the proof of the previous theorem maps each solution of the ring problem to one of the optimal dynamos of Theorem 11.

Theorem 13 *Any ring solution corresponds to an optimal dynamo of Theorem 11.*

Concluding,

Theorem 14 *The optimal dynamos for $C\langle 2, k \rangle$ are exactly the ones of Theorem 11.*

4   Strongly Chorded Rings

A *strongly chorded ring* is a chorded ring $C\langle p, k \rangle$ such that $p \geq \lfloor \frac{k}{2} \rfloor$. In the following let $q = k - p$, so the chordal ring can be also indicated as: $C\langle k - q, k \rangle$. In this section we show that the trivial lower bound $p+1$ (see Corollary 1) on the weight of dynamos is matched by the optimal dynamos for this class of chorded rings. In the following we shall distinguish the two cases: $k > p+1$ and $k \leq p+1$.

4.1   Optimal Dynamos - Case $k > p + 1$

We now consider strongly chorded ring $C\langle p, k \rangle$ where $k > p + 1$.

For the next results, it is useful to decompose the length-k window of a dynamo as follows: Z_c is the part of the window from which any other node of the window can be seen, including the two extreme blacks; Z_l and Z_r are the sets of nodes located respectively at the left and at the right of Z_c, excluding the two extreme blacks (see Figure 1). Observe that Z_c is never empty since

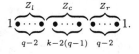

Fig. 1. Decomposition of the length-k of a strongly chorded ring dynamo

$q \leq \lceil \frac{k}{2} \rceil$. Moreover, Z_l, Z_r and the two extreme blacks do not exist when $q = 1$.

Lemma 7 *In a dynamo for* $C\langle p, k \rangle$*, all white nodes of* Z_c *become black in one step.*

Proof. Any white node v of Z_c sees all the window, since $1 + (q - 2) + i \leq k - q$ for all i such that $0 \leq i \leq k - 2(q - 1) - 1$. Thus, v sees at least $p + 1$ black nodes which must be part of the dynamo. Hence, it turns black in one step.

Using the previous lemma we now show a dynamo matching the lower bound.

Theorem 15 *In strongly chorded rings* $C\langle p, k \rangle$ *there exist dynamos of length* k *with weight* $p + 1$.

Proof. We consider two cases depending on whether $|Z_l| + |Z_r| \leq p - 1$ or not.
Case 1, $2(k - p - 2) \leq p - 1$: To obtain a dynamo, we colour black all nodes of $Z_l \cup Z_r$ and arbitrary $p - 1 - 2(k - p - 2)$ nodes of Z_c. In total there are $p + 1$ black nodes. By Lemma 7 the nodes in Z_c become black in one step; so, the window converges and by Property 1 our construction is a dynamo.
Case 2, $2(k - p - 2) \geq p - 1$: To obtain a dynamo, we colour black the $\lfloor \frac{p-1}{2} \rfloor$ nodes to the left of Z_c and the $\lceil \frac{p-1}{2} \rceil$ nodes to the right of Z_c. Also in this case we have a total of $p + 1$ black nodes and, by Lemma 7 the nodes in Z_c become black in one step. Thus, there are at least $3p + 3 - k$ consecutive black nodes after the first step: the $k - 2(k - p - 1)$ nodes of Z_c plus the $\lfloor \frac{p-1}{2} \rfloor$ to its left and the $\lceil \frac{p-1}{2} \rceil$ to its right. Consider the closest white node v to the left of this set S of consecutive nodes with *step* ≤ 1. Node v is at distance at least $q - 2 - \lfloor \frac{p-1}{2} \rfloor$ from the leftmost black node of the window. But it is easy to verify that $q - 2 - \lfloor \frac{p-1}{2} \rfloor \leq p$; hence, v has p black neighbours after the first step on its right and at least one black on its left, thus it becomes black in at most two steps. By symmetry, also the closest white node v' to the right of S becomes black in at most 2 steps. Applying this reasoning inductively, it follows that the entire window converges.

Thus, we have that:

Theorem 16 *A dynamo* X *is optimal for strongly chorded* $C\langle p, k \rangle$ *iff:* $length(X) = k$ *and* $weight(X) = p + 1$.

4.2   Optimal Dynamos - Case $k = p + 1$

In this section we consider strongly chorded ring $C\langle p, k \rangle$ where $k = p + 1$. These chordal rings $C\langle k - 1, k \rangle$ are also called *fan graphs*. Notice that, when $k = \lfloor \frac{n}{2} \rfloor$ such a chordal ring is a complete graph.

We now show that the trivial lower bound on the weigth of an optimal dynamo is matched by exactly one dynamo.

Since the degree of this type of chordal rings is $2k$, the trivial lower bound on the weight of an optimal dynamo is k.

Clearly there exists a dynamo of length and weight k; such a dynamo consists of a sequence of k black nodes (such a configuration is a dynamo by Property 1); and obviously there is no other dynamo with length and weight k. Thus we have:

Theorem 17 *A configuration X is an optimal dynamo for strongly chorded $C\langle k-1, k \rangle$ iff: length$(X) = $ weight$(X) = k$. And there exists exactly one such an optimal dynamo.*

References

1. Z. Agur, A. S. Fraenkel, S. T. Klein. The number of fixed points of the majority rule. *Discrete Mathematics*, 70:295–302, 1988.
2. H. Attiya, J. van Leeuwen, N. Santoro, S. Zaks Efficient elections in chordal ring networks. *Algorithmica*, 4:437-446, 1989.
3. J.C. Bermond, J. Bond, D. Peleg, S. Perennes. Tight bounds on the size of 2-monopolies. In *Proc. 3rd Colloquium on Structural Information and Communication Complexity*, 170–179. 1996.
4. J.-C. Bermond, F. Comellas, D.F. Hsu. Distributed loop computer networks: a survey. *Journal of Parallel and Distributed Computing*, 24:2-10, 1995.
5. J.C. Bermond, D. Peleg. The power of small coalitions in graphs. In *Proc. 2nd Coll. Structural Information and Communication Complexity*, 173–184. 1995.
6. R. De Prisco, A. Monti, L. Pagli. Efficient testing and reconfiguration of VLSI linear arrays. *Theoretical Computer Science*, 197:171-188, 1998.
7. P. Flocchini, F. Geurts, N. Santoro. Optimal Dynamos in Chordal Rings. ULB, Département d'Informatique, Tech. Report 411, http://www.ulb.ac.be/di.
8. P. Flocchini, E. Lodi, F. Luccio, L. Pagli, N. Santoro. Irreversible dynamos in tori. *Proc. EUROPAR 98*, 554-562,1998.
9. P. Flocchini, E. Lodi, F. Luccio, L. Pagli, N. Santoro. Monotone dynamos in tori. In *Proc. 6th Coll. Struct. Information and Communication Complexity*, 152-165, 1999.
10. E. Goles, S. Martinez. *Neural and Automata Networks, Dynamical Behavior and Applications*. Kluwer Academic Publishers, 1990.
11. E. Goles, J. Olivos. Periodic behavior of generalized threshold functions. *Discrete Mathematics*, 30:187–189, 1980.
12. D. Krizanc, F.L. Luccio. Boolean routing on chordal rings. In *Proc. 2nd Coll. Structural Information and Communication Complexity*, 1995.
13. N. Linial, D. Peleg, Y. Rabinovich, M. Sachs. Sphere packing and local majority in graphs In *Proc. 2nd ISTCS*, 141-149, 1993.
14. F. Luccio, L. Pagli, H. Sanossian. Irreversible dynamos in butterflies. In *Proc. 6th Coll. Structural Information and Communication Complexity*, 204-218, 1999.
15. B. Mans. Optimal Distributed Algorithms in Unlabeled Tori and Chordal Rings. *Journal on Parallel and Distributed Computing*, 46(1): 80-90, 1997.
16. G. Moran. The r-majority vote action on 0–1 sequences. *Discrete Mathematics*, 132:145–174, 1994.
17. G. Moran. On the period-two-property of the majority operator in infinite graphs. *Transactions of the American Mathematical Society*, 347(5):1649–1667, 1995.
18. A. Nayak, N. Santoro, and R. Tan. Fault-tolerance of reconfigurable systolic arrays. In *Proc. 20th Int'l Symp. Fault-Tolerant Computing*, 202–209, 1990.
19. Yi Pan. A near-optimal multi-stage distributed algorithm for finding leaders in clustered chordal rings. *Information Sciences*, 76 (1-2):131-140, 1994.
20. D. Peleg. Local majority voting, small coalitions and controlling monopolies in graphs: A review. In *Proc. 3rd Coll. Structural Information and Communication Complexity*, 152–169, 1997.

21. D. Peleg. Size bounds for dynamic monopolies In *Proc. 4th Coll. Structural Information and Communication Complexity*, 151-161, 1997.
22. S. Poljak. Transformations on graphs and convexity. *Complex Systems*, 1:1021–1033, 1987.
23. N. Santoro, J. Ren, A. Nayak. On the complexity of testing for catastrophic faults. In *Proc. 6th Int'l Symposium on Algorithms and Computation*, 188-197, 1995.

Recognizing Bipartite Incident-Graphs of Circulant Digraphs

Johanne Cohen*, Pierre Fraigniaud**, and Cyril Gavoille

[1] LRI, Univ. Paris-Sud, 91405 Orsay cedex, France
Johanne.Cohen@lri.fr
[2] LRI, Univ. Paris-Sud, 91405 Orsay cedex, France
Pierre.Fraigniaud@lri.fr
[3] LaBRI, Univ. Bordeaux I, 33405 Talence cedex, France
gavoille@labri.u-bordeaux.fr

Abstract. Knödel graphs and Fibonacci graphs are two classes of bipartite incident-graph of circulant digraphs. Both graphs have been extensively studied for the purpose of fast communications in networks, and they have deserved a lot of attention in this context. In this paper, we show that there exists an $O(n \log^5 n)$-time algorithm to recognize Knödel graphs, and that the same technique applies to Fibonacci graphs. The algorithm is based on a characterization of the cycles of length six in these graphs (bipartite incident-graphs of circulant digraphs always have cycles of length six). A consequence of our result is that none of the Knödel graphs are edge-transitive, apart those of $2^k - 2$ vertices. An open problem that arises in this field is to derive a polynomial-time algorithm for any infinite family of bipartite incident-graphs of circulant digraphs indexed by their number of vertices.

Keywords: graph isomorphism, circulant graphs, chordal rings, broadcasting, gossiping.

1 Introduction

So-called Knödel graphs and Fibonacci graphs have been used by Knödel [12], and Even and Monien [7] (see also [5,13,18]), respectively, for the purpose of performing efficient communications in networks. More precisely, consider a network of n nodes, and assume that communications among the nodes proceed by a sequence of synchronous calls between neighboring vertices. A *round* is defined as the set of calls performed at the same time. Knödel on one hand, and Even and Monien on the other hand, were interested in computing the minimum number of rounds necessary to perform a all-to-all broadcasting, also called *gossiping*, between the nodes (see [8,9,11] for surveys on gossiping and related problems). The communication constraints assume that a call involves exactly two neighboring nodes, and that a node can communicate to at most one neighbor at a

---

* Additional support by the DRET of the DGA.
** Additional support by the CNRS.

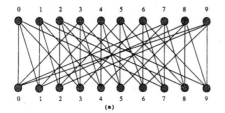 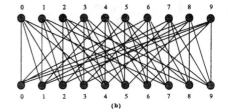

Fig. 1. A Knödel graph of 20 vertices (a), and a Fibonacci graph of 20 vertices (b).

time. Knödel considered the 2-way mode (full-duplex) in which the two nodes involved in the same call can exchange their information in one round, whereas Even and Monien considered the 1-way mode (half-duplex) in which the information can flow in one direction at a time, that is, during the call between x and y, only y can receive information from x, or x can receive information from y, not both. Under these hypotheses, it was shown in [12] that, for n even, one cannot perform gossiping in less that $\lceil \log_2 n \rceil$ rounds in the 2-way mode, and that there are graphs, called here Knödel graphs, that allow gossiping to be performed in $\lceil \log_2 n \rceil$ rounds. Even and Monien have shown in [7] that, for n even, one cannot perform gossiping in less than $2 + \lceil \log_\varrho \frac{n}{2} \rceil$ rounds in the 1-way mode, where $\varrho = \frac{1+\sqrt{5}}{2}$, and that there are graphs, called here Fibonacci graphs, that allow gossiping to be performed in that number of rounds (up to an additive factor of one).

Knödel graphs and Fibonacci graphs are bipartite graphs $G = (V_1, V_2, E)$ of $2n$ vertices. Each partition has n vertices labeled from 0 to $n - 1$. In the Knödel graphs, there is an edge between $x \in V_1$ and $y \in V_2$ if and only if there exists $i \in \{0, 1, \ldots, k\}$ such that $y = x + 2^i - 1 \pmod{n}$, $k = \lfloor \log_2 n \rfloor$. In the Fibonacci graphs, there is an edge between $x \in V_1$ and $y \in V_2$ if and only if there exists $i \in \{0, 1, \ldots, k\}$ such that $y = x + F(i + 1) - 1 \pmod{n}$, $k = F^{-1}(n) - 1$, where $F(i)$ denotes the ith Fibonacci number ($F(0) = F(1) = 1$, and $F(i) = F(i - 1) + F(i - 2)$ for $i \geq 2$) and $F^{-1}(n)$ denotes the integer i for which $F(i) \leq n < F(i + 1)$. Both graphs are Cayley graphs on the dihedral group, and thus they are vertex-transitive. See Figure 1 for an example of a Knödel graph and a Fibonacci graph. Note that graphs on Figure 1 look pretty dense but Knödel graphs and Fibonacci graphs are of degree $O(\log n)$.

Knowing whether a graph G of n nodes allows gossiping to be performed optimally, that is in $\lceil \log_2 n \rceil$ rounds in the 2-way mode, and in (about) $\lceil \log_\varrho n \rceil$ rounds in the 1-way mode, is NP-complete [5]. In particular there are graphs that are not isomorphic to Knödel graphs (resp. Fibonacci graphs), and that allow gossiping to be performed optimally in the 2-way mode (resp. 1-way mode). In this paper, we want to recognize Knödel graphs and Fibonacci graphs. In other words, given a graph G, we want to know whether G is isomorphic to a Knödel graph of the same order, or to know whether G is isomorphic to a Fibonacci graph of the same order.

A closely related topic deals with circulant digraphs. Recall that a digraph is circulant if nodes can be labeled so that the adjacency matrix is circulant, that is node x has $k + 1$ out-neighbors $x + a_i$ (mod n) for $i = 0, \ldots, k$ for some k and some constants a_i's independent of x. Circulant digraphs are Cayley digraphs over $\mathbb{Z}_n$. Ponomarenko has given in [17] a polynomial-time algorithm to decide whether a given tournament is a circulant digraph (a tournament is a digraph obtained by giving an orientation to the edges of a complete graph). More recently, Muzychuk and Tinhofer [15] have shown that one can decide in polynomial-time whether a digraph of prime order is circulant. Deciding whether two circulant digraphs are isomorphic is also a difficult problem. Ádám [1] conjectured that two circulant digraphs are isomorphic if and only if the generators of one digraph can be obtained from the generators of the other digraph via a product by a constant. This conjecture is wrong [6] although it holds in many cases. For instance, Alspach and Parsons [2] have proved that Ádám's conjecture is true for values of n such as the product of two primes (see also [3,14,16]).

Nevertheless, even if they are closely related to circulant digraphs, Knödel graphs and Fibonacci graphs are not circulant graphs but bipartite incident-graphs of circulant digraphs, and they are thus sometimes called bi-circulant graphs. (The bipartite incident-graph of a digraph $H = (V, A)$ is a graph $G = (V_1, V_2, E)$ such that $V_1 = V_2 = V$, and for any $x_1 \in V_1$, and $x_2 \in V_2$, $\{x_1, x_2\} \in E \Leftrightarrow (x_1, x_2) \in A$. Note that two non isomorphic digraphs H and H' can yield isomorphic bipartite incident-graphs, e.g., $H = (\{u, v\}, \{(u, u), (v, v)\})$ and $H' = (\{u, v\}, \{(u, v), (v, u)\})$.) It is unknown whether there exists a polynomial-time algorithm to decide whether a given graph is isomorphic to a given circulant digraph or a given incident-graph of a circulant digraph. This is why we have studied specific algorithms for the case of Knödel graphs and Fibonacci graphs.

The paper is organized as follows. In Section 2, we study the form of solutions of equations involving powers of two. The characterization of these solutions allows us to recognize Knödel graphs in $O(n \log^5 n)$ time, as shown in Section 3. This algorithm is optimal up to a polylogarithmic factor since Knödel graphs have $\Theta(n \log n)$ edges. Section 4 concludes the paper with some remarks on bipartite incident-graphs of circulant digraphs defined by an arbitrary increasing sequence $(g_i)_{i \geq 0}$ of integers, including Fibonacci graphs.

2   Preliminary Results

Let $H = (V, A)$ be a circulant digraph of n vertices and of generators $g_0, \ldots, g_k$, and let $G = (V_1, V_2, E)$ be the corresponding bipartite incident-graph. By 6-*cycle*, we mean an elementary cycle of length six.

Lemma 1. *There is a 6-cycle in G if and only if one can find a sequence of six generators*

$$(g_{i_0}, g_{i_1}, g_{i_2}, g_{i_3}, g_{i_4}, g_{i_5})$$

and $\alpha \in \{0, 1, 2\}$ such that

$$
\begin{cases}
g_{i_0} + g_{i_2} + g_{i_4} = g_{i_1} + g_{i_3} + g_{i_5} + \alpha\, n \\
g_{i_j} \neq g_{i_{j+1}} \text{ (mod 6)} \text{ for any } j \in \{0, 1, 2, 3, 4, 5\}
\end{cases}
\tag{1}
$$

Proof. Let $(u_0, u_1, u_2, u_3, u_4, u_5)$ be a 6-cycle in G (all the u_i's are pairwise distinct), and assume without loss of generality that $u_0 = 0 \in V_1$. We have:

$$
\begin{cases}
u_1 = u_0 + g_{i_0} \\
u_1 = u_2 + g_{i_1} + \alpha_1\, n \\
u_3 = u_2 + g_{i_2} + \alpha_2\, n \\
u_3 = u_4 + g_{i_3} + \alpha_3\, n \\
u_5 = u_4 + g_{i_4} + \alpha_4\, n \\
u_5 = u_0 + g_{i_5}
\end{cases}
$$

where $\alpha_i \in \{-1, 0\}$ and $g_{i_j} \neq g_{i_{j+1}}$ (mod 6) for any $j \in \{0, 1, 2, 3, 4, 5\}$ since the cycle uses six different edges. Therefore we have

$$
g_{i_0} + g_{i_2} + g_{i_4} = g_{i_1} + g_{i_3} + g_{i_5} + \alpha\, n
$$

where $\alpha = (\alpha_1 - \alpha_2) + (\alpha_3 - \alpha_4)$, that is $-2 \leq \alpha \leq 2$ by definition of the α_i's. By possibly swapping even and odd g's, we get the claimed result with $0 \leq \alpha \leq 2$.

Conversely, let $(g_{i_0}, g_{i_1}, g_{i_2}, g_{i_3}, g_{i_4}, g_{i_5})$ be such that

$$
\begin{cases}
g_{i_0} + g_{i_2} + g_{i_4} = g_{i_1} + g_{i_3} + g_{i_5} + \alpha\, n \\
g_{i_j} \neq g_{i_{j+1}} \text{ (mod 6)} \text{ for any } j \in \{0, 1, 2, 3, 4, 5\}
\end{cases}
$$

with $\alpha \in \{0, 1, 2\}$. Then let $u_0 \in V_1$, and let

$$
\begin{cases}
u_1 = u_0 + g_{i_0} \bmod n \\
u_2 = u_1 - g_{i_1} \bmod n \\
u_3 = u_2 + g_{i_2} \bmod n \\
u_4 = u_3 - g_{i_3} \bmod n \\
u_5 = u_4 + g_{i_4} \bmod n \\
u_6 = u_5 - g_{i_5} \bmod n
\end{cases}
$$

We have

$$
u_6 = u_0 + (g_{i_0} + g_{i_2} + g_{i_4}) - (g_{i_1} + g_{i_3} + g_{i_5}) \bmod n.
$$

Since

$$
g_{i_0} + g_{i_2} + g_{i_4} = g_{i_1} + g_{i_3} + g_{i_5} \pmod{n},
$$

we get $u_6 = u_0$, and therefore $(u_0, u_1, u_2, u_3, u_4, u_5)$ is a cycle of length six in G. This cycle is elementary because $g_{i_j} \neq g_{i_{j+1}}$ (mod 6) for any $j \in \{0, 1, 2, 3, 4, 5\}$. $\square$

From Lemma 1, any bipartite incident-graph of a circulant digraph has 6-cycles since $g_{i_0} = g_{i_3}$, $g_{i_1} = g_{i_4}$, and $g_{i_2} = g_{i_5}$ is a solution of Equation 1. We will solve Equation 1 to characterize 6-cycles of Knödel graphs and Fibonacci graphs, and to identify the possible generators of a candidate to be a Knödel graph or a Fibonacci graph.

Let us start first with Knödel graphs. Let $x_0, x_1, x_2, x_3, x_4, x_5$ be six integers in $\{0, \ldots, k\}$, and let n be any integer such that $2^k \leq n < 2^{k+1}$. From Lemma 1, we are interested in computing the solutions of the equation

$$\begin{cases} 2^{x_0} + 2^{x_2} + 2^{x_4} = 2^{x_1} + 2^{x_3} + 2^{x_5} + \alpha\, n \\ x_i \neq x_{i+1} \pmod 6 \text{ for any } i \in \{0, 1, 2, 3, 4, 5\} \end{cases} \tag{2}$$

where $\alpha \in \{0, 1, 2\}$.

Lemma 2. *For $\alpha = 0$, Equation 2 has four types of solutions:*
$(x_0, x_1, x_2, x_3, x_4, x_5) =$

a) $(\gamma, \gamma'', \gamma', \gamma, \gamma'', \gamma')$ $\gamma, \gamma', \gamma'' \in \{0, \ldots, k\}$ $\gamma \neq \gamma', \gamma' \neq \gamma'', \gamma'' \neq \gamma$
b) $(\gamma, \gamma', \gamma, \gamma', \gamma' + 1, \gamma + 1)$ $\gamma, \gamma' \in \{0, \ldots, k-1\}$ $\gamma \neq \gamma'$
c) $(\gamma, \gamma + 1, \gamma, \gamma', \gamma' + 1, \gamma')$ $\gamma, \gamma' \in \{0, \ldots, k-1\}$ $\gamma \neq \gamma'$
d) $(\gamma, \gamma', \gamma, \gamma + 1, \gamma' + 1, \gamma')$ $\gamma, \gamma' \in \{0, \ldots, k-1\}$ $\gamma \neq \gamma'$

up to cyclic permutations[1] of the x_i's.

Proof. The case x_0, x_2, x_4 pairwise distinct generates the first type of solutions. Assume $x_0 = x_2 = \gamma$ and $x_4 = \gamma' \neq \gamma$. There is an impossibility to solve Equation 2 if $\gamma = \gamma' - 1$ because it would imply either $x_0 = x_1$ or $x_0 = x_5$. If $\gamma \neq \gamma' - 1$, we get a solution if two x_{2i+1}'s are both equal to $\gamma' - 1$, and the third one is equal to $\gamma + 1$. This generates the three last types of solutions by changing $\gamma' - 1$ into γ'. □

The solutions of Lemma 2 induces cycles of length 6. These 6-cycles have their edges labeled by *dimensions* as illustrated on Figure 2 (the generator $2^i - 1$ induces edges in dimension i). However, cycle (b) and cycle (d) are isomorphic (just travel (b) clockwise and (d) counterclockwise from the black node, that is the second and the fourth types of solutions induces the same labeled cycle. In the following, only cycles (a), (b), and (c) will be considered.

Notation. The number of blocks of consecutive 1's in the binary representation of n will be denoted by $B_1(n)$.

For instance $B_1((1101100111010)_2) = 4$, $B_1((100)_2) = 1$, $B_1((101)_2) = 2$, and $B_1((0)_2) = 0$. Integers of the form

$$n = (\overline{1}\,\underbrace{00\ldots00}\,\overline{1}\,\underbrace{00\ldots00}\,\overline{0}\,\overbrace{11\ldots11}\,\underline{0}\,\overbrace{11\ldots11}\,\underline{0}\,\overbrace{11\ldots11}\,\underline{1}\,\underbrace{00\ldots00})_2$$

satisfy $B_1(n) = 5$, and there is a solution to Equation 2 for $\alpha = 1$ with x_1, x_3, x_5 equal to the underlined bit-positions, and x_0, x_2, x_4 equal to the over-lined bit-positions. We have the following lemma:

---
[1] A permutation σ of p symbols is a cyclic permutation if $\sigma(x_1, x_2, \ldots, x_p) = (x_2, \ldots, x_p, x_1)$.

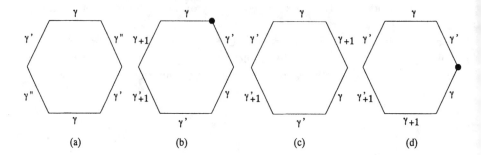

Fig. 2. The four types of solutions of Equation 2 for $\alpha = 0$.

Lemma 3. *If $B_1(n) \geq 6$, then Equation 2 has no solution for $\alpha \neq 0$.*

Proof. If $B_1(n) \geq 6$, then the sum of n and three powers of two cannot result in the sum of three powers of two. Indeed, the binary representation of $2^{x_0} + 2^{x_2} + 2^{x_4}$ has at most three 1-entries, that is $B_1(2^{x_0} + 2^{x_2} + 2^{x_4}) \leq 3$. On the other hand, for every n, and every y, $B_1(n + 2^y) \geq B_1(n) - 1$, and thus $B_1(n + 2^{x_1} + 2^{x_3} + 2^{x_5}) \geq B_1(n) - 3$. Moreover, for n such that $B_1(n) \geq 6$, the binary representation of $n + 2^{x_1} + 2^{x_3} + 2^{x_5}$ has at least four 1-entries. This completes the proof for $\alpha = 1$. The result holds for $\alpha = 2$ too because $B_1(2n) = B_1(n)$. □

Lemma 3 has an important consequence that is, for most of the integers n, Knödel graphs of order $2n$ have 6-cycles only of the form given on Figure 2. There are orders however for which Equation 2 has solutions for $\alpha \neq 0$. This is typically the case for $n = 2^k$. Actually, this special case deserves a particular interest motivated by the simplicity of the solution.

Lemma 4. *If $n = 2^k$ then every 4-cycle of the Knödel graph of order $2n$ is a labeled cycle of type*

$$(k, \gamma - 1, \gamma, \gamma - 1) \ for \ \gamma \in \{1, \ldots, k\}.$$

Proof. By similar arguments as in the proof of Lemma 1, one can check that 4-cycles exist if and only if there exist four generators $g_i = 2^{x_i} - 1$, $0 \leq i \leq 3$, $x_i \neq x_{i+1}$ (mod 4), satisfying $2^{x_0} + 2^{x_2} = 2^{x_1} + 2^{x_3} + \alpha n$ for $\alpha \in \{0, 1\}$. The equation $2^{x_0} + 2^{x_2} = 2^{x_1} + 2^{x_3}$ has no solution for $x_i \neq x_{i+1}$ (mod 4). Therefore, 4-cycles exist only for solutions of $2^{x_0} + 2^{x_2} = 2^{x_1} + 2^{x_3} + 2^k$, that is for $x_0 = k$, and $x_1 = x_3 = x_2 - 1$, $x_2 \in \{1, \ldots, k\}$. Thus the solution is $(x_0, x_1, x_2, x_3) = (k, \gamma - 1, \gamma, \gamma - 1)$, up to a square-cyclic permutation[2] of the x_i's. □

3 Recognizing Knödel Graphs

In order to recognize Knödel graphs, we use the following basic property:

---

[2] A permutation σ of p symbols is a square-cyclic permutation if $\sigma(x_1, x_2, x_3 \ldots, x_p) = (x_3, \ldots, x_p, x_1, x_2)$.

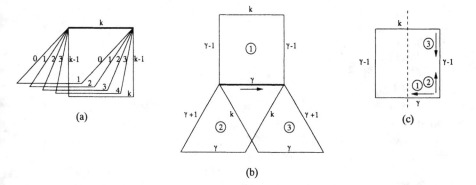

Fig. 3. The 4-cycles in a Knödel graph of order $2 \cdot 2^k$.

Lemma 5. *Let G be a graph whose some edges are colored by 0 or 1. G is a Knödel graph whose edges in dimension 0 and 1 are colored 0 and 1 respectively if and only if:*

1. *A path P using colors 0 and 1, alternatively, from an arbitrary node is Hamiltonian; and*
2. *Assuming the jth node of P is labeled $(j + 1 \bmod 2, \lfloor j/2 \rfloor)$, $j \geq 0$, we have, for every i and x, there is an edge connecting $(0, x)$ with $(1, x + 2^i - 1 \bmod n)$, and no extra edges.*

Proof. Let $G = (V_1, V_2, E)$ be a Knödel graph whose nodes are labeled (i, x), $i = 0, 1$ and $x \in \{0, \ldots, n - 1\}$ so that $V_1 = \{(0, x), x \in \{0, \ldots, n - 1\}\}$, and $V_2 = \{(1, x), x \in \{0, \ldots, n - 1\}\}$. Assume that the edges in dimension 0 and 1 of G are colored 0 and 1 respectively. Then the path P is Hamiltonian. Let u be the starting point of P. Since a Knödel graph is vertex-transitive, one can assume w.l.g. that $u = (1, 0)$. Therefore, the labeling induced by the path corresponds to the connections of a Knödel graph.

Conversely, let G be a graph whose some edges are colored by 0 or 1. Let P be a path using colors 0 and 1, alternatively, from an arbitrary node of G. Assume P is Hamiltonian, label the vertices according to the rule of the second property, and assume the connection rule fulfills. Then G is a Knödel graph by definition. □

Let us start with $n = 2^k$, $k \geq 2$. From Lemma 4, we know that there is only one type of labeled 4-cycle in a Knödel graph of order $2n$, namely $(k, \gamma - 1, \gamma, \gamma - 1)$, for $\gamma \in \{1, \ldots, k\}$. Therefore, for any edge of dimension k, there are k 4-cycles using that edge (see Figure 3(a)). For any edge of dimension $k - 1$, there are two 4-cycles using that edge (see Figure 3(b) where cycle 2 and cycle 3 are the same). For any edge of dimension γ, $\gamma \in \{1, \ldots, k - 2\}$, there are three 4-cycles using that edge (see Figure 3(b)). Finally, for any edge of dimension 0, there are two 4-cycles using that edge (the cycle 1 in Figure 3(b) does not exist for $\gamma = 0$). This counting yields the following corollary of Lemma 4:

Corollary 1. *There exists an $O(n \log^3 n)$-time algorithm to recognize Knödel graphs of order $2n = 2^{k+1}$, $k \geq 1$.*

Proof. Assume $k \geq 4$ (note that for any $k \leq k_0 = O(1)$, recognizing Knödel graphs of order 2^{k+1} can be done in constant time). Given an input graph $G = (V, E)$, we count the number $C_4(e)$ of 4-cycles passing through any edge $e \in E$. From the counting of the number of 4-cycles passing through an edge in a Knödel graph, if there is an edge e such that $C_4(e) \neq 2$, $C_4(e) \neq 3$, and $C_4(e) \neq k$ then G is not a Knödel graph. Otherwise, every edge e with $C_4(e) = 2$ is a candidate to be an edge of dimension 0 or dimension $k - 1$, and every edge e with $C_4(e) = k$ (recall $k \neq 2$ and $k \neq 3$) is a candidate to be an edge of dimension k. From Figure 3(b), edges in dimension $\gamma = k - 1$ are edges e such that $C_4(e) = 2$ and included in a 4-cycle containing an edge of dimension k which is not adjacent to e in this cycle. This allows to distinguish dimensions 0 and $k-1$. From dimensions 0 and k, one can identify dimension 1 by considering all paths of type $0, k, 0$. The end vertices of each such path are connected by an edge of dimension 1 (see Figure 3(a)). Color the edge of dimension 0 and 1 of G by colors 0 and 1, respectively, and check the conditions of Lemma 5.

This algorithm has a cost of $O(n \log^3 n)$ because, for every edge e, counting the number of 4-cycles using that edge takes at most a time of $O(\log^2 n)$ assuming node-adjacency testing in constant time. (The two end-vertices of every edge are both adjacent to $O(\log n)$ nodes. Thus, by testing all the possible edges between these nodes, one can determine the 4-cycles in $O(\log^2 n)$ time.) Checking the conditions of Lemma 5 takes $O(n \log n)$ time. □

Let us carry on our study by considering integers n such that $B_1(n) \geq 6$. In this case, one can apply Lemmas 2 and 3, and we are dealing with the four types of cycles of Figure 2 (recall that cycles (b) and (d) are isomorphic). Figure 2(a) implies that, for any edge of dimension γ, $\gamma \in \{0, \ldots, k\}$, there are $k(k - 1)$ 6-cycles of type (a) using that edge.

The contribution of cycles of type (b) in Figure 2 to each dimension is more difficult to calculate. We proceed as for the 4-cycles studied in the case $n = 2^k$. The counting for $n = 2^k$ can be formalized as follows. Consider again Figure 3. On Figure 3(c), there are four edges (whose one is of a fixed dimension, dimension k), and two possible senses of travel, clockwise and counterclockwise. For any γ such that $1 \leq \gamma \leq k - 1$, there are potentially four positions for γ. However, only three of them are valid because $\gamma \neq k$. Moreover, once the position of γ has been fixed, we have only two ways to travel around the cycle. This fact gives six possible labeled cycles for any edge to belong to. However, only three of these 4-cycles are distinct because there is a symmetry along the axis perpendicular to dimension k. This symmetry reduces the number of travels by a factor of 2: for each travel of Figure 3(c), there is a corresponding travel in Figure 3(b) starting in the direction indicated by the arrow.

Coming back to the 6-cycle of Figure 2(b), there are six edges, and two possible directions (clockwise and counterclockwise). Thus we get twelve possibilities to travel along the edges of a labeled 6-cycle. However, we actually get only

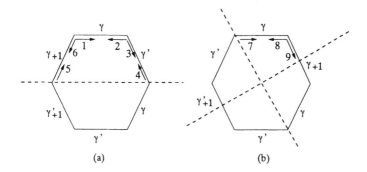

Fig. 4. (a) The six possible travels of the cycle of type (b) in Figure 2. (b) The three possible travels of the cycle of type (c) in Figure 2.

six possibilities for cycle (b) of Figure 2 because of a symmetry along the axis parallel to γ, γ' (see Figure 4(a)).

These six possibilities are:

$$
\left.
\begin{array}{l}
1.\ (\gamma \quad \gamma' \quad \gamma \quad \gamma' \ \ \gamma'+1\ \gamma+1) \\
2.\ (\gamma\ \gamma+1\ \gamma'+1 \quad \gamma' \quad \gamma \quad \gamma') \\
3.\ (\gamma \quad \gamma' \quad \gamma \quad \gamma+1\ \gamma'+1 \quad \gamma') \\
4.\ (\gamma \quad \gamma' \ \ \gamma'+1\ \gamma+1 \quad \gamma \quad \gamma')
\end{array}
\right\} \gamma, \gamma' \in \{0, \ldots, k-1\}, \gamma \neq \gamma';
$$

and

$$
\left.
\begin{array}{l}
5.\ (\gamma\ \gamma-1\ \gamma'\ \gamma-1\ \gamma'\ \gamma'+1) \\
6.\ (\gamma\ \gamma'+1\ \gamma'\ \gamma-1\ \gamma'\ \gamma-1)
\end{array}
\right\} \gamma \in \{1, \ldots, k\}, \gamma' \in \{0, \ldots, k-1\}, \gamma' \neq \gamma-1.
$$

Note that solutions 1 and 3 are the same if $\gamma' = \gamma+1$, solutions 2 and 4 are the same if $\gamma' = \gamma+1$, solutions 3 and 4 are the same if $\gamma' = \gamma-1$, and solutions 5 and 6 are the same if $\gamma' = \gamma-2$.

We do the same analysis for 6-cycles of type (c) in Figure 2. The counting uses the fact that Figure 2(c) is symmetric along the axis perpendicular to the edges $\gamma+1$ and $\gamma'+1$, and along the axis parallel to $\gamma+1, \gamma'+1$ (see Figure 4(b)). Therefore, we get three new possibilities:

$$
\left.
\begin{array}{l}
7.\ (\gamma\ \gamma+1 \quad \gamma \quad \gamma'\ \gamma'+1 \quad \gamma') \\
8.\ (\gamma \quad \gamma' \ \ \gamma'+1\ \gamma' \quad \gamma \quad \gamma+1)
\end{array}
\right\} \gamma, \gamma' \in \{0, \ldots, k-1\}, \gamma \neq \gamma'; \text{ and}
$$

$$
9.\ (\gamma\ \gamma-1\ \gamma'\ \gamma'+1\ \gamma'\ \gamma-1), \gamma \in \{1, \ldots, k\}, \gamma' \in \{0, \ldots, k-1\}, \gamma' \neq \gamma-1.
$$

Note that solutions 1 and 7 are the same if $\gamma' = \gamma+1$, solutions 2 and 7 are the same if $\gamma' = \gamma-1$, and solutions 3 and 7 are the same if $\gamma' = \gamma+1$. Similarly solutions 1 and 8 are the same if $\gamma' = \gamma-1$, solutions 2 and 8 are the same if $\gamma' = \gamma+1$, and solutions 4 and 8 are the same if $\gamma' = \gamma+1$. Finally, solutions 5 and 9 are the same if $\gamma' = \gamma-2$, and solutions 6 and 9 are the same if $\gamma' = \gamma-2$.

Now we can count the number of 6-cycles using a given edge of a Knödel graph.

Lemma 6. *Assume $B_1(n) \geq 6$, and let e be an edge of a Knödel graph of order $2n$ with $k \geq 4$. Let $C_6(e)$ denote the number of 6-cycles using edge e. We have:*

$$C_6(n) = \begin{cases} k^2 + 5k - 10 & \text{if } e \text{ is of dimension } 0 \\ k^2 + 8k - 16 & \text{if } e \text{ is of dimension } 1 \\ k^2 + 8k - 18 & \text{if } e \text{ is of dimension } \gamma, \ 2 \leq \gamma \leq k - 2 \\ k^2 + 8k - 14 & \text{if } e \text{ is of dimension } k - 1 \\ k^2 + 2k - 5 & \text{if } e \text{ is of dimension } k \end{cases}$$

Proof. Let e be an edge of dimension γ, $0 \leq \gamma \leq k$. Let us count the contribution to that edge of the nine travels identified in Figure 4. We get $6k - 10$ cycles for dimension 0, $9k - 16$ cycles for dimension 1, $9k - 18$ cycles for dimension γ, $2 \leq \gamma \leq k - 2$, $9k - 14$ cycles for dimension $k - 1$, and $3k - 5$ cycles for dimension k. Then add $k(k - 1)$ cycles from the solutions of type (a) in Figure 2 to get the claimed result. □

Corollary 2. *There exists an $O(n \log^5 n)$-time algorithm to recognize Knödel graphs of order $2n$, for all n such that $B_1(n) \geq 6$.*

Proof. Assume $k \geq 4$. From Lemma 6, one can identify edges of dimensions 0 and 1 in $O(n \log^5 n)$-time. In time $O(n \log n)$, one can check the conditions of Lemma 5. □

Theorem 1. *There exists an $O(n \log^5 n)$-time algorithm to recognize Knödel graphs of any order.*

Sketch of the proof. (The complete proof is given in the full paper [4].) From Corollaries 1 and 2, the theorem holds for n power of two, or n such that $B_1(n) \geq 6$. Thus, assume that n satisfies $B_1(n) < 6$, $n \neq 2^k$. Assume moreover that $n \neq 2^{k+1} - 1$ (this latter case deserves to be treated separately because the Knödel graph with $n = 2^{k+1} - 1$ is edge-transitive [10]). The key argument is that, for almost all such n, if $C_6(e)$ denotes the number of 6-cycles passing through an edge e of a Knödel graph of order n, then $C_6(e) \neq C_6(e')$ for any e and e' of dimensions 0 and k respectively, and $C_6(e) \neq C_6(e')$ for any e dimension 0 or k, and any e' of dimension i, $i \neq 0$ and $i \neq k$.

The difficulty of the proof comes from the fact that, if $B_1(n) < 6$, then Equation 2 has solutions for $\alpha \neq 0$, and proceeding to a precise counting of the number of 6-cycles passing through the edges of a Knödel graph is tricky. Anyway, we were able to prove that dimension 0 and dimension k can be identified by such counting. From the knowledge of the edges of dimension 0 and k, we have shown that one can determine the set of edges of dimension 1. Then it remains only to check the conditions of Lemma 5.

To summarize, the algorithm is the following:

Algorithm *Recognize.*
Input: a regular graph $G = (V, E)$ of $2n$ vertices, and degree $k = \lfloor \log_2 n \rfloor$;
Output: tell whether or not G is isomorphic to the Knödel graph K of order $2n$.

Phase 1. For every $i \in \{0, \ldots, k\}$, compute p_i = number of 6-cycles passing through any edge of dimension i of K;

Phase 2. For every $e \in E$, compute $C_6(e)$ = number of 6-cycles passing through the edge e of G;

Phase 3. Identify the two sets $S_0 \subset E$, and $S_1 \subset E$, of edges of G that are possibly edges of dimension 0 and 1 in K, respectively;

Phase 4. Check the conditions of Lemma 5.

The first phase has a cost of $O(log^5 n)$, assuming node-adjacency testing in constant time. The second phase has a cost at most $O(n \log^5 n)$ because the degree of every vertex is $O(\log n)$. The third and the fourth phase cost $O(n \log n)$ time.

To prove the correctness of Algorithm *Recognize*, the difficult part is to prove that Phase 3 is doable. This is formaly proved in the complete version of the proof (see [4]). The case of Knödel graphs of order $2^{k+2} - 2$ deserves a particular attention due to the edge-transitivity of such graphs. We have proved that one can identify dimensions 0 and k by counting the number of 6-cycles traversing a path of length 3. □

Corollary 3. *If $n \neq 2^{k+1} - 1$, then Knödel graphs of order $2n$ are non edge-transitive. (From [10], Knödel graphs of order $2^{k+2} - 2$ are edge-transitive.)*

Proof. The proof of correctness of Algorithm *Recognize*, given in [4], uses the fact that, for $n \neq 2^{k+1} - 1$, the number of 6-cycles using edges of different dimensions is not the same. □

4 Conclusion and Further Research

In this paper, we have shown that Knödel graphs can be recognized in $O(n \log^5 n)$ time. The same result holds for Fibonacci graphs (see [4]). The natural question arising in this field is to ask for which sequences g_i the same result holds. Let us formalize this question. Given an infinite and increasing sequence of integers $\Gamma = (g_i)_{i \geq 0}$, consider the sequence $(G_n^\Gamma)_{n \geq 0}$ of circulant digraphs of order n such that, for any $n \geq 0$, G_n^Γ has generators $g_0, g_1, \ldots, g_k$ where k is the largest integer such that $g_k \leq n - 1$ (in other words, $g_k \leq n - 1 < g_{k+1}$). Then let $(H_n^\Gamma)_{n \geq 0}$ be the corresponding sequence of bipartite incident-graphs of order $2n$. The problem is the following:

Problem 1.
INSTANCE: An integer n, and a graph G of $2n$ vertices;
QUESTION: Is G isomorphic to H_n^Γ?

Note that the sequence Γ is fixed in Problem 1, and thus that this problem is different from the problem of deciding whether a graph is isomorphic to the bipartite incident-graph of a circulant digraph, or to decide whether two bipartite incident-graphs of circulant digraphs are isomorphic. These two latter problems are known to be difficult, even if they might be simpler than the problem of deciding whether a graph is a Cayley graph.

We have seen that Problem 1 can be solved polynomially if $\Gamma = (2^i - 1)_{i \geq 0}$ or if $\Gamma = (F(i+1) - 1)_{i \geq 0}$. The question is: for which family the techniques used to solve the problem for Knödel graphs and Fibonacci graphs can be extended, and how? Actually, as soon as we know how to solve Equation 1, then we are able to enumerate the 6-cycles, and to use the same techniques as the techniques used for Knödel graphs and Fibonacci graphs. We let as an open problem the characterization of the sequences Γ for which Problem 1 can be solved using the same techniques as for Knödel graphs and Fibonacci graphs.

Acknowledgments. The authors are thankful to Lali Barrière, Guillaume Fertin, Bernard Mans, and André Raspaud for their valuable helps and comments, and to Jean-Paul Allouche for fruitful discussions. The authors are also thankful to the anonymous referees for many corrections on the original version.

References

1. A. Ádám. Research problem 2-10. *J. Combin. Theory*, 2:393, 1967.
2. B. Alspach and T. Parsons. Isomorphism of circulant graphs and digraphs. *Discrete Mathematics*, 25:97–108, 1979.
3. J-C. Bermond, F. Comellas, and F. Hsu. Distributed loop computer networks: a survey. *Journal of Parallel and Distributed Computing*, 24:2–10, 1995.
4. J. Cohen, P. Fraigniaud, and C. Gavoille. Recognizing bipartite incident-graphs of circulant digraphs. Technical report, Laboratoire de Recherche en Informatique, http://www.lri.fr/~pierre Univ. Paris-Sud, France, 1999.
5. G. Cybenko, D.W. Krumme, and K.N. Venkataraman. Gossiping in minimum time. *SIAM Journal on Computing*, 21(1):111–139, 1992.
6. B. Elspas and J. Turner. Graphs with circulant adjacency matrices. *J. Comb. Theory*, 9:229–240, 1970.
7. S. Even and B. Monien. On the number of rounds necessary to disseminate information. In *First ACM Symposium on Parallel Algorithms and Architectures (SPAA)*, 1989.
8. P. Fraigniaud and E. Lazard. Methods and Problems of Communication in Usual Networks. *Discrete Applied Mathematics*, 53:79–133, 1994.
9. S.M. Hedetniemi, S.T. Hedetniemi, and A. Liestman. A survey of gossiping and broadcasting in communication networks. *Networks*, 18:319–349, 1986.
10. M-C. Heydemann, N. Marlin, and S. Pérennes. Cayley graphs with complete rotations. Technical Report 1155, LRI, Bât. 490, Univ. Paris-Sud, 91405 Orsay cedex, France, 1997. Submitted to the European Journal of Combinatorics.
11. J. Hromković, R. Klasing, B. Monien, and R. Peine. Dissemination of information in interconnection networks (broadcasting and gossiping). In Ding-Zhu Du and D. Frank Hsu, editors, *Combinatorial Network Theory*, pages 125–212. Kluwer Academic, 1995.
12. W. Knodel. New gossips and telephones. *Discrete Mathematics*, 13:95, 1975.
13. R. Labahn and I. Warnke. Quick gossiping by telegraphs. *Discrete Mathematics*, 126:421–424, 1994.

14. B. Mans, F. Pappalardi, and I. Shparlinski. On the Ádám conjecture on circulant graphs. In *Fourth Annual International Computing and Combinatorics Conference (Cocoon '98)*, Lecture Notes in Computer Science. Springer-Verlag, 1998.
15. M. Muzychuk and G. Tinhoffer. Recognizing circulant graphs of prime order in polynomial time. *The electronic journal of combinatorics*, 3, 1998.
16. A. Nayak, V. Accia, and P. Gissi. A note on isomorphic chordal rings. *Information Processing Letters*, 55:339–341, 1995.
17. I. Ponomarenko. Polynomial-time algorithms for recognizing and isomorphisn testing of cyclic tournaments. *Acta Applicandae Mathematicae*, 29:139–160, 1992.
18. V.S. Sunderam and P. Winkler. Fast information sharing in a distributed system. *Discrete Applied Mathematics*, 42:75–86, 1993.

Optimal Cuts for Powers of the Petersen Graph[*]

Sergei L. Bezrukov[1], Sajal K. Das[2], and Robert Elsässer[3]

[1] Dept. of Math. and Computer Science,
University of Wisconsin-Superior, Superior, WI 54880-4500, USA.
[2] Dept. of Computer Science,
University of North Texas, Denton, TX 76203, USA.
[3] Dept. of Math. and Computer Science,
University of Paderborn, 33095 Paderborn, Germany.

Abstract. In this paper we introduce a new order on the set of n-dimensional tuples and prove that this order preserves nestedness in the edge isoperimetric problem for the graph P^n, defined as the nth cartesian power of the well-known Petersen graph. Thus, we show, that there is a graph for which powers the solution of the edge isoperimetric problem preserve nestedness and it is different from the lexicographic order. With respect to this result we determine the cutwidth and wirelength of P^n. These results are then generalized to the cartesian product of P^n and the m-dimensional binary hypercube.

1 Introduction

The subject of the paper is the *edge-isoperimetric problem*, which consists of finding a subset of vertices of a given graph, such that the number of edges separating this subset from its complement, also called edge cut, has minimal size among all subsets of the same cardinality.

For a graph $G = (V_G, E_G)$ with vertex set V_G, edge set E_G and $A \subseteq V_G$ denote

$$\theta_G(A) = \{(u, v) \in E_G \mid u \in A, v \notin A\}.$$
$$\theta_G(t) = \min_{|A|=t} |\theta_G(A)|.$$

Thus, for a given t, where $t = 1, \ldots, |V_G|$, we consider the problem of finding a subset A of vertices of G such that $|A| = t$ and $|\theta_G(A)| = \theta_G(t)$. Such subsets are called *optimal*. We say that optimal subsets are *nested* if there exists a total order $\mathcal{O}$ on the set V_G such that for any $t = 1, \ldots, |V_G|$ the collection of the first t vertices in this order is an optimal subset. In this case the order $\mathcal{O}$ is called an *optimal order*. Edge isoperimetric problems often and naturally arise in various problems on networks such as bisection width, graph embedding or graph partitioning [4,5,16].

---

[*] Partially supported by European Union ESPRIT LTR Project 20244 (ALCOM-IT) and German Research Foundation (DFG) Project SFB-376.

Widmayer et al. (Eds.): WG'99, LNCS 1665, pp. 228–239, 1999.

In this paper we concentrate on the graphs representable as cartesian products. Given two graphs $G = (V_G, E_G)$ and $H = (V_H, E_H)$, their *cartesian product* is defined as a graph $G \times H$ with the vertex-set $V_G \times V_H$ and the edge-set $\{((x, y), (u, v))\}$ iff either $x = u$ and $(y, v) \in E_H$, or $(x, u) \in E_G$ and $y = v$. A graph $G^n = G \times G \times \cdots \times G$ is called the nth *cartesian power* of G.

Well-known examples of product graphs include hypercubes, grids and tori which are used extensively as the interconnection network topologies for multiprocessor architectures. The advantage of constructing a network with the help of the cartesian product operation is that it provides high scalability of the hardware; small diameter, vertex degree and number of edges as compared with the number of vertices; simple message routing algorithms, and good fault-tolerant properties.

Consider the edge isoperimetric problem for the cartesian powers G^n of a graph G. Such problems have been well studied for cliques, i.e., $G = K_p$. Representing the vertices of G^n as n-dimensional tuples, the results of Harper [12] and Lindsey [13] imply that the lexicographic order is an optimal order. Here by the *lexicographic order* we mean the following ordering of n-tuples: we say that a tuple $(x_1, \ldots, x_n)$ is lexicographically larger than $(y_1, \ldots, y_n)$ iff there exists an index i such that $x_j = y_j$ for $1 \le j < i$ and $x_i > y_i$.

These old classical results can be extended to various directions. For instance, taking an (infinite) path instead of a clique leads to a grid. In this case the edge-isoperimetric problem also has nested solutions [1,7]. It is further shown in [3] that the results of [1,7] can be extended to the products of arbitrary trees. The *order*, $\mathcal{G}$, providing the nestedness in this case is much more complicated with respect to the lexicographic order. For the definition of this order and further details, readers are referred to [1,4,7].

The order $\mathcal{G}$ and the lexicographic order are the only known orders, which provide nestedness for products of some graphs in the edge isoperimetric problems. However, as shown in [4] the order $\mathcal{G}$ works for products of trees only. Therefore, two natural questions arise: *(i) for products of which other graphs is the lexicographic order optimal with respect to the edge isoperimetric problems; and (ii) which other optimal orders can one expect?*

In [10], the author considered the cartesian products of k-regular graphs with an even number of vertices p such that $k \ge (3/4)p$. He has shown that for any such graph the size of the edge cut separating a set from its complement is at least as large as for the graph H_p^k obtained from K_p by removing $p - k - 1$ perfect matchings. It turns out that for $k \ge (3/4)p$, the edge isoperimetric problem for cartesian powers of H_p^k has a nested structure of solutions provided by the lexicographic order. Similar results can be derived for powers of complete bipartite graphs with deleted perfect matchings. It is interesting to note that violating the condition $k \ge (3/4)p$ leads to the absence of a nested structure of solutions. The paper [10] also studies the powers of complete p-partite graphs and shows that the lexicographic order still works and provides nestedness. Thus, this extends the result of Ahlswede and Cai [2], who studied powers of complete bipartite graphs.

It is worth mentioning that the lexicographic order yields the so-called *local-global principle* discovered by Ahlswede and Cai [2]. Following the main result of [2], if the lexicographic order is optimal for G^2 then so is for G^n for any $n \geq 3$. The main difficulty in applying this powerful theorem is to establish that the lexicographic order is optimal for the second power of considered graphs. For this no general methods are known yet. See [4] for the local-global principles for some other orders.

In all of the preceding results, the degree of the underlying regular graph is relatively large which is intuitively necessary for the lexicographic order to work. Now the question is what happens if the graphs in the product have smaller degree. For instance, considering the regular graphs of degree 1, we get the hypercubes for which the lexicographic order still works and provides nestedness. For powers of regular graphs of degree 2, e.g. a torus, there is no nested solution in general [11]. Tori are well studied and some isoperimetric inequalities are known for them [6], which are sharp enough for most practical applications.

The next step is to consider the powers of regular graphs of degree 3, for which a huge collection of non-isomorphic graphs exist. However, to the best of our knowledge, none of them (excluding 3-dimensional cube) has been studied with respect to the edge isoperimetric problems. We concentrate on the cartesian powers P^n of the Petersen graph P, which is a regular graph of degree 3 and diameter 2 as shown in Fig. 1a). Note that P is a vertex-symmetric as well as an edge-symmetric graph.

The graphs P^n, known as *folded Petersen networks*, were proposed and extensively studied by Öhring and Das [9,15] as an efficient interconnection network topology for multiprocessors. The folded Petersen network compares favorably with several other product networks in terms of topological properties such as degree, diameter or connectivity. Let us compare, for example, P^n with the $3n$-dimensional cube, Q^{3n}. Both graphs have a vertex degree $3n$. However, P^n has 10^n and Q^{3n} only 8^n vertices. Moreover, the diameter of P^n is $2n$ while that of Q^{3n} is $3n$. For more details, refer to [15]. Multiple disjoint paths between two arbitrary nodes provide a measure of fault-tolerance of a network. It is easily seen that between any two nodes in P there exist three node-disjoint paths. Thus, in P^n there are $3n$ node disjoint paths between any two nodes [15]. Therefore, P^n has best possible fault-tolerant properties among all networks of the same degree.

Interestingly, the bisection width of P^n is known exactly [4]. This fact also stimulates the cut problem having two parts of different cardinalities. It is an important property of a graph from the viewpoint of its minimum layout area in VLSI.

In this paper we answer several questions raised above. More precisely, we introduce a new order $\mathcal{P}^n$ on the set of n-dimensional tuples (which we call the *Petersen order*) and show that this order provides nestedness in the edge isoperimetric problem for P^n, the powers of the Petersen graph. This result allows us to compute the cutwidth and wirelength of P^n exactly, which are respectively defined as the maximum and the mean value of the minimum cut separating the graph into two parts. We extend these results to the product graph $P_m^n = P^n \times Q^m$ where Q^m is the m-dimensional hypercube. The graphs P_m^n, called the *folded Petersen cubes*, have been first studied by Öhring and Das [14]

The paper is organized as follows. In the next section we reformulate the problem and introduce the Petersen order $\mathcal{P}^n$ on the vertex set of the graph, P^n. Section 3 shows that for each t, where $t = 1, \ldots, 10^n$, the set $\mathcal{F}^n(t)$ represented by the initial segment of the order $\mathcal{P}^n$ of length t is an optimal subset. Section 4 is devoted to computing the cutwidth and the wirelength of P^n. Section 5 presents the extensions of these results to the graphs P_m^n.

2   The Petersen Order $\mathcal{P}^n$ and Its Properties

Denote for a Graph $G = (V_G, E_G)$

$$I_G(A) = \{(u,v) \in E_G \mid u, v \in A\}.$$
$$I_G(t) = \max_{|A|=t} |I_G(A)|.$$

It is easily seen that for k-regular graphs holds $\theta_G(A) + 2 \cdot I_G(A) = k|A|$. Thus, we can reformulate the isoperimetric problem stated above as follows: for any $t = 1, \ldots, |V_G|$ find a subset A of vertices of G such that $|A| = t$ and $|I_G(A)| = I_G(t)$.

For regular graphs this problem is equivalent to the edge-isoperimetric problem in the sense that a solution for one also becomes a solution for the other. Hence, optimal subsets w.r.t. θ_G are also optimal w.r.t. I_G.

The order $\mathcal{P}^1$ is shown in Fig. 1a) and it is an easy exercise to ensure oneself that it provides nested optimal subsets of the Petersen graph, P.

Now, by induction on n, we define the order $\mathcal{P}^n$ on the vertex set of P^n for $n \geq 2$. For this purpose let us first define the successor for any vector $(a_1, \ldots, a_n) \in V_{P^n}$ as follows. Denoting $(a_2', \ldots, a_n') = succ(a_2, \ldots, a_n)$ in the order $\mathcal{P}^{n-1}$, we define

$$succ(a_1, \ldots, a_n) = \begin{cases} (a_1 + 1, a_2, \ldots, a_n), & \text{if } a_1 \in \{0, 3, 5, 8\} \\ (a_1 - 1, a_2', \ldots, a_n'), & \text{if } a_1 \in \{1, 4, 6, 9\} \ \& \\ & \text{if } (a_2, \ldots, a_n) \neq (9, \ldots, 9) \\ (a_1, a_2', \ldots, a_n'), & \text{if } a_1 \in \{2, 7\} \ \& \\ & \text{if } (a_2, \ldots, a_n) \neq (9, \ldots, 9) \\ (a_1 + 1, 0, \ldots, 0), & \text{if } a_1 \in \{1, 2, 4, 6, 7\} \ \& \\ & \text{if } (a_2, \ldots, a_n) = (9, \ldots, 9). \end{cases}$$

The order $\mathcal{P}^2$ is illustrated in Fig. 1b). The vertices of the graph $P^2 = P \times P$ are represented as the entries of a 10×10 matrix $\{a_{i,j}\}$, where $i, j = 0, ..., 9$. We assume in this figure that the entry $a_{0,0}$ is in the bottom left corner of the matrix. Furthermore, we assume that the elements $a_{0,0}, ..., a_{9,0}$ of the bottom

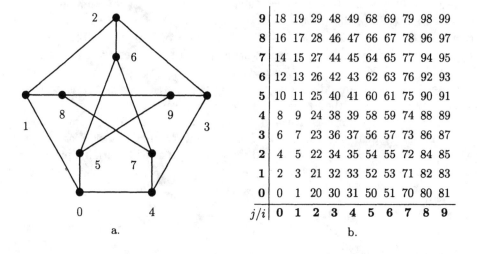

j/i	0	1	2	3	4	5	6	7	8	9
9	18	19	29	48	49	68	69	79	98	99
8	16	17	28	46	47	66	67	78	96	97
7	14	15	27	44	45	64	65	77	94	95
6	12	13	26	42	43	62	63	76	92	93
5	10	11	25	40	41	60	61	75	90	91
4	8	9	24	38	39	58	59	74	88	89
3	6	7	23	36	37	56	57	73	86	87
2	4	5	22	34	35	54	55	72	84	85
1	2	3	21	32	33	52	53	71	82	83
0	0	1	20	30	31	50	51	70	80	81

a.   b.

Fig. 1. (a) The order $\mathcal{P}^1$; (b) the order $\mathcal{P}^2$

row and the elements $a_{0,0}, ..., a_{0,9}$ of the leftmost column represent the vertices of the multiplicands of the product (i.e., vertices of P) taken in the order $\mathcal{P}^1$. The value of the matrix element $a_{i,j}$ is the number of the corresponding vertex of the graph P^2 in the order $\mathcal{P}^2$, as shown in Fig. 1b). With the help of a computer we have verified that the set $\mathcal{F}^2(t)$, which is represented by the initial segment of the order $\mathcal{P}^2$ of length t, is optimal for any t in the range $1 \leq t \leq 100$.

Using induction on n it is easy to show that starting with the vector $(0, \ldots, 0)$ and following the successors, one can achieve any other vector $(a_1, \ldots, a_n) \in V_{P^n}$. Thus, the Petersen order $\mathcal{P}^n$ is well defined.

For $\mathbf{a}, \mathbf{b} \in V_{P^n}$ we write $\mathbf{a} > \mathbf{b}$ if the vertex $\mathbf{a}$ is greater than $\mathbf{b}$ in the order $\mathcal{P}^n$. By analyzing the inductive definition of the order $\mathcal{P}^n$, the following property, called *consistency* in [8], can be verified.

Lemma 1 *Let $\mathbf{a} = (a_1, \ldots, a_n) > (b_1, \ldots, b_n) = \mathbf{b}$ and $a_i = b_i$ for some i. Furthermore, let vectors $\tilde{\mathbf{a}}$ and $\tilde{\mathbf{b}}$ be obtained from $\mathbf{a}$ and $\mathbf{b}$ respectively by omitting their i^{th} entries. Then $\mathbf{a} > \mathbf{b}$ iff $\tilde{\mathbf{a}} > \tilde{\mathbf{b}}$.*

For $A \subseteq V_{P^n}$, $i = 1, \ldots, n$, and $j = 0, \ldots, 9$ let us denote

$$P_i^n(j) = \{(\xi_1, \ldots, \xi_n) \in V_{P^n} \mid \xi_i = j\}$$
$$A_i(j) = A \cap P_i^n(j).$$

We say that A is i-compressed if for any $j = 0, \ldots, 9$ the subset $A_i(j)$ is an initial segment of the set $P_i^n(j)$ in the order $\mathcal{P}^{n-1}$. We call the set A *compressed* if A is i-compressed for $i = 1, \ldots, n$. Standard arguments provide that there is no loss of generality in assuming that an optimal set A is compressed. We used the compression to verify the 2-dimensional solution of our problem with the help of a computer. Larry Harper noted that in this case there are $\binom{20}{10} = 187,756$ compressed sets. The complete choice of such a size is doable by computer but without compression there are $2^{100} \approx 1.3 \times 10^{30}$ possibilities, a prohibitively large number.

3   Proof of the Main Result

Let $\mathbf{a} = (a_1, \ldots, a_n)$ be the largest vertex of the subset $A \subseteq V_{P^n}$ in the order $\mathcal{P}^n$ and let $\mathbf{b} = (b_1, \ldots, b_n)$ be the smallest vertex of the complement $V_{P^n} \setminus A$ in this order. If $A \neq \mathcal{F}^n(t)$ then $\mathbf{a} > \mathbf{b}$. Since A is compressed, $a_i \neq b_i$ for $i = 1, \ldots, n$. Moreover, the sets $A \setminus \mathbf{a}$ and $(A \setminus \mathbf{a}) \cup \mathbf{b}$ are compressed.

For a graph G and $i = 1, \ldots, |V_G|$, let us denote $\delta_G(i) = I_G(i) - I_G(i-1)$ and assume $\delta_G(0) = 0$. By analyzing Fig. 1a) the entries of the following table can be easily verified for the Petersen graph P.

i	0	1	2	3	4	5	6	7	8	9	10
$I_P(i)$	0	0	1	2	3	5	6	8	10	12	15
$\delta_P(i)$	0	0	1	1	1	2	1	2	2	2	3

Table 1: The values of $I_P(i)$ and $\delta_P(i)$

Lemma 2 *With the above notations one has*

$$|I_{P^n}(A)| - |I_{P^n}((A \setminus \mathbf{a}) \cup \mathbf{b})| = \sum_{i=1}^{n}(\delta_P(b_i) - \delta_P(a_i)).$$

The lemma follows from the observation that for a compressed set A it holds:

$$|I_{P^n}(A)| = \sum_{(x_1, \ldots, x_n) \in A} \sum_{i=1}^{n} \delta_P(x_i).$$

Now we are ready to prove the main result.

Theorem 1 *For any $n \geq 1$ and t, $t = 1, \ldots, 10^n$, the set $\mathcal{F}^n(t)$ is optimal w.r.t. edge isoperimetric problem, where $\mathcal{F}^n(t)$ is represented by the initial segment of the order $\mathcal{P}^n$ of length t.*

Proof: We prove the theorem by induction on n. The case $n = 1$ is trivial and the case $n = 2$ follows from the mentioned results based on a computer search. Therefore, let us proceed with $n \geq 3$.

From the definition of the order $\mathcal{P}^n$ it can be concluded that if $\mathbf{a} > \mathbf{b}$, then one of the following five (disjoint) cases occurs:

a. $a_1 - 1 > b_1$;
b. $a_1 - 1 = b_1$ and $b_1 \in \{1, 2, 4, 6, 7\}$;

c. $a_1 - 1 = b_1$, $b_1 \in \{0, 3, 5, 8\}$ and $(a_2, \ldots, a_n) \geq (b_2, \ldots, b_n)$;

d. $a_1 = b_1$ and $(a_2, \ldots, a_n) > (b_2, \ldots, b_n)$;

e. $a_1 + 1 = b_1$, $b_1 \in \{1, 4, 6, 9\}$ and $(a_2, \ldots, a_n) > (b_2, \ldots, b_n)$.

Let A be an optimal compressed set. Following the cases above we can show that in many of them the condition $\mathbf{a} \in A$ implies $\mathbf{b} \in A$ due to the compression. The general strategy to show this is to find a vertex $\mathbf{c}$ satisfying $\mathbf{a} > \mathbf{c} > \mathbf{b}$ such that the vectors $\mathbf{a}, \mathbf{c}$ and $\mathbf{c}, \mathbf{b}$ have an equal entry. On the other hand, if such a vector $\mathbf{c}$ does not exist, using Lemma 2, we show that replacing the vertex $\mathbf{a}$ with $\mathbf{b}$ yields a set B satisfying $|I_{P^n}(B)| \geq |I_{P^n}(A)|$. Clearly, after a finite number of such replacements one can transform A into $\mathcal{F}^n(|A|)$. To prove the theorem, we have to consider each of the above five cases. Due to the space limitation, we present only one of these cases, the other four can be proved in a similar way and will appear in the full version of this paper.

Case e. Assume $a_1 + 1 = b_1$ and $b_1 \in \{1, 4, 6, 9\}$. Denoting $(c_2, \ldots, c_n) = succ(b_2, \ldots, b_n)$, we have $(a_1, c_2, \ldots, c_n) \in A$.

e1. Assume $b_2 \in \{0, 3, 5, 8\}$. Then $succ(b_2, \ldots, b_n) = (b_2 + 1, b_3, \ldots, b_n)$. Since $(a_1, b_2 + 1) > (a_1 + 1, b_2)$, we get

$$\mathbf{a} = (a_1, \ldots, a_n) \geq (a_1, b_2 + 1, b_3, \ldots, b_n) > (a_1 + 1, b_2, b_3, \ldots, b_n) = \mathbf{b}.$$

Therefore, $\mathbf{b} \in A$, a contradiction.

e2. Assume $b_2 \in \{2, 7\}$. If $(b_3, \ldots, b_n) \neq (9, \ldots, 9)$, then let $(d_3, \ldots, d_n) = succ(b_3, \ldots, b_n)$. One has $succ(b_2, \ldots, b_n) = (b_2, d_3, \ldots, d_n)$. Since $(a_1, d_3, \ldots, d_n) > (a_1 + 1, b_3, \ldots, b_n)$, using Lemma 1 we write

$$\mathbf{a} = (a_1, \ldots, a_n) \geq (a_1, b_2, d_3, \ldots, d_n) > (a_1 + 1, b_2, b_3, \ldots, b_n) = \mathbf{b}.$$

This implies $\mathbf{b} \in A$.

Assume $(b_3, \ldots, b_n) = (9, \ldots, 9)$. Then $succ(b_2, \ldots, b_n) = (b_2 + 1, 0, \ldots, 0) \in A$. First assume that $(a_2, \ldots, a_n) \geq succ(succ(b_2, \ldots, b_n)) = (b_2 + 2, 0, \ldots, 0)$. Then $(a_1, b_2 + 2, 0, \ldots, 0) \in A$ since A is 1-compressed. Thus

$$\mathbf{a} \geq (a_1, b_2 + 2, 0 \ldots, 0) > (a_1 + 1, b_2 + 1, 0, \ldots, 0) > (a_1 + 1, b_2, 9, \ldots, 9) = \mathbf{b}.$$

This implies $\mathbf{b} \in A$. If $(a_2, \ldots, a_n) = succ(b_2, \ldots, b_n) = (b_2 + 1, 0 \ldots, 0)$, then $\mathbf{a} = (b_1 - 1, b_2 + 1, 0, \ldots, 0)$ and $\mathbf{b} = (b_1, b_2, 9, \ldots, 9)$. Replacing $\mathbf{a}$ with $\mathbf{b}$ yields a set B such that

$$|I_{P^n}(B)| - |I_{P^n}(A)| = (\delta_P(b_1) + \delta_P(b_2) + (n - 2) \cdot \delta_P(9))$$
$$- (\delta_P(b_1 - 1) + \delta_P(b_2 + 1))$$
$$= (\delta_P(b_1) - \delta_P(b_1 - 1)) + (\delta_P(b_2)$$
$$- \delta_P(b_2 + 1)) + 3(n - 2)$$
$$= 3(n - 2) + 1,$$

since $\delta_P(b_1) - \delta_P(b_1 - 1) = 1$ for $b_1 \in \{1, 4, 6, 9\}$ and $\delta_P(b_2) - \delta_P(b_2 + 1) = 0$ for $b_2 \in \{2, 7\}$.

e3. Assume $b_2 \in \{1, 4, 6, 9\}$ and $(b_3, \ldots, b_n) = (9, \ldots, 9)$. Now $b_2 \neq 9$ since $(a_2, \ldots, a_n) > (b_2, \ldots, b_n)$. Then $succ(b_2, \ldots, b_n) = (b_2 + 1, 0, \ldots, 0)$. First assume

$$(a_2, \ldots, a_n) \geq (e_2, \ldots, e_n) = succ(succ(b_2, \ldots, b_n)).$$

Lemma 1 and the fact that A is 1-compressed imply $(a_1, e_2, \ldots, e_n) \in A$. Now if $b_2 \in \{1, 6\}$ then $(e_2, \ldots, e_n) = (b_2 + 1, 0, \ldots, 0, 1)$. Hence,

$$\mathbf{a} \geq (a_1, b_2+1, 0, \ldots, 0, 1) > (a_1+1, b_2+1, 0, \ldots, 0) > (a_1+1, b_2, 9, \ldots, 9) = \mathbf{b}.$$

If $b_2 = 4$ then $(e_2, \ldots, e_n) = (6, 0, \ldots, 0)$ and

$$\mathbf{a} \geq (a_1, 6, 0, \ldots, 0) > (a_1 + 1, 5, 0, \ldots, 0) > (a_1 + 1, 4, 9, \ldots, 9) = \mathbf{b}.$$

In both cases $\mathbf{b} \in A$ since A is compressed.
If $(a_2, \ldots, a_n) = succ(b_2, \ldots, b_n)$, then $\mathbf{a} = (b_1 - 1, b_2 + 1, 0, \ldots, 0)$ and $\mathbf{b} = (b_1, b_2, 9, \ldots, 9)$. Replacing $\mathbf{a}$ with $\mathbf{b}$ we obtain a set B such that

$$\begin{aligned}
|I_{P^n}(B)| - |I_{P^n}(A)| &= (\delta_P(b_1) + \delta_P(b_2) + (n - 2) \cdot \delta_P(9)) \\
&\quad - (\delta_P(b_1 - 1) + \delta_P(b_2 + 1)) \\
&\geq 3(n - 2) + 1,
\end{aligned}$$

since $\delta_P(b_1) - \delta_P(b_1 - 1) = 1$ for $b_1 \in \{1, 4, 6, 9\}$ and $\delta_P(b_2) \geq \delta_P(b_2 + 1)$ for $b_2 \in \{1, 4, 6\}$.
If $(b_3, \ldots, b_n) \neq (9, \ldots, 9)$, then let us denote $(d_3, \ldots, d_n) = succ(b_3, \ldots, b_n)$. One has $succ(b_2, \ldots, b_n) = (b_2 - 1, d_3, \ldots, d_n)$.
Now assume additionally that

$$(a_2, \ldots, a_n) \geq (e_2, \ldots, e_n) = succ(succ(b_2, \ldots, b_n)).$$

Then $(a_1, e_2, \ldots, e_n) \in A$ similarly to above and

$$(e_2, \ldots, e_n) = (b_2, d_3, \ldots, d_n).$$

Consequently,

$$\mathbf{a} \geq (a_1, b_2, d_3, \ldots, d_n) > (a_1 + 1, b_2 - 1, d_3, \ldots, d_n) > (a_1 + 1, b_2, b_3, \ldots, b_n).$$

Therefore, $\mathbf{b} \in A$.

e4. Lastly, we consider the case $b_2 \in \{1, 4, 6, 9\}$, $(a_2, \ldots, a_n) = succ(b_2, \ldots, b_n)$ and assume

$$(b_3, \ldots, b_n) \neq (9, \ldots, 9).$$

Therefore, $\mathbf{a} = (b_1 - 1, b_2 - 1, a_3, \ldots, a_n)$ and we can apply to the vectors $(a_2, \ldots, a_n)$ and $(b_2, \ldots, b_n)$ the same analysis as those for cases $e1$–$e3$. If in the course of this analysis we will be able to guarantee $\mathbf{b} \in A$ due to the compression, or to replace $\mathbf{a}$ with $\mathbf{b}$ without decreasing the function $I_{P^n}(\cdot)$ then we are done. Otherwise, just one case will give rise to problem again, namely when $b_3 \in \{1,4,6,9\}$, $(a_3, \ldots, a_n) = succ(b_3, \ldots, b_n)$ and $(b_4, \ldots, b_n) \neq (9, \ldots, 9)$.

Continuing this way, the only remaining case left open is the case $\mathbf{a} = succ(\mathbf{b})$ such that

$$\mathbf{a} = (b_1 - 1, b_2 - 1, \ldots, b_{n-1} - 1, b_n + 1), \quad \mathbf{b} = (b_1, \ldots, b_n),$$

where $b_1, \ldots, b_{n-1} \in \{1, 4, 6, 9\}$ and $b_n \neq 9$. However, in this case the replacement of $\mathbf{a}$ with $\mathbf{b}$ leads to a set B with

$$|I_{P^n}(B)| - |I_{P^n}(A)| = \sum_{i=1}^{n-1} (\delta_P(b_i) - \delta_P(b_i - 1)) + (\delta_P(b_n) - \delta_P(b_n + 1)) \geq n - 2,$$

since $\delta_P(b_i) - \delta_P(b_i - 1) = 1$ for $b_i \in \{1, 4, 6, 9\}$ and $\delta_P(b_n) - \delta_P(b_n + 1) \geq -1$.

$\square$

4   Cutwidth and Wirelength of P^n

For $A \subseteq V_{P^n}$ denote

$$\partial(A) = \{(u, v) \in E_{P^n} \mid u \in A, \ v \notin A\}.$$
$$g_n(t) = \min_{|A|=t} |\partial(A)|.$$

Since the graph P^n is regular of degree $3n$, for any $t = 1, \ldots, 10^n$ it holds:

$$2 \cdot I_{P^n}(t) + g_n(t) = 3nt. \tag{1}$$

Therefore, for any initial segment A of the Petersen order $\mathcal{P}^n$, we have $|\partial(A)| = g_n(|A|)$.

For a graph G, let f be a bijective mapping $f : V_G \mapsto \{1, \ldots, |V_G|\}$. Let

$$con_f(i) = |\{(u, v) \in E_G \mid f(u) \leq i, \ f(v) \geq i + 1\}|, \quad \text{for} \quad 1 \leq i < |V_G|,$$

$$cw(G) = \min_f \max_i \{con_f(i)\}, \quad wl(G) = \min_f \{ \sum_{i=1}^{|V_G|-1} con_f(i)\}.$$

The parameters $cw(G)$ and $wl(G)$ are respectively called the *cutwidth* and *wirelength* of the graph G. For $G = P^n$, since the subsets with minimum value of the function $|\partial(\cdot)|$ are nested, it holds (cf. [8]):

$$cw_n = cw(P^n) = \max_t \{g_n(t)\}, \quad wl_n = wl(P^n) = \sum_{t=0}^{10^n} g_n(t).$$

It can be easily shown that $cw_1 = 6$ and $wl_1 = 41$. Now, we are ready to compute the cutwidth and the wirelength of the powers of the Petersen graph. The proofs follow from theorem 1 and will be omitted, due to the space limitation.

Theorem 2 *The cutwidth of the graph P^n is given by*

$$cw_n = \begin{cases} (6.25) \cdot 10^{n-1} + (2^{n-1} - 4)/12, & \text{if } n \text{ is odd} \\ (6.25) \cdot 10^{n-1} + (2^{n-1} - 8)/12, & \text{if } n \text{ is even.} \end{cases}$$

Theorem 3 *The wirelength of P^n is given by*

$$wl_n = \frac{37}{82} \cdot 100^n + \frac{72}{82} \cdot 18^{n-1} - \frac{1}{2} \cdot 10^n.$$

5   Products of Petersen Powers with Hypercubes

Let Q^m be the graph of the m-dimensional hypercube which is the mth cartesian power of the clique with two vertices. Consider the edge-isoperimetric problem on the product graph $P^n_m = P^n \times Q^m$. These families of graphs are called folded Petersen cubes [14] and extensively studied to model multiprocessor interconnection networks.

5.1   Edge-Isoperimetric Problem on P^n_m

We show that the edge-isoperimetric problem on the graph P^n_m has nested solutions provided by the new order, $\mathcal{Q}^n_m$, presented below. We represent the vertices of P^n_m as $(n + m)$-dimensional vectors $(a_1, \ldots, a_n, \alpha_1, \ldots, \alpha_m)$, where $(a_1, \ldots, a_n) \in P^n$ and $(\alpha_1, \ldots, \alpha_m) \in Q^m$.

For vertices $\mathbf{a} = (a_1, \ldots, a_n, \alpha_1, \ldots, \alpha_m)$ and $\mathbf{b} = (b_1, \ldots, b_n, \beta_1, \ldots, \beta_m)$ of $P^n \times Q^m$, we write $\mathbf{a} \succ \mathbf{b}$ in the order $\mathcal{Q}^n_m$ iff

(i) $(a_1, \ldots, a_n) > (b_1, \ldots, b_n)$ in the order $\mathcal{P}^n$,   or
(ii) $(a_1, \ldots, a_n) = (b_1, \ldots, b_n)$ and $(\alpha_1, \ldots, \alpha_m)$ is greater than $(\beta_1, \ldots, \beta_m)$ in the lexicographic order.

It is an easy exercise to ensure oneself that the order $\mathcal{Q}^n_m$ satisfies the consistency property [8] similar to Lemma 1.

We introduce compressions similarly to that in Section 4. The initial segments of the order $\mathcal{Q}^n_m$ of length t will be denoted by the set $\mathcal{F}^n_m(t)$.

Theorem 4 *For any $n \geq 1$, $m \geq 1$ and $t = 1, \ldots, 10^n \cdot 2^m$ the set $\mathcal{F}^n_m(t)$ is optimal, where $\mathcal{F}^n_m(t)$ is represented by the initial segment of the order $\mathcal{Q}^n_m$ of length t.*

The proof of the theorem is similar to the one in theorem 1 and will be omitted due to the space limitation.

5.2   Cutwidth and Wirelength of P_m^n

The simple structure of the order Q_m^n immediately derives the formulas for the cutwidth, $cw_{n,m}$, and the wirelength, $wl_{n,m}$, of the graph P_m^n. We assume that $n \geq 1$ and $m \geq 1$. Furthermore, let

$$q_m(r) = \min_{|A|=r} |\{(u,v) \in E_{Q^m} \mid u \in A, \; v \notin A\}|,$$

where the minimum runs over all subsets of V_{Q^m} of size r. It follows from Theorem 4 that the corresponding function for the graph P_m^n, denoted by $G_{n,m}(s)$, is of the form

$$G_{n,m}(s) = r \cdot g_n(t+1) + (2^m - r) \cdot g_n(t) + q_m(r),$$

where $r = s \bmod 2^m$ and $t = \lfloor s/2^m \rfloor$. In these terms

$$cw_{n,m} = \max_s \{G_{n,m}(s)\} \quad \text{and} \quad wl_{n,m} = \sum_{s=1}^{10^n \cdot 2^m} G_{n,m}(s).$$

Let $cw(Q^m) = \max_r \{q_m(r)\}$ be the cutwidth of the hypercube Q^m.

Now if n is even, then $g_n(t+1) = g_n(t) = cw_n$ for some t (cf. the Claim in the proof of Theorem 3). Hence, $cw_{n,m} = 2^m \cdot cw_n + cw(Q^m)$. If n is odd, then, using the Claim again, at most one of $g_n(t+1)$ and $g_n(t)$ equals cw_n. Therefore, we get

$$G_{n,m}(s) \leq \begin{cases} 2^m(cw_n - 1) + q_m(t) & \text{if } g_n(t+1) < cw_n \;\&\; g_n(t) < cw_n \\ 2^m \cdot cw_n - t + q_m(t) & \text{if } g_n(t+1) < cw_n \;\&\; g_n(t) = cw_n \\ 2^m(cw_n - 1) + t + q_m(t) & \text{if } g_n(t+1) = cw_n \;\&\; g_n(t) < cw_n. \end{cases}$$

It can be shown that the last expression is the largest one if $t \geq 2^{m-1}$.

Note that $\max_{2^{m-1} \leq t \leq 2^m} \{t + q_m(t)\} = cw(Q^{m+1})$. Therefore, the cutwidth of P_m^n is given by

$$cw_{n,m} = \begin{cases} 2^m \cdot cw_n + cw(Q^m), & \text{if } n \text{ is even} \\ 2^m(cw_n - 1) + cw(Q^{m+1}), & \text{if } n \text{ is odd.} \end{cases}$$

Similar arguments provide the wirelength of P_m^n as $wl_{n,m} = 2^m \cdot wl_n + 10^n \cdot wl(Q^m)$.

From these results one can derive formulas for $cw_{n,m}$ and $wl_{n,m}$ in terms of n and m by using Theorems 2 and 3, and known results $cw(Q^m) = (2^{m+1} - 2 + (m \bmod 2))/3$ and $wl(Q^m) = 2^{m-1}(2^m - 1)$ (cf. e.g. [4,12]).

Acknowledgments

The authors would like to thank Professor Larry Harper (University of California at Riverside) whose helpful suggestions and comments greatly improved the paper.

References

1. Ahlswede R., Bezrukov S.L., *Edge Isoperimetric Theorems for Integer Point Arrays*, Appl. Math. Lett. **8** (1995), No. 2, 75–80.
2. Ahlswede R., Cai N., *General Edge-isoperimetric Inequalities, Part II: a Local-Global Principle for Lexicographic Solution*, Europ. J. Combin. **18** (1997), 479–489.
3. Bezrukov S.L., *An Equivalence in Discrete Extremal Problems*, to appear in Discr. Math.
4. Bezrukov S.L., *Edge Isoperimetric Problems on Graphs*, to appear in János Bolyai Math. Series, Budapest.
5. Bezrukov S.L., Chavez J.D., Harper L.H., Röttger M., Schroeder U.-P., *The congestion of n-cube layout on a rectangular grid*, to appear in Discr. Math.
6. Bollobás B., Leader I., *An Isoperimetric Inequality on the Discrete Torus*, SIAM J. Appl. Math. **3** (1990), 32–37.
7. Bollobás B., Leader I., *Edge-isoperimetric Inequalities in the Grid*, Combinatorica **11** (1991), 299–314.
8. Chavez J.D., Harper L.H., *Discrete Isoperimetric Problems and Pathmorphisms*, preprint.
9. Das S. K., Ibel M., Hohndel D., Öhring S., *Efficient Communication in the Folded Petersen Networks*, International Journal of Foundations of Computer Science, **8** (1997), 163–185.
10. Elsässer R., *Kantenseparatoren in Kartesischen Produkten von Graphen*, diploma-thesis, University of Paderborn (1998).
11. Harper L.H., *A Necessary Condition on Minimal Cube Numberings*, J. Appl. Prob. **4** (1967), 397–401.
12. Harper L.H., *Optimal Assignment of Numbers to Vertices*, J. Sos. Ind. Appl. Math. **12** (1964), 131–135.
13. Lindsey II J.H., *Assignment of Numbers to Vertices*, Amer. Math. Monthly **7** (1964), 508–516.
14. Öhring S., Das S.K., *The Folded Petersen Cube Networks: New Competitors for the Hypercube*, in: Proc. 5th IEEE Symp. on Parallel and Distr. Processing, 1993, 582–589.
15. Öhring S., Das S.K., *The Folded Petersen Network: A New Versatile Multiprocessor Interconnection Topology*, in: Proc. Int. Workshop on Graph-Theoretic Concepts in Computer Science (WG'93), Lecture Notes in Computer Science, vol. **790** (1994), 301–314.
16. Rolim J., Sykora O., Vrt'o I., *Optimal Cutwidths and Bisection Widths of 2- and 3-Dimensional Meshes*, in: Proc. Int. Workshop on Graph-Theoretic Concepts in Computer Science (WG'95), Lecture Notes in Computer Science vol **1017** (1995), 252–264.

Dihamiltonian Decomposition of Regular Graphs with Degree Three[*]

Jung-Heum Park[1] and Hee-Chul Kim[2]

[1] School of Computer Science and Engineering
The Catholic University of Korea, Korea
jhpark@tcs.cuk.ac.kr
[2] Department of Computer Science and Engineering
Hankuk University of Foreign Studies, Korea
hckim@maincc.hufs.ac.kr

Abstract. We consider the dihamiltonian decomposition problem for 3-regular graphs. A graph G is *dihamiltonian decomposable* if in the digraph obtained from G by replacing each edge of G as two directed edges, the set of edges are partitioned into 3 edge-disjoint directed hamiltonian cycles. We suggest some conditions for dihamiltonian decomposition of 3-regular graphs: for a 3-regular graph G, it is dihamiltonian decomposable only if it is bipartite, and it is not dihamiltonian decomposable if the number of vertices is a multiple of 4. Applying these conditions to interconnection network topologies, we investigate dihamiltonian decomposition of cube-connected cycles, chordal rings, etc.

1 Introduction

An r-regular graph G is *hamiltonian decomposable* if it is possible to partition the set of edges into k edge-disjoint hamiltonian cycles when r is $2k$ and it is possible to partition the set of edges into k edge-disjoint hamiltonian cycles and a 1-factor when r is $2k + 1$. Here a 1-factor of a graph is a 1-regular spanning subgraph.

The symmetric digraph of an undirected graph G is defined as the digraph obtained from G by replacing each edge (u, v) of G as two directed edges, $\langle u, v \rangle$ and $\langle v, u \rangle$. An r-regular graph G is *dihamiltonian decomposable* if the set of edges in its symmetric digraph can be partitioned into r edge-disjoint directed hamiltonian cycles.

For a graph G to have either a hamiltonian decomposition or a dihamiltonian decomposition, it is necessary that G is loopless, connected, and regular. A hamiltonian decomposable r-regular graph with even r is also dihamiltonian decomposable, since each hamiltonian cycle in the decomposition can be regarded as two directed hamiltonian cycles of opposite direction. When r is odd, it is not always true.

[*] This work was supported by the Korea Science and Engineering Foundation under grant no. 98-0102-07-01-3.

A survey on hamiltonian decomposition of graphs is provided in[3,4]. Much of the focus of research has been directed towards proving that some special cases of Cayley graphs over an abelian group are hamiltonian decomposable, such as the product of any number of cycles, m-cubes, and recursive circulants[5,9]. But still, the current status of the matter lies, for the most part, in the sphere of problems and conjectures. For the dihamiltonian decomposition of graphs, there is only a trivial result such that the hamiltonian decomposable graphs are dihamiltonian decomposable. It remains open whether m-cubes and recursive circulants with an odd degree are dihamiltonian decomposable.

The problem of dihamiltonian decomposition of graphs has an application in the design of reliable communication algorithms such as broadcasting and multicasting under the wormhole routing model. In the way of packing as many edge-disjoint directed hamiltonian cycles as possible, reliable communication algorithms are presented on m-cubes and tori[6], and m-dimensional meshes[7]. Note that the dihamiltonian decomposition of a regular graph results in the maximum number of edge-disjoint directed hamiltonian cycles.

In this paper, we consider the dihamiltonian decomposition of 3-regular graphs. We show that if a 3-regular graph is dihamiltonian decomposable, it is necessarily bipartite. We also show that every 3-regular graph with $n = 4k$ vertices is not dihamiltonian decomposable. Applying these results to interconnection network topologies with degree 3, we investigate dihamiltonian decomposition of cube-connected cycles[8], chordal rings[1], etc.

A 3-regular graph G with multiple edges is not dihamiltonian decomposable, since edge-connectivity of G is less than or equal to 2 and it is not possible for three edge-disjoint directed hamiltonian cycles to pass through the edge cut. Thus we assume a graph is simple. Graph theoretic terms not defined here can be found in[2].

This paper is organized as follows. We show that a dihamiltonian decomposable graph is bipartite in Section 2, and prove that a graph with $n = 4k$ vertices is not dihamiltonian decomposable in Section 3. In Section 4, we consider the dihamiltonian decomposition of the interconnection network topologies, and finally concluding remarks are given in Section 5.

2   Necessary Conditions

In this section we present some properties for dihamiltonian decomposition of 3-regular graphs. Let G be a 3-regular graph with n vertices which is dihamiltonian decomposable. Note that the number of vertices of 3-regular graphs is even. Let G_S denote the symmetric digraph of G. Then the set of edges of G_S is partitioned into 3 directed hamiltonian cycles. Let C, which is one of such cycles, be $(v_{n-1}, v_{n-2}, \cdots, v_1, v_0, v_{n-1})$. Let C_0 and C_1 be the other two directed hamiltonian cycles. For the symmetric digraph G_S of G, $G_S - C$ is defined as the digraph which is obtained by deleting all the edges in C from G_S. From now on, it is assumed that all arithmetic operations on the indices of vertices is performed with mod n.

We call $\langle v_i, v_{i+1} \rangle$, $0 \leq i \leq n-1$, to be a boundary edge w.r.t. C, and $\langle v_i, v_j \rangle$ with $j \neq i+1$ to be a chord edge w.r.t. C(see Figure 1). An edge (v_i, v_j) in G is called a boundary (resp. chord) edge w.r.t. C if $\langle v_i, v_j \rangle$ or $\langle v_j, v_i \rangle$ in G_S is a boundary (resp. chord) edge w.r.t. C. The size of chord edge $\langle v_i, v_j \rangle$ is defined as $\min\{i - j, j - i\}$. We will now investigate the properties of two edge-disjoint hamiltonian cycles C_0 and C_1. In the following, C_0 denotes the hamiltonian cycle which contains $\langle v_0, v_1 \rangle$ unless it is otherwise specified.

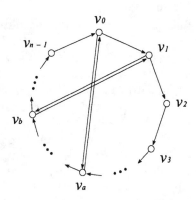

Fig. 1. $G_S - C$

Lemma 1 *(a) In each of C_0 and C_1, the boundary edge and the chord edge w.r.t. C appear alternately. (b) For each boundary edge $\langle v_i, v_{i+1} \rangle$ w.r.t. C in C_0 (resp. C_1), i is even (resp. odd).*

Proof We first show that the lemma is true for C_0. Let $\langle v_i, v_{i+1} \rangle$ be an arbitrary boundary edge w.r.t. C in C_0. Since both $\langle v_{i+1}, v_i \rangle$ and $\langle v_{i+2}, v_{i+1} \rangle$ are in C, the edge which immediately follows the edge $\langle v_i, v_{i+1} \rangle$ on C_0 should be the chord edge for C_1 to be a hamiltonian cycle. For every chord edge $\langle v_i, v_j \rangle$ in C_0, the edge which immediately follows it on C_0 is the boundary edge $\langle v_j, v_{j+1} \rangle$ since $\langle v_j, v_{j-1} \rangle$ is in C. Thus the boundary edge and the chord edge appear alternately in C_0. The proof of part (b) is as follows. Let $\langle v_i, v_{i+1} \rangle$ be an arbitrary boundary edge w.r.t. C in C_0. For each j with $0 \leq j < i$, one of $\langle v_{j-1}, v_j \rangle$ and $\langle v_j, v_{j+1} \rangle$ is in C_0 and the other is not in C_0. Therefore, i should be even since $\langle v_0, v_1 \rangle$ is in C_0. It can be similarly shown that the lemma holds for C_1.   □

Using Lemma 1, we present a necessary condition for dihamiltonian decomposition of 3-regular graphs.

Theorem 1 *If a 3-regular graph G is dihamiltonian decomposable, G is a bipartite graph.*

Proof Let C, C_0, and C_1 be 3 edge-disjoint hamiltonian cycles in G_S. Let $C = (v_{n-1}, v_{n-2}, \cdots, v_1, v_0, v_{n-1})$, and let C_0 be the hamiltonian cycle which contains $\langle v_0, v_1 \rangle$. A vertex v_a is called even vertex if a is even; otherwise it is called an odd vertex. Every boundary edge joins even vertex and odd vertex since n is even. Therefore we have only to show that every chord edge w.r.t. C join an even vertex and an odd vertex. Consider an arbitrary chord edge $\langle v_i, v_j \rangle$ w.r.t. C in C_0. The edge which immediately precedes $\langle v_i, v_j \rangle$ on C_0, is a boundary edge $\langle v_{i-1}, v_i \rangle$ by Lemma 1. Thus $i - 1$ is even, that is, i is odd, by Lemma 1. Since the edge which immediately follows $\langle v_i, v_j \rangle$ on C_0, is a boundary edge $\langle v_j, v_{j+1} \rangle$, j is even. It can be similarly shown that every chord edge w.r.t. C in C_1 joins an even vertex and an odd vertex. Therefore G is a bipartite graph. $\square$

We denote by G_0^C the graph obtained by deleting all the boundary edges w.r.t. C and contracting two vertices $\{v_i, v_{i+1}\}$ in the digraph $G_S - C$ for every even i, $0 \le i \le n - 1$. We denote by G_1^C the graph obtained by deleting all the boundary edges w.r.t. C and contracting two vertices $\{v_i, v_{i+1}\}$ in $G_S - C$ for every odd i, $0 \le i \le n - 1$. Then we have the following properties. The proof is straightforward, and omitted here.

Property 1 *Each of G_0^C and G_1^C is a cycle with length $n/2$.*

Property 2 *A 3-regular graph G is dihamiltonian decomposable if and only if for some hamiltonian cycle C in the symmetric digraph G_S of G, each of G_0^C and G_1^C is a cycle with length $n/2$.*

3   Case of $n = 4k$

Some 3-regular graphs are not dihamiltonian decomposable. In this section, we will show that every 3-regular graphs with $n = 4k$ vertices can not be dihamiltonian decomposable by utilizing the following lemma.

Lemma 2 *If there exists a 3-regular graph with $n \ge 8$ vertices which is dihamiltonian decomposable, then there exists a 3-regular graph with $n - 4$ vertices which is dihamiltonian decomposable.*

Proof Let G be a 3-regular graph with n vertices. Let C be a hamiltonian cycle $(v_{n-1}, v_{n-2}, \cdots, v_1, v_0, v_{n-1})$ in dihamiltonian decomposition of G. Let C_0 and C_1 be two edge-disjoint hamiltonian cycles in $G_S - C$.
   Case 1   There is a chord edge of size 3 w.r.t. C in G.
   Let (v_{i+1}, v_{i+4}) be a chord edge of size 3 w.r.t. C in G, and let v'_{i+2} and v'_{i+3} be the vertices connected with v_{i+2} and v_{i+3} by a chord edge, respectively(see Figure 2). Let C_0 and C_1 be the hamiltonian cycles which contain the edge $\langle v_i, v_{i+1} \rangle$ and $\langle v_{i+1}, v_{i+2} \rangle$, respectively. We have $C_0 = (v_i, v_{i+1}, v_{i+4}, v_{i+5}, P_{1,1}, v'_{i+2}, v_{i+2}, v_{i+3}, v'_{i+3}, P_{1,2}, v_i)$ where $P_{1,1}$ is the path from v_{i+5} to v'_{i+2} and $P_{1,2}$ is the path from v'_{i+3} to v_i. We have $C_1 = (v'_{i+3}, v_{i+3}, v_{i+4}, v_{i+1}, v_{i+2}, v'_{i+2},$

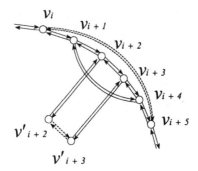

Fig. 2. Illustration of the proof of Lemma 2 (Case 1)

$P_{1,3}, v'_{i+3})$ where $P_{1,3}$ is the path from v'_{i+2} to v'_{i+3} in C_1. We first consider the case that all vertices in $\{v_i, v_{i+1}, v_{i+2}, v_{i+3}, v_{i+4}, v_{i+5}, v'_{i+2}, v'_{i+3}\}$ are distinct.

We define a graph G' with $n-4$ vertices as $G' = (G - \{v_{i+1}, v_{i+2}, v_{i+3}, v_{i+4}\}) \cup \{(v_i, v_{i+5}), (v'_{i+2}, v'_{i+3})\}$ (see Figure 2). We will show that G' is a simple 3-regular graph by proving that (i) (v_i, v_{i+5}) is not a chord edge w.r.t. C in G, and that (ii) (v'_{i+2}, v'_{i+3}) is not a boundary edge w.r.t. C in G.

If (v_i, v_{i+5}) is a chord edge, C_0 becomes a cycle $(v_i, v_{i+1}, v_{i+4}, v_{i+5}, v_i)$ with length 4 by Lemma 1, which is a contradiction to C_0 being a hamiltonian cycle. If (v'_{i+2}, v'_{i+3}) is a boundary edge, then for some k, either $v'_{i+2} = v_k$ and $v'_{i+3} = v_{k+1}$ or $v'_{i+2} = v_{k+1}$ and $v'_{i+3} = v_k$. In the first case, C_1 becomes a cycle $(v_{i+1}, v_{i+2}, v'_{i+2} = v_k, v'_{i+3} = v_{k+1}, v_{i+3}, v_{i+4}, v_{i+1})$ with length 6 by Lemma 1, which is a contradiction. In the second case, C_0 becomes a cycle $(v_{i+2}, v_{i+3}, v'_{i+3} = v_k, v'_{i+2} = v_{k+1}, v_{i+2})$ with length 4, which is also a contradiction.

Using C, C_0, and C_1, we construct 3 edge-disjoint hamiltonian cycles C', C'_0, and C'_1 in the symmetric digraph of G'. Let $C' = (v_{n-1}, v_{n-2}, \cdots, v_{i+5}, v_i, v_{i-1}, \cdots, v_0, v_{n-1})$, and $C'_0 = (v_i, v_{i+5}, P_{1,1}, v'_{i+2}, v'_{i+3}, P_{1,2}, v_i)$ and $C'_1 = (v'_{i+3}, v'_{i+2}, P_{1,3}, v'_{i+3})$. Then C', C'_0 and C'_1 are edge-disjoint hamiltonian cycles in the symmetric digraph of G'.

We consider the case that some vertices in $\{v_i, v_{i+1}, v_{i+2}, v_{i+3}, v_{i+4}, v_{i+5}, v'_{i+2}, v'_{i+3}\}$ are not distinct. Since G is bipartite, either $v'_{i+2} = v_{i+5}$ or $v'_{i+3} = v_i$. If $v'_{i+2} = v_{i+5}$, $P_{1,1}$ is a path with length 0. If $v'_{i+3} = v_i$, $P_{1,2}$ is a path with length 0. It does not hold that both $v'_{i+2} = v_{i+5}$ and $v'_{i+3} = v_i$; otherwise C_0 becomes a cycle with length 6. In either case, C', C'_0 and C'_1 are edge-disjoint hamiltonian cycles in G'.

Case 2   There are no chord edges of size 3 w.r.t. C in G and either C_0 or C_1 has a boundary edge w.r.t. C, $\langle v_i, v_{i+1} \rangle$, $0 \le i \le n-1$, such that when we traverse that hamiltonian cycle starting at $\langle v_i, v_{i+1} \rangle$, $\langle v_{i+4}, v_{i+5} \rangle$ precedes $\langle v_{i+2}, v_{i+3} \rangle$.

Let C_0 and C_1 be the hamiltonian cycles which contain the edge $\langle v_i, v_{i+1} \rangle$ and $\langle v_{i+1}, v_{i+2} \rangle$, respectively(see Figure 3). We have $C_0 = (v_i, v_{i+1}, v'_{i+1}, P_{2,1}, v'_{i+4}, v_{i+4}, v_{i+5}, P_{2,2}, v'_{i+2}, v_{i+2}, v_{i+3}, v'_{i+3}, P_{2,3}, v_i)$ where $P_{2,1}$ is the path from

v'_{i+1} to v'_{i+4} and $P_{2,2}$ is the path from v_{i+5} to v'_{i+2} and $P_{2,3}$ is the path from v'_{i+3} to v_i. We have $C_1 = (v'_{i+1},\ v_{i+1},\ v_{i+2},\ v'_{i+2},\ P_{2,4},\ v'_{i+3},\ v_{i+3},\ v_{i+4},\ v'_{i+4},$ $P_{2,5},\ v'_{i+1})$ where $P_{2,4}$ is the path from v'_{i+2} to v'_{i+3} and $P_{2,5}$ is the path from v'_{i+4} to v'_{i+1}. Since there are no chord edges of size 3 w.r.t. C, all vertices in $\{v_i,$ $v_{i+1},\ v_{i+2},\ v_{i+3},\ v_{i+4},\ v_{i+5},\ v'_{i+1},\ v'_{i+2},\ v'_{i+3},\ v'_{i+4}\}$ are distinct.

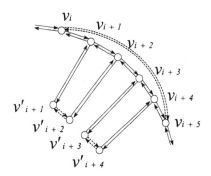

Fig. 3. Illustration of the proof of Lemma 2 (Case 2)

Let G' be the graph such that $G' = (G - \{v_{i+1},\ v_{i+2},\ v_{i+3},\ v_{i+4}\}) \cup \{(v_i, v_{i+5}),\ (v'_{i+1}, v'_{i+2}),\ (v'_{i+3}, v'_{i+4})\}$(see Figure 3). Then we can see that G' is a simple 3-regular graph by proving that (i) (v_i, v_{i+5}) is not a chord edge w.r.t. C, that (ii) (v'_{i+1}, v'_{i+2}) is not a boundary edge w.r.t. C, and that (iii) (v'_{i+3}, v'_{i+4}) is not a boundary edge w.r.t. C.

If (v_i, v_{i+5}) is a chord edge, then C_0 is $(v'_{i+4},\ v_{i+4},\ v_{i+5},\ v_i,\ v_{i+1},\ v'_{i+1},$ $\cdots,\ v'_{i+4})$, that is, C_0 is $(v_i,\ v_{i+1},\ \cdots,\ v_{i+2},\ v_{i+3},\ \cdots,\ v_{i+4},\ v_{i+5},\ v_i)$. Since $\langle v_{i+4}, v_{i+5}\rangle$ appears after $\langle v_{i+2}, v_{i+3}\rangle$ in C_0, it contradicts the assumption of Case 2. If (v'_{i+1}, v'_{i+2}) is a boundary edge, then for some k, either $v'_{i+1} = v_k$ and $v'_{i+2} = v_{k+1}$ or $v'_{i+1} = v_{k+1}$ and $v'_{i+2} = v_k$. In the first case, C_0 is $(v_i,\ v_{i+1},$ $v'_{i+1} = v_k,\ v'_{i+2} = v_{k+1},\ v_{i+2},\ v_{i+3},\ v'_{i+3},\ \cdots,\ v_i)$, which contradicts the assumption of Case 2. In the second case, C_1 becomes a cycle $(v_{i+1},\ v_{i+2},\ v'_{i+2} = v_k,$ $v'_{i+1} = v_{k+1},\ v_{i+1})$ with length 4, which is a contradiction.

If (v'_{i+3}, v'_{i+4}) is a boundary edge, then for some k, either $v'_{i+3} = v_k$ and $v'_{i+4} = v_{k+1}$ or $v'_{i+3} = v_{k+1}$ and $v'_{i+4} = v_k$. In the first case, C_0 is $(v_{i+2},$ $v_{i+3},\ v'_{i+3} = v_k,\ v'_{i+4} = v_{k+1},\ v_{i+4},\ v_{i+5},\ \cdots,\ v_i,\ v_{i+1},\ \cdots,\ v_{i+2})$, that is, C_0 is $(v_i,\ v_{i+1},\ \cdots,\ v_{i+2},\ v_{i+3},\ v'_{i+3},\ v'_{i+4},\ v_{i+4},\ v_{i+5},\ \cdots,\ v_i)$, which contradicts the assumption of Case 2. In the second case, C_1 becomes a cycle $(v_{i+3},\ v_{i+4},$ $v'_{i+4} = v_k,\ v'_{i+3} = v_{k+1},\ v_{i+3})$ with length 4, which is a contradiction.

The dihamiltonian decomposition of G' is as follows. Let C' be $(v_{n-1},\ v_{n-2},$ $\cdots,\ v_{i+5},\ v_i,\ v_{i-1},\ \cdots,\ v_0,\ v_{n-1})$, and C'_0 be $(v_i,\ v_{i+5},\ P_{2,2},\ v'_{i+2},\ v'_{i+1},\ P_{2,1},\ v'_{i+4},$ $v'_{i+3},\ P_{2,3},\ v_i)$ and C'_1 be $(v'_{i+1},\ v'_{i+2},\ P_{2,4},\ v'_{i+3},\ v'_{i+4},\ P_{2,5},\ v'_{i+1})$. Then $C',\ C'_0$ and C'_1 are edge-disjoint hamiltonian cycles in G'.

Case 3 Both Case 1 and Case 2 do not hold.

Let v_i be an arbitrary vertex and let C_0 and C_1 be the disjoint hamiltonian cycles in $G_S - C$ which contain the edge $\langle v_i, v_{i+1} \rangle$ and $\langle v_{i+1}, v_{i+2} \rangle$, respectively(see Figure 4). We first consider the case that all vertices in $\{v_i, v_{i+1}, v_{i+2}, v_{i+3}, v_{i+4}, v_{i+5}, v_{i+6}, v'_{i+1}, v'_{i+2}, v'_{i+3}, v'_{i+4}, v'_{i+5}\}$ are distinct. When we traverse C_0 starting at $\langle v_i, v_{i+1} \rangle$, $\langle v_{i+2}, v_{i+3} \rangle$ precedes $\langle v_{i+4}, v_{i+5} \rangle$ in C_0. Also, when we traverse C_1 starting at $\langle v_{i+1}, v_{i+2} \rangle$, $\langle v_{i+3}, v_{i+4} \rangle$ precedes $\langle v_{i+5}, v_{i+6} \rangle$ in C_1; otherwise it is Case 2. Therefore C_0 and C_1 can be represented as follows: $C_0 = (v_i, v_{i+1}, v'_{i+1}, P_{3,1}, v'_{i+2}, v_{i+2}, v_{i+3}, v'_{i+3}, P_{3,2}, v'_{i+4}, v_{i+4}, v_{i+5}, v'_{i+5}, P_{3,3}, v_i)$ where $P_{3,1}$ is the path from v'_{i+1} to v'_{i+2} and $P_{3,2}$ is the path from v'_{i+3} to v'_{i+4} and $P_{3,3}$ is the path from v'_{i+5} to v_i; $C_1 = (v'_{i+1}, v_{i+1}, v_{i+2}, v'_{i+2}, P_{3,4}, v'_{i+3}, v_{i+3}, v_{i+4}, v'_{i+4}, P_{3,5}, v'_{i+5}, v_{i+5}, v_{i+6}, P_{3,6}, v'_{i+1})$ where $P_{3,4}$ is the path from v'_{i+2} to v'_{i+3} and $P_{3,5}$ is the path from v'_{i+4} to v'_{i+5} and $P_{3,6}$ is the path from v_{i+6} to v'_{i+1}.

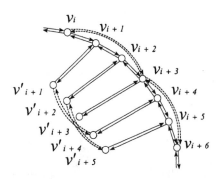

Fig. 4. Illustration of the proof of Lemma 2 (Case 3)

Let G' be the graph such that $G' = (G - \{v_{i+1}, v_{i+2}, v_{i+4}, v_{i+5}\}) \cup \{(v_i, v_{i+3}), (v_{i+3}, v_{i+6}), (v'_{i+1}, v'_{i+4}), (v'_{i+2}, v'_{i+5})\}$(see Figure 4). Then we can see that G' is a simple 3-regular graph by proving that (i) both (v_i, v_{i+3}) and (v_{i+3}, v_{i+6}) are not chord edges, that (ii) (v'_{i+1}, v'_{i+4}) is not a boundary edge, and that (iii) (v'_{i+2}, v'_{i+5}) is not a boundary edge.

Since there are no chord edges of size 3 in Case 3, (i) holds. If (v'_{i+1}, v'_{i+4}) is a boundary edge, then for some k, either $v'_{i+1} = v_k$ and $v'_{i+4} = v_{k+1}$ or $v'_{i+1} = v_{k+1}$ and $v'_{i+4} = v_k$. In the first case, C_0 is $(v_i, v_{i+1}, v'_{i+1} = v_k, v'_{i+4} = v_{k+1}, v_{i+4}, v_{i+5}, \cdots, v_{i+2}, v_{i+3}, \cdots, v_i)$, which is Case 2. In the second case, C_1 is $(v_{i+3}, v_{i+4}, v'_{i+4} = v_k, v'_{i+1} = v_{k+1}, v_{i+1}, v_{i+2}, \cdots, v_{i+5}, v_{i+6}, \cdots, v_{i+3}) = (v_{i+1}, v_{i+2}, \cdots, v_{i+5}, v_{i+6}, \cdots, v_{i+3}, v_{i+4}, v'_{i+4}, v'_{i+1}, v_{i+1})$, which is also Case 2.

If (v'_{i+2}, v'_{i+5}) is a boundary edge, then for some k, either $v'_{i+2} = v_k$ and $v'_{i+5} = v_{k+1}$ or $v'_{i+2} = v_{k+1}$ and $v'_{i+5} = v_k$. In the first case, C_1 is $(v_{i+1}, v_{i+2}, v'_{i+2} = v_k, v'_{i+5} = v_{k+1}, v_{i+5}, v_{i+6}, \cdots, v_{i+3}, v_{i+4}, \cdots, v_{i+1})$, which is Case 2. In the second case, C_0 is $(v_{i+4}, v_{i+5}, v'_{i+5} = v_k, v'_{i+2} = v_{k+1}, v_{i+2}, v_{i+3}, \cdots, v_i, v_{i+1}, \cdots, v_{i+4}) = (v_i, v_{i+1}, \cdots, v_{i+4}, v_{i+5}, v'_{i+5}, v'_{i+2}, v_{i+2}, v_{i+3}, \cdots, v_i)$, which is also Case 2.

The dihamiltonian decomposition C', C_0' and C_1' of G' is similar to the previous cases. Let C' be $(v_{n-1}, v_{n-2}, \cdots, v_{i+6}, v_{i+3}, v_i, v_{i-1}, \cdots, v_0, v_{n-1})$, and C_0' be $(v_i, v_{i+3}, v_{i+3}', P_{3,2}, v_{i+4}', v_{i+1}', P_{3,1}, v_{i+2}', v_{i+5}', P_{3,3}, v_i)$, and C_1' be $(v_{i+1}', v_{i+4}', P_{3,5}, v_{i+5}', v_{i+2}', P_{3,4}, v_{i+3}', v_{i+3}, v_{i+6}, P_{3,6}, v_{i+1}')$. Then C', C_0' and C_1' are edge-disjoint hamiltonian cycles in the symmetric digraph of G'.

It remains to consider the case that some vertices in $\{v_i, v_{i+1}, v_{i+2}, v_{i+3}, v_{i+4}, v_{i+5}, v_{i+6}, v_{i+1}', v_{i+2}', v_{i+3}', v_{i+4}', v_{i+5}'\}$ are not distinct. In case of $v_{i+1}' = v_{i+6}$, $P_{3,6}$ is a path with length 0. In case of $v_{i+5}' = v_i$, $P_{3,3}$ is a path with length 0. In either case, C', C_0' and C_1' are edge-disjoint hamiltonian cycles in G'. $\quad\Box$

Now we are ready to present our main theorem.

Theorem 2 *Every 3-regular graph G with $n = 4k$ vertices is not dihamiltonian decomposable.*

Proof We use induction on the number of vertices, n. When $n = 4$, G is a complete graph. Since G is not a bipartite graph, it is not dihamiltonian decomposable by Theorem 1. Assume that the theorem is true for $n = 4k \geq 4$. We will show that the theorem is true for $n = 4(k + 1)$. Suppose for a contradiction that there exists a 3-regular graph with $n = 4(k + 1)$ vertices which is dihamiltonian decomposable. Then from Lemma 2, there is a 3-regular digraph with $n = 4k$ vertices which is dihamiltonian decomposable. This leads to a contradiction. $\quad\Box$

4 Applications to Interconnection Network Topologies

In this section, we consider the dihamiltonian decomposition problem of interconnection network topologies whose vertex degree is 3 such as CCC(cube-connected cycles)[8], chordal rings[1] and $C_m \oplus C_m$.

4.1 CCC

The m-dimensional cube-connected cycles (CCC) is constructed from the m-dimensional hypercube by replacing each node of the hypercube with a cycle of m nodes in the CCC. The ith dimension edge incident to a node of the hypercube is then connected to the ith node of the corresponding cycle of the CCC. The m-dimensional CCC is a 3-regular graph.

Theorem 3 *Every m-dimensional CCC is not dihamiltonian decomposable, $m \geq 3$.*

Proof The number of vertices in m-dimensional CCC is $m2^m$, which is a multiple of 4. Therefore it is not dihamiltonian decomposable by Theorem 2. $\quad\Box$

4.2 Chordal Ring

For an even n and odd r, chordal ring $CR(n, r)$ is a graph G such that its vertex set is $\{v_0, v_1, \cdots, v_{n-2}, v_{n-1}\}$ and the edge set is $\{(v_i, v_j) \mid j =_{\bmod n} (i + 1), 0 \leq i \leq n - 1\} \cup \{(v_i, v_j) \mid j =_{\bmod n} (i + r), \text{ for even } i, 0 \leq i \leq n - 1\}$. The chordal ring is a 3-regular graph.

Theorem 4 *(a) Every chordal ring $CR(n,r)$ with $n = 4k$ is not dihamiltonian decomposable. (b) Chordal ring $CR(n,r)$ with $n = 4k + 2$ is dihamiltonian decomposable if $n/2$ is relatively prime to both $\lfloor r/2 \rfloor$ and $\lceil r/2 \rceil$.*

Proof The proof of (a) follows from Theorem 2. Let G be a chordal ring $CR(n,r)$ with $n = 4k + 2$. Let C be a hamiltonian cycle $(v_{n-1}, v_{n-2}, \cdots, v_1, v_0, v_{n-1})$ in the symmetric digraph of G. Since G_0^C and G_1^C are cycles with length $n/2$ if and only if $n/2$ is relatively prime to $\lfloor r/2 \rfloor$ and $\lceil r/2 \rceil$ respectively, G is dihamiltonian decomposable by Property 2.   □

$CR(4k + 2, 3)$, $CR(4k + 2, 2k + 1)$, and $CR(4k + 2, 4k - 1)$ are dihamiltonian decomposable by Theorem 4. The condition of Theorem 4 (b) is not a necessary condition. There are some dihamiltonian decomposable chordal rings $CR(4k + 2, r)$ without a dihamiltonian decomposition such that one of three hamiltonian cycles in the decomposition is $(v_{n-1}, v_{n-2}, \cdots, v_1, v_0, v_{n-1})$. $CR(30, 5)$ is such a chordal ring, which has the minimum number of vertices.

4.3 $C_m \oplus C_m$

$C_m \oplus C_m$ is a class of 3-regular graphs such that some edges between the vertices of two cycles with length m are added so that the degree of every vertex is 3. Petersen graph belongs to $C_m \oplus C_m$ where m is 5. The product of C_m and complete graph K_2, $C_m \times K_2$, also belongs to $C_m \oplus C_m$.

Theorem 5 *Every graph in $C_m \oplus C_m$ is not dihamiltonian decomposable, $m \geq 3$.*

Proof   Let G be a graph in $C_m \oplus C_m$. G has $2m$ vertices. If m is odd, G is not dihamiltonian decomposable since it is not a bipartite graph. If m is even, G is not dihamiltonian decomposable since the number of vertices is a multiple of 4.   □

5   Concluding Remarks

The dihamiltonian decomposition of regular graphs can be used to design reliable communication algorithms for broadcasting and multicasting under the wormhole routing model. We suggested some conditions for 3-regular graphs to be dihamiltonian decomposable. For a 3-regular graph G, it is bipartite if it is dihamiltonian decomposable. If the number of vertices is $4k$, then G is not dihamiltonian decomposable. Using these results, we showed that chordal rings with $4k$ vertices, m-dimensional CCCs, and $C_m \oplus C_m$ are not dihamiltonian decomposable. We also proposed a class of chordal rings with $4k + 2$ vertices which are dihamiltonian decomposable. The characterization for 3-regular graphs to have a dihamiltonian decomposition should be further studied.

References

1. B. W. Arden and H. Lee, "Analysis of chordal ring network," *IEEE Trans. Computers* **30**, pp. 291-295, 1981.
2. J. A. Bondy and U. S. R. Murty, *Graph Theory with Applications*, 5th printing, American Elsevier Publishing Co., Inc., 1976.
3. J. Bosák, *Decompositions of Graphs*, Kluwer Academic Publishers, Dordrecht, Netherlands, 1990.
4. S. J. Curran and J. A. Gallian, "Hamiltonian cycles and paths in cayley graphs and digraphs - a survey," *Discrete Mathematics* **156**, pp. 1-18, 1996.
5. G. Gauyacq, C. Micheneau, and A. Raspaud, "Routing in recursive circulant graphs: edge forwarding index and hamiltonian decomposition," in *Proc. of International Workshop on Graph-Theoretic Concepts in Computer Science WG'99*, Smolenice Castle, Slovak Republic, pp. 227-241, 1998.
6. S. Lee and K. G. Shin, "Interleaved all-to-all broadcast on meshes and hypercubes," *IEEE Trans. Parallel and Distributed Systems* **5**, pp. 449-458, 1994.
7. J.-H. Lee, C.-S. Shin, and K.-Y. Chwa, "Directed hamiltonian packing in d-dimensional meshes and its applications," in *Proc. of International Symposium on Algorithms and Computation ISAAC'96*, Osaka, Japan, pp. 295-304, 1996.
8. F. T. Leighton, *Introduction to Parallel Algorithms and Architectures: Arrays · Trees · Hypercubes*, Morgan Kaufmann Publishers, San Mateo, California, 1992.
9. J.-H. Park, "Hamiltonian decomposition of recursive circulants," in *Proc. of International Symposium on Algorithms and Computation ISAAC'98*, Taejon, Korea, pp. 297-306, 1998.

Box-Rectangular Drawings of Plane Graphs
(Extended Abstract)

Md. Saidur Rahman, Shin-ichi Nakano, and Takao Nishizeki

[1] Department of Computer Science and Engineering, Bangladesh University of
Engineering and Technology
Dhaka-100, Bangladesh `saidur@cse.buet.edu`
[2] Department of Computer Science, Gunma University
Kiryo 376-8515, Japan
`nakano@msc.cs.gunma-u.ac.jp`
[3] Graduate School of Information Sciences, Tohoku University
Sendai 980-8579, Japan
`nishi@ecei.tohoku.ac.jp`

Abstract. In this paper we introduce a new drawing style of a plane
graph G, called a "box-rectangular drawing." It is defined to be a drawing
of G on an integer grid such that every vertex is drawn as a rectangle,
called a box, each edge is drawn as either a horizontal line segment
or a vertical line segment, and the contour of each face is drawn as
a rectangle. We establish a necessary and sufficient condition for the
existence of a box-rectangular drawing of G. We also give a simple linear-
time algorithm to find a box-rectangular drawing of G if it exists.
Keywords: Graph, Algorithm, Graph Drawing, Rectangular Drawing,
Box-drawing, Box-rectangular drawing.

1 Introduction

Recently automatic drawings of graphs have created intense interest due to their
broad applications, and as a consequence, a number of drawing styles and corresponding drawing algorithms have come out [DETT94]. Among different drawing
styles, an "orthogonal drawing" has attracted much attention due to its beautiful
applications in circuit layouts, database diagrams, entity-relationship diagrams,
etc [B96, CP98, K96, T87]. An *orthogonal drawing* of a plane graph G is a drawing of G in which each vertex is drawn as a grid point on an integer grid and each
edge is drawn as a sequence of alternate horizontal and vertical line segments
along grid lines as illustrated in Fig. 1(a). Any plane graph with the maximum
degree at most four has an orthogonal drawing. However, a plane graph with a
vertex of degree 5 or more has no orthogonal drawing.

A *box-orthogonal drawing* of a plane graph G is a drawing of G on an integer
grid such that each vertex is drawn as a rectangle, called a *box*, and each edge
is drawn as a sequence of alternate horizontal and vertical line segments along
grid lines, as illustrated in Fig. 1(b). Some of the boxes may be degenerated
rectangles, i.e., points. A box-orthogonal drawing is a natural generalization

Widmayer et al. (Eds.): WG'99, LNCS 1665, pp. 250–261, 1999.

of an ordinary orthogonal drawing, and moreover, any plane graph has a box-orthogonal drawing even if there is a vertex of degree 5 or more. Several results are known for box-orthogonal drawings [BK97, FKK96, PT98].

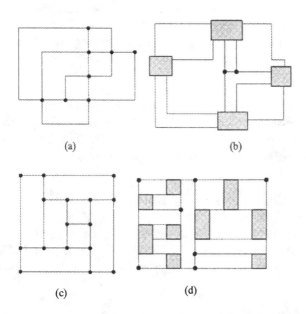

(a)   (b)

(c)   (d)

Fig. 1. (a) An orthogonal drawing, (b) a box-orthogonal drawing, (c) a rectangular drawing, and (d) a box-rectangular drawing.

An orthogonal drawing of a plane graph G is called a *rectangular drawing* of G if each edge of G is drawn as a straight line segment without bends and the contour of each face of G is drawn as a rectangle, as illustrated in Fig. 1(c). Since a rectangular drawing has practical applications in VLSI floorplanning, much attention has been paid to it [KK84, L90]. However, not every plane graph has a rectangular drawing. A necessary and sufficient condition for a plane graph G to have a rectangular drawing is known [T84], and several linear-time algorithms to find a rectangular drawing of G are also known [KH97, RNN98a].

Thus a box-orthogonal drawing is a generalization of an orthogonal drawing, while an orthogonal drawing is a generalization of a rectangular drawing. Hence an orthogonal drawing is an intermediate of a box-orthogonal drawing and a rectangular drawing. In this paper we introduce a new style of drawings as another intermediate of the two drawing styles. The new style is called a *box-rectangular drawing* and is formally defined as follows.

A *box-rectangular drawing* of a plane graph G is a drawing of G on an integer grid such that each vertex is drawn as a (possibly degenerated) rectangle, called a *box*, and the contour of each face is drawn as a rectangle, as illustrated in Fig. 1(d). If G has multiple edges or a vertex of degree 5 or more, then G has no rectangular drawing but may have a box-rectangular drawing. However, not every plane graph has a box-rectangular drawing. We will see in Section 2 that box-rectangular drawings have beautiful applications in floorplanning of MultiChip Modules (MCM) and in architectural floorplanning.

In this paper we establish a necessary and sufficient condition for the existence of a box-rectangular drawing of a plane graph, and give a linear-time algorithm to find a box-rectangular drawing if it exists. The sum of the width and the height of an integer grid required by a box-rectangular drawing is bounded by $m + 2$, where m is the number of edges in a given graph.

The rest of the paper is organized as follows. Section 2 describes some applications of box-rectangular drawings. Section 3 introduces some definitions and presents preliminary results. Section 4 deals with box-rectangular drawings of G for a special case where some vertices of G are designated as corners of the rectangle corresponding to the contour of the outer face. Section 5 deals with the general case where no vertex is designated as a corner.

2   Applications of Box-Rectangular Drawings

In this section we mention some applications of box-rectangular drawings.

As mentioned in Section 1, rectangular drawings have practical applications in VLSI floorplanning. In a VLSI floorplanning problem, an input is a plane graph F as illustrated in Fig. 2(a); F represents the functional entities of a chip, called *modules*, and interconnections among the modules; each vertex of F represents a module, and an edge between two vertices of F represents the interconnections between the two corresponding modules. An output of the problem for the input graph F is a partition of a rectangular chip area into smaller rectangles as illustrated in Fig. 2(d); each module is assigned to a smaller rectangle, and furthermore, if two modules have interconnections, then their corresponding rectangles must be adjacent, that is, must have a common boundary.

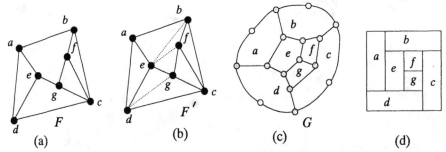

Fig. 2. Floorplanning by a rectangular drawing.

A conventional floorplanning algorithm using rectangular drawings is outlined as follows. First, obtain a graph F' by triangulating all inner faces of F as illustrated in Fig. 2(b), where dotted lines indicate new edges added to F. Then obtain a dual-like graph G of F' as illustrated in Fig. 2(c), where the four vertices of degree 2 drawn by white circles correspond to the four corners of the rectangular chip area. Finally, by finding a rectangular drawing of G, obtain a possible floorplan for F as illustrated in Fig. 2(d).

In the conventional floorplan above, two rectangles are always adjacent if the modules corresponding to them have interconnections. However, two rectangles may be adjacent even if the modules corresponding to them have no interconnections. For example, module e and module f have no interconnection in Fig. 2(a), but their corresponding rectangles are adjacent in the floorplan in Fig. 2(d). Such unwanted adjacencies are not desirable in some other floorplanning problems. In floorplanning of a MultiChip Module (MCM), two chips generating excessive heat should not be adjacent, or two chips operating on high frequency should not be adjacent to avoid malfunctioning due to their interference [S95]. Unwanted adjacencies may cause a dangerous situation in some architectural floorplanning, too [FW74]. For example, in a chemical industry, a processing unit that deals with poisonous chemicals should not be adjacent to a cafeteria.

We can avoid the unwanted adjacencies if we obtain a floorplan for F by using a box-rectangular drawing instead of a rectangular drawing, as follows. First, without triangulating the inner faces of F, find a dual-like graph G of F as illustrated in Fig. 3(b). Then, by finding a box-rectangular drawing of G, obtain a possible floorplan for F as illustrated in Fig. 3(c). In Fig. 3(c) rectangles e and f are not adjacent although there is a dead space corresponding to a vertex of G drawn by a rectangular box. Such a dead space to separate two rectangles in floorplanning is desirable for dissipating excessive heat in an MCM or for ensuring safety in a chemical industry.

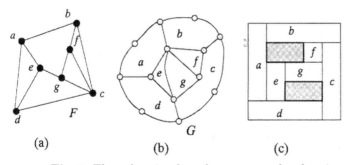

Fig. 3. Floorplanning by a box-rectangular drawing.

3   Preliminaries

In this section we give some definitions and present preliminary results.

Throughout the paper we assume that a *graph* G is a so-called multigraph which has no self loops but may have *multiple edges*, i.e., edges sharing both ends. If G has no multiple edges, then G is called a *simple* graph. We denote the set of vertices of G by $V(G)$, and the set of edges of G by $E(G)$. Let $n = |V(G)|$

and $m = |E(G)|$. The *degree* of a vertex v, denoted by $d(v)$, is the number of edges incident to v in G. We denote the maximum degree of a graph G by $\Delta(G)$ or simply by Δ.

A *cycle* C is an alternating sequence $v_0, e_0, v_1, e_1, \cdots, e_{l-1}, v_l (= v_0)$ of vertices and edges where $e_i = v_i v_{i+1}$ for each i and all edges $e_0, e_1, \cdots, e_{l-1}$ are distinct each other. If no vertex appears twice or more on C, then C is called a *simple cycle*. Thus a pair of multiple edges is a simple cycle. We hereafter call a simple cycle a cycle unless otherwise specified.

A graph is *planar* if it can be embedded in the plane so that no two edges intersect geometrically except at a vertex to which they are both incident. A *plane* graph is a planar graph with a fixed embedding in the plane. A plane graph G divides the plane into connected regions called *faces*. The unbounded region is called the *outer face*. We regard the *contour* of a face as a cycle formed by the edges on the boundary of the face. We denote the contour of the outer face of G by $C_o(G)$.

Let G be a plane connected graph. For a cycle C in G, we denote by $G(C)$ the plane subgraph of G inside C (including C). An edge which is incident to exactly one vertex of a cycle C and located outside of C is called a *leg* of C, and the vertex of C to which the leg is incident is called a *leg-vertex* of C. A cycle C in G is called a k-*legged cycle* if C has exactly k legs. We say that cycles C and C' in a plane graph G are *independent* if $G(C)$ and $G(C')$ have no common vertex. A set S of cycles is *independent* if any pair of cycles in S are independent.

We often use the following operation on a plane graph G [O67]. Let v be a vertex of degree d in a plane graph G, let $e_1 = vw_1, e_2 = vw_2, \cdots, e_d = vw_d$ be the edges incident to v, and assume that these edges $e_1, e_2, \cdots, e_d$ appear clockwise around v in this order. Replace v with a cycle $v_1, v_1 v_2, v_2, v_2 v_3, \cdots, v_d v_1, v_1$, and replace the edges vw_i with $v_i w_i$ for $i = 1, 2, \cdots, d$. We call the operation above the *replacement of a vertex by a cycle*. The cycle $v_1, v_1 v_2, v_2, v_2 v_3, \cdots, v_d v_1, v_1$ in the resulting graph is called the *replaced cycle* corresponding to vertex v of G.

We often construct a new graph from a graph as follows. Let v be a vertex of degree 2 in a connected graph G. We replace the two edges $u_1 v$ and $u_2 v$ incident to v with a single edge $u_1 u_2$, and delete v. We call the operation above *the removal of a vertex of degree 2* from G.

The *width* and the *height* of a rectangular drawing D of G is the width and the height of the rectangle corresponding to $C_o(G)$. The following result on rectangular drawings is known.

Lemma 1. *Let G be a connected plane graph such that all vertices have degree 3 except four vertices of degree 2 on $C_o(G)$. Then G has a rectangular drawing if and only if G has none of the following three types of cycles [T84]: (r1) 1-legged cycles, (r2) 2-legged cycles which contain at most one vertex of degree 2, and (r3) 3-legged cycles which contain no vertex of degree 2. Furthermore one can check in linear time whether G satisfies the condition above, and if G does then one can find a rectangular drawing of G in linear time. The sum of the width and the height of the produced rectangular drawing is bounded by $\frac{n}{2}$, where n is the number of vertices in G [RNN98a].*   □

We now give some definitions regarding box-rectangular drawings. We say that a vertex of graph G is drawn as a *degenerated box* in a box-rectangular drawing D if the vertex is drawn as a point in D. We often call a degenerated box in D a *point* and call a non-degenerated box a *real box*. We call the rectangle corresponding to $C_o(G)$ the *outer rectangle*, and we call a corner of the outer rectangle simply a *corner*. A box in D containing at least one corner is called a *corner box*. A corner box may be degenerated. The *width* of a box-rectangular drawing D is the width of the outer rectangle. The *height* of D is defined in a similar manner.

If $n = 1$, that is, G has exactly one vertex, then the box-rectangular drawing of G is trivial; the drawing is just a degenerated box corresponding to the vertex.

Thus in the rest of the paper, we may assume that $n \geq 2$. We now have the following four facts and a lemma.

Fact 31 *Any corner box in a box-rectangular drawing contains either one or two corners.*   □

Fact 32 *Any box-rectangular drawing has either two, three, or four corner boxes.*
                                                                        □

Fact 33 *In a box-rectangular drawing D of G, any vertex v of degree 2 or 3 satisfies the following (i), (ii) or (iii). (i) Vertex v is drawn as a point containing no corner; (ii) v is drawn as a corner box containing exactly one corner; and (iii) v is drawn as a real (corner) box containing exactly two corners.*   □

Fact 34 *In a box-rectangular drawing D of G, every vertex of degree 5 or more is drawn as a real box.*   □

Lemma 2. *If G has a box-rectangular drawing, then G has a box-rectangular drawing in which every vertex of degree 4 or more is drawn as a real box.*   □

The choice of vertices as corner boxes plays an important role in finding a box-rectangular drawing. For example, the graph in Fig. 4(a) has a box-rectangular drawing if we choose vertices a, b, c and d as corner boxes as illustrated in Fig. 4(g). However, the graph has no box-rectangular drawing if we choose vertices p, q, r and s as corner boxes. If all vertices corresponding to corner boxes are designated for a drawing, then it is rather easy to determine whether G has a box-rectangular drawing with the designated corner boxes. We deal this case in Section 4. In Section 5 we deal with the general case where no vertex of G is designated as corner boxes.

4   Box-Rectangular Drawing with Designated Corner Boxes

In this section we establish a necessary and sufficient condition for the existence of a box-rectangular drawing of a plane graph G when all vertices of G corresponding to corner boxes are designated, and we also give a simple linear-time algorithm to find such a box-rectangular drawing of G if it exists.

By Fact 32, any box-rectangular drawing has either two, three or four corner boxes. In Section 4.1 we consider the case where exactly four vertices are designated as corner boxes, and in Section 4.2 we consider the case where two or three vertices are designated as corner boxes.

4.1   Drawing with Exactly Four Designated Corner Boxes

In this section we assume that exactly four vertices a, b, c and d in a given plane graph G are designated as corner boxes. We construct a new graph G'' from G through an intermediate graph G', and reduce the problem of finding a *box-rectangular drawing* of G with four designated vertices to a problem of finding a *rectangular drawing* of G''.

A plane graph having a vertex of degree 1 has no box-rectangular drawing. Also, a plane graph having a cycle with exactly one leg-vertex has no box-rectangular drawing. Thus we may assume that our input plane graph G has no vertex of degree 1 and has no cycle with exactly one leg-vertex.

We first construct G' from G as follows. If a vertex of degree 2 in G, as vertex d in Fig. 4(a), is a designated vertex, then it is drawn as a corner point in a box-rectangular drawing of G. On the other hand, if a vertex of degree 2, as vertex t in Fig. 4(a), is not a designated vertex, then it is drawn as a point on a vertical or horizontal line segment corresponding to the two edges incident to it, as illustrated in Fig. 4(g). Thus we remove all non-designated vertices of degree 2 one by one from G. The resulting graph is G'. Thus all vertices of degree 2 in G' are designated vertices. Note that some new multiple edges may appear in G'. Clearly G has a box-rectangular drawing with the four designated corner boxes if and only if G' has a box-rectangular drawing with the four designated corner boxes. Fig. 4(a) illustrates a plane graph G with four designated vertices a, b, c and d, and Fig. 4(b) illustrates G'. Fig. 4(f) illustrates a box-rectangular drawing D' of G', and Fig. 4(g) illustrates a box-rectangular drawing D of G.

Since every vertex of degree 2 in G' is a designated vertex, it is drawn as a (corner) point in any box-rectangular drawing of G'. Every designated vertex of degree 3 in G', as vertex a in Fig. 4(b), is drawn as a real box since it is a corner. On the other hand, every non-designated vertex of degree 3 in G' is drawn as a point. These facts together with Lemma 2 imply that if G' has a box-rectangular drawing then G' has a box-rectangular drawing D' in which all designated vertices of degree 3 and all vertices of degree 4 or more in G' are drawn as real boxes.

We now construct G'' from G'. Replace by a cycle each of the designated vertices of degree 3 and the vertices of degree 4 or more, as illustrated in Fig. 4(c). The replaced cycle corresponding to a designated vertex x of degree 3 or more contains exactly one edge, say e_x, on the contour of the outer face, where $x = a, b, c$ or d. Put a dummy vertex x' of degree 2 on e_x. The resulting graph is G''. We let $x' = x$ if a designated vertex x has degree 2. (See Fig. 4(d).) Now G'' has exactly four vertices a', b', c', and d' of degree 2 on $C_o(G'')$, and all other vertices have degree 3.

We now have the following theorem.

Theorem 1. *Let G be a connected plane graph with four designated vertices a, b, c and d on $C_o(G)$, and let G'' be the graph transformed from G as mentioned above. Then G has a box-rectangular drawing with four corner boxes corresponding to a, b, c and d if and only if G'' has a rectangular drawing.*

Proof. Necessity. Assume that G has a box-rectangular drawing with the four designated vertices a, b, c and d as corner boxes. Then by Fact 33 and Lemma 2 G has a box-rectangular drawing D in which all designated vertices of degree 3 and all vertices of degree 4 or more are drawn as real boxes, as illustrated in Fig. 4(g). The drawing D immediately yields a box-rectangular drawing D' of G', and D' immediately gives a rectangular drawing D'' of G''.

Sufficiency. Assume that G'' has a rectangular drawing D'' as illustrated in Fig. 4(e). In D'', each replaced cycle is drawn as a rectangle, since it is a face in G''. Furthermore, the four vertices a', b', c' and d' of degree 2 in G'' are drawn as corners of the rectangle corresponding to $C_o(G'')$. Therefore, D'' immediately gives a box-rectangular drawing D' of G' having the four vertices a, b, c and d as corner boxes, as illustrated in Fig. 4(f). Then, inserting non-designated vertices of degree 2 on horizontal or vertical line segments in D', one can immediately obtain from D' a box-rectangular drawing D of G having the designated vertices a, b, c and d as corner boxes, as illustrated in Fig. 4(g). □

We now have the following theorem.

Theorem 2. *Given a plane graph G with m edges and four designated vertices a, b, c and d on $C_o(G)$, one can determine in $O(m)$ time whether G has a box-rectangular drawing with a, b, c and d as corner boxes, and if G has, then one can find a box-rectangular drawing of G in $O(m)$ time. The sum of the width and the height of a produced box-rectangular drawing of G is bounded by $m + 2$.*

Proof. Time Complexity. Clearly one can construct G'' from G in time $O(m)$. G'' is a connected plane graph such that all vertices have degree 3 except vertices a', b', c' and d' of degree 2. Therefore, by Lemma 1 one can determine in linear time whether G'' has a rectangular drawing or not and find a rectangular drawing D'' of G'' if it exists. One can easily obtain a box-rectangular drawing D of G from D'' in linear time.

Grid size. Let n_2 be the non-designated vertices of degree 2 in G. Let $n' = |V(G')|$ and $m' = |E(G')|$. Then $m' = m - n_2$. We replace some vertices of G' by cycles and add at most 4 dummy vertices to construct G'' from G'. Therefore G'' has at most $2m' + 4$ vertices. From Lemma 1, the sum of the width and the height of the produced rectangular drawing of G'' is bounded by $\frac{2m'+4}{2} = m' + 2$. Now the insertion of a vertex of degree 2 on a horizontal line segment or a vertical line segment increases the width or the height of the box-rectangular drawing by at most one. Thus the sum of the width and the height of the produced box-rectangular drawing of G is bounded by $m' + 2 + n_2 = m + 2$. □

There are infinitely many cycles with four designated vertices for which the sum of the width and the height of any box-rectangular drawing of the cycles is $m - 2$.

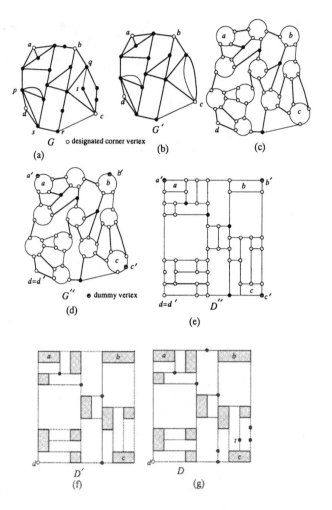

Fig. 4. Illustration of G, G', G'', D'', D' and D.

4.2   Drawing with Two or Three Designated Corner Boxes

If two or three vertices in G are designated as corner boxes and no other vertex can be a corner box, then we can easily reduce this case to the case where exactly four vertices are designated as corner boxes. The details are omitted in this extended abstract.

5   Box-Rectangular Drawing with No Designated Corner Boxes

In this section we consider the general case where no vertex of G is designated as a corner box.

By Fact 32 there are either two, three or four corner boxes in any box-rectangular drawing of G. Therefore, considering all combinations of two, three and four vertices on $C_o(G)$ as corner boxes and applying the algorithm in the previous section for each of the combinations, one can determine whether G has a box rectangular drawing. Such a straightforward method requires time $O(n^5)$ since there are $O(n^4)$ combinations and the algorithm in Section 4 can examine in time $O(n)$ whether G has a box-rectangular drawing for each of them. In this section we first establish a necessary and sufficient condition for the existence of a box-rectangular drawing of G when no vertex is designated as a corner box, and then show that the characterization leads to a linear-time algorithm.

In Section 5.1, we first derive a necessary and sufficient condition for a plane graph G with the maximum degree $\Delta \leq 3$ to have a box-rectangular drawing, and then give a linear-time algorithm to obtain a box-rectangular drawing of such a plane graph G if it exists. In Section 5.2 we give a linear-time algorithm for a plane graph G with the maximum degree $\Delta \geq 4$ by modifying the algorithm in Section 5.1.

5.1   Box-Rectangular Drawing of G with $\Delta \leq 3$.

Let G be a plane graph with the maximum degree at most 3. As in Section 4, we can assume that G is connected and has neither a vertex of degree 1 nor a 1-legged cycle. The following theorem is a main result of Section 5.1.

Theorem 3. *A plane connected graph G with $\Delta \leq 3$ has a box-rectangular drawing if and only if G satisfies the following two conditions: (br1) every 2-legged or 3-legged cycle in G contains an edge on $C_o(G)$; and (br2) any set S of independent cycles in G satisfies $2 \cdot |S_2| + |S_3| \leq 4$, where S_2 is the set of 2-legged cycles in S and S_3 is the set of 3-legged cycles in S.* □

Before proving the necessity of Theorem 3, we observe the following fact.

Fact 51 *In a box-rectangular drawing D of G, any 2-legged cycle of G contains at least two corners, any 3-legged cycle of G contains at least one corner, and any cycle with four or more legs may contain no corner.* □

We now prove the necessity of Theorem 3.

Necessity of Theorem 3. Assume that G has a box-rectangular drawing D. By Fact 51 any 2-legged or 3-legged cycle in D contains a corner, and hence contain an edge on $C_o(G)$.

Let S be any set of independent cycles in G. Then by Fact 51 any 2-legged cycle in S contains at least two corners and any 3-legged cycle in S contains at least one corner. Since all these cycles in S are independent, they are vertex-disjoint each other. Therefore there are at least $2 \cdot |S_2| + |S_3|$ corners in D. Since there are exactly four corners in D, we have $2 \cdot |S_2| + |S_3| \leq 4$. □

We give a constructive proof for the sufficiency of Theorem 3. Our proof, which is omitted in this extended abstract, implies the following corollary.

Corollary 1. *A plane connected graph G with $\Delta \leq 3$ has a box-rectangular drawing if and only if G satisfies the following four conditions: (c1) every 2-legged or 3-legged cycle in G contains an edge on $C_o(G)$; (c2) at most two 2-legged cycles of G are independent each other; (c3) at most four 3-legged cycles of G are independent of each other; and (c4) if G has a pair of independent 2-legged cycles C_1 and C_2, then $\{C_1, C_2, C_3\}$ is not independent for any 3-legged cycle C_3 in G, and neither $G(C_1)$ nor $G(C_2)$ has more than two independent 3-legged cycles of G.*   □

Using a method similar to ones in [RNN98a, RNN98b], one can determine whether a given plane graph with $\Delta \leq 3$ satisfies the conditions in Corollary 1 in time $O(m)$. If G satisfies the conditions in Corollary 1, then one can construct a box-rectangular drawing of G in linear time, following the method described in the constructive proof of the sufficiency of Theorem 3 and using the algorithm in [RNN98a]. (We omit the proof in this extended abstract.) We, thus, have the following theorem.

Theorem 4. *Given a plane graph with the maximum degree $\Delta \leq 3$, one can determine in time $O(m)$ whether G has a box-rectangular drawing or not, and if G has, one can find a box-rectangular drawing of G in time $O(m)$, where m is the number of edges in G. The sum of the width and the height of a produced box-rectangular drawing is bounded by $m + 2$.*   □

5.2   Box-Rectangular Drawings of Graphs with $\Delta \geq 4$.

In this section we give a necessary and sufficient condition for a plane graph with $\Delta \geq 4$ to have a a box-rectangular drawing when no vertex is designated as corner boxes.

Let G be a plane connected graph with $\Delta \geq 4$. We first transform G into a graph H with $\Delta(H) \leq 3$, and then obtain a box-rectangular drawing of G by applying the algorithm in section 5.1 to H with appropriately choosing designated vertices.

We construct H from G by replacing each vertex v of degree four or more in G by a cycle. Each replaced cycle corresponds to a real box in a box-rectangular drawing. We do not replace a vertex of degree 2 or 3 by a cycle since such a vertex may be drawn as a point by Fact 33. Thus $\Delta(H) \leq 3$. We now have the following theorem.

Theorem 5. *Let G be a connected plane graph with no vertex of degree 1, and let H be the graph transformed from G as above. Then G has a box-rectangular drawing if and only if H has a box-rectangular drawing.*   □

It is rather easy to prove the necessity of Theorem 5; one can easily transform any box-rectangular drawing of G to a box-rectangular drawing of H. On the other hand, it is not trivial to prove the sufficiency. However, we give a method to find a box-rectangular drawing of G in linear time if H has a box-rectangular drawing. The detail is omitted in this extended abstract.

References

B96.   T. C. Biedl, *Optimal orthogonal drawings of triconnected plane graphs*, Proc. SWAT'96, Lect. Notes in Compute Science, Springer, 1097 (1996), pp. 333-344.

BK97.   T. C. Biedl and M. Kaufmann, Area-efficient static and incremental graph drawings, Proc. of 5th European Symposium on Algorithms, Lect. Notes in Computer Science, Springer, 1284 (1997), pp. 37-52.

CP98.   T. Calamoneri and R. Petreschi, Orthogonally drawing cubic graphs in parallel, Journal of Parallel and Distributed Comp., 55 (1998), pp. 94-108.

DETT94.   G. de Battista, P. Eades, R. Tamassia and I. G. Tollis, *Algorithms for drawing graphs: an annotated bibliography*, Comp. Geom. Theory Appl., 4 (1994), pp. 235-282.

FKK96.   U. Fößmeier, G. Kant and M. Kaufmann, *2-visibility drawings of plane graphs*, Proc. of Graph Drawing '96, Lect. Notes in Compute Science, Springer, 1190 (1997), pp. 155-168.

FW74.   R. L. Francis and J. A. White, *Facility Layout and Location*, Prentice-Hall, Inc, New Jersey, 1974.

K96.   G. Kant, *Drawing planar graphs using the canonical ordering*, Algorithmica (1996) 16, pp. 4-32.

KH97.   G. Kant and X. He, *Regular edge labeling of 4-connected plane graphs and its applications in graph drawing problems*, Theoretical Computer Science, 172 (1997), pp. 175-193.

KK84.   K. Kozminski and E. Kinnen, *An algorithm for finding a rectangular dual of a planar graph for use in area planning for VLSI integrated circuits*, Proc. of 21st DAC, (1984), pp. 655-656.

L90.   T. Lengauer, *Combinatirial Algorithms for Integrated Circuit Layout*, John Wiley & Sons, Chichester, 1990.

O67.   O. Ore, *The Four Color Problem*, Academic press, New York, 1967.

PT98.   A. Papakostas and I. G. Tollis, *Orthogonal drawings of high degree graphs with small area and few bends*, Proc. of 5th Workshop on Algorithms and Data Structures, Lect. Notes in Computer Science, Springer, 1272 (1998), pp. 354-367.

RNN98a.   M. S. Rahman, S. Nakano and T. Nishizeki, *Rectangular grid drawings of plane graphs*, Comp. Geom. Theo. Appl. 10 (3) (1998), pp. 203-220.

RNN98b.   M. S. Rahman, S. Nakano and T. Nishizeki, *A linear algorithm for optimal orthogonal drawings of triconnected cubic plane graphs*, Proc. of Graph Drawing'97, Lect. Notes in Computer Science, Springer, 1353 (1998), pp. 99-110. Also, Journal of Graph Alg. Appl., to appear.

S95.   N. Sherwani, *Algorithms for VLSI Physical Design Automation*, 2nd edition, Kluwer Academic Publishers, Boston, 1995.

T84.   C. Thomassen, *Plane representations of graphs*, (Eds.) J.A. Bondy and U.S.R. Murty, Progress in Graph Theory, Academic Press Canada, (1984), pp. 43-69.

T87.   R. Tamassia, *On embedding a graph in the grid with the minimum number of bends*, SIAM J. Comput. 16 (1987), pp. 421-444.

A Multi-scale Algorithm for Drawing Graphs Nicely

Ronny Hadany and David Harel

The Weizmann Institute of Science,
Rehovot, Israel
harel@wisdom.weizmann.ac.il

Abstract. We describe a multi-scale approach to the problem of drawing undirected straight-line graphs "nicely". In contrast to conventional global/dynamic algorithms, we employ increasingly coarser-scale representations of the graph, together with a simple local organization scheme, without imposing formal criteria on the quality of the picture at large. Our algorithm can deal easily with very large graphs; some of our examples contain over 1000 vertices.

1   Introduction

A large amount of work on the problem of graph layout has been carried out in recent years, resulting in a number of sophisticated and powerful algorithms. An extensive and detailed survey can be found in [DETT]. One of the most basic problems in this area involves laying out arbitrary undirected graphs, with edges drawn as straight-line segments. The goal is to try to achieve as "nice" a drawing as possible.

A popular approach to this problem is the spring method, first proposed by Eades [E], which likens a graph to a mechanical collection of rings and connecting springs and strives to achieve physical-like equilibrium. Notable work that followed [E] includes the algorithms in [KK] and [FR].

In [DH], we used a different method, whereby simulated annealing was used to maximize a cost function reflecting the aesthetic quality of the drawing. The criteria incorporated into the cost function were: (i) distributing vertices away from each other, (ii) keeping edges short, (iii) minimizing edge crossings, and (iv) keeping vertices from coming too close to edges. The algorithm of [DH] performs well on small graphs, but becomes unsatisfactory when applied to graphs of over 30 vertices or so, especially with respect to minimizing edge crossings. Some of the planar graphs that do not result in planar layouts are particularly annoying. In general, these limitations apply to the other algorithms too.

Later, in [SH], we constructed a system that first carried out a rather complex set of preprocessing steps, designed to produce a topologically good, but not necessarily nice-looking layout, and then subjected it to a downhill-only version of the simulated annealing algorithm of [DH]. The intermediate layout is planar for planar graphs, since the first step of the algorithm in [SH] is to test

Widmayer et al. (Eds.): WG'99, LNCS 1665, pp. 262–277, 1999.

for planarity, producing a planar layout whenever possible. Planar graphs with around 50 vertices, for example, yield excellent results. For 50-vertex graphs that can be made planar by extracting a small number of edges, the results are still good, but for graphs that are far from planar the results can still be worse than manually produced drawings, even for medium sized inputs. For larger graphs, the performance of the algorithm in [SH] further decreases, a fact that is true even for planar graphs, since although the layout will be planar, the downhill annealing finds it increasingly more difficult to remove extreme distortions in the results of the preprocessing stages.

In this paper, we approach the problem in quite a different way. We do not impose any formal criteria on the quality of the final picture, nor do we liken the graph to a physical system; in fact, we do not work on the graph globally at all. Instead, we work locally, in a multi-level (or multi-scale) fashion; see, e.g., [B1,B2]. The algorithm, described in Sections 3 and 4, incorporates a family of natural coarse-scale representations of the graph, together with a simple local organization scheme. In this respect, our work can be viewed as an alternative to the traditional global/dynamic approach that is often used, and which we discuss briefly in Section 2.

The algorithm can deal easily with very large graphs, as illustrated in Section 5, where we apply it to some 400-vertex and 1000-vertex graphs. It also seems to implicitly suggest a different notion of aesthetics for diagrammatic structures, which we discuss briefly in Section 6.

We should remark that in very recent work, joint with Yehuda Koren [HHK], the topological method we use here for coarsening has been changed, which, together with some additional improvements makes it possible to deal easily, and in fact much faster, with graphs of several thousand vertices. These results will be published separately.

2   The Global/Dynamic Approach

Given a graph $G = (V, E)$, with $V = \{v_1, \ldots, v_n\}$, asking for a "nice" drawing of G is an ill-posed problem. Certain regularity can be achieved by reformulating the request as a minimization problem, in the following way. Each vertex v_i is identified with a point in the plane, $r_i = (r_i^1, r_i^2)$, and an edge (v_i, v_j) will be drawn as a straight line from r_i to r_j. Based on the graph structure, one defines an energy functional (or cost function) $H(\bar{r}) = H(r_1, r_2, \ldots, r_n)$, which is designed carefully to capture the interactions of the vertices and the edges in a way that the most pleasant configuration thereof assigns H the minimum value. A global/dynamic graph drawing algorithm is then simply a way of minimizing H and finding its minimum energy configuration $\bar{r}^*$.

A general approach to this search is the *vertex-by-vertex minimization*, or *relaxation*. At each step of the process, the position of only one vertex, say r_i, is changed, keeping the positions of all others fixed. The new value of r_i is chosen so as to reduce $H(\bar{r})$ as much as possible. Repeating this step for all vertices is called a *relaxation sweep*. By performing a sufficiently long sequence of relaxation sweeps, one hopes to be able to get as close to $\bar{r}^*$ as one wishes. This minimization procedure, although seemingly quite reasonable, is actually quite naive, and suffers from the following drawbacks:

- Slow convergence: In each relaxation, only a single vertex is allowed to move, while all others are held fixed. Thus, many relaxations might be needed to obtain a new configuration $\bar{r}$ globally different from the initial configuration $\bar{r}_0$. For graphs with n vertices, convergence to a minimum can take $\Omega(n^2)$ sweeps.
- High sweep cost: The energy functionals in question often contain $\Omega(n^2)$ summands, since they typically consider properties of each pair of vertex locations. This bounds from below the cost of the calculations in each sweep. The cost, combined with the convergence time, yields a total of $\Omega(n^4)$ operations.
- Convergence to local minima: Most kinds of minimization schemes, such as vertex-by-vertex minimization, get easily trapped in local minima, thus failing to lead to true minimal configurations. This phenomenon is especially severe when the energy functional has a multitude of local minima and multi-scale attraction basins. In such cases, simple stochastic minimization schemes, such as the simulated annealing of [DH], do not help either.

The algorithm we propose here does not utilize a cost function that purports to capture aesthetics in a quantitive way. Rather, we use a relatively simple procedure for local organization of neighborhoods, but in a multi-scale fashion, with the scaling itself constituting the subtle part of the algorithm.

3   Multi-scale Representation and Relaxation

We first develop a notion of scale relevant to the graph drawing problem.

A natural strategy for drawing a graph nicely is to first consider an abstraction, disregarding some of the graph's fine details. This abstraction is then drawn, yielding a "rough" layout in which only the general structure is revealed. Then the details are added and the layout is corrected. To employ such a strategy it is crucial that the abstraction retain essential features of the graph. Thus, one has to define the notion of coarse-scale representations of a graph, in which the combinatorial structure is significantly simplified but features important for visualization are well preserved. The drawing process will then "travel" between these representations, and introduce multi-scale corrections.

Assuming we have already defined the multiple levels of coarsening, the general structure of our strategy is as follows:

1. Perform *fine-scale* relocations of vertices that yield a locally organized configuration.
2. Perform *coarse-scale* relocations (through local relocations in the coarse representations), correcting global disorders not found in stage 1.

3. Perform *fine-scale* relocations, correcting local disorders introduced by stage 2.

We now formalize the notion of multi-scale representations, and then devise a natural family of relaxation schemes, determining the rules of interaction between scales. The algorithm itself is presented in Section 4.

Multiple scales

Given a graph $G = (V, E)$, we construct a sequence of graphs $G = G_0 = (V_0, E_0), \ldots, G_k = (V_k, E_k)$, where $|V_{i+1}| = c|V_i|$, for constant $0 < c < 1$. The graph G_{i+1} is derived from G_i by edge contractions, which are defined later, in such a way that the set E_{i+1} of vertices of G_{i+1} is the set $\{[u] | u \in E_i\}$ of equivalence classes, or *clusters*, of the vertices of G_i, by the relation that puts u and v into the same class $[u] = [v]$ whenever $(u, v) \in E_i$ is a contracted edge.

The construction of G_{i+1} from G_i is carried out iteratively: Taking G_i as the initial graph $G^0 = G_i$, in iteration l an edge $(u, v) \in G^l$ is contracted, forming a new graph G^{l+1}. The edge (u, v) is chosen to be optimal with respect to a cost $\mu = \mu(u, v)$, which is a convex combination of the following three objectives:

- The *cluster number:* $Cl(u, v) = |C_{G^l}(u)| + |C_{G^l}(v)|$
- The *degree number:* $D(u, v) = \max(deg_{G^l}(u), deg_{G^l}(v))$
- The *homotopic number:* $Ho(u, v) = |\{w \in G^l | (w, u), (w, v) \in E_{G^l}\}|$

The cost μ is devised so that the graph G_{i+1}, although being combinatorially much simpler than G_i (containing significantly less vertices), captures some basic topological properties. Minimizing the cluster number encourages uniform covering of G_i by G_{i+1} with small clusters, thus preserving graph proportions. Namely, for every pair of vertices $u, v \in V_i$, the distances satisfy $d_{G_i}(u, v) \approx d_{G_{i+1}}([u], [v])$. Minimizing the degree number encourages vertices in G_i with high degree to be transformed unaltered to G_{i+1}; that is, for such a u we might have $[u] = \{u\}$.

Minimizing the homotopic number restricts G_{i+1} so as not to "oversimplify" G_i. In fact, it aims at preserving the Euler characteristic of G_i. The Euler characteristic of a graph $G = (V, E)$, denoted by $\chi(G) = |V| - |E|$, determines (rather surprisingly) a rather strong topological invariant of G, namely, its fundamental group. This is a free group over a set of generators Y, with $|Y| = 1 - \chi(G)$. Informally, one can view the fundamental group of a graph as characterizing its cellular structure (or its structure as a labyrinth). Minimizing the reduction of the value $1 - \chi$ (a contraction step can only decrease $1 - \chi$) helps preserve the cellular structure of the graph.

The sequence $G_0, \ldots, G_k$ constitutes a multi-scale representation of the graph G. We refer to graph G_i as the *scale i representation of G*.

Energy functionals

In each scale i the representation G_i is associated with a function $X_i : V_i \longrightarrow R^2$, which we refer to as the *scale i configuration* of G. The relaxation scheme in scale i acts on X_i, and is based on the heuristic that "nice" drawing relates to "good" isometry. More precisely, a configuration $X : V \longrightarrow R^2$ that preserves combinatorial distances, i.e., for all $u, v \in V$, $d_G(u, v) \sim \|X(u) - X(v)\|$, is claimed to be "nice". Accordingly, in every scale i we utilize an energy functional of the form

$$H_i = \sum_{u,v \in V_i} w_{uv}(\|X_i(u) - X_i(v)\| - hl_{uv})^2$$

Here h is a scaling factor, and the numbers w_{uv} and l_{uv} are defined as follows. In the initial scale (scale 0) l_{uv} is the combinatorial distance $d_{G_0}(u,v)$ between the vertices u and v in G_0, and $w_{uv} = 1/d_{G_0}(u,v)$ for all $u,v \in V_0$. The w_{uv} and l_{uv} for H_{i+1} are obtained from those of H_i as follows. (For clarity we use A and B to denote vertices of G_{i+1}.)

$$H_{i+1} = \sum_{A,B \in V_{i+1}} w_{AB}(\|X_{i+1}(A) - X_{i+1}(B)\| - hl_{AB})^2$$

where

$$w_{AB} = \sum_{u \in A\ v \in B} w_{uv} \quad \text{and} \quad l_{AB} = \frac{\sum_{u \in A\ v \in B} w_{uv}l_{uv}}{\sum_{u \in A\ v \in B} w_{uv}}$$

Now, from H_i we derive a family $\{H_i^j \mid j = 1,\ldots,diam(G_i)\}$ of functionals:

$$H_i^j = \sum_{u \in V_i} \sum_{d_{G_i}(u,v) \leq j} w_{uv}(\|X_i(u) - X_i(v)\| - hl_{uv})^2$$

Each functional H_i^j is a localization of the functional H_i, in the following sense. If one thinks of l_{uv}, for $u,v \in V_i$, as a metric defined on G_i (except for scale 0, l_{uv} is not necessarily a true metric), then the minimum configuration X^* with respect to H_i is the best approximation of an isometry of G_i into R^2. According to this interpretation, it follows that the minimum configuration X_i^{j*} of H_i^j is the best approximation of a local isometry of G_i into R^2. This minimum X_i^{j*} is local, in the sense that it approximates an isometry only when restricted to neighborhoods of radius $\leq j$ in G_i.

Relaxation scheme

For each scale i, we define an (i,j) relaxation sweep as follows. Visit the vertices of G_i in some order, and for each vertex $u \in V_i$ employ a steepest descent minimization procedure with respect to the functional H_i^j. Namely, for every vertex u, and on the basis of the current configuration X_i, first compute the partial derivative of H_i^j with respect to $X_i(u)$:

$$\frac{\partial H_i^j}{\partial X_i(u)} = \sum_{d_{G_i}(u,v) \leq j} \frac{\|X_i(u) - X_i(v)\|/h - l_{uv}}{\|X_i(u) - X_i(v)\|/h}(X_i(u) - X_i(v))$$

Next, compute the unit vector

$$D_u = -\frac{\partial H_i^j}{\partial X_i(u)} \bigg/ \left\|\frac{\partial H_i^j}{\partial X_i(u)}\right\|$$

Finally, change $X_i(u)$ in the direction D_u as long as the energy H_i^j is reduced.

4 The Multi-scale Algorithm

The algorithm has two phases. The first contains preprocessing operations. It constructs the representations $G_0, \ldots, G_k$ and the functionals H_i^j. The second phase is the drawing process, and consists of a recursive procedure with following schematic structure:

1. pre-relaxation
2. gravitation
3. recursive call
4. interpolation
5. post-relaxation

At level i of the recursion, the algorithm acts on the representation G_i given with an initial configuration X_i^0. Step 1 is a sequence of (i, j) relaxation sweeps for $j = j_1^0, j_2^0, \ldots, j_l^0$, where $j_1^0 \le j_2^0 \le \ldots \le j_l^0$, yielding a configuration X_i^l. This X_i^l introduces a local improvement to X_i^0, namely all neighborhoods of radius $\le j_l$ are better organized. In step 2, X_i^l is transformed into a configuration X_{i+1}^0 of the graph G_{i+1}, by assigning $X_{i+1}^0(u)$, for $u \in V_{i+1}$, the center of gravity of all $X_i^l(v)$, for $v \in [u]$ (hence the name "gravitation"):

$$X_{i+1}^0(u) = \sum_{v \in [u]} X_i^l(v)$$

This X_{i+1}^0 serves as an initial configuration for the next recursive level, carried out on the graph G_{i+1}. The recursion in step 3 returns with a configuration X_{i+1}^f of G_{i+1}. In step 4, this X_{i+1}^f is transformed by interpolation into a configuration X_i^{l+1} of the graph G_i, by the translation

$$X_i^{l+1}(u) = X_i^l(u) + (X_{i+1}^f([u]) - X_{i+1}^0([u]))$$

We might say that a vertex $u \in V_i$ inherits the movement of its parent $[u] \in G_{i+1}$. The configuration X_i^{l+1} introduces an improvement to X_i^l in a global sense, but it might contain local disorders due to the interpolation. These are fixed in step 4, by employing another sequence of (i, j) relaxation sweeps, this time for $j = j_1^1, j_2^1, \ldots, j_{l'}^1$, where $j_1^1 \le j_2^1 \le \ldots \le j_{l'}^1$, yielding the configuration $X_i^{l+l'+1}$. The hope is that $X_i^{l+l'+1}$ will introduce both local and global improvements to X_i^0.

The complexity of the preprocessing phase of the algorithm depends heavily on the implementation, but in any case it is dominated by the time taken to construct the H_i^j functionals. The construction of H_i^j involves the calculation of all pairs of shortest distances in G. If carried out naively, this can take $O(VE)$, where we write V and E for their sizes, $|V|$ and $|E|$, respectively. Actually, the algorithm works well with approximated shortest distances too, which is an easier problem that can be solved in time $O(V^2)$. Thus, a sophisticated implementation can execute the first phase of the algorithm in $O(V^2)$.

As to the second phase, the complexity of an (i, j) relaxation sweep turns out to be $O(\sum_{u \in V_i} |\Gamma_j(u)|)$, where $\Gamma_j(u) = \{v \in V_i \mid d_{G_i}(u, v) \leq j\}$. Thus, $|\Gamma_j(u)|$ is the number of operations in the calculation of the partial derivative $\frac{\partial H_i^j}{\partial X_i(u)}$. If we require G_i to be of bounded degree Δ_i, we have

$$O(\sum_{u \in V_i} |\Gamma_j(u)|) \leq O(|V_i||\Delta_i|^j) = O(V_i)$$

Hence, in such a case, at level i of the recursion, steps 1 and 5 together contribute $O(l|V_i||\Delta_i|^l) + O(l'|V_i||\Delta_i|^{l'}) = O(V_i)$. Applying gravitation to the configuration of G_i to obtain that of G_{i+1} (step 2), costs $O(\sum_{u \in V_{i+1}} |[u]|)$, where $|[u]|$ is the number of operations in the calculation of $X_{i+1}(u)$ from $X_i(v)$, for $v \in [u]$. If, in the construction of G_{i+1}, we assume that the cluster sizes $|[u]|$, for $u \in V_{i+1}$, are bounded, the contribution of the gravitation step will be $O(V_i)$. Similar considerations show that the cost of the interpolation (step 4) will also be $O(V_i)$.

We conclude that the total cost of stage i in the recursion is $O(V_i)$. Since we employ a fixed number of scales, the overall complexity of the recursive procedure is dominated by the cost of the first stage of the recursion, which is $O(V)$. Thus, the preprocessing is quadratic in $|V|$, whereas the drawing process is linear in $|V|$.

5   Examples

We illustrate our approach with several examples. The algorithm was implemented in C, and runs on a Silicon Graphics workstation. The drawing process for the 400-vertex graphs, for example, take less than 5 seconds. The preprocessing takes longer, of course. For each example, we show a number of final results of the algorithm run with a number of varying levels of coarsening. For some we also show the representations involved, i.e., the graph on a given scale after carrying out the reduction in the number of vertices. In all examples, the initial configuration was a random scattering.

The construction of the coarse-scale representations is carried out with the cost

$$\mu(u, v) = Cl(u, v) + Ho(u, v) + 2D(u, v)$$

We have set things up so that step 1 of the algorithm involves ten (i, j) sweeps, for $j = 1, 2$ and 4, and step 5 involves ten (i, j) sweeps, for $j = 4$.

First we deal with one of the simplest kinds of graphs — a circle. Fig. 1 shows the results obtained for a 160-vertex circle. The left-hand column shows the representations and the right-hand column shows the final outcomes, for 2, 3 and 4 levels of coarsening. Observe the improvement in the output as more scales enter the game. This improvement becomes more global and the resemblance between the output and a perfect circle is increased. Another way of viewing the results is to think of the initial configuration (a random scatter) as suffering from multi-scale disorders. Associated with each disorder is a natural scale, in which the relaxation scheme 'sees' the disorder and deals with it effectively.

The graph in Fig. 2 consists of 40 circles of 10 vertices each, concatenated in a grand circle to form a 400-vertex torus. Again, note the gradual globalization of the solution as more scales are involved. The middle solution, using three coarsenings, has the torus 'twisted' in 3D space, although obviously the algorithm knows nothing at all about 3D. Also, notice how the representations retain the torus-like topology, even the coarsest one in the left-hand bottom part of the figure.

Figs. 3 and 4 show the algorithm's results on a 1000-vertex torus (100 circles of 10 vertices each) and a 1024-vertex torus (64 circles of 16 vertices each), respectively. We have not included the representations here, just the final outcomes.

This might be a good place to point out that we could have subjected the outcomes shown in the paper to some relatively straightforward stochastic global scheme (e.g., the simulated annealing algorithm of [DH]), in order to remove minor distortions from the final layout. We have not done so, however, and show the raw results as produced by the multi-scale method on its own.

We now turn to grids. These too are amenable to "multi-scale aesthetics". Fig. 5 shows the results obtained for a 400-vertex 20×20 grid, together with the representations. Fig. 6 shows a 1024-vertex 32×32 grid without the representations.

Fig. 7 shows a 15×15 grid, in which the two pairs of opposite corners have been connected by an edge. We show only the final result, obtained with three levels of coursening. Such a graph, although combinatorially similar to a simple grid, causes the algorithm to display more complicated behavior. The connected corners interfere with the movement of other vertices, requiring more work in the organization process. This is less significant on coarser scales, so that fortunately it is overcome quite naturally by the multi-scale algorithm. Another consequence of such a combinatorial structure is the possible appearance of a "deep" local optimum "trap". Again, this phenomenon is less significant as the number of vertices is reduced and the combinatorial structure simplified, so that it too is handled well by the algorithm when operating on coarse-scale representations of the graph.

6   Discussion

The approach we describe in this paper is novel in several ways.

On the operational level, it is capable of handling large graphs containing thousands of vertices, yielding interesting layouts — perhaps the "right" layouts, and this in very acceptable time limits. For example, we are aware of no other algorithm that can deal nearly as well with laying out the large torus and grid graphs of Figs. 3, 4 and 6. (See also the remark at the end of the Introduction about the very recent [HHK].)

On the conceptual level, our algorithm deals with a real-life, ill-defined problem, of the kind that seems to require carefully wrought heuristics. However, we

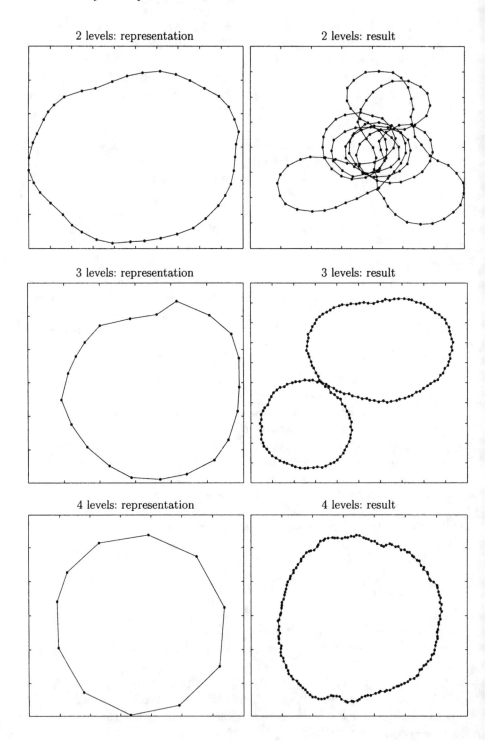

Fig. 1. 160-vertex circle

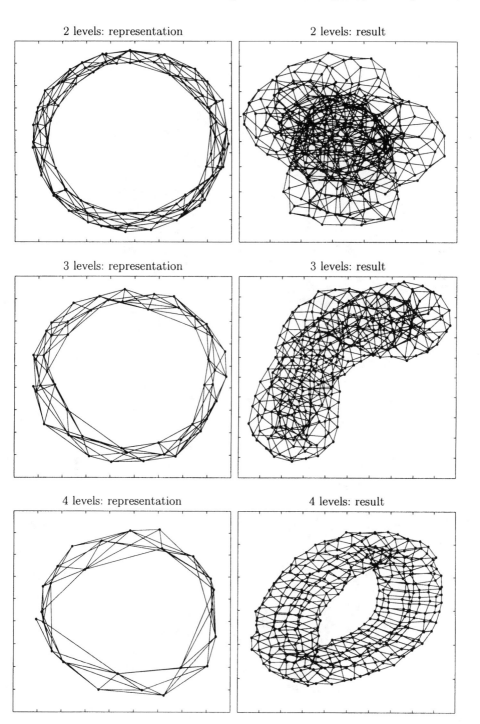

2 levels: representation

2 levels: result

3 levels: representation

3 levels: result

4 levels: representation

4 levels: result

Fig. 2. 400-vertex torus (40×10)

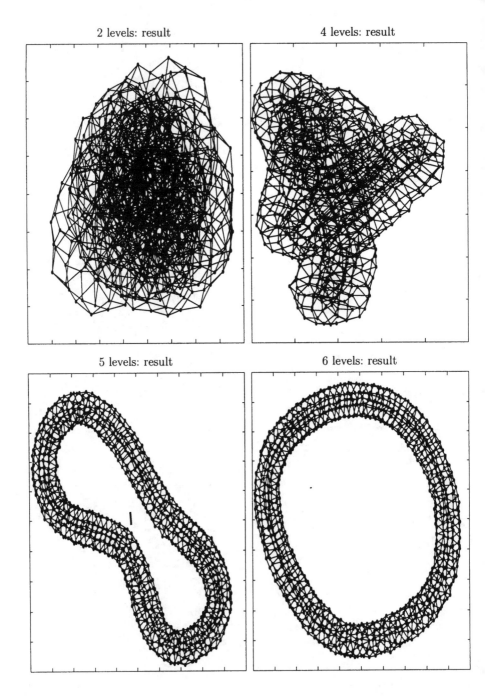

Fig. 3. 1000-vertex torus (100×10)

2 levels: result

3 levels: result

4 levels: result

5 levels: result

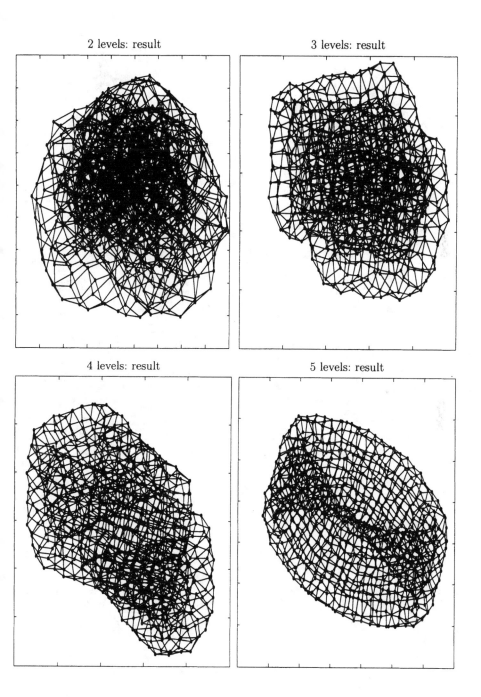

Fig. 4. 1024-vertex torus (64×16)

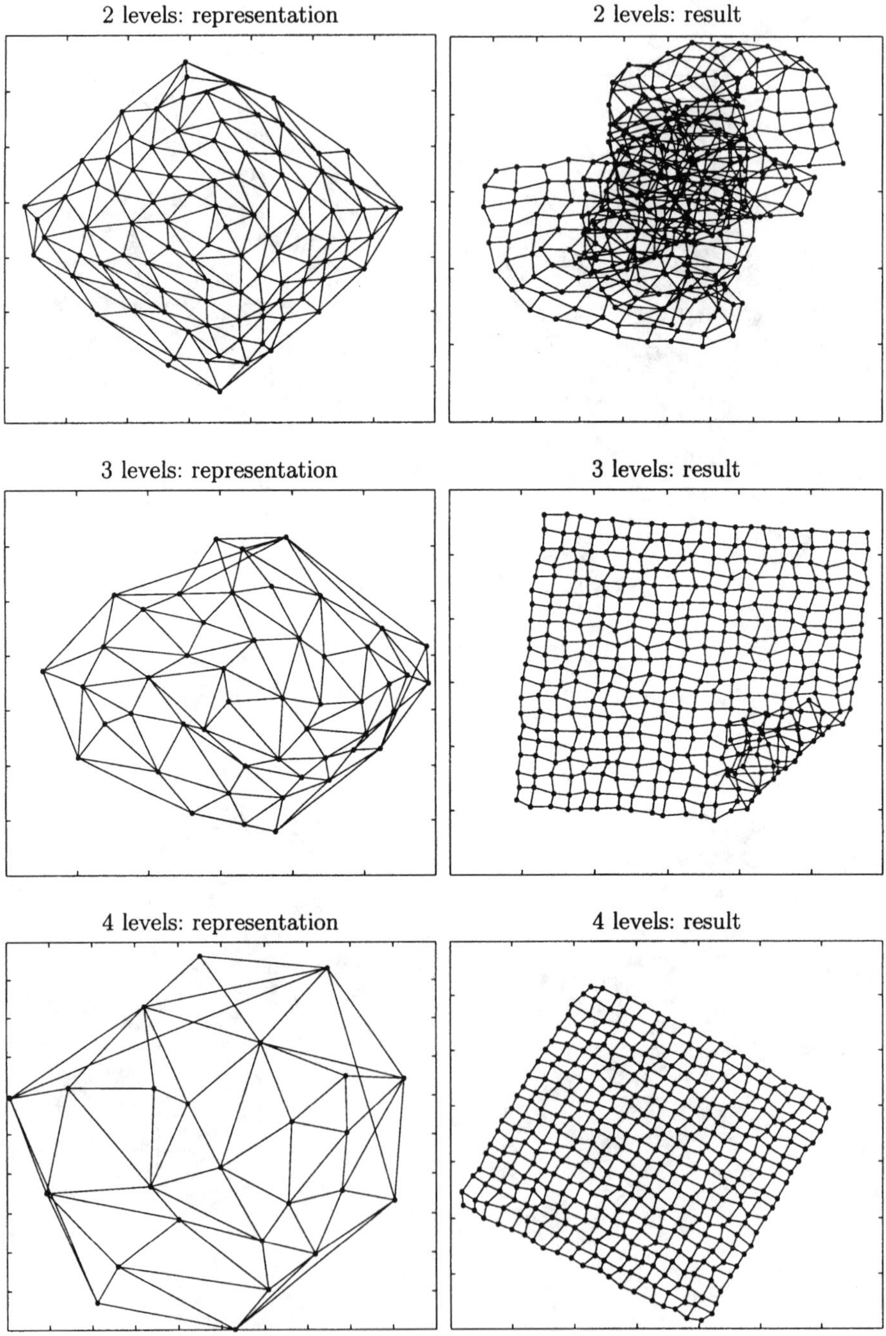

Fig. 5. 400-vertex grid (20×20)

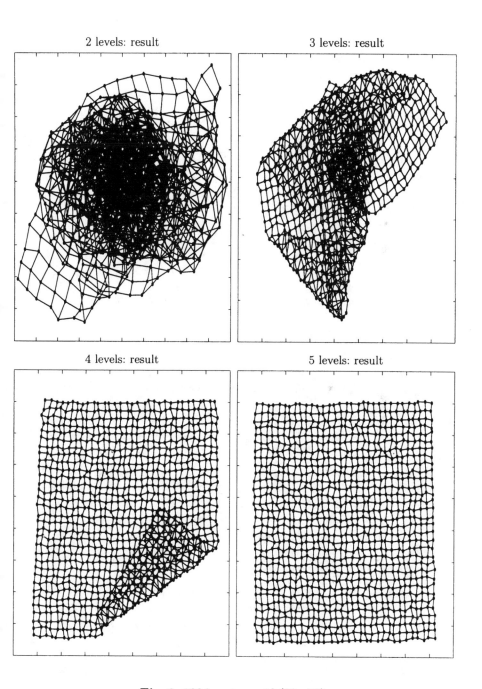

Fig. 6. 1024-vertex grid (32×32)

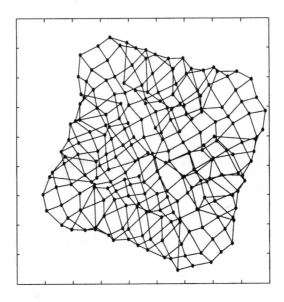

Fig. 7. 225-vertex grid with knotted corners

may argue that the algorithm deals with this problem in a heuristic free fashion, in that it does not depend critically on a heuristic definition of "nice". The drawing process consists of relaxation steps and coarsening steps. In a relaxation step, a local relaxation scheme is employed to organize small neighborhoods, and when dealing with small enough neighborhoods it makes no real difference which scheme is used. We can choose any of the the relaxation schemes in [E], [FR], or [KK], for example, or even a stochastic relaxation scheme, such as simulated annealing [DH]. The main argument is that in our approach the role of the relaxation steps is not to draw the entire graph, but merely to organize small neighborhoods therein. The full picture "emerges" from the coarse-scales down to the fine ones. Perhaps a good way to summarize this is to say that local neighborhoods in coarser scales encapsulate global information about the finer scales. And the algorithm is particularly suited to graphs that have this hard-to-define multi-scale aesthetics property.

On the other hand, our work is far from being perfect, and should be extended and strengthened in several ways. For example, the construction of the coarse-scale representations of the graph is carried out through a sequence of edge contractions determined by specific criteria. Here we have used very basic criteria, taking care of only the most vital topological properties that should be preserved, namely, cluster size, vertex degree, and homotopy. Devising more detailed criteria, that integrate higher level visual properties of the graph, will surely improve the results for additional kinds of graphs. An example of a more elaborate criterion has to do with edges connecting distant sections of the graph; that is, edges whose removal from the graph will significantly increase the distance between its two endpoint vertices. It might be wise not to contract such edges. In fact, in [HHK] we use a simpler global criterion to achieve a rather spectacular improvement.

Experimental work should be carried out in optimizing the performance of the algorithm. One could try to distribute wisely the amount of relaxation work between the scales, investing more relaxation sweeps in coarser scales. An intersting parameter is the recursion index, namely, the number of recursive calls on each level. Different values of this parameter determine the work distribution between the scales. Simple recursion (i.e., a single recursive call), distributes the work evenly between the scales. As the recursion index grows, the work is concentrated in the coarser scales.

Another direction that should be looked into is the incorporation of stochastics. One approach, hinted at earlier, would be to develop a hybrid system, in the spirit of [SH]. It would link a deterministic multi-scale algorithm (such as ours) to achieve a good first approximation of an optimal drawing with a stochastic fine-tuning algorithm, such as simulated annealing with detailed beauty criteria. Another approach would be to try to develop a "true" multi-scale simulated annealing algorithm, in which the stochastic part is used throughout.

Acknowledgements

We thank Achi Brandt for many helpful discussions.

References

B1.   Brandt, A., "Multilevel computations of integral transforms and particle interactions with oscillatory kernels", *Computer Physics Communications* **65** (1991), 24–38.

B2.   Brandt, A., "Multigrid Methods in Lattice Field Computations", *Nuclear Physics B (Proc. Suppl.)* **26** (1992), 137–180.

DH.   Davidson, R., and D. Harel, "Drawing Graphs Nicely Using Simulated Annealing", *ACM Transactions on Graphics* **15** (1996), 301–331. (Preliminary version: Technical Report, The Weizmann Institute of Science, 1989.)

DETT. Di Battista, G., P. Eades, R. Tamassia and I. G. Tollis, *Graph Drawing: Algorithms for the Visualization of Graphs*, Prentice Hall, NJ, 1999.

E.   Eades, P., "A Heuristic for Graph Drawing", *Cong. Numer.* **42** (1984), 149–160.

FR.   Fruchterman, T.M.G., and E. Reingold, "Graph Drawing by Force-Directed Placement", *Software-Practice & Experience* **21** (1991), 1129–1164.

HHK.   Hadany, R., Harel, D. and Y. Koren, in preparation.

SH.   Harel, D., and M. Sardas, "Randomized Graph Drawing with Heavy-Duty Preprocessing", *J. Visual Lang. and Comput.* **6** (1995), 233–253.

KK.   Kamada, T., and S. Kawai, "An Algorithm for Drawing General Undirected Graphs", *Inf. Proc. Lett.* **31** (1989), 7–15.

All Separating Triangles in a Plane Graph Can Be Optimally "Broken" in Polynomial Time

Anna Accornero[1], Massimo Ancona[2], and Sonia Varini[1]

[1] Dipartimento di Matematica,
annaa@spawn.disi.unige.it
varini@fourier.dima.unige.it
soniav@spawn.disi.unige.it

[2] Dipartimento di Informatica e Scienza dell' Informazione, Università di Genova,
Via Dodecaneso 35, 16146 Genova, Italy
ancona@disi.unige.it
http://www.disi.unige.it/person/AnconaM/

Abstract. Lai and Leinwand have shown that an arbitrary *plane* (i.e., embedded planar) graph G can be transformed, by adding *crossover* vertices, into a new plane graph G' admitting a *rectangular dual*. Moreover, they conjectured that finding a *minimum set* of such crossover vertices is an NP-complete problem. In this paper it is shown that the above problem can be resolved in polynomial time by reducing it to a *graph covering* problem, and an efficient algorithm for finding a minimum set of edges on which to insert the crossover vertices is also presented.

1 Introduction

Rectangular dualization is a method for VLSI floorplanning which is concerned with the allocation of space to component modules and their interconnections. A floorplanning can be obtained by constructing the *rectangular dual* of the *circuit structure graph*[1]. Each planar graph admits a *dual* graph: i.e. a graph composed by non overlapping regions of the plane such that each region corresponds to a vertex of the original graph and two regions are neighbors (i.e. share a common boundary) if and only if the corresponding vertices in the original graph are adjacent. When all regions are rectangles[2], and their union is a rectangle too, then the dual is called a rectangular dual. However, not all planar graphs *admit* a rectangular dual.

The problem was studied in [2,3,6] by providing both conditions guaranteeing the existence of a rectangular dual and algorithms for its construction. We suppose to start from a plane, i.e. planar embedded, and *triangulated* graph. The problem of triangulating a planar graph has been studied in[6,4]. Each cycle of a plane graph surrounds a region of the plane. An elementary region (i.e., not including in its interior any vertex of G) is called a *face*.

---

[1] A graph theoretic description of the circuit.

[2] Four of them never meeting in a single point.

Widmayer et al. (Eds.): WG'99, LNCS 1665, pp. 278–290, 1999.

In a planar graph, each cycle of length three which is not a face is called a *separating triangle* (ST hereafter). Each graph containing an ST does not admit a rectangular dual. Lai and Leinwand [6] suggested to modify the graph by *breaking* all STs by adding new *crossover* vertices on an edge of each ST. This breaking adds new areas to the floor plan that are used only for interconnections, thus this waste of space should be kept as low as possible. Lai and Leinwand conjectured that the problem adding a *minimum* number of such crossover vertices is an NP-complete problem, in this paper we show that the above problem can be solved in polynomial time.

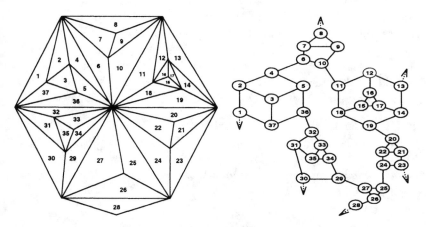

Fig. 1. A triangulated graph $G_\triangle$ which does not have RDG; and its dual graph G_T.

2   Background

A planar graph $G = (V, E)$ is a graph that can be drawn on the plane without crossing edges. A *plane* graph is *planar embedding* of a planar graph[3]. A plane graph is composed of *faces*. The exterior face is unbounded and called the outerface. If all the internal faces of G are triangles the graph is said to be *triangulated*. A *Rectangular Dual* (RDG) of a n-vertex graph $G = (V, E)$ is a rectangle R partitioned into n non-overlapping rectangles with the following properties:

- each vertex $i \in V$ corresponds to a distinct rectangle i in the RDG;
- (i, j) is an edge of E iff i and j are adjacent in R (that is they have either an edge or part of it in common).

It is easy to see that some graphs do not admit RDG and for yet others it is not unique. Kozminski and Kinnen determined the conditions to be fulfilled

---

[3] We suppose that all edges incident to each vertex are listed in clockwise order.

by a biconnected plane graph in order to admit an RDG[2] and developed an algorithm for its construction.

Theorem 1. *A biconnected plane graph G that (a) has all faces of degree 3, (b) has all internal vertices with degree at least 4, and (c) all cycles in G that are not faces have degree ≥ 4, has an RDG iff it has no more than 4 PICs[4].*

A cycle in G of length 3, which is not a face, is called a *Separating Triangle* (an ST): *a plane graph containing an ST does not admit an RDG* [2]. Lai and Leinwand were the first to propose to modify G by adding new vertices so as to obtain a new graph G' which does not contain STs [6]. They defined an algorithm which finds and breaks each ST T in G by adding a new vertex to an edge of T.

In the next sections we show that an optimal breaking of all the STs in a plane triangulated graph can be found in polynomial time. Our purpose is to modify the graph as little as possible in order to obtain a graph which has an RDG. We have found a polynomial algorithm which determines the minimum number of vertices and their exact position. Our method is based on graph covering:

Definition 1. *Given a simple non-directed graph $G = (V, E)$, a* matching *is a subset of edges $M \subset E$ such that no two edges of M are adjacent. A* covering *is a subset of edges $C \subset E$ such that every vertex is the endpoint of at least one edge of C. A* maximum matching (minimum covering) *is a matching (covering) of maximum (minimum) cardinality.*

3   The Method

Our method takes as input a plane triangulated graph $G_\triangle$ (Figure 1 shows a triangulated plane graph). The algorithm is divided into 3 steps:

Step 1: Dualization
Construction of the `geometrical` dual G_T of $G_\triangle$. The vertices of G_T correspond to the internal faces of $G_\triangle$. There is an edge between two vertices a and b of G_T iff the corresponding faces in $G_\triangle$ are adjacent, i.e. they share an edge.

Figure 1b shows the dual graph G_T of the graph of Figure 1a. We describe the algorithm referring to Figure 1, by starting from a triangulated plane graph $G_\triangle$, i.e. a graph with all faces of length three. The construction of the geometrical dual G_T of $G_\triangle$ is straightforward and it will not be described here: it is implicitly performed in Step 2.

Step 2: Structuration
In this phase all STs are identified, and their geometrical relationship (i.e., if they are adjacent or nested) is encoded. The associated *structured graph* $\mathcal{G}_T = (M, N, E)$ of G_T is constructed (Algorithm 1), by recursively *shrinking* each subgraph surrounded by an ST in $G_\triangle$ into a single *macrovertex* (or macro for short) of $\mathcal{G}_T$. The shrinking is recursively repeated until no more ST to be

---

[4] A PIC, i.e. *Corner Implying path*, is a segment $v_1, v_2, ..., v_k$ of the outermost cycle U of G such that $(v_1, v_k) \in E - U$ and $(v_i, v_j) \notin E - U$ $1 < i, j < k$.

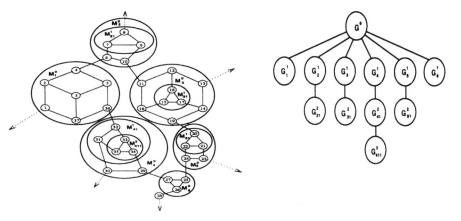

Fig. 2. The structured graph $\mathcal{G}_T$ computed from Figure 1 and its hierarchical structure.

contracted are found. Figure 2a shows the structured graph $\mathcal{G}_T$ relative to Figure 1; for example the ST $1 - 4 - 36$ is shrunk into macro M_1^0. The main problem is to identify all STs and their mutual relationship; in fact, some of them are adjacent, i.e. they share an edge, some are nested and eventually also adjacent. In Figure 1a ST $T_1 = \{20, 21, 22\}$ is nested and adjacent to $T_2 = \{23, 24, T_1\}$, while $T_3 = \{25, 26, 27\}$ is adjacent to T_2. Finally ST $T_4 = \{15, 16, 17\}$ is nested, but not adjacent into $\{11, 12, 13, ...17, 18, 19\}$. Note that two adjacent STs may be broken by a single crossover vertex. $\mathcal{G}_T$ encodes the relationship existing among the STs of G_T. More formally, $\mathcal{G}_T$ is obtained by recursively *shrinking* a subset of vertices $W \subset V$ into a single new vertex w. The condensing operation produces a new graph, called the *reduced* graph, whose set of vertices is $V - W \cup \{w\}$ (all vertices of W are omitted and a new vertex w added to replace W) and where the edges are of the following form:

- (u, v) if $u \in V - W$ and $v \in V - W$ and $(u, v) \in E$;
- (u, w) if $u \in V - W$ and there is a $v \in W$ such that $(u, v) \in E$ (all edges with both endpoints in W are omitted). Vertex v is called a *port* of w, while edges (u, v) (u, w) are identified by saying that edge $(u, v) \equiv (u, w)$ is incident to macro w into its port v[5].

Intuitively a structured graph is a hierarchy of graphs connected by a *parent-child* relation defined by the shrinking operation above[6]; formally a structured graph can be recursively defined as follows:

Definition 2. *A structured graph is a pair $\mathcal{G} = (\mathcal{T}, G)$ where G is a graph and $\mathcal{T}$ a tree of subgraphs of G such that: $G^0 = (M \cup N, E)$ is a simple graph called the*

---

[5] We draw (u, v) as a line entering w and terminating on v (see, for example, edge $32 - 36$ in Figure 2).

[6] The parent graph is the reduced graph and the child the shrunk graph.

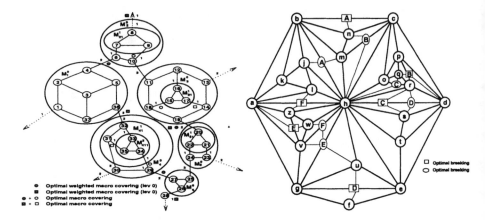

Fig. 3. Optimal Macro coverings on $\mathcal{G}_T$ and corresponding breakings of $G_\triangle$.

restriction of $\mathcal{G}$ to level-0, or the root graph of $\mathcal{G}$. N is a set of simple vertices, while M is the set of level-0 macrovertices of $\mathcal{G}$. Each macro $m \in M$ is the collapsing of a structured subgraph $\mathcal{G}_m$ of $\mathcal{G}$. We say that $\mathcal{G}_m = Exp(m)$ is the (structured) expansion of m. We also denote $G_m = \mathcal{G}_m[0] = Exp_0(m)$ the root graph[7] of $\mathcal{G}_m$.

The following definitions introduce some concepts about structured graphs that will be used hereafter:

Definition 3. Given a structured graph $\mathcal{G} = (T, G)$ its restriction at level n is the structured graph $\mathcal{G}^{(n)}$ whose macrovertices of level $\ell = n$ are considered simple vertices (their internal structure is discarded). Note that $\mathcal{G}[0]$, also denoted G^0, is always a simple graph.

Definition 4. A vertex v (simple or macro) is at level 0 if it belongs to G^0, it is at level $n > 0$ if it belongs to the level 0 expansion of a macrovertex at level $n - 1$. The level of a vertex v is denoted $\ell(v)$.

For example (Figure 2) M_{411}^2 is at level 2, while M_2^0 is at level 0[8]. A more accurate description of structured graph is reported in [1].

Now we find an ST in $G_\triangle$ and collapse it into a macrovertex M of $\mathcal{G}_T$ by obtaining a structured graph $\mathcal{G}_T^1$. By collapsing a new ST of $G_\triangle$ we obtain a second structured graph $\mathcal{G}_T^2$. By recursively collapsing all the ST of $G_\triangle$ we transform it into $\mathcal{G}_T$. The condensing operation is *consistent* for each $\mathcal{G}_T^i$ because if $ST_1 = \{t_1, t_2, .., t_k\}$ and $ST_2 = \{t_1', t_2', ..., t_l'\}$ are two STs such that $ST_1 \cap ST_2 \neq \emptyset$ then $ST_1 \subseteq ST_2$ or $ST_2 \subseteq ST_1$. In other words the construction of $\mathcal{G}_T$ does not depend on the order of finding and shrinking the STs. Each macrovertex contains exactly three ports[9], and three adjacent edges (Figure 5).

---

[7] Also called the level-0 restriction of $\mathcal{G}_m$.

[8] We use symbol M_j^i to denote the j macro belonging to level i.

[9] A port may be a macrovertex.

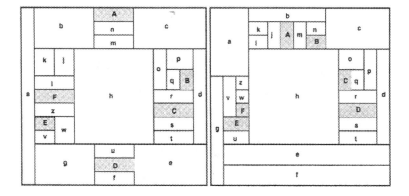

Fig. 4. Two RDGs from graph of Figure 3

Figure 5a shows an ST of $G_\triangle$ while 5b its shrinking into a macrovertex M_{abc} of $\mathcal{G}_T$. We now detail how to find all STs of $G_\triangle$ and collapse them into a macrovertex.

Definition 5. *Two faces (i.e., elementary triangles) t_1 and t_2 of $G_\triangle$ (vertices of G_T) are said to be incident iff they share at least a vertex in $G_\triangle$. A path (cycle) of faces is a sequence $t_1, \ldots, t_n$ of faces such that t_i is incident to t_{i+1} for all $i = 1, \ldots, n-1$ $(t_1 = t_n)$.*

Definition 6. *Two faces t_1 and t_2 are said to be adjacent if they share one edge in $G_\triangle$.*

In Figure 5a triangles $t_1 = A$ and $t_2 = b$ are incident, while A and a are adjacent. Note that two adjacent faces are also incident, but two incident faces are not necessary adjacent. It is immediate to show that:

Theorem 2. *A cycle of three faces t_1, t_2 and t_3 define a separating triangle $\{v_1, v_2, v_3\}$ in $G_\triangle$ if there exist another cycle of three faces T_1, T_2 and T_3[10] such that T_1 is adjacent to t_1, T_2 is adjacent to t_2 and T_3 is adjacent to t_3. In other words T_1, T_2 and T_3 define the same ST $\{v_1, v_2, v_3\}$ defined by t_1, t_2 and t_3. Moreover, if all vertices and adjacencies in $G_\triangle$ are listed in clockwise order, then three of the six triangles can be classified as* internal *triangles and three as* external *triangles (see Figure 5).*

From the above theorem we derive an algorithm for computing $\mathcal{G}_T$ from $G_\triangle$. The algorithm (procedure B_STG) performs a double visit of the adjacency list $L(v)$ of each vertex $v \in G_\triangle$ finding all cycles of length three through v. When a cycle is found, if it is a face then a vertex t of $\mathcal{G}_T$ is created otherwise the cycle

---

[10] Some of the T_i or t_i could be degenerated, i.e., coincident with the outer unbounded infinite face.

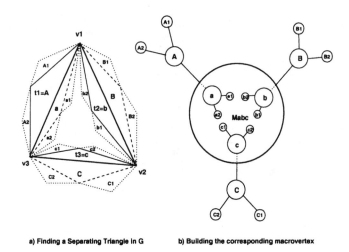

a) Finding a Separating Triangle in G   b) Building the corresponding macrovertex

Fig. 5. Building $\mathcal{G}_T$ from G_T

is an ST and the subgraph surrounded by it in $G_\triangle$ originates a macro of $\mathcal{G}_T$ (see Figure 5b and procedure Create_Macro).

The algorithm runs in time $O(n^2)$[11], where n is the number of vertices of $G_\triangle$.

Algorithm 1. *Let* a,b,c *and* A,B *and* C *the internal and the external triangles of an ST* {v1,v2,v3} *defined by Theorem 2 (see Figure 5).*

```
procedure Create_Macro(M: Macro;a,b,c,A,B,C: Face;v1,v2,v3: Vertex);
begin
    New_Macro(M,a,b,c); (* M is the created macro *)
    (* a,b,c are the ports of M, A,B,C become adjacent to M *)
    substitute A,B,C with NULL in a,b,c respectively;
    (*disconnect the inner graph ports from the outer graph*)
    substitute M with a,b,c in A,B,C respectively;
    (* connect the outer graph to M *)
end Create_Macro;
procedure B_STG(v: vertex);
var w,z: vertex;
begin
    for each w ∈ L(v) do
        for each z ∈ L(v) − {w} adjacent to w do;
        if is_face(v,w,z) then Create_Vertex(t,v,w,z)
            else Create_Macro(M,a,b,c,A,B,C,v,w,z)
        end (* if *)
    end (* inner for *)
    end (* outer for *)
end B_STG.
```

---

[11] $O(n^3)$ when applied to all vertices of $G_\triangle$.

While procedure `Create_Macro` turn an ST in $G_\triangle$ into a macrovertex of $\mathcal{G}_T$ (Figure 5).

Step 3: Optimal Breaking

Each edge e of $G_\triangle$ corresponds to a single edge e_T of $\mathcal{G}_T$. Finding a minimum set of additional vertices to break all separating triangles is equivalent to the problem of finding a *minimum covering* of the subset of macrovertices of $\mathcal{G}_T$. This problem is called *macro-covering* of $\mathcal{G}_T$ and is formulated and resolved in the next session.

4   Covering in Structured Graphs

An optimal breaking of $G_\triangle$ is now reduced to a covering problem in $\mathcal{G}_T$, i.e. to the problem of finding a minimum number of edges incident to all macrovertices of $\mathcal{G}_T$. Figure 3 shows two optimal breakings, derived from the two minimum sets of edges that are incident to all macrovertices and Figure 4 shows the rectangular duals derived from coverings and breakings in Figure 3. Finally, we will show that a minimum macro-covering can be efficiently constructed by computing a sequence of minimum weighted covering problems on each simple graph of $\mathcal{G}_T$ by visiting in postorder its hierarchy.

Definition 7. *Given an undirected structured graph* $\mathcal{G} = (T, G)$ *then a* macro-covering *on* $\mathcal{G}$ *is a collection* $\mathcal{K}$ *of edges of* $\mathcal{G}$, *such that there is an edge* $e \in \mathcal{K}$ *incident to each macrovertex* m *of* $\mathcal{G}$, *i.e.:*

- *If* m *is at level* 0, *i.e.* $\mathcal{G}[0] = (M \cup N, E)$ *and* $m \in M$, *then there exists an edge* $e \in E \cap \mathcal{K}$ *incident to* m;
- *If* m *is at level* n, *then there exists a macrovertex* u *at level* $n' < n$ *such that* $G_u^0 = Expo(u) = (M_u \cup N_u, E_u)$, $m \in M_u$ *and there exists an edge* $e = (m, x) \in E_u \cap \mathcal{K}$ *with* $x \in N_u \cup M_u$;

Definition 8. *Given a (weighted) graph* G; *we define* minimum (weighted) covering *of* G *constrained to* $e = (u, v) \in E$ *the problem of finding a minimum (weighted) covering in* G *containing edge* e. *Intuitively we want to choose a minimum (weighted) covering between the set of coverings which include the edge* $e = (u, v)$. *We also say that the above minimum covering is* constrained *to* $e = (u, v)$.

Note that, in a structured graph, some edges are incident to more that two vertices (when their endpoints are macrovertices). Such edges are *hyper-edges* showing that structured graphs are a special case of hypergraphs. *Moreover, the minimum covering problem is NP-complete in general hypergraphs, while it runs in polynomial time in simple graphs. What happens for our structured graphs?* A minimum macro-covering can be conveniently reduced to a minimum weighted covering problem in simple graphs: with a bottom-up visit of the hierarchy of $\mathcal{G}_T$ weights are assigned to the three arcs incident to each macrovertex. Thus,

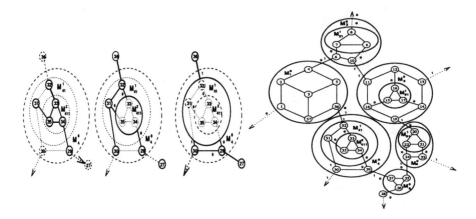

Fig. 6. Bottom-up computation of weights of macro M_4^0 of Figure 2; weights on $\mathcal{G}_T$.

macro-covering can be solved in polynomial time. The weight assigned to an edge $e = (u, v)$ is equal to *the minimum number of edges to be added to $Exp_0(u)$ and $Exp_0(v)$ if e is included into a minimum cover of $\mathcal{G}_T$*[12]. Such a number could be computed by resolving a weighted minimum covering *constrained* to e, i.e. a covering on $Exp_0(u)$ after removing all macrovertices covered by e[13].

Algorithm 2. *Computation of weights in $\mathcal{G}_T$. Method: visit in post-order the hierarchy of $\mathcal{G}_T$ and assign to each edge $e = (u, v)$ incident to at least a macro (u or v) the cost $w_u + w_v + 1$, where w_u (w_v) is the solution of the minimum weighted macro-covering problem on $G_u^{(e)} = Exp_0(u)$ ($G_v^{(e)} = Exp_0(v)$), $w_u = 0$ ($w_v = 0$) for simple vertices. $G_u^{(e)}$ is obtained from $G_u = Exp_0(u)$ by deleting all macros covered by edge e, thus exposing some inner macrovertices with corresponding weights.*

For example, in Figure 6a.2, $u = M_{41}^1$ is covered by edge $e = (36, 32)$ and $G_u^{(e)} = (\{29, 30, 31, 32\} \cup \{z\}, \{(29, z), (31, 32), (31, z)\})$ where $z = M_{411}^2$. Procedures `Prune` and `Visit` implement the algorithm. Weight w_u and w_v of edge $e = (u, v)$ are denoted $e.w1$ and $e.w2$ respectively (initially $e.w1 = 0$ and $e.w2 = 0$). Procedure `Cons_Weight` computes a minimum weighted covering on F, by using appropriate weights $W(e) = e.w1 + e.w2 + 1$.

The covering is restricted to macrovertices, thus procedure `Prune` computes $G_m^{(e)}$ from $G_m = Exp_0(m)$ by including all macros and all adjacent simple vertices with incident edges. The resulting graph may become disconnected, in this case `Cons_Weight` has to be activated on all the connected components. Figure 6 (example of Figure 2) shows the bottom-up computation of weights on M_{411}^2, M_{41}^1 and M_4^0. The complete set of weights of $\mathcal{G}_T$ are shown in Figure 3.

---

[12] $Exp_0(u) = \emptyset$ if u is a simple vertex.
[13] Actually, a single (free) covering on $Exp_0(u)$ combined with weights of nested macros is sufficient to this aim.

```
procedure Prune(var F:Graph; N: Macro);
var x:vertex; G: Graph; V: SetOfVertex; G: SetOfEdge;
begin
    let G = (N ∪ M, E) =Exp₀(N);
    V:=M ∪ Ports(N);(* start with macros and ports*)
    for each m ∈ M do
        for each v ∈ L(m) do
            V:=V ∪ {v}; E ∪ {(m,v)}
        end for; (* inner*)
    end for;(* outer*)
    V:=V ∪ {P₁,P₂,P₃};(* add external ports*)
    E:=E∪{e₁,e₂,e₃};(* and entering edges*)
    F:=(V,E);
end Prune;
```

```
procedure Visit(G: ST_Graph);
var w:integer;F,G : Graph;
begin
    let G:=G⁰; (* restriction to level 0 *)
    for each macro N ∈ G do
        Visit(Exp₀(G)); (* expansion of N *)
        Prune(F,N);
        for each edge e incident to N do (* exactly 3 edges *)
            G:=F^(e);(* delete macros broken by e*)
            w:=Cons_Weight(G); (*W(e) = e.w1 + e.w2 + 1*)
        end for;
        if e = (N,v) then e.w1:=w else e.w2:=w endif
    end for (* macro *);
end Visit;
```

Cons_Weight computes a weighted optimal graph covering in polynomial time $(O(n^3)$ [8]). Let n_i be the number of vertices of each macro, for $i = 0, \dots, m$, where m is the number of macros in $\mathcal{G}_T$. It can be shown that $\sum_0^m n_i = n + m$, thus $\sum_0^m n_i^3 << (n + m)^3$ (n_0 is the number of vertices of the root graph). It follows that our method is faster than computing a single minimum covering on a single graph of the size of $\mathcal{G}_T$.

Algorithm 3. *Computation of the minimum macro-covering Γ of $\mathcal{G}_T$. Let Γ^0 a solution of the minimum weighted macro-covering problem on G^0 restriction of $\mathcal{G}_T$ at level 0, computed by Algorithm 2. Then $W(\Gamma^0)$ is the minimum number of edges covering $\mathcal{G}_T$ such that $|\Gamma| = W(\Gamma^0)$. Method:*

a) *Let $\Gamma := \Gamma^0$;*

b) *Visit in preorder the hierarchy of $\mathcal{G}_T$ and for each macro M of $\mathcal{G}_T$ let $e \in \Gamma$ be an edge of minimum weight $w(e)$ incident into M. Then let $\Gamma := \Gamma \cup \Gamma_e(M)$, where $\Gamma_e(M)$ is the minimum weighted covering of $Exp_0(M)$ (the expansion graph of M) constrained to e.*

```
procedure M_Covering(Var Γ: Set of edge;M: Macro-vertex);
begin
    for each macro N ∈ Exp₀(M) do
        let e ∈ Γ an edge incident to M;
        Γ := Γ ∪ Γₑ(M);
        M_Covering(Γ,N);
    end for;
end M_Covering;
```

It is immediate to verify that the above algorithm runs in polynomial time.

Theorem 3. *The cardinality of macro-covering Γ on $\mathcal{G}_T$, computed by the above algorithm, is minimum.*

Proof. The proof can be obtained inductively from the leaves to the root on the hierarchy of $\mathcal{G}_T$ by observing that:

a) From the definition of weight it follows that, for each macro M at level l, the weight $w(e)$ of each edge e incident to a macro $N \in Exp(M)$ (or two macros N_1 and N_2) represents *the number of edges to be added to each local optimal solution S at level $l+1$ in $Exp(N)$ (or $Exp(N_1)$ and $Exp(N_2)$), when e is* included *into S (e itself is included in $w(e)$).*

b) The minimum and unconstrained weighted covering on the root graph $\mathcal{G}_T[0]$ represents the minimum number of edges belonging to each optimal macro-covering of $\mathcal{G}_T$.

c) the optimal covering $\mathcal{C}$ may be constructed by collecting in a top-down visit of the hierarchy, all the edges, belonging to the local solutions constrained to each edge inserted in S, during the visit of the parent graph.

Definition 9. *A breaking of a plane graph $G = (V, E)$ is subset Γ of E such that each ST in G has at least an edge in Γ.*

The following Lemma asserts that each optimal breaking of $G_\triangle$ is equivalent to a macro-covering of $\mathcal{G}_T$ and viceversa.

Lemma 1. *There is a 1-1 correspondence $\wp$ mapping each breaking Γ of G into a macro-covering Θ of $\mathcal{G}_T$. Moreover, if Γ is optimal then also Θ is optimal, i.e. $|\Theta|$ is minimum and viceversa.*

Proof. For each $e \in \Gamma$ let $\wp(e) = (t_1, t_2)$ where t_1 and t_2 are the two triangles mutually adjacent in G through e. Obviously, $\wp$ is 1-1 because to each edge of E there correspond a unique pair of triangles in $G_\triangle$ and G_T adjacent through it. ($\Rightarrow$) *Let π be the 1-1 mapping defined by Algorithm 1 and mapping each ST S in G into a macro $M = \pi(S)$ of $\mathcal{G}_T$, and Γ a breaking of G. Then, for each $S = \{v_1, v_2, v_3\}$ in G there is an edge of S, let it $e = (v_1, v_2)$, belonging to Γ. Let (t, T) respectively the internal and external triangles mutually adjacent via e. Then t is a port and T the adjacent external vertex of the macrovertex $M = \pi(S)$*

constructed from S. It follows that there is an edge $(T, M) \equiv (T, t)$ incident to M in $\mathcal{G}_T$, i.e. $\Theta = \{e'|e' = \wp(e)\&e \in \Gamma\}$ is a macro-covering of $\mathcal{G}_T$.

($\Leftarrow$)Let Θ be a macro-covering of $\mathcal{G}_T$ and $\theta \in \Theta$ and edge $\theta = (t_0, M)$ where M is a macrovertex of $\mathcal{G}_T$. Let $S = \pi^{-1}(M)$ the ST of G from which M has been built. Them M has a port p_0 on which θ is incident. Both t_0 and p_0 could be macros. In this case they have a port t_1 and p_1 that, in turn, may be a macro. By iterating this reasoning we obtain two simple vertices in $\mathcal{G}_T$ t_l and p_k internal to two sequences of macros $t_0, t_1, ..., t_{l-1}$ and $p_0, p_1, ..., p_{k-1}$ respectively. From Algorithm 1 it follows that t_l and p_k are two faces of G such that t_l is an external face of all STs $\pi^{-1}(M)$ and $\pi^{-1}(p_i), i = 0, ..., k-1$. Then, t_l and p_k are adjacent and their common edge in G say (v, w) is a breaking edge for all the above STs. Moreover, by exchanging the role of p_k with that of t_l we can show that (v, w) is a breaking edge for all STs $\pi^{-1}(t_i), i = 0, ..., l-1$. Thus, we can say that each $\theta \in \Theta$ there corresponds a unique edge $\wp^{-1}(\theta)$ in G that breaks all STs associated by π^{-1} to all macros on which θ is incident. It follows that $\Gamma = \wp^{-1}(\Theta)$ is a breaking of G.

Optimality

($\Rightarrow$)The optimality of $\Theta = \wp < \Gamma >= \{\theta|\theta = \wp(\gamma) \quad \gamma \in \Gamma\}$ follows from the optimality of Γ: otherwise there would exist a macro-covering Ψ of $\mathcal{G}_T$ such that $|\Psi| < |\Theta|$. This implies that $\Delta = \wp^{-1}(\Psi)$ is a breaking of G such that $|\Psi| < |\Gamma|$ that is a contradiction.

($\Leftarrow$)The optimality of $\Gamma = \wp^{-1} < \Theta >$ follows by that of Θ in a similar way, for the injectivity of $\wp$.

5   Conclusions

In this paper we have shown that all the STs in a plane and triangulated graph can be optimally broken in polynomial time. The problem is reduced to the solution an optimal weighted covering of a hierarchy of simple graphs. Even if the complexity of the available algorithms for resolving the latter problem is $O(n^3)$ (where $n = |V|)^{14}$ [8], the *divide and conquer* approach proposed here, partitions the graph into a hierarchy of *small* graphs, making the method effectively usable on very large graphs like that used in VLSI design.

References

1. M. Ancona, L. De Floriani, J. S. Deogun, Path Problems in Structured Graphs,The Computer Journal, V. 29, N. 6, 1986, pp. 553-563.
2. K. Kozminski, E. Kinnen, Rectangular dual of planar graphs, Networks 5(1985), pp. 145-157.
3. J. Bhasker, S. Sahni, A linear algorithm to check for the existence of a rectangular dual of a planar triangulated graph, Networks 7(1987), pp. 307-317.

---

[14] Or $O(m \, n \, log \, n)$ where $m = |E|$.

4. T. Biedl, G. Kant, M. Kaufmann, On Triangulating Planar Graphs under the Four-Connectivity Constraint, Proc. of SWAT94, LNCS Vol. 824, Springer Verlag, 1994, pp. 83-94.
5. M.Gondran, M.Minoux, *Graphs and Algorithms*, John Wiley-Interscience Pubblication, 1984
6. Y. Lai, M. Leinwand, Algorithms for Floorplan Design Via Rectangular Dualization, IEEE Transactions on Computer-Aided Design, V. 7, N. 12, De. 1988, pp. 1278-1289.
7. G. Kant, X. He, Two Algorithms for Finding Rectangular Duals of Planar Graphs, Procs. WG'93, LNCS Vol. 790, Springer Verlag 1994, pp. 396-410.
8. Z. Galil, Efficient Algorithms for Maximum Matching in Graphs, ACM Computing Surveys, Vol. 18, N. 1, pp. 23-38, March 1986.

Linear Orderings of Random Geometric Graphs[*]

Josep Díaz[1], Mathew D. Penrose[2], Jordi Petit[1], and María Serna[1]

[1] Departament de Llenguatges i Sistemes Informàtics
Universitat Politècnica de Catalunya
Campus Nord C6. Jordi Girona 1-3. 08034 Barcelona. Spain
{diaz,jpetit,mjserna}@lsi.upc.es
[2] Department of Mathematical Sciences, University of Durham
South Road, Durham DH1 3LE, England
Mathew.Penrose@durham.ac.uk

Abstract. In *random geometric graphs*, vertices are randomly distributed on $[0,1]^2$ and pairs of vertices are connected by edges whenever they are sufficiently close together. Layout problems seek a linear ordering of the vertices of a graph such that a certain measure is minimized. In this paper, we study several layout problems on random geometric graphs: *Bandwidth, Minimum Linear Arrangement, Minimum Cut, Minimum Sum Cut, Vertex Separation* and *Bisection*. We first prove that some of these problems remain **NP**-complete even for geometric graphs. Afterwards, we compute lower bounds that hold with high probability on random geometric graphs. Finally, we characterize the probabilistic behavior of the lexicographic ordering for our layout problems on the class of random geometric graphs.

1 Introduction

Several well-known optimization problems on graphs can be formulated as Layout Problems. Their goal is to find a linear ordering (layout) of the nodes of an input graph such that a certain measure is minimized. Graph layout problems are an important class of problems with many different applications in Computer Science [4], Biology [14], Archaeology [3] and Linear Algebra [19]. Finding an optimal layout is **NP**-hard in general, and therefore it is natural to develop and analyze efficient methods that give good approximations in practice. However, evaluating heuristics as simulated annealing, greedy algorithms or spectral methods is a hard task [17,18].

A standard way of analyzing the efficiency of an heuristic algorithm is to evaluate its performance on random instances. Two classes of random instances have been widely used in the literature to enable comparisons of algorithms for layout and partitioning problems: *random graphs* and *random geometric graphs*. We denote the former class by $\mathcal{G}_{n,p}$, where n represents the number of nodes and p is the probability of the existence of each possible edge. Random graphs

[*] This research was partially supported by ESPRIT LTR Project no. 20244 — ALCOM-IT, CICYT Project TIC97-1475-CE, and CIRIT project 1997SGR-00366.

$\mathcal{G}_{n,p}$ have received much attention and together with the probabilistic method have become a powerful tool in combinatorics (see e.g. [1]). The approximation properties of sparse random graphs for different layout problems are considered in [8,19] and partitioning algorithms for random graphs are studied in [5,6]. On the other hand, we denote the class of random geometric graphs as $\mathcal{G}_n(r)$, where n is the number of vertices and r is called the radius. The vertices of a random geometric graph correspond to n points randomly distributed on the unit square. Each of its possible edges appears if and only if the distance between their two end-points is at most r. Random geometric graphs are considered a relevant abstraction to model graphs that occur in practice in real applications, such as finite element graphs, VLSI circuits, and communication graphs [11,12]. Moreover, since for many problems $\mathcal{G}_{n,p}$ random graphs do not serve to differentiate good from bad heuristics [8,6,19], random geometric graphs offer a good alternative. Even though many empirical studies have used random models of geometric graphs [11,18,12], its theoretical study has mainly focussed on parameters as their clique number or chromatic number, or in their connectivity properties [16].

In this paper, we are concerned with bounds for several layout measures on random geometric graphs. The layout problems that we consider are: *Bandwidth*, *Minimum Linear Arrangement*, *Minimum Cut*, *Minimum Sum Cut*, and *Vertex Separation*. We also consider the *Bisection* problem, which is a partitioning problem, but can be also treated as a layout problem. All these problems, formally defined in Section 2, are **NP**-complete. In Section 3, we prove that some of them remain **NP**-complete even for geometric instances. In Section 4, we compute lower bounds that hold with high probability on random geometric graphs. Afterwards, we obtain tight bounds on the cost of the *projection ordering* that is obtained by the projection of each node of a given random geometric graph into the x-axis. Section 5 analyzes this ordering. Our main result is the fact that the projection ordering is, with high probability, a constant approximation algorithm for our layout problems on the class of random geometric graphs considered here.

2   Definitions

We always consider undirected graphs without self loops. A *layout* φ on a graph $G = (V, E)$ is a one-to-one function $\varphi : V \to [n] = \{1, \ldots, n\}$ with $n = |V|$. Given a graph G and a layout φ on G, let us define:

$$L(i, \varphi, G) = \{u \in V(G) \ : \ \varphi(u) \leq i\}$$
$$R(i, \varphi, G) = \{u \in V(G) \ : \ \varphi(u) > i\}$$
$$\theta(i, \varphi, G) = \{uv \in E(G) \ : \ u \in L(i, \varphi, G) \wedge v \in R(i, \varphi, G)\}$$
$$\delta(i, \varphi, G) = \{u \in L(i, \varphi, G) \ : \ \exists v \in R(i, \varphi, G) : uv \in E(G)\}$$
$$\lambda(uv, \varphi, G) = |\varphi(u) - \varphi(v)| \quad \text{where } uv \in E(G).$$

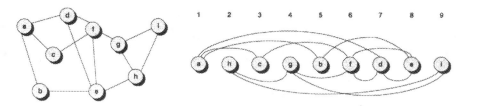

Fig. 1. A graph and a layout.

The problems we consider and their associated measures are:

- Bandwidth (BANDWIDTH): Given a graph $G = (V, E)$, find $\mathrm{MINBW}(G) = \min_\varphi \mathrm{BW}(\varphi, G)$ where $\mathrm{BW}(\varphi, G) = \max_{uv \in E} \lambda(uv, \varphi, G)$.
- Minimum Linear Arrangement (MINLA): Given a graph $G = (V, E)$, find $\mathrm{MINLA}(G) = \min_\varphi \mathrm{LA}(\varphi, G)$ where $\mathrm{LA}(\varphi, G) = \sum_{uv \in E} \lambda(uv, \varphi, G) = \sum_{i=1}^{n} |\theta(i, \varphi, G)|$.
- Minimum Cut Width (MINCUT): Given a graph $G = (V, E)$, find $\mathrm{MINCUT}(G) = \min_\varphi \mathrm{CUT}(\varphi, G)$ where $\mathrm{CUT}(\varphi, G) = \max_{i=1}^{n} |\theta(i, \varphi, G)|$.
- Vertex Separation (VERTSEP): Given a graph $G = (V, E)$, find $\mathrm{MINVS}(G) = \min_\varphi \mathrm{VS}(\varphi, G)$ where $\mathrm{VS}(\varphi, G) = \max_{i=1}^{n} |\delta(i, \varphi, G)|$.
- Minimum Sum Cut (MINSUMCUT): Given a graph $G = (V, E)$, find $\mathrm{MINSC}(G) = \min_\varphi \mathrm{SC}(\varphi, G)$ where $\mathrm{SC}(\varphi, G) = \sum_{i=1}^{n} |\delta(i, \varphi, G)|$.
- Bisection (BISECTION): Given a graph $G = (V, E)$, find $\mathrm{MINBIS}(G) = \min_\varphi \mathrm{BIS}(\varphi, G)$ where $\mathrm{BIS}(\varphi, G) = |\theta(\lfloor n/2 \rfloor, \varphi, G)|$.

It is well known that all the above problems are **NP**-complete for general graphs [9,10,13].

We introduce now several classes of geometric graphs on the plane. These graphs depend on which kind of norm is used to measure distances. Under the l_2 norm (the Euclidean norm), the distance between two points (x_1, y_1) and (x_2, y_2) is $((x_1 - x_2)^2 + (y_1 - y_2)^2)^{1/2}$. Under the l_∞ norm, their distance is $\max\{|x_1 - x_2|, |y_1 - y_2|\}$.

A graph is a *unit disk graph* if each vertex can be mapped to a closed, unit diameter disk in the plane such that two vertices are adjacent (in the graph) if and only their corresponding disks intersect (in the plane). A graph is a *grid graph* if it is a node-induced finite subgraph of the infinite grid. Observe that grid graphs are unit disk graphs both in l_2 and l_∞: it suffices to associate each node of the grid with a disk or a square (see Figure 2).

We define the class of *random geometric graphs* $\mathcal{G}_n(r_n)$ as the graphs of n nodes that can be obtained from the following experiment: Let the set $\mathcal{X}_n$ consist of n points sampled uniformly and independently at random from the unit square $([0, 1]^2)$; the nodes of the graph correspond to those points, and the edges of the graph connect pairs of distinct points whose distance is at most r_n (a random geometric graph is shown in Figure 3). Random geometric graphs

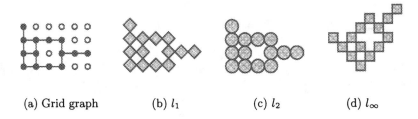

(a) Grid graph   (b) l_1 (c) l_2 (d) l_∞

Fig. 2. Any grid graph is a unit disk graph in l_1, l_2 and l_∞.

induce a probability distribution on unit disk graphs. Observe that, under this distribution, grid graphs have some positive probability.

All through this paper, we use the l_∞ norm. Furthermore, in the following we restrict our attention to the case

$$r_n = \sqrt{\frac{a_n}{n}} \quad \text{where} \quad a_n = b_n \log n \quad \text{with} \quad b_n \to \infty \quad \text{and} \quad b_n = O\left(\sqrt{\log n}\right).$$

It is important to remark that through this choice, the construction of sparse but connected graphs is guaranteed: Define the connectivity distance ρ_n of a random geometric graph by $\rho_n = \inf\{r \mid G \in \mathcal{G}_n(r) \text{ is connected}\}$. It is known [2] that as $n \to \infty$, $\left(\sqrt{n/\log n}\right)\rho_n$ converges to $\frac{1}{2}$ almost surely.

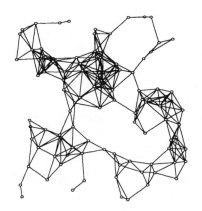

Fig. 3. A random geometric graph.

3 Complexity Results

In this section we will consider the decisional counterparts of the optimization problems previously defined. Let us show now that some of the layout problems we consider are still hard to solve efficiently, even when restricted to geometric instances.

Theorem 1. BANDWIDTH, MINCUT *and* VERTSEP *remain **NP**-complete even when restricted to grid graphs (and therefore, even when restricted to unit disk graphs).*

We could not obtain similar results for MINSUMCUT, MINLA and BISECTION. However, for the BISECTION problem, we are able to give a weak result:

Theorem 2. *If the* BISECTION *problem is **NP**-complete even when restricted to planar graphs with maximum vertex degree 4, then* BISECTION *is **NP**-complete even when restricted to unit disk graphs.*

The proof of these results will be given in the full version of this paper. Remark that Papadimitriou and Sideri [15] conjecture the hypothesis of Theorem 2, which is an important open problem.

4   Lower Bounds

In this section we find asymptotic lower bounds for the optimum cost of our various layout problems. As said, we take $r_n = \sqrt{a_n/n}$ where $a_n = b_n \log n$ with $b_n \to \infty$ as $n \to \infty$ and $b_n = O\left(\sqrt{\log n}\right)$. We consider a collection $\mathcal{X}_n$ of n points independently uniformly distributed in the unit square $[0,1]^2$. Consider $\lfloor 2/r_n \rfloor^2$ little boxes of size $\frac{1}{2}r_n \times \frac{1}{2}r_n$ placed packed in $[0,1]^2$ starting at $(0,0)$. Notice that, by construction, any two points of $\mathcal{X}_n$ in neighboring boxes (including diagonal neighbors) will be connected by an edge in the geometric graph induced by $\mathcal{X}_n$.

Definition 1. *Given $\epsilon \in \left(0, \frac{1}{2}\right)$, let us say that a configuration of n points in the unit square is ϵ-nice if every box has at least $\frac{1}{4}(1-\epsilon)a_n$ points and at most $\frac{1}{4}(1+\epsilon)a_n$ points.*

Using Chernoff's bounds and Boole's inequality, it is possible to show that, given any $\epsilon \in \left(0, \frac{1}{2}\right)$, $\mathbf{Pr}\left[\mathcal{X}_n \text{ is } \epsilon\text{-nice}\right] \to 1$ as n tends to infinity.

Proposition 1. *Let $\epsilon \in \left(0, \frac{1}{2}\right)$. Then for all large enough n, for any ϵ-nice geometric graph G with n nodes, and any layout φ of G, and any integer $i \in \left\{\lceil \frac{1}{4}n \rceil, \ldots, \lfloor \frac{3}{4}n \rfloor \right\}$, we have $|\theta(i, \varphi, G)| \geq \frac{3}{64}(1-\epsilon)^3 n^{1/2} a_n^{3/2}$.*

Proof. Consider n points in an ϵ-nice configuration, and take an arbitrary ordering φ of the n points. Let $\alpha = i/n$ and assume $\alpha \in \left[\frac{1}{4}, \frac{3}{4}\right]$. Call the first i points in the ordering "red" and the others "green". Let R_n be the set of boxes containing at least $\frac{1}{4}\beta a_n$ red points (red boxes), and let G_n be the set of boxes containing fewer than $\frac{1}{4}\beta a_n$ red points (green boxes) with $\beta = \frac{1}{2}(1-\epsilon)$. Define u, v, w and t such that un is the number of green points in red boxes, vn is the number of red points in red boxes, wn is the number of red points in green boxes, and tn is the number of green points in green boxes. According to these definitions, $v + w = \alpha$ and $u + t = 1 - \alpha$. Moreover, as G is ϵ-nice, each box contains $\frac{1}{4}(1-\epsilon)a_n$ or more points. Therefore, each green box contains at least $\frac{1}{4}(1 - \epsilon - \beta)a_n = \frac{1}{4}\beta a_n$ green points.

Observe that $|\theta(i, \varphi, G)|$ is the total number of edges between opposite color points. Let us refer to such edges as "within-box" if the points in question lie in the same box, or "between-box" if not. Let C_1 (respectively C_2) denote the contribution to $|\theta(i, \varphi, G)|$ from within-box edges that are within green boxes (respectively, red boxes). Each red point in a green box is connected, at least, to all the green points of its own box, and there are wn red points in all the green boxes. As a consequence, $C_1 \geq \frac{1}{4}wn\beta a_n$. Similarly, $C_2 \geq \frac{1}{4}un\beta a_n$.

Let C_3 denote the contribution to $|\theta(i, \varphi, G)|$ of between-box edges. As, by ϵ-niceness, no box can contain more than $\frac{1}{4}(1+\epsilon)a_n$ points and there are vn red points in red boxes and tn green points in green boxes, we have

$$|R_n| \geq \frac{vn}{\frac{1}{4}(1+\epsilon)a_n} = \frac{4v}{(1+\epsilon)r_n^2} \quad \text{and} \quad |G_n| \geq \frac{tn}{\frac{1}{4}(1+\epsilon)a_n} = \frac{4t}{(1+\epsilon)r_n^2}.$$

Let ∂G denote the number of pairs of neighbor boxes of opposite colors in G. We have

$$C_3 \geq \partial G(\tfrac{1}{4}\beta a_n)(\tfrac{1}{4}\beta a_n) = \partial G\tfrac{(1-\epsilon)^2}{64}a_n^2.$$

The following isoperimetric inequality, giving a lower bound for ∂G, can be proved along the following lines. If G_n includes and entirely green row of boxes, and R_n includes an entirely red row of boxes, then each column includes a red-green neighbor pair of boxes, which contributes at least 3 to ∂G (remember diagonal neighbors are counted) except for the pair in the right-most column which contributes 1, so that $\partial G \geq 3\lfloor 2/r_n \rfloor - 2$. If R_n containes no entirely red row or column, and more rows than columns have non-empty intersection with R_n, then there are at least $\sqrt{|R_n|}$ such rows, and each contains a red-green neighbor pair which contributes at least 3 to ∂G, so that $\partial G \geq 3\sqrt{|R_n|}$. Combining these and analogous cases we have

$$\partial G \geq \min\left\{3\sqrt{|R_n|},\ 3\sqrt{|G_n|},\ 3\left\lfloor\frac{2}{r_n}\right\rfloor - 2\right\} \geq 3\min\left\{\sqrt{|R_n|},\ \sqrt{|G_n|}\right\}$$

$$\geq \frac{6}{r_n}\min\left\{\sqrt{\frac{v}{1+\epsilon}},\ \sqrt{\frac{t}{1+\epsilon}}\right\}.$$

Using the results obtained so far, we obtain

$$|\theta(i, \varphi, G)| \geq C_1 + C_2 + C_3 \geq A_n(u+w) + B_n\min\left\{\sqrt{v},\ \sqrt{t}\right\}$$

where $A_n = \frac{1-\epsilon}{8}na_n$ and $B_n = \frac{3}{32}a_n^{3/2}n^{1/2}(1-\epsilon)^2(1+\epsilon)^{-1/2}$.

Remember that $u+t = 1-\alpha$, $v+w = \alpha$ and $\alpha \in \left[\frac{1}{4}, \frac{3}{4}\right]$. When $t < (1-\alpha)(1-\epsilon)$ we have $u \geq \frac{1}{4}\epsilon$. When $v < \alpha(1-\epsilon)$, we have $w \geq \frac{1}{4}\epsilon$. In both cases

$$|\theta(i, \varphi, G)| \geq A_n(u+w) \geq \tfrac{1}{4}\epsilon A_n.$$

Finally, when $v \geq \alpha(1-\epsilon)$ and $t \geq (1-\alpha)(1-\epsilon)$, we have

$$|\theta(i, \varphi, G)| \geq B_n\min\left\{\sqrt{v},\ \sqrt{t}\right\} \geq B_n\min\left\{\sqrt{\alpha},\ \sqrt{(1-\alpha)}\right\}\sqrt{1-\epsilon}.$$

Since we assume $\alpha \in \left[\frac{1}{4}, \frac{3}{4}\right]$, we have $\min\left\{\sqrt{\alpha}, \sqrt{(1-\alpha)}\right\} \geq \frac{1}{2}$. Hence, as A_n grows faster than B_n since $a_n/n \to 0$, joining these three cases we get for n big enough that

$$|\theta(i, \varphi, G)| \geq \min\left\{\tfrac{1}{4}\epsilon A_n, \tfrac{1}{2}B_n\sqrt{1-\epsilon}\right\} \geq \tfrac{3}{64}a_n^{3/2}n^{1/2}(1-\epsilon)^3. \qquad \square$$

Theorem 3 (Lower bounds). *Let $\epsilon \in (0, \frac{1}{2})$. Then for n big enough, the following lower bounds hold for any ϵ-nice geometric graph G with n vertices:*

$$\frac{\text{MINBIS}(G)}{n^{1/2}a_n^{3/2}} \geq \tfrac{3}{64}(1-\epsilon)^{-3} \tag{lb1}$$

$$\frac{\text{MINCUT}(G)}{n^{1/2}a_n^{3/2}} \geq \tfrac{3}{64}(1-\epsilon)^{-3} \tag{lb2}$$

$$\frac{\text{MINLA}(G)}{n^{3/2}a_n^{3/2}} \geq \tfrac{3}{128}(1-\epsilon)^{-3}(1 - 4n^{-1}) \tag{lb3}$$

$$\frac{\text{MINVS}(G)}{n^{1/2}a_n^{1/2}} \geq \tfrac{3}{400}(1-\epsilon)^{-4} \tag{lb4}$$

$$\frac{\text{MINSC}(G)}{n^{3/2}a_n^{1/2}} \geq \tfrac{3}{800}(1+\epsilon)^{-4}(1 - 4n^{-1}) \tag{lb5}$$

$$\frac{\text{MINBW}(G)}{n^{1/2}a_n^{1/2}} \geq \tfrac{3}{400}(1-\epsilon)^{-4} \tag{lb6}$$

Proof. The proof of (lb1) and (lb2) is directly obtained from Proposition 1. To prove (lb3), take any layout φ; using Proposition 1, we have:

$$\text{LA}(\varphi, G) \geq \sum_{i=\lceil n/4 \rceil}^{\lfloor 3n/4 \rfloor} |\theta(i, \varphi, G)| \geq \left(\tfrac{1}{2}n - 2\right)\tfrac{3}{64}(1-\epsilon)^{-3}n^{1/2}a_n^{3/2}.$$

To prove (lb4), let Δ be the degree of the graph (i.e. the maximum degree of its vertices). Then, for any layout φ and any $i \in [n]$, we have $|\delta(i, \varphi, G)| \geq |\theta(i, \varphi, G)|/\Delta$. For any ϵ-nice graph, $\Delta \leq \frac{25}{4}(1+\epsilon)a_n$. Therefore, by Proposition 1, for any layout φ and any i with $\frac{1}{4}n \leq i \leq \frac{3}{4}n$, we have $|\delta(i, \varphi, G)| \geq \frac{3}{400}(1-\epsilon)^{-4}n^{1/2}a_n^{1/2}$, implying (lb4), and also (lb5) since

$$\text{SC}(\varphi, G) \geq \sum_{i=\lceil n/4 \rceil}^{\lfloor 3n/4 \rfloor} |\delta(i, \varphi, G)| \geq \left(\tfrac{1}{2}n - 2\right)\tfrac{3}{400}(1-\epsilon)^{-4}n^{1/2}a_n^{1/2}.$$

Finally, let us prove that (lb6) holds. Before the node at position i, $\frac{1}{4}n \leq i \leq \frac{3}{4}n$, there have to be at least $|\delta(i, \varphi, G)|$ nodes, all of them connected with some other nodes located after the position i. So, the first of these $|\delta(i, \varphi, G)|$ nodes must have an edge that jumps at least $|\delta(i, \varphi, G)|$ nodes. In other words, for any layout φ there is an edge $uv \in E(G)$ with $\lambda(uv, \varphi, G) \geq |\delta(i, \varphi, G)|$. Thus (lb6) follows from (lb4).

5   The Projection Ordering

In this section, we characterize the behavior of the projection ordering. Recall that the projection ordering is obtained through the projection of the nodes on the x-axis. Another way to see this ordering is to sweep a vertical line starting from $x = 0$ to $x = 1$, numbering vertices in the order the line touches them. As in the previous section, we consider a collection $\mathcal{X}_n$ of n points independently and uniformly distributed in the unit square $[0, 1]^2$ and work in the case $r_n = \sqrt{a_n/n}$ where $a_n = b_n \log n$ with $b_n \to \infty$ and $b_n = O\left(\sqrt{\log n}\right)$. The coordinates of a point u are denoted $x(u)$ and $y(u)$. We dissect the unit square in boxes of size $\gamma r_n \times \gamma r_n$ with $\gamma = 1/k$ for some large enough integer k. Let $t = n/(\gamma^2 a_n)$ denote the total number of boxes. Without loss of generality, we will suppose that $1/(\gamma r_n)$ and t are integers.

Definition 2. *A set $\mathcal{X}_n$ of n points in $[0, 1]^2$ is said to be γ-good if every box contains no less than $p_- = (1-\gamma)\gamma^2 a_n$ points and no more than $p_+ = (1+\gamma)\gamma^2 a_n$ points. In this case, the random geometric graph induced by $\mathcal{X}_n$ is also said to be γ-good.*

Later we will prove that with high probability, random geometric graphs are γ-good for any $\gamma \in (0, \frac{1}{2})$. The behavior of the Projection ordering on γ-good graphs is characterized by the following result:

Theorem 4. *Let G_n be a sequence of γ-good graphs with n vertices, and let π be the projection layout on G_n. Then, for any $\epsilon \in (0, \frac{1}{2})$ and for any measure $f \in \{\mathrm{BW}, \mathrm{VS}, \mathrm{SC}, \mathrm{CUT}, \mathrm{BIS}, \mathrm{LA}\}$, we have*

$$1 - \epsilon \leq \left(\lim_{n\to\infty} \frac{f(\pi, G_n)}{A_f} \right) \leq 1 + \epsilon$$

where

$$A_{\mathrm{BW}} = n^{1/2} a_n^{1/2}, \qquad A_{\mathrm{VS}} = n^{1/2} a_n^{1/2}, \qquad A_{\mathrm{SC}} = n^{3/2} a_n^{1/2},$$
$$A_{\mathrm{CUT}} = n^{1/2} a_n^{3/2}, \qquad A_{\mathrm{BIS}} = n^{1/2} a_n^{3/2}, \qquad A_{\mathrm{LA}} = n^{3/2} a_n^{3/2}.$$

5.1   Upper Bounds on the Projection Ordering

Definition 3. *Consider the geometric graph G induced by $\mathcal{X}_n$. Given a node u from G, let $\theta(u)$ denote the cut induced by the projected layout π on u, that is, the number of edges vw such that $x(v) \leq x(u)$ and $x(u) < x(w)$. Given an edge uv from G, let $\lambda(uv)$ denote the length induced by the π on uv, that is, the number of nodes w such that $x(u) < x(w)$ and $x(w) < x(v)$.*

Lemma 1. *For any node u and for any edge uv of any γ-good graph, we have that*

$$\theta(u) \leq c_\theta(\gamma) \cdot n^{1/2} a_n^{3/2} \quad and \quad \lambda(uv) \leq c_\lambda(\gamma) \cdot n^{1/2} a_n^{1/2}$$

where $\lim_{\gamma\to 0} c_\theta(\gamma) = 1$ and $\lim_{\gamma\to 0} c_\lambda(\gamma) = 1$.

Proof. Every possible edge is between boxes with centers at distance at most r_n. Thus,

$$\theta(u) \leq \sum_{i=0}^{1/\gamma} \left(\frac{1}{\gamma} - i + 1\right) \left(\frac{2}{\gamma} + 1\right) \frac{1}{\gamma r_n} \cdot p_+^2 \leq c_\theta(\gamma) \cdot n^{1/2} a_n^{3/2}.$$

On the other hand, $\lambda(uv)$ is bounded above by the number of possible nodes in the columns of boxes between the column of u and the column of v. Thus,

$$\lambda(uv) \leq p_+ \left(\frac{1}{\gamma} + 2\right) \frac{1}{\gamma r_n} \leq c_\lambda(\gamma) \cdot n^{1/2} a_n^{1/2}.$$

Corollary 1. *For any γ-good graph G with n nodes, the following upper bounds on the cost of the projected layout π of G hold:*

$$\text{CUT}(\pi, G) \leq c_\theta(\gamma) \cdot n^{1/2} a_n^{3/2} \tag{ub1}$$

$$\text{BIS}(\pi, G) \leq c_\theta(\gamma) \cdot n^{1/2} a_n^{3/2} \tag{ub2}$$

$$\text{BW}(\pi, G) \leq c_\lambda(\gamma) \cdot n^{1/2} a_n^{1/2} \tag{ub3}$$

$$\text{LA}(\pi, G) \leq c_\theta(\gamma) \cdot n^{3/2} a_n^{3/2} \tag{ub4}$$

$$\text{VS}(\pi, G) \leq c_\lambda(\gamma) \cdot n^{1/2} a_n^{1/2} \tag{ub5}$$

$$\text{SC}(\pi, G) \leq c_\lambda(\gamma) \cdot n^{3/2} a_n^{3/2} \tag{ub6}$$

Proof. Bounds (ub1), (ub2) and (ub3) follow directly from Lemma 1. Bounds (ub4), (ub5) and (ub6) hold because for any layout φ, we have $\text{LA}(\varphi, G) \leq n \cdot \text{CUT}(\varphi, G)$, $\text{VS}(\varphi, G) \leq \text{BW}(\varphi, G)$ and $\text{SC}(\varphi, G) \leq n \cdot \text{VS}(\varphi, G)$.

5.2   Lower Bounds on the Projection Ordering

Lemma 2. *For any sequence G_n of γ-good graphs with n vertices, the following lower bounds on the cost of the projected layout π of G_n hold:*

$$\frac{\text{BW}(\pi, G_n)}{n^{1/2} a_n^{1/2}} \geq c_1(\gamma) \tag{lb1}$$

$$\frac{\text{VS}(\pi, G_n)}{n^{1/2} a_n^{1/2}} \geq c_2(\gamma) \tag{lb2}$$

$$\liminf_{n \to \infty} \frac{\text{SC}(\pi, G_n)}{n^{3/2} a_n^{1/2}} \geq c_3(\gamma) \tag{lb3}$$

$$\liminf_{n \to \infty} \frac{\text{CUT}(\pi, G_n)}{n^{1/2} a_n^{3/2}} \geq c_4(\gamma) \tag{lb4}$$

$$\liminf_{n \to \infty} \frac{\text{BIS}(\pi, G_n)}{n^{1/2} a_n^{3/2}} \geq c_5(\gamma) \tag{lb5}$$

$$\liminf_{n \to \infty} \frac{\text{LA}(\pi, G_n)}{n^{3/2} a_n^{3/2}} \geq c_6(\gamma) \tag{lb6}$$

where $c_i(\gamma)$ are functions that only depend on γ and such that $\lim_{\gamma \to 0} c_i(\gamma) = 1$.

Proof. Let G be a γ-good graph with n vertices. Let us prove (lb1). Consider a node u far enough from the square boundaries (u exists because of goodness). This node will be connected with some other node v which is located $k-1$ columns away from the column of u (v also exists because of goodness). The length of the edge uv in the projected layout π is certainly larger than the total number of nodes located at columns between the column of u and the column of v:

$$\lambda(\pi, uv, G) \geq p_- \cdot (k-2) \cdot \frac{1}{\gamma r_n} \geq (\gamma-1)(2\gamma-1) \cdot a_n^{1/2} n^{1/2}.$$

As $\mathrm{BW}(\pi, G)$ is the maximal edge length, we obtain the claimed bound.

Let us give a proof of (lb2). Consider any node u far enough of the square boundaries. All the nodes in the $k-2$ columns preceding the columns of u must be connected to some node in the next column after the column of u. Therefore,

$$\mathrm{VS}(\pi, G) \geq p_- \cdot (k-2) \cdot \frac{1}{\gamma r_n} \geq (\gamma-1)(2\gamma-1) \cdot a_n^{1/2} n^{1/2}.$$

Let us prove (lb3). We can extend the previous proof to all the points which are away from the left and the right borders of the unit square:

$$\mathrm{SC}(\pi, G) \geq p_- \left(\frac{1}{\gamma r_n}\right)\left(\frac{1}{\gamma r_n} - 2k\right)\left(p_- \cdot (k-2) \cdot \frac{1}{\gamma r_n}\right).$$

In this case,

$$\lim_{n \to \infty} \frac{\mathrm{SC}(\pi, G)}{n^{3/2} a_n^{1/2}} \geq 1 - 4\gamma + 5\gamma^2 - 2\gamma^3.$$

We prove now (lb4) and (lb5). Take any node u in the central part of $[0,1]^2$. We have

$$\mathrm{CUT}(\pi, G) \geq \sum_{i=1}^{k-2} p_-^2 (k-i-1) \left(\frac{1}{\gamma r_n} - 2k\right) 2k.$$

Therefore,

$$\lim_{n \to \infty} \frac{\mathrm{CUT}(\pi, G)}{n^{1/2} a_n^{3/2}} \geq 1 - 5\gamma + 9\gamma^2 - 7\gamma^3 + 2\gamma^4.$$

As the $\lceil n/2 \rceil$-th node of the projected layout must be in the central part of $[0,1]^2$, we have also that $\lim_{n \to \infty} \frac{\mathrm{BIS}(\pi, G)}{n^{1/2} a_n^{3/2}} \geq 1 - 5\gamma + 9\gamma^2 - 7\gamma^3 + 2\gamma^4$.

Finally, let us prove (lb6). We can extend the cut width proof to all the points which are away from the left and the right borders of the unit square:

$$\mathrm{LA}(\pi, G) \geq p_- \left(\frac{1}{\gamma r_n}\right)\left(\frac{1}{\gamma r_n} - 2k\right) \cdot \sum_{i=1}^{k-2} p_-^2 (k-i-1) \left(\frac{1}{\gamma r_n} - 2k\right) 2k.$$

In this case,

$$\lim_{n \to \infty} \frac{\mathrm{LA}(\pi, G)}{n^{3/2} a_n^{3/2}} \geq 1 - 6\gamma + 14\gamma^2 - 16\gamma^3 + 9\gamma^4 - 2\gamma^5$$

5.3 Approximability of the Projection Ordering

In Section 4, we have given lower bounds that hold with high probability for all the considered problems on ϵ-nice graphs. Theorem 4 characterizes the behavior of the Projection ordering on γ-good graphs. Using Chernoff's bounds and Boole's inequality, one can show that the probability of a random geometric graph to be both ϵ-nice and γ-good tends to one as n tends to infinity. Therefore, we have the following result:

Theorem 5. *The projection ordering is an approximation algorithm with high probability for the Bandwidth, Minimum Linear Arrangement, Minimum Cut, Minimum Sum Cut, Vertex Separation and Bisection problems on the class $\mathcal{G}_n(r)$ with $r_n = \sqrt{a_n/n}$ where $a_n = b_n \log a_n$ with $b_n \to \infty$ and $b_n = O\left(\sqrt{\log n}\right)$.*

6 Conclusion

In this paper we have presented upper and lower bounds for different measures of vertex orderings. We have also shown that the projection ordering is able to deliver with high probability solutions whose cost is not more than a constant times bigger the optimum on a particular class of random geometric graphs for several layout problems. Given the importance of the considered problems and the intensive use of these graphs in experimental papers, our result fills an important gap that existed between theory and practice.

We have considered only the two-dimensional geometric graphs as most real instances belong to that case, but we think that similar results will also hold on d-dimensional spaces. Our current work is trying to generalize the results on other models of random geometric graphs. For instance, it would be interesting to understand how the optimal costs of our problems change for different radii.

On the other hand, in [7] layout problems for lattice graphs and random lattice graphs are studied. Our main result in that paper is a convergence theorem (analoguous to the Beardwood, Halton and Hammersley theorem for the Euclidian TSP) for the optimal cost of the MINLA and MINSUMCUT problems, for the case where the underlying graph is obtained through a subcritical site percolation process.

References

1. N. Alon, J. H. Spencer, and P. Erdős. *The probabilistic method.* Wiley-Interscience, New York, 1992.
2. M. J. Appel and R. P. Russo. The connectivity of a graph on uniform points in $[0,1]^d$. Technical Report #275, Department of Statistics and Actuarial Science, University of Iowa, 1996.
3. J.E. Atkins, E.G. Boman, and B. Hendrickson. A spectral algorithm for seriation and the consecutive ones problem. *SIAM J. Comput.*, 28(1):297–310, 1999.
4. S. N. Bhatt and F. T. Leighton. A framework for solving VLSI graph layout problems. *Journal of Computer and System Sciences*, 28:300–343, 1984.

5. R. Boppana. Eigenvalues and graph bisection: An average case analysis. In *Proc. on Foundations of Computer Science*, pages 280–285, 1987.

6. T. Bui, S. Chaudhuri, T. Leighton, and M. Sipser. Graph bisection algorithms with good average case behavior. *Combinatorica*, 7:171–191, 1987.

7. Josep Díaz, Mathew D. Penrose, Jordi Petit, and María Serna. Layout problems on lattice graphs. In T. Asano, H. Imai, D.T. Lee, S. Nakano, and T. Tokuyama, editors, *Computing and Combinatorics — 5th Annual International Conference, COCOON'99, Tokyo, Japan, July 26-28, 1999*, number 1627 in Lecture Notes in Computer Science. Springer–Verlag, Berlin, July 1999.

8. J. Díaz, J. Petit, M. Serna, and L. Trevisan. Approximating layout problems on random sparse graphs. Technical Report LSI 98-44-R, Universitat Politècnica de Catalunya, 1998.

9. M. R. Garey and D. S. Johnson. *Computers and Intractability: A Guide to the Theory of NP-Completeness*. Freeman, San Francisco, 1979.

10. F. Gavril. Some NP-complete problems on graphs. In *Proc. 11th. Conference on Information Sciences and Systems*, pages 91–95, John Hopkins Univ., Baltimore, 1977.

11. D. S. Johnson, C. R. Aragon, L. A. McGeoch, and C. Schevon. Optimization by simulated annealing: an experimental evaluation; part I, graph partitioning. *Operations Research*, 37(6):865–892, November 1989.

12. K. Lang and S. Rao. Finding Near-Optimal Cuts: An Empirical Evaluation. In *Fourth Annual ACM-SIAM Symposium on Discrete Algorithms*, pages 212–221, 1993.

13. T. Lengauer. Black-white pebbles and graph separation. *Acta Informatica*, 16:465–475, 1981.

14. G. Mitchison and R. Durbin. Optimal numberings of an $n \times n$ array. *SIAM Journal on Discrete Mathematics*, 7(4):571–582, 1986.

15. C. H. Papadimitriou and M. Sideri. The bisection width of grid graphs. In *First ACM-SIAM Symposium on Discrete Algorithms*, pages 405–410, San Francisco, 1990.

16. M. D. Penrose. The longest edge of the random minimal spanning tree. *The Annals of Applied Probability*, 7:340–361, 1997.

17. J. Petit. Combining Spectral Sequencing with Simulated Annealing for the MINLA Problem: Sequential and Parallel Heuristics. Technical Report LSI-97-46-R, Departament de Llenguatges i Sistemes Informàtics, Universitat Politècnica de Catalunya, 1997.

18. J. Petit. Approximation Heuristics and Benchmarkings for the MINLA Problem. In R. Battiti and A. Bertossi, editors, *Alex '98 — Building bridges between theory and applications*, http://www.lsi.upc.es/~jpetit/MinLA/Benchmark, February 1998. Università di Trento.

19. J. S. Turner. On the probable performance of heuristics for bandwidth minimization. *SIAM Journal on Computing*, 15:561–580, 1986.

Finding Smallest Supertrees under Minor Containment*

Naomi Nishimura[1], Prabhakar Ragde[1], and Dimitrios M. Thilikos[2]

[1] Department of Computer Science, University of Waterloo,
Waterloo, Ontario, Canada, N2L 3G1. FAX (519) 885-1208
{nishi,pragde}@plg.uwaterloo.ca
[2] Departament de Llenguatges i Sistemes Informàtics,
Universitat Politècnica de Catalunya,
Campus Nord – Mòdul C5, c/Jordi Girona Salgado, 1-3.
E-08034, Barcelona, Spain
sedthilk@lsi.upc.es

Abstract. The diversity of application areas relying on tree-structured data results in a wide interest in algorithms which determine differences or similarities among trees. One way of measuring the similarity between trees is to find the smallest common superstructure or supertree, where common elements are typically defined in terms of a mapping or embedding. In the simplest case, a supertree will contain exact copies of each input tree, so that for each input tree, each vertex of a tree can be mapped to a vertex in the supertree such that each edge maps to the corresponding edge. More general mappings allow for the extraction of more subtle common elements captured by looser definitions of similarity. We consider supertrees under the general mapping of minor containment. Minor containment generalizes both subgraph isomorphism and topological embedding; as a consequence of this generality, however, it is NP-complete to determine whether or not G is a minor of H, even for general trees. By focusing on trees of bounded degree, we obtain an $O(n^3)$ algorithm which determines the smallest tree T such that both of the input trees are minors of T, even when the trees are assumed to be unrooted and unordered.

1 Introduction

The breadth of algorithmic research on trees stems from both the simplicity of the structure and the variety of application domains. When information about a data set can be derived from its tree structure, comparisons among two or more data sets can entail determining similarities among two or more trees. Algorithms of this type have been developed in areas such as compiler design, structured text databases, theory of natural languages, computer vision [18], and

* Research supported by the Natural Sciences and Engineering Research Council of Canada. The research of the third author was partially supported by the Ministry of Education and Culture of Spain – Grant no MEC-DGES SB97 0K148809.

computational biology (the reader is directed to a previous paper on trees [10] for further references).

Comparisons of trees range from the classical tree pattern matching problem (finding an exact copy of one tree in another) to numerous variants, including problems on multiple trees and inexact matches. Each problem can be viewed as finding a way to relate trees by mappings, where trees are related if it is possible to map vertices to sets of vertices and edges to sets of edges subject to certain constraints. Researchers have considered different types of trees (ordered, unordered, labeled, unlabeled) and different mappings between pairs of trees (exact matching, approximate matching, subgraph isomorphism, topological embedding, minor containment) [5,13,14,3,9]. In addition, researchers have measured the similarity between trees by finding the largest common subtree or smallest common supertree under various constraints [1,4,7,8,10,12,19].

In this paper we consider the problem of finding the smallest common supertree under minor containment. Concisely, a graph G is a minor of a graph H if it is possible to map all the vertices in G to mutually disjoint connected subgraphs in H and there exists a bijection, from the edges of G to the edges of H that are not in any of these subgraphs, such that the images of the endpoints of any edge e in G contain the endpoints of the image of e through this bijection; equivalently we can view the mapping as taking edges to paths. Minor containment is of interest due to its generality; it encompasses both subgraph isomorphism and topological embedding and is fundamental in the work of Robertson and Seymour on graph minors [17]. However, due in large part to the generality, many problems which are tractable under subgraph isomorphism and topological embedding are NP-complete for minor containment. In particular, it is NP-complete to determine whether or not one tree is a minor of another [6], but this can be determined in polynomial time when there is a degree bound of $O(\log n/\log \log n)$ [9]. We thus restrict our attention to trees of bounded degree, noting that the resultant supertree will also be of bounded degree (in contrast, a common subtree of two bounded degree trees may not have bounded degree).

Interest in supertrees under minor containment arises from their applications to editing, image clustering, genetics, chemical structure analysis, and evolution [12,19]. Previous algorithms to find supertrees have been limited to special cases: in ordered minor containment, there is an order imposed on the children of each node in each input tree, and this order must be preserved by the mapping [12]; for evolutionary trees, the leaves have distinct labels and are constrained to map to other leaves [19].

2   Preliminaries

Each input to our algorithm is a bounded-degree tree (an undirected graph with no cycles). $V(T)$ denotes the vertices of T and $E(T)$ the edges of T. A tree T may be rooted at a distinguished vertex r; in this case we can view the rooted tree as a directed graph, with children and parents defined as in standard graph-theoretic references [2]. When processing rooted trees we will consider a *subtree*

T_v of T, defined to be the subgraph of T induced by v and all its descendants. More generally, for A a subset of the children of some node v, we define T_A to be the subgraph induced by v, the vertices in A, and all descendants of nodes in A. For A an arbitrary subset of vertices, $T[A]$ is defined to be the subgraph of T induced by A.

Given input trees Q and R, we wish to find a tree T such that both Q and R are minors of T and T is as small as possible. There are several equivalent definitions of minors; the most relevant one for our purposes is given below. Intuitively, a graph G is a minor of a graph H (or H is a major of G) if H can be obtained from G by a series of vertex and edge deletions and edge contractions, where a contraction of an edge (u, v) in G is the operation that replaces u and v by a new vertex whose neighbors are the vertices that were adjacent to u or v. We will use the notation $G \leq H$ ($G \geq H$) to denote that G is a minor (major) of H. It is not difficult to see that the following definition is equivalent:

Definition 1. *A tree Q is a* minor *of a tree T if and only if there exists a surjection $f : V(T) \to V(Q)$ such that*

1. *for each $a \in V(Q)$, $T[f^{-1}(a)]$ is connected;*
2. *for each pair $a, b \in V(Q)$, $f^{-1}(a) \cap f^{-1}(b) = \phi$; and*
3. *for $S = \{(u, v) \in E(T) \mid f(u) \neq f(v)\}$, there exists a bijection $\xi : S \to E(Q)$ such that for each $e = (s, t) \in S, \xi(e) = (f(s), f(t))$.*

We call f a minor embedding *of T into Q. Intuitively, $f^{-1}(a)$ is the set of vertices of T contracted into a; (2) captures the notion that each vertex of T corresponds to exactly one vertex of Q; and (3) captures the notion that uncontracted edges of T are preserved in Q.*

The problem we wish to solve is that of determining the smallest common acyclic major of Q and R, henceforth called the *smallest common tree major*. For sctmj(Q, R) the minimum number of vertices in a common tree major of Q and R, it is not difficult to see that $\max\{|V(Q)|, |V(R)|\} \leq$ sctmj$(Q, R) \leq |V(Q)| + |V(R)|$. We observe that sctmj$(Q, R) = |Q|$ if and only if $R \leq Q$. In [6], Duchet proved that it is NP-complete to determine whether one tree is a minor of another. It is now easy to prove that deciding whether sctmj$(Q, R) \leq k$ for two general trees Q, R is NP-complete. In view of this, we will restrict our attention to the case where the input graphs are both trees with maximum degree bounded by a fixed constant.

In the remainder of the paper we will make use of the following notational conventions. Since we will be finding a graph T such that Q and R are both minors of T, we will use f to denote the minor embedding of T into Q and g to denote the minor embedding of T into R. We will use letters near the beginning of the alphabet for vertices of Q and letters near the end of the alphabet for vertices of R.

3   Expansions

To facilitate understanding of the algorithm, it is beneficial to consider the mappings between Q, R, and a common tree major T. The edges of T correspond to edges in the input trees Q and R; we distinguish between *strong* edges, which correspond to edges in both Q and R, and *weak* edges, each of which corresponds to an edge in only one of Q and R. For f and g the minor embeddings of T into Q and R, respectively, $f^{-1}(a)$ and $g^{-1}(u)$ describe connected subgraphs of T. Since each vertex in $f^{-1}(a)$ is in $g^{-1}(u)$ for some $u \in V(R)$, we can associate a with a set of vertices in $V(R)$ with overlapping preimages. This notion is formalized in a graph called an *expansion* of Q and R consisting of edges between associated vertices. More formally:

Definition 2. *For Q and R trees on disjoint sets of vertices, an* expansion *of Q and R is a bipartite graph $\mathcal{E} = (V(\mathcal{E}), E(\mathcal{E}))$ with bipartition $(V(Q), V(R))$ such that*

1. *the neighborhood in $\mathcal{E}$ of any vertex of $V(R)$ (respectively, $V(Q)$) induces a connected subgraph of Q (respectively, R);*
2. *$\mathcal{E}$ has no isolated vertices;*
3. *the neighborhoods in $\mathcal{E}$ of two vertices in $V(Q)$ (respectively, $V(R)$) intersect in at most one vertex; and*
4. *for every edge (a, b) in $E(Q)$, either there are edges (a, u) and (b, u) in $\mathcal{E}$ for some $u \in V(R)$, or there are edges (a, u) and (b, v) in $\mathcal{E}$ for some edge $(u, v) \in E(R)$ (and symmetrically for edges in R).*

Given an expansion $\mathcal{E}$ of Q and R, we define $T_\mathcal{E}$ to be a graph whose vertices are edges in $\mathcal{E}$ and whose edges are formed by condition 4 in the definition above. For an edge $(a, b) \in E(Q)$, if there are edges (a, u) and (b, u) in $\mathcal{E}$, then $\{(a, u), (b, u)\}$ is an edge in $T_\mathcal{E}$, and if there are edges (a, u) and (b, v) in $\mathcal{E}$ for some $(u, v) \in E(R)$, and neither (a, v) nor (b, u) is in $\mathcal{E}$, then $\{(a, u), (b, v)\}$ is in $T_\mathcal{E}$. Edges $(u, v) \in E(R)$ define edges in $T_\mathcal{E}$ in a similar fashion. In the former case we call the edge (a, b) *weak*; in the latter case, (a, b) and (u, v) are *strong*. Note that there is a natural bijection $f_\mathcal{E}$ between strong edges in $E(Q)$ and strong edges in $E(R)$.

The proof of the following lemma is an easy consequence of the definition of an expansion.

Lemma 1. *For any expansion $\mathcal{E}$ of Q and R, if $(a_i, u_i) \in \mathcal{E}, i = 1, 2, 3$, and a_2 is on the path connecting a_1 and a_3 in Q, then u_2 is on the path connecting u_1 and u_3 in R.*

Lemma 2. *If $\mathcal{E}$ is an expansion of two trees Q and R, then $T_\mathcal{E}$ is isomorphic to a common tree major of Q and R.*

To prove Lemma 2, we can show that $T_\mathcal{E}$ is connected and contains exactly $|E(\mathcal{E})| - 1$ edges, forming a tree. Furthermore, if we alter $T_\mathcal{E}$ by contracting the edges resulting from $E(Q)$ ($E(R)$, respectively) we obtain a graph that is isomorphic to R (Q, respectively).

Lemma 3. *For T a smallest common tree major of Q and R, there exists an expansion $\mathcal{E}$ such that $T_{\mathcal{E}}$ is isomorphic to T.*

Proof: Given minor embeddings f and g of T into Q and R, for each $a \in V(Q)$ and each $u \in V(R)$, $|f^{-1}(a) \cap g^{-1}(u)| \leq 1$, since otherwise the minor of T obtained after contracting the edges in the graph induced by $\{f^{-1}(a) \cap g^{-1}(u)\}$ would be a smaller common tree major of Q and R. We define the expansion $\mathcal{E}$ to be the set $\{(a, u) \, : \, |f^{-1}(a) \cap g^{-1}(u)| = 1\}$. It is straightforward to verify the claim that $\mathcal{E}$ is an expansion of Q and R.   Q.E.D.

As a corollary of Lemmata 2 and 3, we can conclude that $\mathsf{sctmj}(Q, R)$ is the number of edges in the minimum expansion of Q and R. It is now easy to prove the following.

Lemma 4. *For trees Q and R and for any $a \in V(Q)$, $\mathsf{sctmj}(Q, R)$ is the minimum over all $u \in V(R)$ of the number of edges in the smallest expansion $\mathcal{E}$ of Q and R such that (a, u) is an edge in $\mathcal{E}$.*

4   Smallest Common Tree Major Algorithm

4.1   Algorithm Overview

For algorithmic convenience, we construct a rooted tree major, where any node of either input tree could be associated with the root. We fix a root for one tree and then try all possible rootings of the other tree; the following description concerns one possible choice of a root.

Our algorithm proceeds by dynamic programming, at each stage building tree majors of various subtrees of our inputs. After topologically sorting each tree with respect to the chosen root, we process each vertex a in $V(Q)$ in order from leaves to root, pairing a with each u in $V(R)$ in order from leaves to root.

For a given pair (a, u) we wish to determine the size of the largest common tree major T such that $Q_a \leq T$ and $R_u \leq T$ where for r the root of T, $f(r) = a$ and $g(r) = u$. We solve this problem using subproblems involving children of a and u, where in each subproblem we specify not only the roots of the subtrees of Q and R, but also the subsets of the children included thus far in the mapping.

Expansions, as defined in the previous section, give a convenient framework for expressing the progress of the algorithm, where expansions involving subgraphs of Q and R are augmented to form expansions of larger subgraphs of Q and R. The dynamic programming formulation of the problem relies on a set of subproblems at $a \in V(Q)$ and $u \in V(R)$, where each subproblem corresponds to one choice of how the children of a and the children of u are related, assuming that (a, u) is to be an edge in the expansion and that all subproblems rooted at children have already been solved.

4.2   Technical Lemmas

When processing (a, u), we are assuming that $(a, u) \in E(\mathcal{E})$ and attempting to see where subsets of the children of a and u can map. To build our intuition, we consider the process from the point of view of Q (viewing from R is symmetric and hence the reasoning identical). Each child b of a must eventually be involved in $\mathcal{E}$. There are four different cases for a child b of a, reflecting four different possible smaller expansions involving subtrees rooted at the children of a and u (for an illustration of the case analysis that follows, see Figure 1).

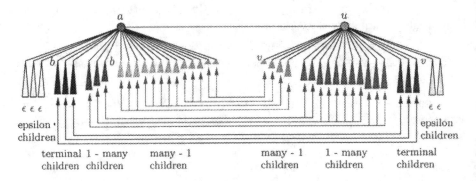

Fig. 1. The different ways possible smaller expansions involving subtrees rooted at the children of a and u can be combined to a general expansion.

1. (epsilon child) The subtree rooted at b is not involved in any previous expansion. It will be included by creating an edge in $\mathcal{E}$ from each vertex in the subtree to u.
2. (terminal child) The subtree rooted at b has been mapped to a subtree rooted at a child v of u, where (a, v) is not an edge in any previous expansion. In this case the edges (a, b) and (u, v) will be corresponding strong edges.
3. (one-many child) The subtree rooted at b is mapped to subtrees rooted at a set of children of u, where (b, u) is an edge in a previous expansion. In this case (a, b) is a weak edge.
4. (many-one child) A set of subtrees rooted at children of a is mapped to a subtree rooted at a child v of u, where (a, v) is an edge in a previous expansion. In this case (u, v) is a weak edge.

We formalize the possible associations of children by a tuple for each possible pair of subsets A of children of a and X of children of u and each possible mapping among vertices.

Definition 3. *Given two sets A, X we define $\Pi(A, X)$ as the set containing all tuples*

$$(\{A^e, A^t, A^o, A^m\}, \{X^e, X^t, X^o, X^m\}, \tau, \alpha, \chi)$$

that satisfy the following properties.

1. *$\{A^e, A^t, A^o, A^m\}$ is a partition of A;*
2. *$\{X^e, X^t, X^o, X^m\}$ is a partition of X;*
3. *$\tau : A^t \to X^t$ is a bijection;*

4. $\alpha : X^m \rightarrow A^o$ *is a surjection; and*
5. $\chi : A^m \rightarrow X^o$ *is a surjection.*

The proof of the following lemma uses Lemma 1.

Lemma 5. *For any expansion $\mathcal{E}$ of two trees Q and R and any $(a, u) \in \mathcal{E}$, there exists a tuple $(\{A^e, A^t, A^o, A^m\}, \{X^e, X^t, X^o, X^m\}, \tau, \alpha, \chi)$ in $\Pi(\text{children}(a), \text{children}(u))$, such that the following hold:*

1. *there are no strong edges in Q_{A^e} or R_{X^e};*
2. *all edges from a to vertices in A^t and from u to vertices in X^t are strong;*
3. *all edges from a to vertices in A^o and from u to vertices in X^o are weak;*
4. *for all $b \in A^t$, $v \in X^o$, and $c \in A^o$, $f_{\mathcal{E}}$ maps (a, b) to $(u, \tau(b))$, the strong edges in Q_b to the strong edges in $R_{\tau(b)}$, the strong edges in R_v to the strong edges in $Q_{\alpha(v)}$, and the strong edges in Q_c to the strong edges in $R_{\tau(c)}$.*

In order to define the recurrence for our dynamic programming algorithm, we need to be able to decompose a minimum size expansion of two trees into minimum size expansions of pairs of subtrees. The following two lemmas do this for the subtrees needed when considering decompositions induced by removal of a strong edge and a weak edge, respectively.

Lemma 6. *If $\mathcal{E}$ is a minimum size expansion of Q and R, (a, b) is a strong edge of Q, and $(u, v) = f_e((a, b))$ is the corresponding strong edge of R, then $\mathcal{E}$ is the union of a minimum size expansion of $Q \setminus Q_b$ and $R \setminus R_v$ containing (a, u) and a minimum size expansion of Q_b and R_v containing (b, v).*

Lemma 7. *If $\mathcal{E}$ is a minimum size expansion of Q and R, (a, b) is a weak edge of Q, and (a, u) and (b, u) are edges of $\mathcal{E}$, then there exists a subset X of the children of u such that $\mathcal{E}$ is the union of a minimum size expansion of Q_b and $R_u \setminus (R_{\text{children}(u) \setminus X} \setminus \{u\})$ containing (b, u) and a minimum size expansion of $Q \setminus (Q_a \setminus \{a\})$ and $R \setminus (R_X \setminus \{u\})$ containing (a, u).*

The following lemma is proved by applying Lemma 5, followed by repeated applications of Lemma 6 and Lemma 7.

Lemma 8. *For any expansion $\mathcal{E}$ of two trees Q and R with minimum size and any $(a, u) \in \mathcal{E}$, there exists a tuple $(\{A^e, A^t, A^o, A^m\}, \{X^e, X^t, X^o, X^m\}, \tau, \alpha, \chi)$ in $\Pi(\text{children}(a), \text{children}(u))$ such that $(\mathcal{E}_e \cup \mathcal{E}_t \cup \mathcal{E}_a \cup \mathcal{E}_u)$ is a partition of $\mathcal{E}$ where*

1. *A^e and X^e define epsilon children. That is, $\mathcal{E}_e = \{(a, z) \mid z \in V(R_{A^e})\} \cup \{(c, u) \mid c \in V(Q_{X^e})\}$.*
2. *A^t and X^t define terminal children. That is, $\mathcal{E}_t = \bigcup_{b \in A^t} \mathcal{E}_{t,b}$ where, for any vertex $b \in A^t$, $\mathcal{E}_{t,b}$ is a minimum expansion of Q_b and $R_{\tau(b)}$ that contains $(b, \tau(b))$.*
3. *A^o defines one-many children. That is, $\mathcal{E}_a = \bigcup_{b \in A^o} \mathcal{E}_{a,b}$ where, for any vertex $b \in A^o$, $\mathcal{E}_{a,b}$ is a minimum expansion of Q_b and $R_{\alpha^{-1}(b)}$ that contains (b, u).*
4. *X^o defines many-one children. That is, $\mathcal{E}_u = \bigcup_{v \in X^o} \mathcal{E}_{u,v}$ where, for any vertex $v \in X^o$, $\mathcal{E}_{u,v}$ is a minimum expansion of $Q_{\chi^{-1}(v)}$ and R_v that contains (a, v).*

4.3   Algorithm Details

Procedure Expansion(Q, R, a, u)
Input: Two trees Q, R and two vertices $a \in V(Q), u \in V(R)$.
Output: $\min\{|\mathcal{E}| \; : \; \mathcal{E}$ is an expansion of Q and R and $(a, u) \in \mathcal{E}\}$.

1: Root Q and R at a and u respectively.
2: Topologically sort $V(Q)$, giving $L_Q := \{a_1, \ldots, a_{|V(Q)|}\}$ where $a = a_{|V(Q)|}$.
3: Topologically sort $V(R)$, giving $L_R := \{u_1, \ldots, u_{|V(R)|}\}$ where $u = u_{|V(R)|}$.
4: **for** $i := 1 \ldots |V(Q)|$ **do**
5:    **for** $j := 1 \ldots |V(R)|$ **do**
6:       **if** a_i and u_j are leaves **then** $I(a_i, u_j, \phi, \phi) := 1$
7:       **else**
8:          **for all** $X \subseteq$ children(u_j) and $A \subseteq$ children(a_i) **do**
9:             $x := |V(Q)| + |V(R)|$
10:            **for all** $(\{A^e, A^t, A^o, A^m\}, \{X^e, X^t, X^o, X^m\}, \tau, \alpha, \chi) \in \Pi(A, X)$ **do**
11:               $x := \min\{\, x, \; |V(Q_{A^e})| + |V(R_{X^e})| - 1 +$ (i)
                       $\sum_{b \in A^t} I(b, f_t(b), \text{children}(b), \text{children}(\tau(b))) +$ (ii)
                       $\sum_{b \in A^o} I(b, u_i, \text{children}(b), \alpha^{-1}(b)) +$ (iii)
                       $\sum_{v \in X^o} I(a_i, v, \chi^{-1}(v), \text{children}(v)) \,\}$ (iv)
12:        $I(a_i, u_i, A, X) := x$
13: **return** $I(a, u, \text{children}(a), \text{children}(u))$

Theorem 1. *For any trees Q and R rooted at a and u respectively, Expansion (Q, R, a, u) returns the minimum number of edges in any expansion $\mathcal{E}$ containing (a, u).*

Proof: We prove that for Q and R rooted at a and u respectively, for any $c \in V(Q)$, $z \in V(R)$, and any $A \subseteq$ children(c) and $X \subseteq$ children(z), the quantity $I(c, z, A, X)$ computed by the algorithm is the minimum number of edges over all expansions $\mathcal{E}$ of Q_A and R_X, where $(c, z) \in \mathcal{E}$. The proof is by induction on the order of computation.

Consider the computation of $I(c, z, A, X)$. As L_Q and L_R are topological sortings of $V(Q)$ and $V(R)$ respectively, we can conclude that $I(d, y, A_d, X_y)$ has already been computed in the following three cases, which cover the expressions on the right-hand side of step 11.

1. $d \in$ children(c), $y \in$ children(z), $A_d \subseteq$ children(d), and $X_y \subseteq$ children(y).
2. $d \in$ children(c), $y = z$, $A_d \subseteq$ children(d), and $X_y \subseteq$ children(z).
3. $d = c$, $y \in$ children(z), $A_d \subseteq$ children(c), and $X_y \subseteq$ children(y).

If we assume by the inductive hypothesis that the values $I(d, y, A_d, X_y)$ are correct, then by Lemma 8 there is a choice of $(\{A^e, A^t, A^o, A^m\}, \{X^e, X^t, X^o, X^m\}, \tau, \alpha, \chi)$ that results, at step 11, in x taking on the minimum number of edges in an expansion $\mathcal{E}$ of Q_A and R_X containing (c, z), as required. Q.E.D.

Theorem 2. *For any pair of trees Q and R of bounded degree, sctmj(Q, R) can be computed in $O(n^3)$ time where $n = \max\{|V(Q)|, |V(R)|\}$.*

Proof: The if-statement at step 6 is invoked $O(n^2)$ times, and because the maximum degrees of Q and R are bounded by a constant, the loops at steps 8 and 10 result in a constant number of iterations of step 11.   Q.E.D.

5   Conclusions and Further Work

We have shown an $O(n^3)$ algorithm for finding the smallest common tree major of two trees Q and R, where both Q and R are unrooted and undirected, and have degree bounded by a fixed constant. The degree restriction can be relaxed to maximum degree $O(\log n/\log\log n)$ while keeping the running time of the algorithm polynomial, since the multiplicative factor is in $O(2^{2d}d^d)$ for trees of maximum degree d. Our algorithm can be generalized to the problem of determining the edit distance (under the operations of edge contraction, vertex expansion, and relabeling) of a pair of a edge-labeled, unrooted, unordered trees, by incorporating labels into the definition of the expansion. Both of these can be solved by NC algorithms using the technique of Brent restructuring [9]. Our work is also related to work on intertwines [15]: the value sctmj(Q, R) is the minimum size of an acyclic intertwine of Q and R.

Although the NP-completeness of minor containment for general trees suggests the intractability of finding the largest common subgraph under minors, there is hope for solving other related problems. The problem of determining whether or not G is a minor of H is solvable in polynomial time for G and H both bounded-degree partial k-trees [16] or for G and H both k-connected k-paths [11]; solving the largest common supergraph problem for each of these graph classes would be an obvious extension to our work. It is an open question whether or not the ideas in our algorithm can be extended to solve the largest common tree major problem for three or more input trees.

References

1. A. Amir and D. Keselman. Maximum agreement subtree in a set of evolutionary trees: metrics and efficient algorithms. *SIAM Journal on Computing*, 26(6):1656–1669, December 1997.
2. J. A. Bondy and U.S.R. Murty. *Graph Theory with Applications*. North-Holland, 1976.
3. M. J. Chung. $O(n^{2.5})$ time algorithms for the subgraph homeomorphism problem on trees. *Journal of Algorithms*, 8:106–112, 1987.
4. Richard Cole and Ramesh Hariharan. An $O(n\log n)$ algorithm for the maximum agreement subtree problem for binary trees. In *Proceedings of the Seventh Annual ACM-SIAM Symposium on Discrete Algorithms*, pages 323–332, 1996.
5. M. Dubiner, Z. Galil, and E. Magen. Faster tree pattern matching. In *Proceedings of the 31st Annual Symposium on Foundations of Computer Science*, pages 145–150, 1990.
6. P. Duchet. Tree minors. Presentation at *AMS-IMS-SIAM Joint Summer Research Conference on Graph Minors*, 1991 (personal communication, A. Gupta).

7. M. Farach, T. Przytycka, and M. Thorup. On the agreement of many trees. *Information Processing Letters*, 55(6):297–301, 1995.
8. M. Farach and M. Thorup. Fast comparison of evolutionary trees. In *Proceedings of the Fifth Annual ACM-SIAM Symposium on Discrete Algorithms*, pages 481–488, 1994.
9. A. Gupta and N. Nishimura. The parallel complexity of tree embedding problems. *Journal of Algorithms*, 18(1):176–200, 1995.
10. A. Gupta and N. Nishimura. Finding largest subtrees and smallest supertrees. *Algorithmica*, 21:183–210, 1998.
11. A. Gupta, N. Nishimura, A. Proskurowski, and P. Ragde. Embeddings of k-connected graphs of pathwidth k. Manuscript.
12. T. Jiang, L. Wang, and K. Zhang. Alignment of trees – an alternative to tree edit. In *Combinatorial Pattern Matching*, pages 75–86, 1994.
13. P. Kilpeläinen and H. Mannila. Ordered and unordered tree inclusion. *SIAM Journal on Computing*, 24(2):340–356, 1995.
14. S. R. Kosaraju. Efficient tree pattern matching. In *Proceedings of the 30th Annual Symposium on Foundations of Computer Science*, pages 178–183, 1989.
15. J. Lagergren. The size of an intertwine. In *Proceedings of the 23rd International Colloquium on Automata, Languages, and Programming*, volume 820 of *Lecture Notes in Computer Science*, pages 520–531, 1994.
16. J. Matoušek and R. Thomas. On the complexity of finding iso- and other morphisms for partial k-trees. *Discrete Mathematics*, 108:343–364, 1992.
17. N. Robertson and P. Seymour. Graph minors II. Algorithmic aspects of tree-width. *Journal of Algorithms*, 7:309–322, 1986.
18. K. Siddiqi, A. Shokoufandeh, S. Dickinson, and S. Zucker. Shock graphs and shape matching. *International Journal of Computer Vision*, to appear.
19. T. Warnow. Tree compatibility and inferring evolutionary history. In *Proceedings of the Fourth Annual ACM-SIAM Symposium on Discrete Algorithms*, pages 382–391, 1993.

Vertex Cover:
Further Observations and Further Improvements

Jianer Chen[1], Iyad A. Kanj[1], and Weijia Jia[2]

[1] Dept. of Computer Science, Texas A&M University,
College Station, Texas , USA
{chen,iakanj}@cs.tamu.edu

[2] Dept. of Computer Science, City University of HongKong,
HongKong SAR, China
wjia@cs.cityu.edu.hk

Abstract. Recently, there have been increasing interests and progresses in lowering the worst case time complexity for well-known NP-hard problems, in particular for the VERTEX COVER problem. In this paper, new properties for the VERTEX COVER problem are indicated and several new techniques are introduced, which lead to a simpler and improved algorithm of time complexity $O(kn + 1.271^k k^2)$ for the problem. Our algorithm also induces improvement on previous algorithms for the INDEPENDENT SET problem on graphs of small degree.

1 Introduction

The current paper was motivated by two lines of research on algorithms for NP-hard optimization problems. The first is the recent progress on parameterized algorithms for the VERTEX COVER problem (given a graph G and a parameter k, deciding if G has a vertex cover of k vertices), which is central in the study of fixed-parameter tractability theory [4] and has important applications in fields such as computational biochemistry [7]. Buss developed the first fixed-parameter tractable algorithm of running time $O(kn + 2^k k^{2k+2})$ for the problem (see [2]), which was later improved to $O(kn + 2^k k^2)$ by Downey and Fellows [5]. More recently, parameterized algorithms for the VERTEX COVER problem have further drawn researchers' attention, and continuous improvements on the problem have been developed. Balasubramanian et al. [1] first broke the bound 2 barrier in the base of the exponential term and developed an $O(kn + 1.324718^k k^2)$ time algorithm for the problem, which was then slightly improved to an $O(kn + 1.31951^k k^2)$ time algorithm by Downey et al. [6]. This algorithm was further improved most recently by Niedermeier and Rossmanith by their $O(kn + 1.29175^k k^2)$ time algorithm [10].

The other motivation is from the research on worst case analysis of algorithms for the INDEPENDENT SET problem. Since the initialization by Tarjan and Trojanowski [14] with an $O(2^{n/3})$ time algorithm for the INDEPENDENT SET problem, there have been continuous improved algorithms for the problem

Widmayer et al. (Eds.): WG'99, LNCS 1665, pp. 313–324, 1999.

[8,12,13]. The best of these algorithms is due to Robson [12], whose algorithm solves the INDEPENDENT SET problem in time $O(2^{0.276n})$.

In the present paper, we develop a further improved parameterized algorithm for the VERTEX COVER problem, starting with two simple but important observations. Our first observation is on the size for the problem kernel [6] and is based on a theorem by Nemhauser and Trotter [11]. We show that in order to decide if a graph has a vertex cover of k vertices, we essentially only need to concentrate on graphs of at most $2k$ vertices. This observation makes it become possible for us to apply dynamic programming techniques to further improve parameterized algorithms for the VERTEX COVER problem. Our second observation is a new, but simple technique to deal with degree-2 vertices, which greatly simplifies the case by case combinatorial analysis. We have also developed a new technique called "iterative branching" which is used to maintain a special structure for a graph so that an efficient branching search procedure is always applicable.

Using the new techniques and the observations, we are able to develop improved parameterized algorithms for the VERTEX COVER problem. More precisely, we present an $O(kn + 1.271^k k^2)$ time algorithm for the VERTEX COVER problem. This greatly improves the previous best parameterized algorithm of running time $O(kn + 1.29175^k k^2)$ by Niedermeier and Rossmanith [10].

We further indicate that our improved parameterized algorithm for the VERTEX COVER problem can be employed to develop improved exponential time algorithm for the INDEPENDENT SET problem on graphs of small degree. For example, our algorithm induces an $O(1.161^n)$ time algorithm for the INDEPENDENT SET problem on graphs of degree bounded by 3, which significantly improves the previously best algorithm of running time $O(2^{0.276n}) \approx O(1.211^n)$ by Robson [12].

2   On Problem Kernel and Degree-2 Vertices

Let G be a graph and V' be a subset of vertices in G. In the rest of this paper, we will denote by $G(V')$ the subgraph induced by the vertex set V'.

Suppose (G, k) is an instance for the VERTEX COVER problem, where G is a graph of n vertices and k is an integer. By *reduction to problem kernel*, we mean we apply an efficient preprocessing procedure on the instance (G, k) to construct another instance (G_1, k_1), where G_1 is a smaller graph (the kernel) and $k_1 \leq k$, such that the graph G_1 has a vertex cover of k_1 vertices if and only if the original graph G has a vertex cover of k vertices. Buss [2] explained a simple algorithm of running time $O(kn)$ that reduces an instance (G, k) for the VERTEX COVER problem to another instance (G_1, k_1), where the graph G_1 has at most k^2 edges and $k_1 \leq k$. This result has been extensively used in the latter improved parameterized algorithms for the VERTEX COVER problem [1,6,10].

We show that the size of the kernel can be further reduced. This is based on a theorem due to Nemhauser and Trotter [11], as stated in the following proposition.

Proposition 1. [NT-Theorem].   *There is an $O(\sqrt{n}m)$ time algorithm that, given a graph G_1 of n vertices and m edges, constructs two disjoint subsets C_0 and V_0 of vertices in G_1 such that*

(1). *Every minimum vertex cover of $G_1(V_0)$ plus C_0 forms a minimum vertex cover for G_1;*

(2). *A minimum vertex cover of $G_1(V_0)$ contains at least $|V_0|/2$ vertices.*

Now we explain how Proposition 1 is used to obtain a smaller kernel. Given an instance (G, k) for the VERTEX COVER problem, let (G_1, k_1) be the instance constructed by Buss' algorithm, where G_1 has at most $\leq k^2$ edges and $k_1 \leq k$. We apply Proposition 1 to the graph G_1 to construct the two subsets C_0 and V_0, as described in Proposition 1. Now the graph G_1 has a vertex cover of size k_1 if and only if the induced subgraph $G_1(V_0)$ has a vertex cover of size $k_1 - |C_0|$. Since the minimum vertex cover of the graph $G_1(V_0)$ consists of at least $|V_0|/2$ vertices, a necessary condition for the graph $G_1(V_0)$ to have a vertex cover of size $k_1 - |C_0|$ is that the number of vertices $|V_0|$ of the graph $G_1(V_0)$ is bounded by $2k_1 - 2|C_0| \leq 2k$. Let $G_2 = G_1(V_0)$ and $k_2 = k_1 - |C_0|$, then we have constructed an instance (G_2, k_2) for the VERTEX COVER problem, where G_2 has at most $2k_2$ vertices and $k_2 \leq k$, such that the graph G_2 has a vertex cover of k_2 vertices if and only if the graph G_1 has a vertex cover of k_1 vertices, which in consequence is a sufficient and necessary condition for the original graph G to have a vertex cover of k vertices.

Since the graph G_1 has $O(k^2)$ vertices and $O(k^2)$ edges, according to Proposition 1, the instance (G_2, k_2) can be constructed from the instance (G_1, k_1) in time $O(k^3)$. Summarizing these discussions, we conclude

Theorem 1. *There is an algorithm of running time $O(kn + k^3)$ that given an instance (G, k) for the VERTEX COVER problem constructs another instance (G_2, k_2), where the graph G_2 contains at most $2k_2$ vertices and $k_2 \leq k$, such that the graph G has a vertex cover of k vertices if and only if the graph G_2 has a vertex cover of k_2 vertices.*

Our second observation is on the processing of degree-2 vertices during the branching search for a vertex cover of size k. First note that unless a graph G has less than k vertices, which can be easily checked, the graph G has a vertex cover of size exactly k if and only if it has a vertex cover of size bounded by k. Therefore, for each instance (G, k) of the VERTEX COVER problem, we only need to decide if the graph G has a vertex cover of size bounded by k.

There have been several methods proposed in dealing with degree-2 vertices in parameterized algorithms for the VERTEX COVER problem [1,6,10]. Most of those methods consider the combinatorial structures case by case and apply different operations according to the combinatorial structures. Here we propose a new method which is more uniform and simpler, and seems more convenient in processing degree-2 vertices.

Suppose v is a degree-2 vertex in the graph G with two neighbors u and w such that u and w are not adjacent to each other. We construct a new graph G' as follows: remove the vertices v, u, and w and introduce a new vertex v_0 that is adjacent to all neighbors of the vertices u and w in G (of course except the vertex v). We say that the graph G' is obtained from the graph G by *folding the vertex v*. See Figure 1 for an illustration of this operation.

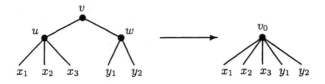

Fig. 1. Vertex folding

Lemma 1. *Let G' be a graph obtained by folding a degree-2 vertex v in a graph G, where the two neighbors of v are not adjacent to each other. Then the graph G has a vertex cover of size bounded by k if and only if the graph G' has a vertex cover of size bounded by $k - 1$.*

The following two trivial situations will be ignored in our rest discussion. If the graph G has a degree-1 vertex v, we can simply include the neighbor of v in the minimum vertex cover "for free". If the graph G has a degree-2 vertex v whose two neighbors u and w are adjacent to each other, then we can simply include the vertices u and w in the minimum vertex cover "for free".

3   On Graphs of Degree Bounded by 3 and by 4

We first describe an algorithm for the VERTEX COVER problem on graphs of degree bounded by 3. This algorithm well illustrates the power of the vertex folding techniques. For a vertex v, we denote by $N(v)$ the set of neighbors of v.

Let G be a graph in which every vertex is of degree bounded by 3. We also assume that the graph G contains at least one vertex of degree at most 2: if G has no such a vertex, we subdivide an edge in G by two degree-2 vertices. According to Lemma 1, the resulting graph G' has a vertex cover of size bounded by $k + 1$ if and only if the graph G has a vertex cover of size bounded by k. We then instead work on the instance $(G', k + 1)$. Consider the algorithm given in Figure 2.[1]

Since the graph given to each stage is always a proper subgraph of the original graph G of degree bounded by 3, we ensure that the graph at the beginning of each stage always has a vertex of degree 2. The new vertex v_0 is always immediately eliminated in a stage if it has degree larger than 2.

---

[1] For simplicity, by "including a vertex v in the vertex cover C" we mean "adding v to C, removing v and the edges incident on v from the graph G, and then also removing all vertices of degree 0 in G". Similarly, by "branching at a vertex v_0", we mean we branch into two search paths, one includes the vertex v_0 in the vertex cover C then works recursively on the resulting graph, and the other includes all neighbors of v_0 in C, then works recursively on the resulting graph. These conventions will be used throughout this paper.

VC-Degree-3
Input: A graph G and an integer k
1. $C = \emptyset$;
2. **while** $|C| < k$ **and** G is not empty **do**
2.1. pick a degree-2 vertex v in G;
2.2. fold v;
2.3. **if** the new vertex v_0 has degree larger than 2 **then** branch at v_0.

Fig. 2. An algorithm for vertex cover on graphs of degree bounded by 3

We consider the running time of the algorithm **VC-Degree-3**. In the rest of this paper, we always let $C(k)$ be the number of search paths in the search tree of our algorithm searching for a vertex cover of size bounded by k. In Step 2.2 of the algorithm **VC-Degree-3**, folding the vertex v reduces the parameter k by 1 (according to Lemma 1). When we branch at v_0 in Step 2.3, we further reduce the parameter k either by 1 (by including v_0 in C), or by $|N(v_0)| \geq 3$ (by including all neighbors of v_0 in C). Therefore, we always branch with

$$C(k) \leq C(k-1) \quad \text{or} \quad C(k) \leq C(k-2) + C(k-4)$$

It is easy to verify that $C(k) = \alpha^k$, where $\alpha = 1.272\cdots$ is the root of the polynomial $x^4 - x^2 - 1$. Finally, according to Theorem 1, we can assume that the graph G has at most $2k$ vertices (thus $O(k)$ edges). Therefore, each search path takes time $O(k)$. In conclusion, we have

Theorem 2. *The* VERTEX COVER *problem on graphs of degree bounded by 3 can be solved in time* $O(kn + 1.273^k k)$.

The time bound in Theorem 2 can be further improved using dynamic programming techniques. We will describe this improvement in section 5.

Now we consider graphs of degree bounded by 4.

Lemma 2. *Let v be a vertex of degree 3 in a graph G. Then there is a minimum vertex cover of G that contains either all three neighbors of v or at most one neighbor of v. Moreover, if all neighbors of v have degree at least 3 and there is an edge between two neighbors of v, then we can branch with $C(k) \leq 2C(k-3)$.*

Based on Lemma 2 and the vertex folding technique, we derive the following theorem (a proof can be found in [3]).

Theorem 3. *The* VERTEX COVER *problem on graphs of degree bounded by 4 can be solved in time* $O(1.277^k k + kn)$.

4   Iterative Branching

In this section we consider graphs of degree bounded by 5. ¿From the previous sections, we have seen great advantages for always having a vertex of degree bounded by 3. This condition can be easily preserved while we are dealing with graphs of degree bounded by 4 since every proper subgraph of such a graph has at least one vertex of degree bounded by 3. However, the condition is no longer guaranteed for graphs of degree bounded by 5. In the following, we first introduce a technique, the *iterating branching method*, that imposes this condition on graphs of degree bounded by 5.

Let G be a graph whose vertices have degree either 4 or 5. We also assume that the graph G is not regular. Consider the algorithm in Figure 3.

Iterative-Branch
while G has a vertex of degree 5 and no vertex of degree ≤ 3 **do**
    pick a vertex v of degree 4 that is adjacent to a vertex u of degree 5;
    branch with (1) including u in the vertex cover, and **STOP**;
              (2) including $N(u)$ in the vertex cover.

Fig. 3. The iterative branch algorithm

In the algorithm **Iterative-Branch** the process at the two branches at the vertex u is *asymmetric*: in case we include u in the vertex cover, the vertex v becomes of degree 3 so we *stop* the process; while in case we include $N(u)$ in the vertex cover, we continue the process by working on another pair of degree-4 and degree-5 vertices. Therefore, when the process stops, we should end either with a graph with a vertex of degree bounded by 3, or with a graph in which all vertices have degree less than 5. In the latter case, we can apply Theorem 3.

Before we formally present our algorithm, we give an intuitive illustration for our algorithm. At each stage of our algorithm, we assume that the given graph satisfies the following degree condition:

Degree-Assumption.   All vertices in the graph have degree bounded by 5 and at least one vertex has degree bounded by 3.

We branch at a vertex of degree bounded by 3 according to its combinatorial structures. In case there is no vertex of degree bounded by 3 left in the graph, we apply the iterative branching process to reinstate the Degree-Assumption so that the graph is applicable for the next stage.

We let $C(k)$ be the number of search paths in the search tree of our algorithm to construct a vertex cover of at most k vertices in a general graph of degree bounded by 5. Moreover, we let $C_0(k)$ be the number of search paths in the search tree of our algorithm to construct a vertex cover of at most k vertices in a graph that satisfies the Degree-Assumption. We will concentrate on the analysis of the function $C_0(k)$. Due to the space limit, we only present the detailed analysis for a few cases for illustration. The detailed analysis for other cases can be found in [3]. Let G be a graph satisfying the Degree-Assumption.

Case 1. The graph G has a vertex v of degree 1. We branch with

$$C_0(k) \leq C(k-1) \tag{1}$$

Case 2. The graph G has a vertex v of degree 2. We branch with

$$C_0(k) \leq C(k-2) \quad \text{or} \quad C_0(k) \leq C(k-1) \quad \text{or} \quad C_0(k) \leq C(k-2)+C(k-6) \quad (2)$$

Excluding cases 1-2, we can assume in the following discussion that all vertices of the graph G have degree at least 3. Let v be a vertex of degree 3 in G with neighbors x, y, and z, with z of the largest degree among x, y, and z. Without less of generality, we also assume that the degree of z is larger than 3.

Case 3. The degree of z is 5.

We branch at the vertex z. The branch including $N(z)$ in the vertex cover reduces the parameter k by 5. In the branch that includes z in the vertex cover, the vertex v becomes of degree 2, thus we fold v. If the new vertex v_0 has degree larger than 5, we further branch at v_0. Thus in the branch including z, we either reduce the parameter k by 2, or further branch and reduce the parameter k by either 3 or 8. This gives

$$C_0(k) \leq C(k-5)+C(k-2) \quad \text{or} \quad C_0(k) \leq C(k-5)+C(k-3)+C(k-8) \quad (3)$$

Now we can assume that all vertices x, y, and z have degree at most 4.

Case 4. There are two edges among x, y, and z. We branch with

$$C_0(k) \leq C(k-1) \quad (4)$$

Case 5. There is exactly one edge among x, y, and z. We branch with

$$C_0(k) \leq C(k-3) + C_0(k-3) \quad (5)$$

we now assume that there is no edge among the vertices x, y, and z.

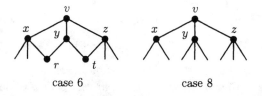

case 6   case 8

Fig. 4. Vertex cover for a graph of degree bounded by 5

Case 6. There are two vertices r and t, other than v, that are adjacent to more than one vertex in $\{x, y, z\}$. (see Figure 4, case 6).

By Lemma 2, there is a minimum vertex cover S that either contains all x, y, and z, or contains at most one of them. In the latter case, all three vertices v, r, and t must be in S. Therefore, we can branch by either including x, y, and z in S, or including v, r, and t in S. Note that in both cases, the resulting graph contains a vertex of degree bounded by 3. This gives

$$C_0(k) \leq 2C_0(k-3) \quad (6)$$

Case 7. There is exactly one vertex t, other than v, that is adjacent to more than one vertex in $\{x, y, z\}$. The recurrence relations are

$$C_0(k) \leq C(k-2) + C(k-5) \quad \text{or}$$
$$C_0(k) \leq C(k-3) + C(k-8) + C(k-6) + C(k-11) \quad \text{or}$$
$$C_0(k) \leq C(k-3) + C(k-8) + C(k-5) \quad \text{or}$$
$$C_0(k) \leq C(k-2) + C(k-6) + C(k-11) \tag{7}$$

Excluding cases 1-7, now we can assume that v is the only vertex that is adjacent to more than one vertex in $\{x, y, z\}$.

Case 8. A vertex in $\{x, y, z\}$, say x, has degree 3.

Then we branch at z. In case we include z, the vertex v becomes of degree 2 (see Figure 4, case 8), so we fold v, totally reducing the parameter k by 2. The new vertex v_0 has degree at most 5, so we do not need to further branch at it.

In case we include $N(z)$, the vertex x becomes of degree 2, so we fold x and if the new vertex has degree larger than 5, we further branch at the new vertex. Thus, in this branch, we either reduce the parameter k by 5, or further branch at the new vertex and reduce the parameter k totally by either 6 or at least 11. Thus, the recurrence relation is

$$C_0(k) \leq C(k-2) + C(k-5) \quad \text{or} \quad C_0(k) \leq C(k-2) + C(k-6) + C(k-11) \tag{8}$$

Excluding cases 1-8, we can assume all x, y, z have degree exactly 4.

Case 9. A neighbor t of one of x, y, and z has degree 5. We branch with

$$C_0(k) \leq C(k-3) + C(k-6) + C(k-6) \quad \text{or}$$
$$C_0(k) \leq C(k-3) + C(k-6) + C(k-7) + C(k-12) \quad \text{or} \tag{9}$$
$$C_0(k) \leq C(k-3) + C(k-7) + C(k-12) + C(k-7) + C(k-12) \quad \text{or}$$
$$C_0(k) \leq C(k-3) + C(k-7) + C(k-12) + C(k-6)$$

Case 10. All neighbors of x, y, z have degrees at most 4. We branch with

$$C_0(k) = C_0(k-2) + C_0(k-4) \quad \text{or} \quad C_0(k) = C_0(k-3) + C_0(k-8) + C_0(k-4) \tag{10}$$

This completes our construction for all possible cases. Note that it is easy to verify that in all cases, the resulting graph has all its vertices of degree bounded by 5 and contains at least one vertex of degree bounded by 4.

The function $C(k)$ can be expressed by the function $C_0(h)$ by applying the **Iterative Branch** algorithm on a general graph of degree bounded by 5:

$$C(k) = C_0(k-1) + C_0(k-6) + \cdots + C_0(k-5h+4) + C_0(k-5h)$$

for some integer $h \leq \lfloor k/5 \rfloor$. Using this formula in the recurrences (1)-(10), we can prove (see [3] for a proof)

Lemma 3. *The recurrences (1)-(10) have a solution $C_0(k) \leq 1.2852^k$.*

Theorem 4. *The* VERTEX COVER *problem on graphs of degree bounded by 5 can be solved in time $O(1.2852^k k + kn)$.*

5   Putting All Together and Further Improvements

Now we are ready to describe the entire algorithm.

Given a general graph G, we first branch at each vertex of degree larger than 5. The number $C(k)$ of search paths in the search tree of our algorithm satisfies the recurrence relation:

$$C(k) \le C(k-1) + C(k-6)$$

Note that the function $C(k) = 1.2852^k$ satisfies this recurrence relation.

After eliminating all vertices of degree larger than 5, we get a graph of degree bounded by 5. Now combining this with Theorem 4, which gives an $O(1.2852^k k)$ time algorithm for the VERTEX COVER problem on graphs of degree bounded by 5, we conclude with an $O(kn + 1.2852^k k^2)$ time algorithm for the VERTEX COVER problem on general graphs.

Theorem 5. *The* VERTEX COVER *problem on general graphs can be solved in time* $O(kn + 1.2852^k k^2)$.

Theorem 5 is a clear improvement over the best previous algorithm of time $O(kn + 1.29175^k k^2)$ for the VERTEX COVER problem [10].

Further improvement of the algorithm can be obtained by using a technique of dynamic programming, which is similar to the one used by Robson [12] for solving the INDEPENDENT SET problem.

Let G_6 be a graph of degree bounded by 6. Pick a constant $0 < \alpha < 1$. We first count how many subsets V' of vertices of G are there such that the induced subgraph $G(V')$ is connected and has a vertex cover of size bounded by αk. According to Theorem 1, we can assume that the subset size $|V'|$ is bounded by $2\alpha k$. Each induced connected subgraph $G(V')$ of at most $2\alpha k$ vertices can be represented by a 5-ary tree as follows: order the edges of G in an arbitrary order. Pick any vertex v in $G(V')$ and any edge e incident to v (e does not have to be in $G(V')$). Label the root of the 5-ary tree by v (and e). Now the edges incident to v correspond to the children of v in the 5-ary tree. If a vertex w adjacent to v is in $G(V')$, then the corresponding child is labeled by w, and if the vertex w adjacent to v is not in $G(V')$, then the corresponding child is a leaf. It is easy to see that given a tree and a vertex v and an edge e, a connected induced subgraph is uniquely determined. Therefore, the number of connected subgraphs induced by at most $2\alpha k$ vertices is bounded by $2k \cdot 3k \cdot \beta_5 = 6k^2 \beta_5$, where β_5 is the number of different 5-ary trees of $2\alpha k$ vertices. It is known [9] that

$$\beta_5 = \binom{5 \cdot 2\alpha k}{2\alpha k} \frac{1}{4 \cdot 2\alpha k + 1} \le \binom{10\alpha k}{2\alpha k} \frac{1}{8\alpha k}$$

Using Stirling approximation [9], we have

$$\beta_5 \le c_5 (10^{10}/(4 \cdot 8^8))^{\alpha k}$$

where c_5 is a small constant. Therefore, the number of connected subgraphs induced by at most $2\alpha k$ vertices is bounded by $\sqrt{k}(10^{10}/(4 \cdot 8^8))^{\alpha k}$. Now for each such induced subgraph, we apply our algorithm to check whether it has a vertex cover of size bounded by αk. If so, we record it. For each such an induced subgraph, checking this takes time $O(1.2852^{\alpha k}k)$. Therefore, the total time for checking this for all connected subgraphs induced by a vertex subset of at most $2\alpha k$ vertices is bounded by

$$\sqrt{k}(10^{10}/(4 \cdot 8^8))^{\alpha k}1.2852^{\alpha k}$$

Now suppose we have decided the value of α. Then we modify our algorithm for general graphs as follows. First we branch at each vertex of degree at least 7 with the recurrence relation

$$C(k) \leq C(k-1) + C(k-7)$$

Note that the function $C(k) = 1.271^k$ satisfies this recurrence relation. After eliminating all vertices of degree larger than 6, we obtain a graph G_6 of degree bounded by 6. Now we enumerate all connected subgraphs of G_6 with at most $2\alpha k$ vertices and check if each of them has a vertex cover of size bounded by αk. By the above discussion, this can be done in time $\sqrt{k}(10^{10}/(4 \cdot 8^8))^{\alpha k}1.2852^{\alpha k}$. Finally, we apply the algorithm in Theorem 5 to the graph G_6 *but to find at most $k - \alpha k$ vertices in the vertex cover*. Once we have found at least $k - \alpha k$ vertices in the vertex cover, we check directly whether the subgraph consisting of the rest of the vertices has a vertex cover of size bounded by αk. In this case, the running time of our algorithm can be given by

$$O(kn + 1.2852^{k-\alpha k}k^2 + \sqrt{k}(10^{10}/(4 \cdot 8^8))^{\alpha k}1.2852^{\alpha k})$$

Choosing α so that

$$1.2852^{k-\alpha k}k^2 = \sqrt{k}(10^{10}/(4 \cdot 8^8))^{\alpha k}1.2852^{\alpha k}$$

we obtain $\alpha \approx 0.04557$ and the running time of our algorithm is reduced to

$$O(kn + 1.2852^{k-0.04557k}k^2) \leq O(kn + 1.271^k k^2)$$

Theorem 6. *The* VERTEX COVER *problem is solvable in time* $O(kn+1.271^k k^2)$.

Using the same technique, we can reduce the running time for our algorithms on graphs of degree bounded by 3 and by 4.

Theorem 7. *The* VERTEX COVER *problem on graphs of degree bounded by 3 can be solved in time* $O(kn + 1.2497^k k)$.

Theorem 8. *The* VERTEX COVER *problem on graphs of degree bounded by 4 can be solved in time* $O(kn + 1.2588^k k)$.

6   Improving Algorithms for INDEPENDENT SET

The problem of MAXIMUM INDEPENDENT SET problem — given a graph, find a maximum independent set — has been playing a major role in the study of optimization problems. Initialized by Tarjan and Trojanowski's algorithm of running time $O(2^{n/3})$ [14], efficient exponential time algorithms for the MAXIMUM INDEPENDENT SET problem have been investigated for more than two decades. Jian [8] refined Tarjan and Trojanowski's algorithm and presented an algorithm of running time $O(2^{0.304n})$, and Shindo and Tomita [13] developed more recently a simpler algorithm of running time $O(2^{n/2.863})$. The best algorithm for the MAXIMUM INDEPENDENT SET problem is due to Robson [12] whose algorithm for solving the MAXIMUM INDEPENDENT SET problem runs in time $O(2^{0.276n})$. This bound has stood as the best for about two decades. All these algorithms for the MAXIMUM INDEPENDENT SET problem use similar approach as the one recently used in the development of parameterized algorithms for the VERTEX COVER problem, i.e., they are based on exhaustive combinatorial search. Note that according to the theory of fixed-parameter tractability [4,5], it is very unlikely that we can develop algorithms of running time $O(c^k p(n))$ for constructing an independent set of k vertices, where c is a constant and $p(n)$ is a polynomial of n.

We show a different approach to solving the MAXIMUM INDEPENDENT SET problem via the VERTEX COVER problem. We show that using the algorithms we have developed for the VERTEX COVER problem, we can obtain improved algorithms for the MAXIMUM INDEPENDENT SET problem on graphs of small degree. We first prove a lemma.

Lemma 4. *Let G be a graph of degree bounded by d and having n vertices. Then a minimum vertex cover of G contains at most $n(d-1)/d$ vertices.*

Proof. It is easy to construct an independent set of at least n/d vertices in G. Therefore, a maximum independent set of G contains at least n/d vertices. Consequencely, a minimum vertex cover contains at most $n(d-1)/d$ vertices.   □

We first show our improved algorithm for graphs of degree bounded by 3.

Theorem 9. *For graphs of degree bounded by 3, there is an algorithm of running time $O(1.161^n)$ that solves the MAXIMUM INDEPENDENT SET problem.*

Proof. Given a graph of degree bounded by 3, instead of finding a maximum independent set, we try to construct a vertex cover of k vertices, starting from $k = 1, 2, 3, \ldots$. At the first k for which we are able to construct a vertex cover of k vertices, we know this vertex cover is a minimum vertex cover. Thus, the complement of these vertex cover gives a maximum independent set. As shown in Lemma 4, we must have $k \leq 2n/3$. Therefore, the running time of this algorithm is bounded by $O(1.2497^{2n/3}) = O(1.161^n)$.   □

Similarly, we can prove

Theorem 10. *There is an algorithm of running time $O(1.18842^n)$ that solves the* MAXIMUM INDEPENDENT SET *problem on graphs of degree bounded by 4.*

Note that both Theorem 9 and Theorem 10 significantly improve the previous best algorithm by Robson [12], whose algorithm runs in time $O(2^{0.276n}) \approx O(1.211^n)$.

References

1. Balasubramanian, R., Fellows, M. R., and Raman, V.: An Improved Fixed Parameter Algorithm for Vertex Cover. Information Processing Letters **65** (1998) 163-168
2. Buss, J. F. and Goldsmith, J.: Nondeterminism within P. SIAM Journal on Computing **22** (1993) 560-572
3. Chen, J., Kanj, I. A., and Jia, W.: Vertex Cover: Further Observation and Further Improvements. Tech. Rep., Dept. of Computer Science, Texas A&M University (1999)
4. Downey, R. G. and Fellows, M. R.: Fixed-Parameter Tractability and Completeness I: Basic Results. SIAM Journal on Computing **24** (1995) 873-921
5. Downey, R. G. and Fellows, M. R.: Parameterized Computational Feasibility, In: Clote P. and Remmel J. (eds.): Feasible Mathematics II. Boston, Birkhauser (1995) 219-244
6. Downey, R. G., Fellows, M. R., and Stege, U.: Parameterized Complexity: A Framework for Systematically Confronting Computational Intractability. In: Roberts, F., Kratochvil, J., and Nesetril, J. (eds.): Contemporary Trends in Discrete Mathematics: From DIMACS and DIMATIA to the Future. AMS-DIMACS Proceedings Series **49** (1999) 49-99
7. Hallett, M., Gonnet, G., and Stege, U.: Vertex Cover Revisited: A Hybrid Algorithm of Theory and Heuristic. Manuscript. (1998)
8. Jian, T.: An $O(2^{0.304n})$ Algorithm for Solving Maximum Independent Set Problem. IEEE Trans. Comput. **35** (1986) 847-851
9. Knuth, D. E.: The Art of Computer Programming, Vol. 1. Addison-Wesley, Reading, Mass. (1968)
10. Niedermeier, R. and Rossmanith, P.: Upper Bounds for Vertex Cover Further Improved. Proc. of the 16th Symposium on Theoretical Aspects of Computer Science (STACS'99). Lecture Notes in Computer Science **1563** (1999) 561-570
11. Nemhauser, G. L. and Trotter, L. E.: Vertex Packing: Structural Properties and Algorithms. Mathematical Programming **8** (1975) 232-248
12. Robson, J. M.: Algorithms for Maximum Independent Sets. Journal of Algorithms **7** (1986) 425-440
13. Shindo, M. and Tomita, E.: A Simple Algorithm for Finding A Maximum Clique and Its Worst-Case Time Complexity. Sys. and Comp. in Japan **21** (1990) 1-13
14. Tarjan, R. E. and Trojanowski, A. E.: Finding A Maximum Independent Set. SIAM Journal on Computing **6** (1977) 537-546

On the Hardness of Recognizing Bundles in Time Table Graphs*

Annegret Liebers, Dorothea Wagner, and Karsten Weihe

University of Konstanz
Faculty of Mathematics and Computer Science
Fach D 188, D-78457 Konstanz, Germany
Annegret.Liebers@uni-konstanz.de
Dorothea.Wagner@uni-konstanz.de
Karsten.Weihe@uni-konstanz.de

Abstract. In a cooperation with the national German railway company, we construct a directed graph from a set of train time tables where train stations correspond to vertices, and where pairs of consecutive stops of trains correspond to edges. We consider the problem of locating vertices of this time table graph that intuitively correspond to train stations where the underlying railroad network branches into several directions, and that induce a partition of the edge set into *bundles*.

We formulate this problem as a graph theoretic optimization problem, and show for two versions of the problem that they are NP-hard. For the first version we show that it is even NP-complete to decide whether any other solution besides the trivial one exists.

1   Introduction

We are given train time tables of long distance, regional, and local trains consisting of more than 140 000 trains. Together, they stop at about 25 000 train stations all over Europe. Each time table contains a list of consecutive stops for one particular train, indicating the times of arrival and departure for each stop. Such a set of train time tables induces a directed *time table graph* as follows: Each train station appearing in some time table becomes a vertex of the graph, and if there is a train leaving station r and having station s as its next stop, then there is a directed edge (r, s) in the graph. By definition, a time table graph has no multiple edges.

The original problem to be solved as posed by the TLC^1 is now the following: Decide, solely on the basis of the train time tables, for each edge of the time table graph, whether trains traveling along it pass through other train stations on the way without stopping there. If so, the sequence of these other train stations is also sought. In other words, the edges have to be classified as representing

---

* This work was partially supported by the DFG under grant Wa 654/10-2.
1 *Transport-, Logistik- und Consulting GmbH/EVA-Fahrplanzentrum*, the subsidiary of the *Deutsche Bahn AG* that is our partner in this cooperation

Widmayer et al. (Eds.): WG'99, LNCS 1665, pp. 325–337, 1999.

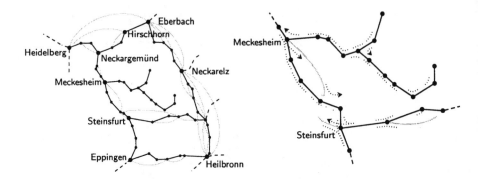

Fig. 1. *Left:* An example of the underlying undirected graph of a time table graph. Real edges are drawn in black, transitive edges are drawn in grey. *Right:* Line graph edges induced by the trains operating on this part of the time table graph are drawn in dotted lines. Edges shown undirected mean that there is a directed edge in either direction.

segments of the physical railroad network (*real* edges) or not (*transitive* edges), and for each transitive edge (s, t), an appropriate s-t-path consisting of only real edges is sought. Figure 1 shows an example. See also [4, Sect. 2.3] for a discussion of the application background and the modeling aspects of this edge classification problem.

Our approach to this problem is based on the following observation: Often there are bundle-like edge sets along a line of railroad tracks that are delimited by the branching points of the physical railroad network (in Fig. 1 for example between Heilbronn and Eppingen, or between Neckarelz and Eberbach). If we can find sets of edges that form such bundles, then we can, within each bundle, classify the edges: The physical railroad network appears as a (unique) Hamilton path of the bundle. Notice that since a time table graph is a directed graph, along one line of railroad tracks there are usually two bundles, one in each direction.

Suppose that we could guess the branching points of the physical railroad network. Then this vertex set together with the train time tables induces a partition of the edges of the time table graph as follows: If a train travels along the edge (r, s) and then along the edge (s, t), and if s is not a branching point, then (r, s) and (s, t) belong to the same edge set. Ideally, these edge sets form the bundles we are looking for. Edges connecting end points of bundles like (Heidelberg, Eberbach) in Fig. 1 will never be put together into a bundle with any other edge, so they will form a singleton in the induced edge partition. The edges in these singletons are exactly the edges of a time table graph that remain unclassified by this approach.

So our goal is to find a vertex subset of a time table graph that induces a partition of the edge set into bundles, and we want there to be few singletons in the partition. Sect. 2 will formally define this graph theoretic optimization problem by formalizing the notion of a bundle. Notice that the vertex set we are looking for is not necessarily exactly the set of branching points of the physical railroad network, because then trains that travel through a branching point without stopping there would cause us to unite edge sets that we would want to regard as two separate bundles. The edge (Heidelberg, Hirschhorn) is one of several examples for such a situation in Fig. 1.

Two optimization criteria will be considered: The first one is minimizing the cardinality of the vertex subset that induces an edge partition into bundles. The second one is finding a vertex subset inducing an edge partition into bundles, so that the number of singletons in the partition is minimized.

Section 3 constitutes the main part of the paper. It first provides the following NP-hardness result for each of the two optimization criteria: We show that it is even NP-complete to decide whether there is any other solution besides the trivial one. This is a strong result in comparison to other graph theoretic problems such as VERTEX COVER or FEEDBACK VERTEX SET [2, Problems GT1 and GT7], which are NP-hard to solve to optimality, but for which it is polynomial to find a vertex set that is a solution and that is minimal with respect to set inclusion. Intuitively, the crucial difference seems to be that if a vertex set V' is not a solution for VERTEX COVER or FEEDBACK VERTEX SET, then no subset $V'' \subset V'$ is, whereas this is not true for the problems discussed in this paper.

For the NP-completeness proof we will construct instances where one bundle is usually opposite (as defined below) to more than one other bundle. Demanding that a bundle has at most one opposite bundle yields a second version of the problem formulation. We will show that it is again NP-hard to solve this problem to optimality, but we do not know what the complexity is of deciding whether a solution other than the trivial one exists. The NP-hardness result holds again for both optimization criteria mentioned above.

2 Formalizing the Bundle Recognition Problem

Recall the following definition:

Definition 1. *Given a directed graph $G = (V, A)$, its* line graph $L = (A, A_L)$ *is the directed graph in which every edge of G is a vertex, and in which there is an edge (a, b) if and only if there are three vertices r, s, and t of G so that $a = (r, s)$ and $b = (s, t)$ are edges in G.*

A set of train time tables induces the time table graph $G = (V, A)$ together with a subset A'_L of its line graph edge set: If there is a train with train stations r, s, and t as consecutive stops, in that order, then the line graph edge $\big((r, s), (s, t)\big)$ is in A'_L.

Definition 2. *Given a directed graph $G = (V, A)$ and a subset A'_L of its line graph edges, a vertex subset $V' \subseteq V$ induces a partition of A by the following procedure:*

1. *Every edge of A is its own set.*
2. *For every line graph edge $\big((r, s), (s, t)\big) \in A'_L$, if s is not in V', then unite the edge sets that (r, s) and (s, t) belong to.*

Fig. 2. *Left:* $V' = \{o, s\}$ would induce two edge sets, one consisting of the four short edges, and one consisting of the two long edges — note that there are no line graph edges $\big((p, q), (q, s)\big)$ or $\big((o, q), (q, r)\big)$. Classifying edges along Hamilton paths within bundles as real would lead to classifying all six edges as real. $V' = \{o, q, s\}$, however, induces four edge sets, among them two singletons, resulting in four edges classified as real and two edges remaining unclassified. *Right:* $V' = \{o, s\}$ induces two edge sets that are opposite of each other.

We still need to specify how to recognize whether the edge sets of a directed graph $G = (V, A)$ induced by a line graph edge subset A'_L and a vertex subset $V' \subseteq V$ form bundles that are suitable for our edge classification approach. Since we classify edges along Hamilton paths within bundles as real, a situation as depicted on the left side of Fig. 2 would lead to wrongly classifying two edges as real. We can avoid such wrong classifications by demanding that two different bundles do not share vertices outside V', unless, of course, the two bundles are *opposite* of each other as defined below and as illustrated on the right side of Fig. 2.

Definition 3. *For a directed acyclic graph G, define a partial order "$\leq$" on its vertex set by $r \leq s$ for two (not necessarily distinct) vertices of G if and only if there is a directed path (possibly of length zero) from r to s in G. If neither $r \leq s$ nor $s \leq r$ for two vertices r and s, we write $r \| s$.*

We are now ready to cast the intuitive notion of opposite edge sets into a formal definition, and to state the bundle recognition problem as a graph theoretic problem:

Definition 4. *For a directed graph $G = (V, A)$ and a subset $A' \subseteq A$ of its edge set, let $G[A']$ denote the directed graph induced by A'. Given two disjoint edge subsets $A_1 \subseteq A$ and $A_2 \subseteq A$ of G such that $G[A_1]$ and $G[A_2]$ are acyclic, A_1 and A_2 are called* opposite *if the following two conditions hold:*

1. *There is an edge $(r, s) \in A_1$ such that $(s, r) \in A_2$.*
2. *If r and s are two distinct vertices belonging both to $G[A_1]$ and to $G[A_2]$, and if $r \leq s$ in $G[A_1]$, then either $s \leq r$ or $s \| r$ in $G[A_2]$.*

Bundle Recognition Problem. *Given a directed graph $G = (V, A)$ and a subset A'_L of its line graph edge set, find a vertex set $V' \subseteq V$ such that the sets of the induced edge partition fulfill the following three conditions:*

1. *Each edge set induces a directed acyclic graph.*
2. *This graph contains a Hamilton path.*
3. *Two distinct edge sets do not have vertices outside V' in common, except possibly if the edge sets are opposite.*

Note that there is always a trivial solution for the bundle recognition problem, namely $V' = V$. But this solution is useless because it induces an edge partition consisting only of singletons, and our goal is to minimize the number of induced singletons. Since adding a vertex to a solution of the bundle recognition problem cannot reduce the number of induced singletons, minimizing the cardinality of V' seems to be a reasonable optimization criterion as well. Note, however, that a minimum cardinality set V' solving the bundle recognition problem does not necessarily yield the smallest possible number of induced singletons. In fact, there are instances where all vertex sets solving the bundle recognition problem and inducing the minimum number of singletons possible contain more vertices than a vertex set of minimum cardinality. We will consider both optimization criteria below.

3   NP-Hardness Results

We will now see that finding a solution for the bundle recognition problem that contains a small number of vertices or that induces a small number of singletons is NP-hard in a very strong sense: It is already NP-complete to decide whether there is any other solution besides the trivial one.

Theorem 1a. *Given a directed graph $G = (V, A)$ and a subset A'_L of its line graph edge set, it is NP-complete to decide whether there is any proper subset $V' \subset V$ solving the bundle recognition problem.*

To prove the theorem, we will make intensive use of the basic arguments illustrated in Fig. 3.

Definition 5. *Given a directed graph $G = (V, A)$ and a subset A'_L of its line graph edge set, a vertex set $V'' \subseteq V$ is an* all-or-none-set *if for every solution $V' \subseteq V$ of the bundle recognition problem, either $V'' \subseteq V'$, or $V'' \cap V' = \emptyset$.*

Proof (of Theorem 1a). Clearly the bundle recognition problem is in NP. We use a reduction from the problem 3-SATISFIABILITY (3SAT; see [1] and [2, Problem LO2]) to see that it is NP-complete. Given a set U of Boolean variables and a set C of clauses over U, we may assume that U contains at least two variables. We construct an instance $(G = (V, A), A'_L)$ of the bundle recognition problem as follows:

For each variable $u \in U$ there are two *literal chains*, one for the positive literal (also denoted by u) and one for the negative literal (denoted by $\bar{u}$), as defined by Fig. 4. Furthermore, the two literal chains belonging to a variable are connected by a *variable cycle* as depicted in Fig. 5. For each clause $c \in C$ there is a *clause cycle* as defined by Fig. 5, connecting the literal chains corresponding to the three literals forming c.

If U consists of the n variables $u_1, \ldots, u_n$, there is a *forcing component* as defined by Fig. 6 from the two literal chains corresponding to u_i to the two literal chains corresponding to u_{i+1} for each $1 \leq i \leq n-1$, and there is a

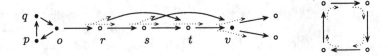

Fig. 3. Given this directed graph (*black and white vertices and black edges*) and this set of line graph edges (*dotted*), assume a vertex subset V' is a solution to the bundle recognition problem. Then $\{o, p, q\} \subseteq V'$ because of the third condition in the bundle recognition problem. If $r \in V'$, then the edges (r, t) and (s, t) belong to different sets of the induced edge partition, so again because of the third condition in the bundle recognition problem, $r \in V'$ implies $t \in V'$. The second condition implies that $v \in V'$, and that if $t \in V'$, then $s \in V'$. The first condition implies that at least two of the four white vertices of the cycle must belong to V'.

So we have that $r \in V'$ implies $t \in V'$, $t \in V'$ implies $s \in V'$, and $s \in V'$ implies $r \in V'$. Together, these implications mean that either all of r, s, and t belong to V', or none of them does. Therefore, we call $\{r, s, t\}$ an *all-or-none-set*.

Vertices that must belong to every solution of the bundle recognition problem will always be drawn in black, while vertices that may or may not belong to a solution will be drawn in white. Line graph edges will always be dotted.

forcing component from the literal chains corresponding to u_n to the literal chains corresponding to u_1.

We will make use of the following lemmas, the proofs of which are omitted:

Lemma 1. *In this instance of the bundle recognition problem there are $18 \cdot |U|$ vertices (the black vertices of the literal chains and of the forcing components in Fig. 4 and 6) that every solution V' of the bundle recognition problem must contain.*

Lemma 2. *Within every literal chain, the set of white vertices forms an all-or-none-set.*

Lemma 3. *If for a forcing component attached to the literal chain x by the vertices a_x and b_x as shown in Fig. 6, a solution V' of the bundle recognition problem contains a_x or b_x, then V' contains all three vertices a_x, b_x, and c_x.*

Lemma 4. *Consider a variable u and the vertices c_u and $c_{\overline{u}}$ of the corresponding forcing component as shown in Fig. 6. If both c_u and $c_{\overline{u}}$ are contained in a solution V' of the bundle recognition problem, then the vertices d_u, e_u, f_u, g_u, $d_{\overline{u}}$, $e_{\overline{u}}$, $f_{\overline{u}}$, and $g_{\overline{u}}$ all belong to V' as well.*

We need to show that there is a fulfilling truth assignment for the variables of the 3SAT instance if and only if there is a solution V' for the bundle recognition

Fig. 4. An example of a literal chain as used in the construction for the proof of Theorem 1a. For a literal appearing in ℓ clauses, the corresponding literal chain is constructed to contain 6 black vertices, $6 + 2\ell$ white vertices, $18 + 4\ell$ edges, and $12 + 4\ell - 1$ line graph edges. Each literal chain is vertex and edge disjoint from every other literal chain.

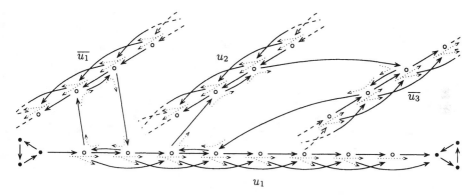

Fig. 5. For each variable u in the construction for the proof of Theorem 1a, the two literal chains for u and $\overline{u}$ are connected by a cycle of four edges and four line graph edges, called a *variable cycle*. The picture shows the variable cycle for u_1. Each variable cycle is vertex and edge disjoint from every other variable cycle.

For each clause c in the construction for the proof of Theorem 1a, the literal chains corresponding to the three literals making up c are connected by a *clause cycle* of six edges and six line graph edges. The picture shows the clause cycle for a clause $c = (u_1 \vee u_2 \vee \overline{u_3})$. Each clause cycle is vertex and edge disjoint from every other clause cycle and from every variable cycle.

problem that is a proper subset of V. First assume we have a fulfilling truth assignment for the 3SAT instance. We construct a solution V' for the bundle recognition problem as follows. Include all black vertices in V'. For each literal x that the truth assignment sets true, include the white vertices of its literal chain in V', and include in V' the vertex c_x of the corresponding forcing component. Clearly, this set V' is a proper subset of V. We may convince ourselves that it is also a solution to the bundle recognition problem.

Now assume we are given a solution V' to the bundle recognition problem that is a proper subset of V. We claim that such a set V' yields a fulfilling truth assignment for the 3SAT instance by setting each variable to true for which the white vertices of the positive literal chain are contained in V', and by setting every variable to false for which the white vertices of the negative literal chain are contained in V'.

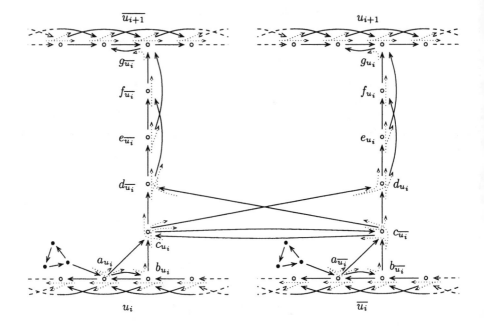

Fig. 6. For each variable in the construction for the proof of Theorem 1a, there is a forcing component connecting the two literal chains of the variable to the literal chains of another variable. Each of the $|U|$ forcing components consists of six black vertices, eight white vertices (besides the white vertices of literal chains that are part of the forcing component), 32 edges, and 28 line graph edges. The line graph edges at vertex c_{u_i} are $\big((c_{\overline{u_i}}, c_{u_i}), (c_{u_i}, d_{\overline{u_i}})\big)$, $\big((b_{u_i}, c_{u_i}), (c_{u_i}, c_{\overline{u_i}})\big)$, and $\big((b_{u_i}, c_{u_i}), (c_{u_i}, d_{u_i})\big)$. The line graph edges at vertex $d_{\overline{u_i}}$ are $\big((c_{u_i}, d_{\overline{u_i}}), (d_{\overline{u_i}}, e_{\overline{u_i}})\big)$, $\big((c_{u_i}, d_{\overline{u_i}}), (d_{\overline{u_i}}, f_{\overline{u_i}})\big)$, $\big((c_{\overline{u_i}}, d_{\overline{u_i}}), (d_{\overline{u_i}}, e_{\overline{u_i}})\big)$, and $\big((c_{\overline{u_i}}, d_{\overline{u_i}}), (d_{\overline{u_i}}, f_{\overline{u_i}})\big)$. The situation at $c_{\overline{u_i}}$ and d_{u_i} is symmetric.

Every forcing component is vertex and edge disjoint from every other forcing component, and from every variable and clause cycle.

We first need to show that for each variable, exactly one of its two literal chains has its white vertices contained in V'. First recall that by Lemma 2, for each literal chain, either all its white vertices are in V' or none is. Now assume there is a variable u for which neither of its two literal chains has its white vertices in V'. But then the four edges of the variable cycle for u form an induced edge set that is cyclic, contradicting that V' is a solution to the bundle recognition problem. So for each variable, at least one of its literal chains has its white vertices in V'. Now assume there is a variable u for which both of its literal chains have their white vertices in V'. Applying Lemmas 3, 4, and 2 repeatedly yields that all white vertices of all literal chains and all white vertices of all forcing components must be in V'. But then $V' = V$, contradicting that V' is a proper subset of V. So we have that for each variable, exactly one of its literal chains has their white vertices in V', and thus V' indeed induces a truth assignment for the variables.

Now assume that this truth assignment leaves a clause c with only false literals. Then the six edges of the clause cycle corresponding to c form an induced edge set that is cyclic, contradicting that V' is a solution to the bundle recognition problem.   □

Notice that in the instances of the bundle recognition problem constructed for this proof, there is a solution $V' \neq V$ if and only if there is a solution so that the induced edge sets form less than $|A|$ singletons. So we have as a corollary of Theorem 1a:

Theorem 1b. *Given a directed graph $G = (V, A)$ and a subset A'_L of its line graph edge set, it is NP-complete to decide whether there is any vertex set $V' \subseteq V$ solving the bundle recognition problem and inducing less than $|A|$ singletons.*

The proof of Theorem 1a makes extensive use of situations where for one edge set A', there are several edge sets to which A' is opposite. Considering the motivation of the bundle recognition problem through time table graphs, one might argue that this situation is very artificial and far removed from the original problem of finding bundles along a line of railroad tracks. So let us consider a version of the bundle recognition problem where we do not allow an edge set to have more than one opposite edge set:

Modified Bundle Recognition Problem. *Given a directed graph $G = (V, A)$ and a subset A'_L of its line graph edge set, find a vertex set $V' \subseteq V$ such that the sets of the induced edge partition fulfill the following four conditions:*

1. *Each edge set induces a directed acyclic graph.*
2. *This graph contains a Hamilton path.*
3. *Two distinct edge sets do not have vertices outside V' in common, except possibly if the edge sets are opposite.*
4. *No edge set has more than one opposite edge set.*

We will see that it is still NP-hard to minimize the number of vertices in a solution V', or to minimize the number of singletons induced by a solution V':

Theorem 2a. *Given a directed graph $G = (V, A)$, a subset A'_L of its line graph edge set, and an integer K, it is NP-complete to decide whether there is a vertex set $V' \subseteq V$ with $|V'| \leq K$ that solves the modified bundle recognition problem.*

Proof. The modified bundle recognition problem is again in NP. We use a reduction from the following variation of ONE-IN-THREE 3SAT: Given a set U of Boolean variables and a set C of clauses over U where each clause contains exactly three positive literals, is there a truth assignment for the variables so that there is exactly one true literal in each clause ? [3] shows that ONE-IN-THREE 3SAT is NP-complete, and [2, Problem LO4] observes that it remains NP-complete even if there are no negative literals. This can be easily seen.

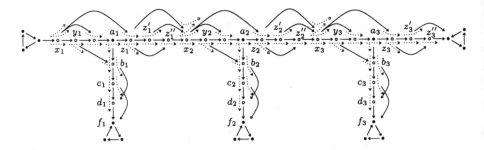

Fig. 7. An example of a variable component as used for the proof of Theorem 2a. If a variable appears ℓ times in clauses, then the corresponding variable component has ℓ *arms* and consists of $6 + 4\ell$ black vertices, 9ℓ white vertices, $6 + 22\ell$ edges, and $22\ell - 1$ line graph edges. The picture shows an example with three arms.

Given U and C, we now construct an instance $\big(G = (V, A), A'_L, K\big)$ of the problem in Theorem 2a as follows: For each variable in U the positive literal of which appears ℓ times in some clause of C there is a *variable component* with ℓ arms as defined by Fig. 7. For each clause c in C there is a *clause component* as defined by Fig. 8. It connects three arms of variable components corresponding to the three literals making up c. Set $K = 6 \cdot |U| + 28 \cdot |C|$.

We will make use of the following lemmas, the proofs of which are again omitted:

Lemma 5. *Every solution to the modified bundle recognition problem must contain the $6 \cdot |U| + 12 \cdot |C|$ black vertices of the variable components.*

Lemma 6. *If a variable component has ℓ arms, then the vertex set $\{x_1, z_1, z'_1, z''_1, \ldots, x_\ell, z_\ell, z'_\ell, z''_\ell\}$ forms an all-or-none-set.*

Lemma 7. *If a variable component has ℓ arms, then for each $i \in \{1, \ldots, \ell\}$, the vertex set $\{y_i, b_i, c_i, d_i\}$ forms an all-or-none-set.*

Lemma 8. *In each clause component and for each $i \in \{1, 2, 3\}$, the vertex set $\{p_i, q_i, r_i, s_i\}$ forms an all-or-none-set.*

To show that there is a truth assignment for the variables in U fulfilling the ONE-IN-THREE 3SAT instance if and only if there is a solution to the modified bundle recognition problem with at most K vertices, first assume there is a fulfilling truth assignment. Construct a solution V' to the modified bundle recognition problem as follows. Let every black vertex be in V'. For each variable that is set to true, let the all-or-none-set containing x_1 be in V'. For every variable that is set to false, let the all-or-none-sets of the arms of the variable component be in V'. For each clause component, let only those four vertices be in V' that

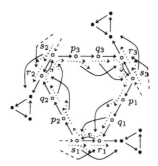

Fig. 8. For each clause, there is a clause component connecting three arms of the variable components corresponding to the literals making up the clause. A clause component contains 6 white vertices (besides the white vertices that are part of the arms of variable components), 18 edges, and 18 line graph edges. Note that each vertex labeled r_i for some $i \in \{1, 2, 3\}$ is called d_j for some integer $j > 0$ in an arm of a variable component. Likewise, each s_i has another label c_j.

will have to belong to it because of Lemma 8. We may convince ourselves that this set V' contains exactly K vertices and that it is a solution for the modified bundle recognition problem.

Now assume we are given a vertex set V' with at most K vertices that solves the modified bundle recognition problem. All the black vertices must then belong to V', and we have the following lemmas:

Lemma 9. *If the all-or-none-set containing x_1 of a variable component with ℓ arms is not in V', then every all-or-none-set corresponding to an arm of the variable component must be in V'. Thus, there are at least 4ℓ vertices of each variable component in V'.*

Lemma 10. *For each clause component, at least two of its all-or-none-sets must belong to V'.*

The number of vertices that must belong to V' according to the above lemmas is exactly K. Therefore, if for a variable component its all-or-none-set containing x_1 belongs to V', then no all-or-none-sets corresponding to arms of the variable component may be in V', and we have a truth assignment for the variables by setting a variable to true if and only if the all-or-none-set containing x_1 belongs to V'. This truth assignment fulfills the ONE-IN-THREE 3SAT instance because due to the limit of K vertices for V' no clause component can have all three of its all-or-none-sets belong to V'. □

As with Theorems 1a and 1b, we can again obtain a result corresponding to Theorem 2a where the optimization criterion is the number of induced singletons rather than the number of vertices in the vertex set V':

Theorem 2b. *Given a directed graph $G = (V, A)$, a subset A'_L of its line graph edge set, and an integer K, it is NP-complete to decide whether there is a vertex set $V' \subseteq V$ that induces at most K singletons, and that solves the modified bundle recognition problem.*

Proof (idea). The proof uses the same ideas as the proof of Theorem 2a but with a slightly altered construction of the instance of the modified bundle recognition problem:

For every variable component with two or more arms, prolong the first arm by inserting an extra white vertex e_1: Replace the edge (d_1, f_1) by two edges (d_1, e_1) and (e_1, f_1), replace the edge (c_1, f_1) by two edges (c_1, e_1) and (d_1, f_1), and add two line graph edges so that $\{y_1, b_1, c_1, d_1, e_1\}$ forms an all-or-none-set.

For every variable component with $\ell \geq 3$ arms, add to every "middle" arm i with $2 \leq i < \ell$ one edge (b_i, f_i), and add one line graph edge $\big((a_i, b_i), (b_i, f_i)\big)$ so that $\{y_i, b_i, c_i, d_i\}$ is still an all-or-none-set. If a variable component has only one arm, then do the same thing to this arm.

If K_F, K_M, and K_L are the number of first, middle, and last arms, counted over all variable components, then set $K = 6 \cdot |U| + 11 \cdot |C| + 12 \cdot K_F + 11 \cdot K_M + 10 \cdot K_L$. □

4   Conclusion

We have considered two versions of a graph theoretic optimization problem modeling the problem of recognizing bundles in time table graphs. For the first version, we have shown that it is even NP-complete to decide whether there is any other solution besides the trivial one. For the proof we constructed instances where one bundle was usually opposite to more than one other bundle. Demanding that a bundle has at most one opposite bundle yields the second version of the problem formulation. We were able to show that it is still NP-hard to solve this problem to optimality. For each of the two versions, we have shown the NP-completeness result for two different optimization criteria.

Algorithmic results as to how to solve the bundle recognition problem in an approximative or heuristic way are now needed. Alternatively, further modifications of the graph theoretic formulation of the problem, for example by demanding that two opposite edge sets actually share all their vertices, may be promising.

References

1. Stephen A. Cook. The complexity of theorem-proving procedures. In *Proceedings of the 3rd Annual ACM Symposium on Theory of Computing, STOC'71*, pages 151–158, 1971.
2. Michael R. Garey and David S. Johnson. *Computers and Intractability: A Guide to the Theory of $\mathcal{NP}$-Completeness*. W H Freeman & Co Ltd, 1979.

3. Thomas J. Schaefer. The complexity of satisfiability problems. In *Proceedings of the 10th Annual ACM Symposium on Theory of Computing, STOC'78*, pages 216–226, 1978.
4. Karsten Weihe, Ulrik Brandes, Annegret Liebers, Matthias Müller-Hannemann, Dorothea Wagner, and Thomas Willhalm. Empirical Design of Geometric Algorithms. In *Proceedings of the 15th ACM Symposium on Computational Geometry, SCG'99*, pages 86–94, 1999.

Optimal Solutions for Frequency Assignment Problems via Tree Decomposition

Arie M.C.A. Koster, Stan P.M. van Hoesel, and Antoon W.J. Kolen

Department of Quantitative Economics, Maastricht University, P.O. Box 616, 6200
MD Maastricht, The Netherlands
{a.koster,s.vanhoesel,a.kolen}@ke.unimaas.nl
http://www.unimaas.nl/~akoster/

Abstract. In this paper we describe a computational study to solve
hard frequency assignment problems (FAPs) to optimality using a tree
decomposition of the graph that models interference constraints. We
present a dynamic programming algorithm which solves FAPs based on
this tree decomposition. With the use of several dominance and bounding techniques it is possible to solve small and medium-sized real-life
instances of the frequency assignment problem to optimality. Moreover,
with an iterative version of the algorithm we obtain good lower bounds
for large-sized instances within reasonable time and memory limits.

1 Introduction

The Frequency Assignment Problem (FAP) in its most basic form has two structural properties: the limited availability of frequencies to be assigned to wireless
connections, and signal interference between connections for some combinations
of frequencies. The diversity of practical applications, ranging from military
communication, television broadcasting, and (the most popular example) mobile telephone communication has not only resulted in many different models,
but also in many different types of instances. We consider a model that is fairly
general in the sense that most variants of the FAP can be translated to this
model: Given is a set of antennae that are all to be assigned a frequency. For
each antenna the available set of frequencies, the domain, is known, though not
necessarily the same for all antennae. In some applications certain frequencies
are favored over others. For instance, if there is an existing frequency plan in
which changes are to be reduced to a minimum. We model this by introducing a
penalty on each frequency from the domain of an antenna, such that the smaller
the penalty the more favorable the frequency is. For pairs of antennae specific
combinations of frequencies may interfere, resulting in loss of quality of the re-
ception of the signals. This loss of quality is measured and penalized with an
amount related to the quality loss. For each pair of antennae we are given the
penalties for all combinations of frequencies possible, the penalty matrix. The
penalty matrices have a structure that can be used in solution methods, namely
that frequencies that are close or even equal have a high penalty, because of
high interference levels, and frequencies at larger distances have no penalty. We

Widmayer et al. (Eds.): WG'99, LNCS 1665, pp. 338–349, 1999.

model this problem on graphs as follows: the vertices represent the antennae; each antenna pair with a nonzero penalty matrix is connected by an edge. This graph is called the constraint graph. The objective is to find a frequency plan that minimizes the sum of the vertex and edge penalties.

The FAP is, in most of its variants, hard to solve, due to its close relation to the vertex coloring problem. Therefore, many heuristic approaches have been suggested using all of the known methods in operations research and artificial intelligence, like simulated annealing (cf. Tiourine, Hurkens and Lenstra [10]), tabu search (cf. Castelino, Hurley and Stephens [4]) and genetic algorithms (Kolen [5]). A comparison of these techniques on a specific set of data can be found in [9]. In this paper we concentrate on finding exact solutions, or second best on finding good lower bounds for the FAP. To obtain lower bounds for this problem, Tiourine, Hurkens and Lenstra [10] use nonlinear programming techniques. Non-trivial lower bounds are only obtained for very special cost structures and fairly simple constraint graphs. An exact solution technique has been studied by Koster, van Hoesel and Kolen [6]. They investigate the polyhedral approach (which can also be used to obtain lower bounds). It only works within reasonable time for problems with a very limited number of frequencies available for every antenna.

Forced by the limited success of the exact solution methods so far, we tried to exploit the structure of the constraint graph more directly in our approach. Instances of the FAP have a geographical nature, since each antenna is placed in a two-dimensional map. Moreover, this geography influences interference, since pairs of antennae have no interference if their distance is far enough. Finally, concentrations of antennae are found in densely populated areas. These areas are connected with one another with a limited number of edges. This led us to believe that many instances have a constraint graph with a *tree-like structure*, and thus may be solved using a *tree decomposition* of the constraint graph with small *treewidth*. The theory on tree decompositions and tree-width has been developed by Robertson and Seymour [8]. Problems on graphs with small tree-width can usually be solved using dynamic programming (see Bodlaender [2]). We used these ideas, together with sophisticated processing techniques, on a set of instances for which the previous techniques generated only few significant results, i.e., for a small set of instances reasonable lower bounds were computed. We are now able to solve many of these instances to optimality, and we found that the instances with known lower bounds are in fact the ones with (after preprocessing) a very small treewidth. Moreover, in an iterative version of our algorithm we are able to generate good lower bounds on the very difficult instances fairly quickly. The algorithm is applicable on all instances with limited tree-width. Finally, the FAP is a partial constraint satisfaction problem with binary relations (PCSP). It seems likely due to the generic nature of the FAP, that our techniques are also applicable to other PCSPs.

In Section 2 and 3 we respectively model the FAP in detail, and introduce the notions tree decomposition and treewidth of a graph. In Section 4 we propose the dynamic programming algorithm based on the tree decomposition of the constraint graph. We present an iterative extension of the algorithm that provides lower bounds for the original problem in Section 5. The computational results obtained with these methods are the topic of Section 6. For reasons of brevity, additional reduction techniques and a heuristic for the tree decomposition of a graph are omitted. For a complete presentation, we refer to [7].

2   Problem Description

The FAP can be formulated in a mathematical form by a quadruple (G, D, p, q). Here, $G = (V, E)$ is the so-called constraint graph. All n vertices $v \in V$ in this graph correspond to an antenna to which we have to assign a frequency. The set $D = \{D_v : v \in V\}$ consists of a collection of sets of frequencies D_v for every vertex $v \in V$. The set D_v is also called the domain of vertex $v \in V$. The last two components of a FAP are given by the functions p and q. The function p gives the level of interference, given an edge $\{v, w\} \in E$ and two domain elements $d_v \in D_v$, $d_w \in D_w$ for the vertices incident to that edge, whereas the function q gives the level of preference, given a domain element $d_v \in D_v$ for a vertex $v \in V$. The function p is also called the edge-penalty function, whereas the function q is also called the vertex-penalty function. Now, the problem is to select from every domain D_v exactly one domain element in such a way that the total sum of both the edge- and vertex-penalties is minimized.

The above described FAP can be formulated as a binary linear programming problem using the following binary variables for all $v \in V$, $d_v \in D_v$

$$y(v, d_v) \qquad = \begin{cases} 1 & \text{if } d_v \in D_v \text{ is selected} \\ 0 & \text{otherwise} \end{cases}$$

and for all $\{v, w\} \in E$, $d_v \in D_v$, $d_w \in D_w$

$$z(v, d_v, w, d_w) \qquad = \begin{cases} 1 & \text{if } (d_v, d_w) \in D_v \times D_w \text{ is selected} \\ 0 & \text{otherwise} \end{cases}$$

Now, if $q(v, d_v)$ and $p(v, d_v, w, d_w)$ denote the vertex- and edge penalties, respectively, a $\{0, 1\}$ linear programming formulation of the frequency assignment problem is given by

$$\min \quad \sum_{\{v,w\} \in E} \sum_{d_v \in D_v} \sum_{d_w \in D_w} p(v, d_v, w, d_w) z(v, d_v, w, d_w)$$

$$+ \sum_{v \in V} \sum_{d_v \in D_v} q(v, d_v) y(v, d_v) \tag{1}$$

$$\text{s.t.} \quad \sum_{d_v \in D_v} y(v, d_v) = 1 \qquad \forall v \in V \tag{2}$$

$$\sum_{d_w \in D_w} z(v, d_v, w, d_w) = y(v, d_v) \qquad \forall \{v, w\} \in E, d_v \in D_v \tag{3}$$

$$z(v, d_v, w, d_w) \in \{0, 1\} \qquad \forall \{v, w\} \in E, d_v \in D_v, d_w \in D_w \tag{4}$$

$$y(v, d_v) \in \{0, 1\} \qquad \forall v \in V, d_v \in D_v \tag{5}$$

Constraints (2) model the fact that exactly one value in the domain of a vertex should be selected. Constraints (3) enforce that the combination of values selected for an edge should be consistent with the values selected for the vertices of that edge. In Koster, van Hoesel and Kolen [6] it is proved with a reduction from Maximum Satisfiability that the FAP is NP-hard, even if all domains have size 2.

Finally, we introduce some additional notation which we will use in the sequel of this paper. Let $N(v) = \{w \in V | \{v, w\} \in E\}$ denote the set of vertices adjacent to $v \in V$, whereas $N(S) = \{w \in V \setminus S | \exists v \in S : \{v, w\} \in E\}$ denotes the neighbors of the vertices in the subset $S \subseteq V$. Let $E(S, T)$ denote the set of all edges between the vertices in $S \subseteq V$ and $T \subseteq V$, i.e., $E(S, T) = \{\{v, w\} \in E : v \in S, w \in T\}$. With $E[S]$ we denote all edges between the vertices in $S \subseteq V$, i.e., $E[S] = E(S, S)$. By $G[W] = (W, E[W])$ we denote the subgraph of $G = (V, E)$ induced by W.

3   The Tree Decomposition of a Graph

In the early 1980s Robertson and Seymour introduced the notions tree decomposition and treewidth:

Definition 1 (Robertson and Seymour [8]). *Let $G = (V, E)$ be a graph. A tree-decomposition is a pair $(T, \mathcal{X})$, where $T = (I, F)$ is a tree, and $\mathcal{X} = \{X_i | i \in I\}$ is a family of subsets of V, one for each node of T, such that*

(i). $\bigcup_{i \in I} X_i = V$,

(ii). for every edge $\{v, w\} \in E$, there is an $i \in I$ with $v \in X_i$ and $w \in X_i$, and

(iii). for all $i, j, k \in I$, if j is on the path from i to k in T, then $X_i \cap X_k \subseteq X_j$.

The width of a tree decomposition is $max_{i \in I} |X_i| - 1$. The treewidth of a graph G, denoted by $tw(G)$, is the minimum width over all possible tree decompositions of G.

The problem 'Given a graph $G = (V, E)$ and an integer k, is the treewidth of G at most k' is NP-complete. However, if the integer k is a constant that is not part of the input of the problem, the problem can be solved in polynomial time. These results are due to Arnborg, Corneil and Proskurowski [1]. An algorithm that solves the problem in linear time for constant k is given by Bodlaender [3]. However, this algorithm is exponential in k, and is therefore impractical for graphs with larger treewidth. Therefore, we use a heuristic to construct a tree decomposition with small treewidth (see [7]).

4   Dynamic Programming Algorithm

The algorithm that solves the FAP in polynomial time (given that the treewidth is at most a constant k) is based on the following idea. Suppose the optimal solution is given for all vertices except for one, let say v. Then, the optimal choice for v only depends on the domain elements assigned to the neighbors of v, $N(v)$. In other words, the solution does not depend on the assignment for the vertices not in the neighborhood of v. In general, let $S \subset V$ be a vertex separating set of G with $G[V \setminus S] = G[V_1] \cup G[V_2]$. Then the optimal assignment in both V_1 and V_2 only depends on the assignment in S. So, given an assignment S the problem decomposes in two FAPs on $G[V_1]$ and $G[V_2]$. Thus, the FAP can be solved by solving the two FAPs on $G[V_1]$ and $G[V_2]$ for all $\Pi_{v \in S} |D_v|$ different assignments in S. The problems on $G[V_1]$ and $G[V_2]$ can be solved in the same way.

The idea described above can be formulated as a dynamic programming algorithm using a tree decomposition of the graph. Without loss of generality we assume that the tree decomposition is rooted, and binary. Moreover, we assume that every non-leaf node $i \in I$, X_i is a separating set, which implies that given an assignment for X_i, the FAP decomposes in smaller FAPs for every branch in the tree. First, we introduce some additional notation. Let $Y_i = \{v \in V : \exists j \in I, j$ descendant of i and $v \in X_j\}$ denote the set of vertices that is represented by the subtree rooted at node i. Given a subset $S \subseteq V$, we denote with $d_S = (d_v)_{v \in S}$ an assignment of domain elements $d_v \in D_v$ for every vertex $v \in S$. The total penalty involved by this assignment in the induced subgraph $G[S]$ is given by

$$q(S, d_S) = \sum_{v \in S} q(v, d_v) + \sum_{\{v,w\} \in E[S]} p(v, d_v, w, d_w)$$

This value is a lower bound on the total penalty involved in any complete assignment based on the partial assignment d_S. So, if $q(S, d_S) > u$, where u is the best known value, then this partial assignment cannot be extended to an optimal complete assignment. An assignment will be called non-redundant if $q(S, d_S) < u$. Finally, let D_S denote the complete set of all non-redundant assignments for a set S.

Now, we can describe the dynamic programming algorithm as follows. In a bottom-up way we compute for every node $i \in I$ all non-redundant assignments for the subset Y_i, D_{Y_i}. Starting with a leaf $i \in I$ of the tree, the algorithm stores all non-redundant assignments for the vertices in X_i. The computation of all non-redundant assignments takes $\mathcal{O}(\Pi_{v \in X_i} |D_v|) = \mathcal{O}(d^{|X_i|})$ time, where $d = \max_{v \in V} |D_v|$. Next, given all non-redundant assignments for two nodes $j, k \in I$ with common predecessor $i \in I$, we can compute all non-redundant assignments Y_i by combining every assignment of Y_j, every assignment of Y_k that has the same assignment for the vertices in $X_j \cap X_k$, and every assignment of domain elements to the vertices in $X_i \setminus (X_j \cup X_k)$. However, since X_i is a vertex separating set in the graph, we do not have to store all non-redundant assignments for the vertices in Y_i, but only the assignments that differ for the vertices in X_i. Given an assignment for the vertices in X_i, we only have to store the best assignment for the vertices in $Y_i \setminus X_i$. In other words, we have to store at most $\Pi_{v \in X_i} |D_v|$ assignments for node $i \in I$ instead of $\Pi_{v \in Y_i} |D_v|$

assignments to obtain the overall optimal solution. The computation of these assignments can be done in $\mathcal{O}(\Pi_{v \in X_i \cup X_j \cup X_k}|D_v|) = \mathcal{O}(d^{|X_i|+|X_j|+|X_k|})$. Finally, for the root node $r \in I$ of the tree T, $Y_r = V$, and so we only have to store one solution which gives the desired optimal solution for the problem. The overall computation time of this algorithm is given by $\mathcal{O}(nd^{3k})$, where k is the width of the tree decomposition $(T, \mathcal{X})$ of G that is used. So, for graphs with treewidth bounded by a constant k, this algorithm solves the FAP in polynomial time, but is exponential in k.

In Figure 1 the above described algorithm is represented in a flowchart, where we assume that the nodes are numbered $1, \ldots, |I|$ in a topological order from top to bottom. The performance of the algorithm highly rely on the exact order in which the nodes are numbered and additional techniques to reduce the size of the sets of non-redundant assignments D_{Y_i} (see [7]).

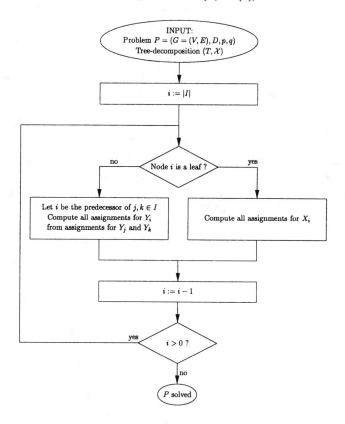

Fig. 1. Dynamic programming algorithm

5   Iterative Version of the Dynamic Programming Algorithm

Both time and memory are insufficient to solve large instances by the dynamic programming algorithm described in Section 4, even if we use several reduction techniques. Therefore, we present in this section an iterative version of the dynamic programming algorithm which provides lower bounds for these instances. The dynamic programming algorithm is used as a subroutine to solve FAPs with substantially smaller domains. Contrary to the original FAP, time and memory are sufficient to solve these FAPs. The consecutive FAPs are constructed in such a way that they provide a sequence of non-decreasing lower bounds for the original problem.

The idea of this iterative method is based on the special structure of the edge-penalties. For example, consider the matrix of edge-penalties given in Figure 2(a). The level of interference on this edge is 10 if the difference between the frequencies is less than the value 2. If we divide the frequencies in two groups $\{1,2\}$, and $\{3,4\}$, we obtain 4 blocks in the table of edge-penalties with the same or almost the same values. In most cases there is no difference between the penalties as long as the pairs of frequencies are in the same block. Therefore, let us relax the FAP to a new FAP in which we have to assign either the subset $\{1,2\}$ or the subset $\{3,4\}$ to the vertices. The edge-penalties in this new FAP are given by the minimum of the values in each block (see Figure 2(b)). Solving this substantially smaller problem provides a lower bound for the optimal value of the original problem. The quality of the lower bound depends on the size of the blocks: many small blocks will provide a better lower bound than a small number of large blocks. In most real-life instances the block structure of the penalty matrices arises naturally by the fact that the available frequencies for an antenna can be divided in groups of frequencies that are in the same part of the spectrum.

d_v, d_w	1	2	3	4
1	10	10	0	0
2	10	10	10	0
3	0	10	10	10
4	0	0	10	10

(a) original penalty matrix

d_v, d_w	$\{1,2\}$	$\{3,4\}$
$\{1,2\}$	10	0
$\{3,4\}$	0	10

(b) new penalty matrix

Fig. 2. Example to illustrate the idea behind the iterative algorithm

We can extend this idea to the following algorithm which provides a sequence of non-decreasing lower bounds for the original FAP. We start with the original problem $P = (G, D, p, q)$ and we partition for every vertex $v \in V$ the domain D_v in an initial number of n_v subsets $D_v^1, \ldots, D_v^{n_v}$. This partition is, for example, based on a natural partition of the frequencies in groups of frequencies that are in the same part of the spectrum.

Next, we construct a new FAP $P' = (G', D', p', q')$, with

- graph $G' = G = (V, E)$,
- domains $D_v' = \{1, \ldots, n_v\}$ for all vertices $v \in V$,
- vertex-penalties $q'(v, i) = \min_{d_v \in D_v^i} q(v, d_v)$ for every vertex $v \in V$, $i \in D_v'$, and
- edge-penalties $p'(v, i, w, j) = \min_{d_v \in D_v^i} \min_{d_w \in D_w^j} p(v, d_v, w, d_w)$ for every edge $\{v, w\} \in E$, $i \in D_v'$, $j \in D_w'$.

So, P' is defined on the same graph as P, and the domains of P' correspond with the subsets D_v^i, $i = 1, \ldots, n_v$. Since the vertex and edge-penalties in P' are the minimum of the penalties in the corresponding subset(s), the optimal value of the problem P' provides a lower bound for the optimal value of the original problem P. Moreover, if this lower bound is equal to the value of the best known solution for the original problem, we have proved that the best known solution for the original problem is optimal. If there is still a gap between the bounds, several ways exist to reduce this gap. On the one hand, the optimal solution of P' can be used to obtain a new upper bound for P. On the other hand, a refinement of the domain-subsets may result in a better lower bound.

Given the optimal solution of P', a new upper bound for the original problem P can be obtained in the following way. The optimal solution corresponds to the assignment of a domain-subset to every vertex. As the vertex and edge-penalties within these subsets do not differ that much, an assignment of frequencies in these subsets may lead to an improved upper bound for the original instance. In other words, in order to obtain a new upper bound, we apply either a heuristic or an exact approach to the restricted problem in which the domain of every vertex is restricted to the domain-subset that is assigned in the optimal solution of P'.

Our way to obtain better lower bounds is based on a refinement of the domain-subsets. A partition $\tilde{D}_v^1, \ldots, \tilde{D}_v^m$ of a domain D_v is called a refinement of another partition $\bar{D}_v^1, \ldots, \bar{D}_v^n$, if for every subset $\tilde{D}_v^i$, $i = 1, \ldots, m$, there exists a subset $\bar{D}_v^j$, $j \in \{1, \ldots, n\}$ in the second partition for which $D_v^i \subseteq D_v^j$. Similarly, a partition of all domains is called a refinement of another partition of all domains, if for all vertices the former partition of its domain is a refinement of the latter partition of its domain. If $\tilde{P}$ and $\bar{P}$ are FAPs corresponding to these partitions, then the value of the optimal solution of $\tilde{P}$ will be at least as high as the value of the optimal solution of $\bar{P}$, which implies that $\tilde{P}$ provides a lower bound that is greater or equal than the lower bound provided by $\bar{P}$.

So, the lower bound given by P' can be improved by a refinement of the partition of the domains. For the refined partition we again construct a FAP which hopefully provides us with a better lower bound. We can repeat the refinement of the partition and solving a new FAP as long as the lower bounds are smaller than the upper bound for P and the efforts to solve P' are reasonable in both time and memory. The way we refine the partition of the subsets is described in [7]. The problems P' can be solved with any exact algorithm. However, the use of the dynamic programming algorithm of Section 4 has an important advantage that reduce the computational effort. With the dynamic programming algorithm information of the previous problem P' can be used to solve the new problem P'. More specific, during the computation of the optimal solution of a previous problem P', we obtain for all $i \in I$ a lower bound l_i^+ for the penalty involved by the induced subgraph $G[Y_i]$ and the edges $E(Y_i, V \setminus Y_i)$. These values are also lower bounds on the penalty in the new problem P', which implies that we can compute upper bounds $u(V \setminus Y_i) = u' - l_i^+$ for all $i \in I$. Here, u' is a general upper bound for the new problem P' which can be computed by one of the heuristics available for the FAP, and $u(V \setminus Y_i)$ is an upper bound for the penalty involved by the induced subgraph $G[V \setminus Y_i]$. If the increase of u' is not too large for two consecutive problems P', then the upper bounds for the subsets are often relatively strong compared with other more difficult to compute upper bounds. Figure 3 shows a flowchart of the above described algorithm.

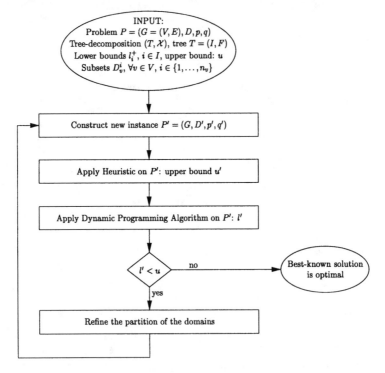

Fig. 3. Iterative version of the algorithm

6   Computational Results

In this section we report on the results we have obtained using the approach described in the previous sections. We tested the methods described in this paper on real-life instances obtained from the CALMA-project. In the CALMA (Combinatorial ALgorithms for Military Applications)-project researchers from England, France, and the Netherlands tested different combinatorial algorithms on the same set of frequency assignment problems. The set consists of two parts. The CELAR instances are real-life problems from a military application. The GRAPH instances are randomly generated problems by Delft University of Technology and have the same characteristics (size of the graphs, size of the domains and structure of penalty-matrices) as the CELAR instances. Results of the CALMA-project as well as all test problems are available by anonymous ftp from `ftp.win.tue.nl` in the directory `/pub/techreports/CALMA`. In this section we prove for 7 out of the 11 penalty-instances that the best known solution is optimal, whereas we obtain very good lower bounds for the other instances. Before this study non-trivial lower bounds were only available for 4 of the instances. All implementations have been carried out in C++. The programs were running on a DEC 2100 A500MP workstation with 128Mb internal memory.

The solution procedure can be divided in three parts. First of all, we apply several preprocessing rules that reduce both the constraint graph and the size of the domains. Secondly, we apply a heuristic to construct a tree decomposition of the preprocessed graph. Next, we apply the dynamic programming algorithm of Section 4.

Tables 1 and 2 show the results for a number of instances. Several instances can already be solved by preprocessing. The instances CELAR 09, GRAPH 06 and GRAPH 12 can be solved very efficiently with the dynamic programming algorithm. The optimal value for all these instances is equal to the best known value. The instance CELAR 06 is more difficult to solve. After more than 7.5 hours the algorithm was able to prove that the best known solution was optimal for this instance as well. Mainly due to the large treewidth we obtained with our heuristic and the limitations in computer memory, we are not able to solve the other instances.

Table 2 shows the results of the heuristic for a tree-decomposition and the results for the dynamic programming algorithm. A lower bound on the treewidth is given by the maximum clique number minus one. For the smaller instances, the gap between lower bound and heuristicly computed width is small; for the larger instances, the gap is substantially larger. The dynamic programming algorithm is not able to solve several CALMA instances. For these problems we apply the iterative version of the algorithm described in Section 5. Before we start our computations we have to partition all domains in an initial number of subsets. In our experiments we start with either 2 or 4 subsets for every vertex. The partition of the subsets is based on a natural partition of the frequencies in the radio spectrum. In each iteration of the algorithm, first a heuristic have to be applied to obtain an upper bound for the new instance P'. In our computational experiments we used the genetic algorithm developed by Kolen [5]. After

instance	before preprocessing			after preprocessing				best
	$|V|$	$|E|$	$|D_v|$	$|V|$	$|E|$	$|D_v|$	fixed	known
CELAR 06	100	350	39.9	82	327	39.9	0	3389
CELAR 07	200	817	39.9	162	764	34.6	0	343592
CELAR 08	458	1655	39.5	365	1539	39.4	0	262
CELAR 09	340	1130	39.5	67	165	35.6	11391	15571
CELAR 10	340	1130	39.5	0	0	-	31516	31516
GRAPH 05	100	416	37.1	0	0	-	221	221
GRAPH 06	200	843	37.7	119	348	16.2	4112	4123
GRAPH 07	200	843	36.7	0	0	-	4324	4324
GRAPH 11	340	1425	37.7	340	1425	32.6	2553	3080
GRAPH 12	340	1255	37.6	61	123	15.3	11496	11827
GRAPH 13	458	1877	38.4	456	1874	38.1	8676	10110

Table 1. Statistics and pre-processing.

instance	tree decomposition		optimal value	cpu-time
	width	max clique - 1	tree decomposition	(sec)
CELAR 06	11	10	3389	27102
CELAR 07	17	10	-	-
CELAR 08	18	10	-	-
CELAR 09	7	7	15571	23
CELAR 10	solved by preprocessing			
GRAPH 05	solved by preprocessing			
GRAPH 06	17	5	4123	29
GRAPH 07	solved by preprocessing			
GRAPH 11	104	7	-	-
GRAPH 12	4	4	11827	11
GRAPH 13	133	6	-	-

Table 2. Treewidth heuristic and dynamic programming algorithm

the calculation of an upper bound for the new instance, we apply the dynamic programming algorithm in the same way as in the previous subsection. If the optimal value obtained by the dynamic programming algorithm is equal to the general upper bound, then we have proved that this upper bound is optimal. Otherwise, the next step is to find a new (better) upper bound via the restricted problem, that can be constructed from the optimal solution of P'. We skip this step in our implementation of the algorithm, because experiments show that it is very difficult to improve the best known value. As last step in an iteration, we refine the partition of the subsets.

In Table 3 we report the results obtained in this way for the instances that we could not solve with the original dynamic programming algorithm. We also applied the iterative version to the instance CELAR 06. If we either start with an initial number of subsets of 2 (or 4), we obtain a lower bound that is one

instance	initial $|D_v|$	lower bound	percentage	upper bound	CPU-time (sec)
CELAR 06	2	3388	99.9	3389	13733.9
CELAR 06	4	3388	99.9	3389	9429.0
CELAR 07	4	290000	84.4	343592	231328.0
CELAR 08	4	87	33.2	262	313168.0
GRAPH 11	4	2553	82.9	3080	0
GRAPH 13	4	8676	85.8	10110	0

Table 3. Computational results iterative version of the algorithm

away from optimal in third the time (or half the time) that was needed to solve the problem to optimality with the original dynamic programming algorithm. In case we start with 2 subsets per vertex we need 38 iterations to achieve the lower bound of 3388, whereas if we start with 4 subsets per vertex we need 35 iterations.

Acknowledgement

The authors would like to thank Rudolf Müller for his suggestions which resulted in the iterative algorithm.

References

1. S. Arnborg, D.G. Corneil, and A. Proskurowski. Complexity of finding embeddins in a k-tree. *SIAM Journal on Algebraic and Discrete Methods*, 8:277–284, 1987.
2. H.L. Bodlaender. A tourist guide through treewidth. *Acta Cybernetica*, 11(1-2):1–21, 1993.
3. H.L. Bodlaender. "A linear time algorithm for finding tree-decompositions of small treewidth". *SIAM Journal on Computing*, 25(6):1305–1317, 1996.
4. D.J. Castelino, S. Hurley, and N.M. Stephens. A tabu search algorithm for frequency assignment. *Annals of Operations Research*, 63:301–319, 1996.
5. A. Kolen. A genetic algorithm for frequency assignment. Technical report, Maastricht University, 1999. Available at http://www.unimaas.nl/~akolen/.
6. A.M.C.A. Koster, C.P.M. van Hoesel, and A.W.J. Kolen. The partial constraint satisfaction problem: Facets and lifting theorems. *Operations Research Letters*, 23(3-5):89–97, 1998.
7. A.M.C.A. Koster, C.P.M. van Hoesel, and A.W.J. Kolen. Solving frequency assignment problems via tree-decomposition. Technical Report RM 99/011, Maastricht University, 1999. Available at http://www.unimaas.nl/~akoster/.
8. N. Robertson and P.D. Seymour. Graph minors. II. algorithmic aspects of tree-width. *Journal of Algorithms*, 7:309–322, 1986.
9. S. Tiourine, C. Hurkens, and J.K. Lenstra. "An overview of algorithmic approaches to frequency assignment problems". In *Calma Symposium on Combinatorial Algorithms for Military Applications*, pages 53–62, 1995. Available at http://www.win.tue.nl/math/bs/ comb_opt/hurkens/calma.html.
10. S. R. Tiourine, C.A.J. Hurkens, and J.K. Lenstra. Local search algorithms for the radio link frequency assignment problem. *Telecommunication systems*, to appear, 1999.

Fixed-Parameter Complexity of λ-Labelings

Jiří Fiala[1*], Ton Kloks[2**], and Jan Kratochvíl[1]

[1] Department of Applied Mathematics and DIMATIA, Charles University
Malostranské nám. 25, 11800 Praha 1, Czech Republic
{honza,fiala}@kam.ms.mff.cuni.cz
[2] Department of Mathematics and Computer Science, Vrije Universiteit
De Boelelaan 1081 A, 1081 HV Amsterdam, The Netherlands
kloks@cs.vu.nl

Abstract. A λ-labeling of a graph G is an assignment of labels from the set $\{0, \ldots, \lambda\}$ to the vertices of a graph G such that vertices at distance at most two get different labels and adjacent vertices get labels which are at least two apart. We study the minimum value $\lambda = \lambda(G)$ such that G admits a λ-labeling. We show that for every fixed value $k \geq 4$ it is NP-complete to determine whether $\lambda(G) \leq k$. We further investigate this problem for sparse graphs (k-almost trees), extending the already known result for ordinary trees.

In a generalization of this problem we wish to find a labeling such that vertices at distance two are assigned labels that differ by at least q and the labels of adjacent vertices differ by at least p (where p and q are given positive integers). We denote the minimum number of labels by $L(G; p, q)$ (hence $\lambda(G) = L(G; 2, 1)$). We show several hardness results for $L(G; p, q)$ including that for any $p > q \geq 1$ there is a $\lambda = \lambda(p, q)$ such that deciding if $L(G; p, q) \leq \lambda(p, q)$ is NP-complete.

1 Introduction

In many practical cases one wishes to assign frequencies such that the difference between frequencies assigned to two transmitters depends on the distance between the transmitters. In this paper we assume that the distances between pairs of transmitters can be modeled by the distance between the corresponding vertices of a graph. In this framework the problem is to find a non-negative real-valued function f on vertices of a graph such that if u and v are vertices at distance i, the values $f(u)$ and $f(v)$ should differ by at least a prescribed integer p_i. The smallest possible maximal value of such a function f over all vertices of the graph G is denoted by $L(G; p_1, p_2, ..., p_k)$.

In [4,10] it was shown that determining $\lambda(G)$ is an NP-complete problem even for graphs G with diameter two. In this paper we study the fixed parameter tractability. Answering a question of [10], we show that it is NP-complete to

* Research supported in part by the Czech Research Grants GAUK 158/99 and GAČR 201/1996/0194.
** Supported by DIMATIA and GAČR 201/99/0242

determine if $\lambda(G) \leq k$ for every fixed integer $k \geq 4$. A fortiori, the problem is not fixed parameter tractable. It is easy to see that $\lambda(G) \leq 3$ if and only if G is a disjoint union of paths of length at most 3. Much less is known for the more general $L(p,q)$-problem. Special instances such as the $L(0,1)$-problem and the $L(1,1)$-problem were already considered in [10]. We show the first complexity results for this problem.

2 Preliminaries

Definition 1. *Let G be a graph and let $d_1, \ldots, d_k$ be a collection of integers. A λ-labeling is an assignment of labels from the set $\{0, \ldots, \lambda\}$ to the vertices of G. We say that the labeling satisfies constraints $(d_1, \ldots, d_k)$ if each pair of vertices at distance i in the graph gets labels that differ by at least d_i. In such a case we talk about $\lambda_{(d_1,\ldots,d_k)}$-labelings. The minimum value λ for which G admits a $\lambda_{(d_1,\ldots,d_k)}$-labeling is denoted by $L(G; d_1, \ldots, d_k)$. We refer to the problem of deciding the existence a $\lambda_{(d_1,\ldots,d_k)}$-labeling as the $L(d_1, \ldots, d_k)$-problem.*

For the sake of history (and also for convenience) we write $\lambda(G)$ instead of $L(G; 2, 1)$.

Definition 2. *A λ-labeling is called* consecutive *if the labels that are used are consecutive.*

Clearly, a consecutive λ-labeling does not always exist. Consecutive labelings were studied in [9]. In general, it can be shown [9] that $G = (V, E)$ has a Hamiltonian path if and only if $\overline{G}$ (the complement of G) has a consecutive $\lambda_{(2,1)}$-labeling with $\lambda \leq |V| - 1$. In this paper we show that the fixed parameter variant with $\lambda \geq 4$ is equally difficult. Most of our NP-hardness results are obtained via graph covers.

Definition 3. *Let G be a graph and H be a multigraph. The H-COVER problem asks for a local isomorphism $f : G \to H$, i.e., a homomorphism that maps edges incident with every vertex x isomorphically to edges incident with $f(x)$.*

The study of the complexity of the H-COVER problem for particular parameter multigraphs H was initiated in [1] and carried on in a series of papers [5,6,7,8]. In particular, the K_t-COVER problem is shown NP-complete for every fixed $t \geq 4$ in [5]. Note that a $(t-1)$-regular graph G covers the complete graph K_t if and only if the vertices of G can be assigned t different colors in such a way that closed neighborhood of each vertex is assigned all t colors, i.e., if and only if $L(G, 1, 1) = t - 1$. The immediate consequence for the complexity of the $L(1,1)$ problem follows: The problem if $L(G; 1, 1) \leq \lambda$ is NP-complete for every fixed $\lambda \geq 3$.

Note also that this problem asks for the chromatic number of the square G^2 of G.

We find convenient to reformulate in terms of graph coloring a class of particular graph covering problems which will be used several times in the hardness proofs:

Definition 4. *Let $k \geq 3$ be an integer. The $BW(k)$–problem is the following 'black and white coloring' problem:*

   Instance: *A k-regular graph G.*

   Question: *Is there a coloring of the vertices of G with labels black and white such that every vertex has exactly two neighbors with the same color and $k - 2$ neighbors with opposite colors.*

(The $BW(k)$–problem is equivalent to the H-COVER problem for H being the multigraph with two vertices of degree k, joined by $k - 2$ parallel edges and having a loop at each of the two vertices.) The following theorem was proved in [7].

Proposition 1. *The $BW(k)$–problem is NP-complete for every $k \geq 3$.*

The paper is organized as follows. In Section 3 we present the positive result, namely a polynomial algorithm for sparse graphs (k-almost trees). The algorithm is polynomial both in λ and the size of the input graph, i.e., the maximum label bound λ is regarded as part of the input. In Section 4 we present the hardness results for fixed λ. Section 5 is devoted to the general $L(p, q)$ problem. Here we only state our results and the proofs are postponed to the Appendix.

3 $L(2, 1)$ for Sparse Graphs

For trees T, $\lambda(T)$ can be computed efficiently [3]. We show how this algorithm can be used to solve the problem for a slightly wider class of graphs, namely k-almost trees for fixed values of k. (A k-almost tree is a connected graph G with $n + k - 1$ edges. Clearly, k-almost trees can be recognized in linear time, cf. e.g., [2]. Let us remark in this connection that the $L(2, 1)$ problem seems to be harder than many other graph-theoretical problems, at least as concerns tree-like graphs. Many generally difficult problems can be solved in polynomial time for graphs of bounded tree-width, whereas the complexity of $L(2, 1)$ for such graphs is unknown (when λ is part of the input, as assumed in this section).

Theorem 1. *Let G be a k-almost tree with n vertices. Then $L(G; 2, 1) \leq \lambda$ can be tested in $O(\lambda^{2k+\frac{9}{2}} n)$ time.*

Proof. Let G be a k-almost tree. Choose a spanning tree of G and let $e_1, e_2, ..., e_k$ be the edges of G that do not belong to the spanning tree.

   In order to use the tree algorithm of Chang and Kuo, we build a new tree T by modifying the spanning tree. For each edge $e_i = (u_i, v_i)$, add two leaves $e'_i = (u_i, x_i)$ and $e''_i = (v_i, y_i)$ to the spanning tree, so that x_i and y_i are new leaves, and denote the new tree by T. Note that identifying vertices u_i with y_i and v_i with x_i gives the original graph G. Clearly T has a $\lambda_{(2,1)}$-labeling such that the labels of u_i and y_i and of v_i and x_i are equal for all $i = 1, \ldots, k$, if and only if the original graph G has a proper $\lambda_{(2,1)}$-labeling.

To test whether T has a $\lambda_{(2,1)}$-labeling we use the following procedure: Fix a pre-labeling on the endvertices of all the edges of $\{e'_i, e''_i \mid 1 \leq i \leq k\}$ such that u_i and y_i and also v_i and x_i have the same labels and test whether there exists a $\lambda_{(2,1)}$-labeling of the tree T compatible with this pre-labeling. Repeat this step for all possible pre-labelings. (Note that there are $O(\lambda^{2k})$ different pre-labelings.)

It is easy to see that the algorithm of [3] works also in case some nodes are pre-labeled.   □

4   Fixed Parameter λ-Labeling is NP-Complete

Griggs and Yeh showed in [4] that it is NP-complete to determine whether $\lambda(G) \leq |V(G)|$ for a general graph G. In this section we focus on the fixed parameter case.

Theorem 2. *For each $\lambda \geq 4$, it is NP-complete to decide if the input graph allows a $\lambda_{(2,1)}$-labeling.*

Proof. We prove our result by induction on λ. The base cases are $\lambda = 4$ and $\lambda = 5$. For these cases we use a reduction from the H-cover problem for multigraphs H_1 and H_2 respectively (see Fig. 2 in the Appendix), which are equivalent to the $BW(3)$ and $BW(4)$ coloring problems and are both NP-complete (Prop. 1).

$\underline{\lambda = 4:}$ We use a reduction from $BW(3)$. Let G be a 3-regular graph for which we want to find a black and white coloring such that every white vertex has exactly two white and one black neighbor and every black vertex has one white and two black neighbors.

We claim that the graph G' obtained by subdividing each edge of G by two new vertices of degree 2, has a $4_{(2,1)}$-labeling if and only if G covers H_1 (cf. an illustrative picture in Fig. 3 in the Appendix).

First assume that G' has a $4_{(2,1)}$-labeling. The original vertices of G have degree 3, so they must be labeled either by number 0 or by 4. An easy case analysis shows that the only possible labelings of the paths of length 3 replacing the edges of G are $(0, 2, 4, 0)$, $(0, 3, 1, 4)$, $(4, 0, 2, 4)$ and the reverse 4-tuples.

Consider a vertex u of G in G'. Say u is labeled 0. Its neighbors (in G') have to be assigned different labels, hence each of the labels $2, 3, 4$ is used on exactly one of the neighbors. Therefore, the three paths leaving u are labeled $(0, 2, 4, 0)$, $(0, 3, 1, 4)$ and $(0, 4, 2, 0)$, and two of them lead to vertices labeled 0 and the third one to a vertex labeled 4. Similarly, if u is labeled 4 then two of the paths starting in u lead to vertices labeled 4 and one to a vertex labeled 0. Thus coloring every vertex labeled 0 with white color (in G), and coloring every vertex labeled 4 with black color translates the given 4-labeling into a black and white coloring of the original graph, satisfying $BW(3)$.

Now assume that a black and white coloring of G is given (satisfying $BW(3)$). We can partition the edges of G into three sets: edges with two white end-vertices, edges with two black end-vertices and "black-white edges". In G', we assign label

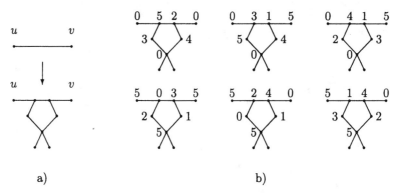

Fig. 1. Replacement for $\lambda = 5$ case and its 5-labeling

0 to every white vertex of G, and label 4 to every black vertex of G. It remains to label the intermediate vertices on the paths replacing edges of G. Each path corresponding to a black-white edge in G, will be labeled by $(0, 3, 1, 4)$. The paths corresponding to white edges in G induce a disjoint union of cycles in G' (every white vertex has two white neighbors in G). On each of these cycles, we use consecutively the scheme $(0, 2, 4, 0)$. Thus every vertex labeled 0 has exactly one neighbor labeled 2, one neighbor labeled 3 and one neighbor labeled 4, i.e., common neighbors of a vertex are assigned different labels. Similarly, the paths corresponding to the black edges of G induce a disjoint union of cycles, and as such will be consecutively labeled $(4, 0, 2, 4)$. Since adjacent vertices (in G') are always assigned labels which are at least 2 apart, this gives a $4_{(2,1)}$-labeling of G'.

$\lambda = 5$: We reduce from the $BW(4)$–coloring problem.

Let $G = (V, E)$ be a 4-regular graph. Replace each edge in E by the graph depicted in Figure 1 a). An easy case analysis shows that this auxiliary graph allows 6 combinations depicted in Figure 1 b), subject to the condition that vertices u and v are labeled 0 or 5 (this must be the case in any $5(2, 1)$-labeling of G' since these vertices have degree 4).

Call the new graph G'. We claim that G' has a $5_{(2,1)}$-labeling if and only if G has an H_2-cover.

Consider a $5_{(2,1)}$-labeling of G'. The four neighbors of a vertex labeled 0 must be labeled $2, 3, 4, 5$, and so the paths starting in such a vertex are labeled $(0, 2, 5, 0)$, $(0, 3, 1, 5)$, $(0, 4, x, 5)$ and $(0, 5, 2, 0)$ (here x is either 1 or 2). Hence two of these paths lead to vertices labeled 0 and the other two lead to vertices labeled 5. Regarding label 0 as white color and label 5 as black color, we see that the restriction of the labeling to the original vertices of G solves the $BW(4)$–problem for G in the affirmative way.

Now assume G covers H_2-cover, i.e., we assume a black-white coloring of G which satisfies $BW(4)$. The edges of G can be partitioned into edges with two white end-vertices, edges with two black end-vertices, and black-white edges. Similarly as in the case $\lambda = 4$, the white-white edges induce a disjoint union of cycles and the corresponding paths can be labeled consecutively $(0, 5, 2, 0)$. The paths corresponding to black-black edges will be labeled $(5, 0, 3, 5)$. The black-white edges induce a 2-regular bipartite graph, whose edges can be partitioned into two perfect matchings. Label the paths corresponding to edges of one of the matchings by $(0, 3, 1, 5)$, and the paths corresponding to the other matching by $(0, 4, 2, 5)$. (Note that the two rightmost labelings of the Figure 1 are not used.) In this way we obtain a $5_{(2,1)}$-labeling for G'.

Induction step: Assume that $G = (V, E)$ is a graph whose $\lambda_{(2,1)}$-labeling is the question. We construct a graph G' which can be $(\lambda + 2)_{(2,1)}$-labeled if and only if G can be $\lambda_{(2,1)}$-labeled.

First construct a complete binary tree T with at least $|V(G)|$ leaves, all of them in the bottom layer. Call the layers, from the bottom to the top, $L_1, \ldots, L_k$.

Add a layer L_0 with $|L_1|$ vertices, and connect these vertices by a perfect matching to the vertices of L_1. Subdivide each edge of T (including the edges between L_0 and L_1) by a new extra vertex. Connect each vertex of G to a unique vertex in L_0. Finally, add leaves to vertices in layers L_i so that their degrees add up to $\lambda + 1$. This completes the construction of G'.

Assume that G' is $(\lambda + 2)_{(2,1)}$-labeled. Vertices from layers L_i have degree $\lambda + 1$ and therefore each of them is labeled either by 0 or by $\lambda + 2$. Furthermore all vertices in the same layer have the same label. Assume without loss of generality that the vertices of L_0 are labeled with label $\lambda + 2$. Then all vertices of G have labels in $\{0, \ldots, \lambda\}$ and we obtain a $\lambda_{(2,1)}$-labeling of G.

Now assume that G can be $\lambda_{(2,1)}$-labeled. Color the vertices from layers L_i with an even index i with $\lambda + 2$ and those from layer with an odd index with 0. Color the vertices between layers L_i and L_{i+1}, $i \geq 1$ arbitrary (but feasible). A vertex between layers L_0 and L_1 has at most 6 forbidden labels: $0, 1, \lambda + 1, \lambda + 2$, and two more labels of vertices at distance two: of the vertex in G and of the vertex between L_1 and L_2 (see Fig. 4 in appendix). Since $\lambda \geq 4$ at least one label is available. Finally, label the leaves at the vertices from layers L_i. Since such a vertex is labeled with 0 or with $\lambda + 2$ and since such a vertex has degree $\lambda + 1$, there is always an available label for every leaf. This gives a $(\lambda + 2)$-labeling of G'. □

As we mentioned in Section 2, sometimes it is important to require that the labels actually used in a labeling are consecutive. An easy argument shows that consecutive $\lambda_{2,1}$-labelings are not easier.

Corollary 1. *The problem whether there exists a consecutive $\lambda_{(2,1)}$-labeling for a given graph G is NP-complete.*

Proof. Let H_λ be the tree containing two vertices of degree $\lambda - 1$ and with $2(\lambda - 2)$ leaves.

For a graph G, consider the disjoint union $G' = G + H_\lambda$. Obviously G' admits a consecutive $\lambda_{(2,1)}$-labeling if and only if G admits a $\lambda_{(2,1)}$-labeling. Hence, the existence of consecutive $\lambda_{(2,1)}$-labeling is also NP-complete for every fixed $\lambda \geq 4$. □

5   Complexity of the Generalized Distance Two Labeling

Recall that the $L(p,q)$ problem asks for a λ-labeling such that adjacent vertices receive labels which are at least p apart, and vertices at distance 2 receive labels which differ by at least q. This is already the general frequency assignment problem restricted to relevant distance 2 (in the $(p_1, p_2, \ldots, p_k)$ condition). Nobody would expect the $L(p,q)$-labeling problem for $p > q \geq 1$ to be easier than $L(2,1)$, however, the actual NP-hardness proofs seem tedious and not easily achievable in full generality. We conjecture:

Conjecture 1. For any $p \geq q \geq 1$ there is a $\lambda(p,q)$ such that deciding if $L(G; p,q) \leq \lambda$ is NP-complete for every fixed $\lambda \geq \lambda(p,q)$.

In this section we gather results which support our conjecture, the proofs are mostly technical and left to the Appendix. We also discuss the complexity of $L(p,q)$ for trees and almost trees.

We first aim at restricting the range of parameters p, q. The following easy generalization of the special case $p = 2, q = 1$ proven in [4] shows that we may restrict our attention to the case of p, q being relatively prime.

Observation 3 *Let p, q, and k be integers. Then $L(G; kp, kq) = kL(G; p, q)$. Hence $L(p,q)$ and $L(kp, kq)$ are polynomially equivalent.*

The case $p = q$ is thus fully understood (NP-complete for $\lambda \geq 3p$ and polynomial for $\lambda < 3p$. So is now the case $p = 2q$ (NP-complete for $\lambda \geq 4q$ and polynomial for $\lambda < 4q$).

For general p, q, we can prove that there is at least one NP-complete fixed parameter instance:

Theorem 4. *For all fixed $p > q \geq 1$, it is NP-complete to decide whether $L(G; p,q) \leq p + q\lceil \frac{p}{q} \rceil$.*

And under slightly less general conditions we can prove that there are infinitely many hard instances:

Theorem 5. *The problem whether $L(G; p,q) \leq p + kq$ is NP-complete for any fixed $k \geq \frac{p}{q}$ and $p > 2q$.*

It follows that for $q = 1$ (and more generally p divisible by q), there are only finitely many polynomial instances (unless of course P=NP), namely if $p > 2$ then the $L(p, 1)$-problem is NP-complete for every fixed $\lambda \geq 2p$. In this case we are able to prove a little more:

Theorem 6. *For every $p > 2$, the $L(p, 1)$-problem is NP-complete for $\lambda \geq p + 5$ and polynomially solvable for $\lambda \leq p + 2$.*

For $p \geq 5$, this result leaves the cases $\lambda = p+3$ and $\lambda = p+4$ as the last open cases for the fixed parameter complexity of $L(p, 1)$ (for $p \leq 4$ the bound $\lambda \geq 2p$ is better than $\lambda \geq p+5$, and $\lambda = 7$ is the only open case for $p = 4$, while $L(3, 1)$ is fully understood being polynomial for $\lambda \leq 5$ and NP-complete for $\lambda \geq 6$).

Finally we comment on the complexity of the $L(p, q)$-problem on trees. Obviously, for fixed λ, the problem is solvable in linear time. When λ is part of the input and $q = 1$, then the same algorithm as we described in Section 3 can be applied to trees and k-almost trees (in the recursive step we only consider labelings (a, b) of the currently processed edge xy such that $|a - b| \geq p$). Hence we get:

Corollary 2. *The $L(p, 1)$-problem is polynomially solvable for k-almost trees for every fixed k.*

However, this algorithm is not of immediate help in case of $q > 1$. In this case we would need to be able to decide the existence of a system of *distant* (rather than distinct) representatives (in the language of the proof of Theorem 1, we would need to decide if the set system M_z, z children of y, allows representatives $a(z) \in M_z$ such that $|a(z) - a(z')| \geq q$ for any two distinct children z, z' of y). Unfortunately, this variant of bipartite matching is NP-complete (easy reduction from SAT). Therefore the complexity of $L(p, q)$ for trees remains (for $q > 1$) open.

References

1. Abello J., M.R. Fellows and J.C. Stillwell, On the complexity and combinatorics of covering finite complexes, *Australasian Journal of Combinatorics* **4** (1991), 103-112
2. Brandstädt, Andreas, Special graph classes – A survey, Schriftenreihe des Fachsbereichs Mathematik, SM-DU-199 (1991), Universität Duisburg Gesamthochschule.
3. Chang, G. J. and D. Kuo, The $L(2, 1)$-labeling problem on graphs, *SIAM J. Disc. Math.* **9**, (1996), pp. 309–316.
4. Griggs, J. R. and R. K. Yeh, Labelling graphs with a condition at distance 2, *SIAM J. Disc. Math.* **5**, (1992), pp. 586–595.
5. Kratochvíl, Jan, Regular codes in regular graphs are difficult, *Discrete Math.* **133** (1994), 191-205
6. Kratochvíl, Jan, Andrzej Proskurowski, and Jan Arne Telle, Covering regular graphs *Journal of Combin. Theory Ser. B* 71 (1997), 1-16
7. Kratochvíl, Jan, Andrzej Proskurowski, and Jan Arne Telle, Covering directed multigraphs I. Colored directed multigraphs, In: *Graph-Theoretical Concepts in Computer Science, Proceedings 23rd WG '97, Berlin, Lecture Notes in Computer Science 1335*, Springer Verlag, (1997), pp. 242-257.
8. Kratochvíl, Jan, Andrzej Proskurowski, and Jan Arne Telle, Complexity of graph covering problems *Nordic Journal of Computing* 5 (1998), 173-195
9. Liu, D. D.-F. and R. K. Yeh, On distance two labellings of graphs, *ARS Combinatorica* **47**, (1997), pp. 13–22.
10. Yeh, Kwan-Ching, Labeling graphs with a condition at distance two, Ph.D. Thesis, University of South Carolina, 1990.

A   Appendix

H_1 H_2

Fig. 2. Multigraphs H_1 and H_2.

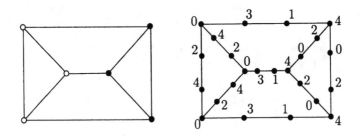

Fig. 3. $BW(3)$-coloring of the prism and labeling of its prime-graph.

Proof of Theorem 4

Theorem: *It is NP-complete to decide whether $L(G; p, q) \leq p + q\lceil \frac{p}{q} \rceil$ for all $p > q \geq 1$.*

Proof. We use a reduction from the $BW(k+2)$–problem, i.e., that of finding a black and white coloring of the vertices of a $(k+2)$-regular graph, such that every vertex has two neighbors with the same color and k neighbors with opposite color.

Let $k = \lceil \frac{p}{q} \rceil - 1$ and let G be a $(k+2)$-regular graph for which we want to find a black and white coloring. Replace each edge of G with a P_4. Call this new graph G'. Assume we have a $\lambda_{(p,q)}$-labeling for G' with $\lambda = p + (k+1)q$. Notice that $\lambda < 2p + q$. The original vertices must be labeled by 0 or λ and their neighbors have labels $p, p+q, \ldots, p+kq, \lambda$ or $0, q, 2q, \ldots, \lambda - p$.

We show that a P_4 starting at a vertex with label 0 and ending in vertex labeled by 0 or by λ can be labeled only by one of the following schemes $(0, p, \lambda, 0)$, $(0, p+iq, jq, \lambda)$ for $1 \leq i \leq k$ and $1 \leq j \leq i$, and $(0, \lambda, p, 0)$.

Suppose the path has vertices w, x, y, z and suppose $f(w) = 0$, where f denotes the labeling function. Clearly $f(x) \geq p$ and $f(y) \geq q$.

colors: $c(v)$ $\lambda + 2$ 0 $c(v')$

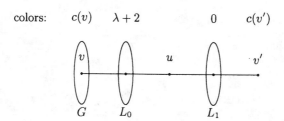

Fig. 4. Induction step of the $L(G, 2, 1)$-problem. Labeling the vertex u, forbidden labels are $\{0, 1, c(v), c(v'), \lambda + 1, \lambda + 2\}$

First assume $f(x) = p$. Then $f(y) \geq 2p$. If $f(z) = \lambda$ then $f(y) \leq \lambda - p < p + q < 2p$ which is a contradiction. Hence $f(z) = 0$. Now $f(y) \in \{p, p+q, \ldots, p + kq, \lambda\}$ and $f(y) \geq 2p$ imply $f(y) = \lambda$.

Now assume that $f(x) = p + iq$ for some $1 \leq i \leq k$. Then $q \leq f(y) \leq iq$. Now, $f(z) = 0$ would imply $f(y) = p + jq$ for some $0 \leq j \leq k + 1$. But then $|f(x) - f(y)| = |i - j| q \leq kq < p$ which is a contradiction. Hence $f(z) = \lambda$ and $f(y) = jq$ for some $1 \leq j \leq i$.

Because the degree of an original vertex is $k + 2$, two of the above mentioned paths lead to a vertex with the same label and the other paths to vertices with complementary label.

Now assume a black and white coloring of G is given. This coloring can be easily extended to an $\lambda_{(p,q)}$-labeling of G'. The subgraph induced by white vertices is a set of cycles and the corresponding P_4's can be colored by $(0, p, \lambda, 0)$ consecutively along the cycles. Similarly for the subgraph induced by black vertices. Edges joining black and white vertices form a bipartite k-regular graph. Hence this graph is k-edge colorable. Fix a coloring by colors $1, 2, \ldots, k$ and use the sequence $(0, p + iq, iq, \lambda)$ on the path corresponding to an edge labeled by i.

$\square$

Proof of Theorem 5

Theorem: *The problem whether $L(G; p, q) \leq p + kq$ is NP-complete for all $p > 2q$ and any integer $k \geq \frac{p}{q}$.*

Proof. We consider two cases.

Case $2q < p \leq 3q$ We use reduction from the $BW(3)$–problem.

Let G be an input graph for the black and white coloring problem. Replace each edge by the graph H in Figure 5 and add leaves to the original vertices to add up their degree to $k + 1$.

The case analysis shows that on the path (z, a, b, x) only the following sequences of colors can appear: $(0, p, \lambda, 0)$, $(0, \lambda, p, 0)$, $(0, \lambda - q, q, \lambda)$, $(\lambda, 0, \lambda - p, \lambda)$ and $(\lambda, \lambda - p, 0, \lambda)$.

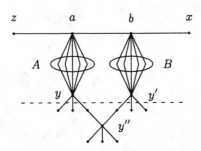

Fig. 5. Edge gadget for the $L(p, q)$-labeling problem

Similarly as in the proof of Theorem 2, it follows that a proper black and white coloring exists if and only if the modified graph allows a $\lambda_{(p,q)}$-labeling.

Case $p > 3q$ In this case we use a reduction from $BW(4)$.

Let G be a 4-regular graph. Use the same edge replacement as in the above case, also add leaves to the original vertices to degrees $k + 1$.

In this case a case analysis gives more possible label schemes of (z, a, b, x), namely $(0, p, \lambda, 0)$, $(0, p + iq, q, \lambda)$, $(0, \lambda - q, iq, \lambda)$, $(0, \lambda, p, 0)$ and their reversals with vertex z labeled by λ. In the above schemes i belongs to the interval $[1, \lfloor \frac{p}{q} \rfloor - 1]$.

Suppose that a $\lambda_{(p,q)}$-labeling of G' exists. Let the original vertices labeled by 0 have white color and those with label λ be colored black.

W.l.o.g. we suppose that the white vertices form a partition of bigger or equal size, compared to set of black vertices. Each white vertex has at most two white neighbors, so it has at least two black neighbors, at most one edge to the black neighbor corresponds to the path labeled $(0, \lambda - q, iq, \lambda)$ and the remaining edges (at least one) correspond to the scheme $(0, p + iq, q, \lambda)$. But the number of paths labeled $(*, *, q, \lambda)$, cannot be greater then number of black vertices, so the number of white vertices is exactly same as the number of black ones. Hence, the given $\lambda_{(p,q)}$-labeling of G' gives a proper $BW(4)$-coloring.

On the other hand $BW(4)$-coloring divides the edges of the graph G into the sets of white cycles, black cycles and cycles with alternating black and white label. By using patterns $(0, p, \lambda, 0)$, $(\lambda, 0, \lambda - p, \lambda)$ and $(0, \lambda - 2q, q, \lambda)(\lambda, 2q, \lambda - q, 0)$ consecutively on these cycles followed by completing colors of added leaves we obtain a proper $\lambda_{(p,q)}$-labeling of the modified graph G'.   □

The Case Analysis for the Labeling of Fig. 5

Here we describe the case study for the theorem 5. We show that under assumption $2q < p$ only schemes $(0, p, \lambda, 0)$, $(0, p + iq, q, \lambda)$, $(0, \lambda - q, iq, \lambda)$, $(0, \lambda, p, 0)$ and their mirrors are applicable to the path (z, a, b, x) on the graph H.

Let us investigate properties of the graph H in Figure. 5.

Sets A and B have the size $|A| = |B| = k - \lceil \frac{p}{q} \rceil$.

If $k = 0$ we obtain exactly result of the theorem 4 so no case analysis is needed.

The vertices z, x, y, y' and y'' have full degree, so they can be labeled by 0 or λ. W.l.o.g we suppose that z is labeled by 0. It implies that $c(a) \in \{p, p+q, p+2q, ..., \lambda - q, \lambda\}$.

If the vertices y and y' have different labels there is no way to label the vertex y'', so $c(y) = c(y')$. If the part of H over the dashed line is pre-labeled with respect to above mentioned condition, the labeling is always extendible to the bottom part of H.

We divide the proof into several cases.

Case A: $c(z) = 0, c(a) = p$ Then labels of the set $A \cup \{b\}$ could be chosen from the interval $[2p, \lambda]$ which contains at most $|A| + 1$ numbers at distance q. Maximum label of A is greater than $\lambda - 2q > \lambda - p \Rightarrow c(y) = c(y') = 0$. If the label of b belongs to the interval $[2p, \lambda - q]$ then the size of the set of possible labels of B is $\lfloor \frac{\lambda - 2p - q - 1}{q} \rfloor + 1 < |B|$. So $c(b) \in [\lambda - q + 1, \lambda] \Rightarrow c(x) = 0 \Rightarrow c(b) = \lambda$.

Case B: $c(z) = 0, c(a) = p + iq, 1 \le i < k, c(y) = 0$ If $1 < i < k - 1$ then in the intervals $[p, iq] \cup [2p + iq, \lambda]$ there are not enough numbers at distance q to label the set A.

Subcase Ba: $i = 1 \Rightarrow c(b) = q \Rightarrow c(x) = \lambda$. The set B can be easily labeled from the interval $[p + 2q, \lambda - q]$.

Subcase Bb: $i = k - 1$ then all possible labels at distance q from the interval $[p, (k-1)q]$ are used in the set A. It gives that $c(b) < p \Rightarrow c(x) = \lambda$ and $c(b)$ is multiple of q. The labels of the set B are chosen from the interval $[c(b) + p, \lambda - 2q]$ which has sufficient size if $c(b) \le (\lfloor \frac{p}{q} \rfloor - 1)q$.

Case C: $c(z) = 0, c(a) = p + iq, 1 \le i < k, c(y) = \lambda$ For b we have $c(b) \ge q$ and labels of B belong to the interval $[q + p, \lambda - p]$. In the case that $c(b) \ge 2q$ there are not enough points to label the set B. Thus $c(b) < 2q < p \Rightarrow c(x) = \lambda \Rightarrow c(b) = q$. The rest of this case should be reduced to the mirror of the case B. (By interchanging the roles of z, a, A and x, b, B.)

Case D: $c(z) = 0, c(a) = \lambda \Rightarrow c(x) = 0$ The only solution is described in the case A.

Proof of Theorem 6

Theorem: *The $L(G; p, 1) \le \lambda$ is an NP-complete problem for $p + 5 \le \lambda < 2p$ and polynomially solvable for $\lambda \le p + 2$.*

Proof. We prove the polynomial and NP-complete parts separately.

The NP-complete cases We are proving that $L(G; p, q) \le \lambda$ is NP complete for $p + 5 \le \lambda < 2p$. Note that in these cases the input graph is bipartite and (assuming the input graph is connected) labels in one class of the bipartition are less than or equal to $\lambda - p$, and labels in the other class greater than or equal to p.

We reduce the 3-colorability of a $(\lambda - p - 1)$-regular $(\lambda - p - 1)$-edge colored graph to our $L(p, 1) \le \lambda$-problem and then prove that this special coloring problem is also NP-complete.

Let G be the graph which is to be 3-colored. Subdivide each edge by one new extra vertex and ask for a $\lambda_{(p,1)}$-labeling of this graph G'.

The graph G' is bipartite and the newly introduced vertices form one class of the bipartition. W.l.o.g we assume that labels from the interval $[p, \lambda]$ in this class are used. We claim that the original vertex has label at most 2, since all its neighbors have distinct labels and one of them should be less than or equal to $p + 2$. Using these labels as colors for the original graph we obtain a proper coloring, since adjacent vertices in G are at distance two in G' and thus have different labels.

In the opposite way a vertex 3-coloring of G together with an edge $(\lambda - p - 1)$-coloring gives us a proper $\lambda_{(p,1)}$-labeling of G' by using the same label from the interval $[p + 2, \lambda]$ for the vertices added to edges with the same color.

The NP-completeness of the special 3-colorability We reduce the 3-SAT problem where each variable has at most two positive and at most two negative occurences.

We give a modification of the classical reduction from 3-SAT. Let F be the given formula with n variables and m clauses. Construct the graph G as follows: For each variable x_i put into G vertices x_i and $\bar{x}_i$ and join them by an edge. Construct a gadget as on the Fig. 6 a) with $2n + 1$ vertices u and join vertices u_{2i-1} and u_{2i} to vertices x_i and $\bar{x}_i$. (This step replaces the vertex of high degree in the original proof.)

For each clause $(l \wedge l' \wedge l'')$ of F add into G the "clause" gadget depicted in the Fig. 6 b) where edges e_0, e_1 and e_2 lead to the vertices corresponding to the literals l, l' and l''.

Construct a gadget with $m + 1$ vertices u' and identify m of them with the lowest vertices of "clause" gadgets. Finally connect vertices u_{2n+1} and u'_{m+1} by an edge.

Assume that G is 3-colored. All vertices u have the same color hence on vertices x_i and $\bar{x}_i$ both of two remaining colors are used. All vertices u' have also the same color different to the color of vertices u. Let this color be viewed as the "false" assignment. The "clause" gadgets guarantee that in each clause at least one literal has "truth" assignment, hence if a 3-coloring exists, the formula F has an satisfying assignment. In vice versa any satisfying assignment can be extended to a 3-coloring of G.

As depicted on the Figures 6 a) and b) the graph G is 4-edge colorable and contains vertices of degree 3 or 4. To have the four-regular graph use two copies of G and join corresponding vertices of degree 3 by a matching. Now the graph

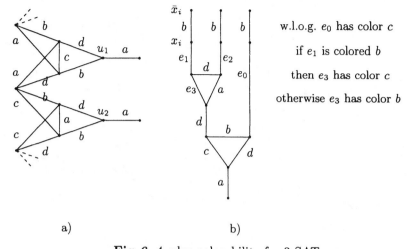

a)   b)

Fig. 6. 4-edge-colorability for 3-SAT

is four-regular, 4-edge-colorable and any coloring of one copy can be extended to a coloring of whole graph by cyclic exchange of colors on the other copy.

Finally we prove by induction the NP-completeness of the 3-coloring of a k-regular k-edge-colorable graph for $k \geq 5$. Two copies of a $(k-1)$-regular $(k-1)$-edge-colorable graph joined by the matching gives us the hardness result similarly as in the above case.

The polynomially solvable cases In the case that a $\lambda_{(p,1)}$-labeling of G exists and $\lambda < p$ then the input graph G has no edges. Similarly concerning a $p_{(p,1)}$-labeling and a $(p+1)_{(p,1)}$-labeling respectively the input graph contains the disjoint set of paths of length at most 3.

Only subgraphs of a (infinite) path $(v_0, v_1, \ldots)$ with leaves added to vertices v_{4k} and v_{4k+1} allows a $(p+2)_{(p,1)}$-labeling. Such a graph is recognizable in the linear time. $\square$

Linear Time Algorithms for Hamiltonian Problems on (Claw,Net)–Free Graphs[*]
(Extended Abstract)

Andreas Brandstädt[1], Feodor F. Dragan[1], and Ekkehard Köhler[2]

[1] Fachbereich Informatik, Universität Rostock,
A. Einstein Str. 21, D-18051 Rostock, Germany.
{ab,dragan}@informatik.uni-rostock.de
[2] Fachbereich Mathematik, Technische Universität Berlin,
Straße des 17. Juni 136, D-10623 Berlin, Germany.
ekoehler@math.TU-Berlin.DE

Abstract. We prove that claw-free graphs, containing an induced dominating path, have a Hamiltonian path, and that two-connected claw-free graphs, containing an induced doubly dominating cycle or a pair of vertices such that there exist two internally disjoint induced dominating paths connecting them, have a Hamiltonian cycle. As a consequence, we obtain linear time algorithms for both problems if the input is restricted to (claw,net)-free graphs. These graphs enjoy those interesting structural properties.

1 Introduction

Hamiltonian properties of claw-free graphs have been studied extensively in the last couple of years. Different approaches have been made and a couple of interesting properties of claw-free graphs have been established (see [1,2,3,4,11,12,13,14,19,20,22,23]). The purpose of this work is to consider the algorithmic problem of finding a Hamiltonian path or a Hamiltonian cycle efficiently. It is not hard to show that both the Hamiltonian path problem and the Hamiltonian cycle problem are NP-complete, even when restricted to line graphs [25]. Hence, it is quite reasonable to ask whether one can find interesting subclasses of claw-free graphs for which efficient algorithms for the above problems exist.

Already in the eighties, Duffus et al. [10] defined the class of CN-free graphs, i.e. graphs that contain neither an induced claw nor an induced net (see Figure 1). Although this definition seems to be rather restrictive, the family of CN-free graphs contains a couple of graph families, that are of interest in their own right. Examples of those families are unit interval graphs, claw-free AT-free graphs and proper circular arc graphs. In their paper [10] Duffus et al. showed that

---

[*] Research of the second author supported by the DFG. Research of the third author supported by the graduate program 'Algorithmic Discrete Mathematics', grant GRK 219/2-97 of the German National Science Foundation (DFG).

this class of graphs has the nice property that every connected CN-free graph contains a Hamiltonian path and every two-connected CN-free graph contains a Hamiltonian cycle. Later, Shepherd [24] proved that there is an $O(n^6)$ algorithm for finding such a Hamiltonian path/cycle in CN-free graphs. Note also that CN-free graphs are exactly the Hamiltonian-hereditary graphs [8], i.e. the graphs for which every connected induced subgraph contains a Hamiltonian path.

In this paper we give a constructive existence proof and present linear time algorithms for the Hamiltonian path and Hamiltonian cycle problems on CN-free graphs. The important structural property that we exploit for this, is the existence of an induced *dominating path* in every CN-free graph (Theorem 1). The concept of a dominating path was first used by Corneil et al. [6] in the context of AT-free graphs. They also developed a simple linear time algorithm for finding such a path in every AT-free graph [5]. As we show in Theorem 1, for the class of CN-free graphs, a linear time algorithm for finding an induced dominating path exists as well. This property is of interest for our considerations since we prove that all claw-free graphs that contain an induced dominating path have a Hamiltonian path (Theorem 2). The proof implies that, given a dominating path, one can construct a Hamiltonian path for a claw-free graph in linear time.

For two-connected claw-free graphs we show, that the existence of a dominating pair is sufficient for the existence of a Hamiltonian cycle (a *dominating pair* is a pair of vertices, such that every induced path connecting them is a dominating path). Again, given a dominating pair, one can construct a Hamiltonian cycle in linear time (Theorem 8). This already implies, for example, a linear time algorithm for finding a Hamiltonian cycle in claw-free AT-free graphs, since every AT-free graph contains a dominating pair and it can be found in linear time [7]. Unfortunately, CN-free graphs do not always have a dominating pair. For example, an induced cycle with more than six vertices is CN-free but does not have such a pair of vertices. Nevertheless, two-connected CN-free graphs have another nice property: they have a good pair or an induced doubly dominating cycle. An *induced doubly dominating cycle* is an induced cycle such that every vertex of the graph is adjacent to at least two vertices of the cycle. A *good pair* is a pair of vertices, such that there exist two internally disjoint induced dominating paths connecting these vertices. We prove that the existence of an induced doubly dominating cycle or a good pair in a claw-free graph is sufficient for the existence of a Hamiltonian cycle (Theorems 6 and 7). Moreover, given an induced doubly dominating cycle or a good pair of a claw-free graph, a Hamiltonian cycle can be constructed in linear time. In Section 4 we present an $O(m + n)$ time algorithm which, for a given CN-free graph, finds either a good pair or an induced doubly dominating cycle.

For terms not defined here, we refer to [9,15]. In this paper we consider finite connected undirected graphs $G = (V, E)$ without loops and multiple edges. The cardinality of the vertex set is denoted by n, whereas the cardinality of the edge set is denoted by m.

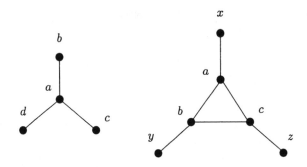

Fig. 1. The claw $K(a; b, c, d)$ and the net $N(a, b, c; x, y, z)$.

A *path* is a sequence of vertices $(v_0, \ldots, v_l)$ such that all v_i are distinct and $v_i v_{i+1} \in E$ for $i = 0, \ldots, l-1$; its *length* is l. An *induced path* is a path where $v_i v_j \in E$ iff $i = j-1$ and $j = 1, \ldots, l$. A *cycle* (k-*cycle*) is a path $(v_0, \ldots, v_k)$ ($k \geq 3$) such that $v_0 = v_k$; its *length* is k. An *induced cycle* is a cycle where $v_i v_j \in E$ iff $|i - j| = 1$ (modulo k). A *hole* H_k is an induced cycle of length $k \geq 5$.

The *distance* $dist(v, u)$ between vertices v and u is the smallest number of edges in a path joining v and u. The *eccentricity* $ecc(v)$ of a vertex v is the maximum distance from v to any vertex in G. The *diameter* $diam(G)$ of G is the maximum eccentricity of a vertex in G. A pair v, u of vertices of G with $dist(v, u) = diam(G)$ is called a *diametral pair*.

For every vertex we denote by $N(v)$ the set of all neighbors of v, $N(v) = \{u \in V : dist(u, v) = 1\}$. The *closed neighborhood* of v is defined by $N[v] = N(v) \cup \{v\}$. For a vertex v and a set of vertices $S \subseteq V$, the minimum distance between v and vertices of S is denoted by $dist(v, S)$. The *closed neighborhood* $N[S]$ of a set $S \subseteq V$ is defined by $N[S] = \{v \in V : dist(v, S) \leq 1\}$.

We say that a set $S \subseteq V$ *dominates* G if $N[S] = V$, and S *doubly dominates* G if every vertex of G has at least two neighbors in S. An induced path of G which dominates G is called an *induced dominating path*. A shortest path of G which dominates G is called a *dominating shortest path*. Analogously one can define an *induced dominating cycle* of G. A *dominating pair* of G is a pair of vertices $v, u \in V$ such that every induced path between v and u dominates G. A *good pair* of G is a pair of vertices $v, u \in V$, such that there exist two internally disjoint induced dominating paths connecting v and u.

The *claw* is the induced complete bipartite graph $K_{1,3}$ and for simplicity, we refer to it by $K(a; b, c, d)$ (see Figure 1). The *net* is the induced six-vertex graph $N(a, b, c; x, y, z)$ shown in Figure 1. A graph is called *CN-free*, or equivalently *(claw, net)–free* if it contains neither a claw nor a net. An *asteroidal triple* of G is a triple of pairwise nonadjacent vertices, such that for each pair of them there exists a path in G that does not contain any vertex in the neighborhood of the third one. A graph is called *AT-free* if it does not contain an asteroidal triple. Finally, a *Hamiltonian path* or *Hamiltonian cycle* of G is a path or cycle, respectively, containing all vertices of G.

2 Induced Dominating Path

In this section we give a constructive proof for the property that every CN-free graph contains an induced dominating path. In fact, we show that there is an algorithm that finds such a path in linear time. To prove the main theorem of this section we will need the following two lemmas (*proofs are omitted*).

Lemma 1. *Let $P = (x_1, \ldots, x_k)$ be an induced path of a CN-free graph G and v be a vertex of G such that $dist(v, P) = 2$. Then any neighbor y of v with $dist(y, P) = 1$ is adjacent to x_1 or to x_k.*

Lemma 2. *Let P be an induced path connecting vertices v and u of a CN-free graph G. Let also s be a vertex of G such that $s \notin N[P]$ and $dist(v, s) \leq dist(v, u)$. Then*

1. *for every shortest path P' connecting v and s, $P' \cap P = \{v\}$ holds, and*
2. *if there is an edge xy of G such that $x \in P \setminus \{v\}$ and $y \in P' \setminus \{v\}$, then both x and y are neighbors of v.*

Theorem 1. *Every CN-free graph G has an induced dominating path and such a path can be found in $O(n + m)$ time.*

Proof. Let G be a CN-free graph. One can construct an induced dominating path in G as follows. Take an arbitrary vertex v of G. Using Breadth First Search (BFS), find a vertex u with the largest distance from v and a shortest path P connecting u with v. Check whether this path P dominates G. If so, we are done. Now, assume that the set $S = V \setminus N[P]$ is not empty. Again, using BFS find a vertex s in S with largest distance from v and a shortest path P' connecting v with s. Create a new path P^* by joining P and P' in the following way: $P^* = (P \setminus \{v\}, P' \setminus \{v\})$, if there is a chord xy between the paths P and P' (see Lemma 2), and $P^* = (P \setminus \{v\}, P')$, otherwise. By Lemma 2, the path P^* is induced. It remains to show that this path dominates G.

Assume there exists a vertex $t \in V \setminus N[P^*]$. First we claim that t is dominated neither by P nor by P'. Indeed, if $t \in (N[P] \bigcup N[P']) \setminus N[P^*]$, then necessarily $tv \in E$ and $v \notin P^*$, i.e. neighbors $x \in P$ and $y \in P'$ of v are adjacent. Therefore, we get a net $N(v, y, x; t, s', u')$, where s' and u' are the vertices at distance two from v on paths P' and P, respectively. Note that vertices s', u' exist because $dist(v, s) \geq 2$.

Thus, t is dominated neither by P nor by P'. Moreover, from the choice of u and s we have $2 \leq dist(v, t) \leq dist(v, s) \leq dist(v, u)$. Now let P'' be a shortest path, connecting t with v, and z be a neighbor of v on this path. Applying Lemma 2 twice (to P, P'' and to P', P''), we obtain a subgraph of G

depicted in Figure 2. We have three shortest paths P, P', P'', each of length at least 2 and with only one common vertex v. These paths can have only chords of type zx, zy, xy. Any combination of them leads to a forbidden claw or net. This contradiction completes the proof of the theorem. Evidently, the method described above can be implemented to run in linear time.   □

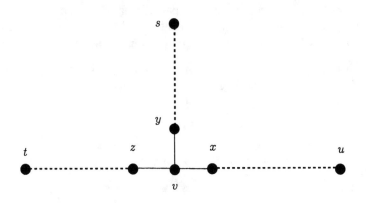

Fig. 2.

It is not clear whether CN-free graphs can be recognized efficiently. But, to apply our method for finding an induced dominating path in these graphs, we do not need to know in advance that a given graph G is CN-free. Actually, our method can be applied to any graph G. It either finds an induced dominating path or returns either a claw or a net of G, showing that G is not CN-free.

Corollary 1. *There is a linear time algorithm that for a given (arbitrary) graph G either finds an induced dominating path or outputs a claw or a net of G.*

Proof is omitted.

3   Hamiltonian Path

In what follows we show that for claw-free graphs the existence of an induced dominating path is a sufficient condition for the existence of a Hamiltonian path. The proof for this result is constructive, implying that, given an induced dominating path, one can find a Hamiltonian path efficiently.

Theorem 2. *Every connected claw-free graph G containing an induced dominating path has a Hamiltonian path. Moreover, given an induced dominating path, a Hamiltonian path of G can be constructed in linear time.*

Proof. Let $G = (V, E)$ be a connected claw-free graph and let $P = (x_1, \ldots, x_k)$, $(k \geq 1)$ be an induced dominating path of G. If $k = 1$, vertex x_1 dominates G

and, since G is claw-free, there are no three independent vertices in $G - \{x_1\}$ (by $G - \{x_1\}$ we denote a subgraph of G induced by vertices $V \setminus \{x_1\}$). If $G - \{x_1\}$ is not connected then, again because G is claw-free, it consists of two cliques C_0, C_1 and a Hamiltonian path of G can easily be constructed. If $G - \{x_1\}$ is connected we can construct a Hamiltonian path as follows. First we construct a maximal path $P_1 = (y_1, \ldots, y_l)$, i.e. all vertices that are not in P_1 are neither connected to y_1 nor to y_l. Let R be the set of all remaining vertices. If $R = \emptyset$ we are done. If there is any vertex in R, it follows that $y_1 y_l \in E$ since otherwise there are three independent vertices in $G - \{x_1\}$. Furthermore, any two vertices of R are joined by an edge, since otherwise they would form an independent triple with y_1 (and with y_l as well). Hence, R induces a clique. Since $G - \{x_1\}$ is connected, there has to be an edge from a vertex $v_R \in R$ to some vertex $y_i \in P_1$ ($1 < i < l$). Now we can construct a Hamiltonian path P of G: $P = (x_1, y_{i+1}, y_{i+2}, \ldots, y_l, y_1, y_2, \ldots, y_i, v_R, \tilde{R})$, where $\tilde{R}$ stands for an arbitrary permutation of the vertices of $R \setminus \{v_R\}$.

For $k \geq 2$ we first construct a Hamiltonian path P_2 for $G' = G(\mathrm{N}[x_1] \setminus \{x_2\})$ as described above, using x_1 as the dominating vertex. At least one endpoint of P_2 is adjacent to x_2 since if $G' - \{x_1\}$ is not connected, x_2 has to be adjacent to all vertices of either C_0 or C_1 (otherwise there is a claw in G) and if $G' - \{x_1\}$ is connected the construction gives a path ending in x_1 which is, of course, adjacent to x_2. To construct a Hamiltonian path for the rest of the graph we define for each vertex x_i ($i \geq 2$) of P a set of vertices $C_i = \mathrm{N}(x_i) \setminus \bigcup_{j=1}^{i-1} \mathrm{N}[x_j]$. Each set C_i forms a clique of G since if two vertices $u, v \in C_i$ are not adjacent then the set u, v, x_i, x_{i-1} induces a claw. Hence we can construct a path $P^* = (P_2, x_2, P_2^C, x_3, P_3^C, x_4, \ldots, x_{k-1}, P_{k-1}^C, x_k, P_k^C)$, where P_i^C stands for an arbitrary permutation of the vertices of $C_i \setminus \{x_{i+1}\}$. This path P^* is a Hamiltonian path of G because it obviously is a path and, since P is a dominating path, each vertex of G has to be either on P, P_2 or in one of the sets C_i.

For the case $k = 1$ both finding the connected components of $G - \{x_1\}$ and constructing the path P_1 can easily be done in linear time. For $k \geq 2$ we just have to make sure that the construction of the sets C_i can be done in $O(n + m)$ and this can be realized easily within the required time bound. $\qquad \Box$

Theorem 3. *Every CN-free graph G has a Hamiltonian path and such a path can be found in $O(n + m)$ time.*

Corollary 2. *There is a linear time algorithm that for a given (arbitrary) graph G either finds a Hamiltonian path or outputs a claw or a net of G.*

4 Induced Dominating Cycle, Dominating Shortest Path, or Good Pair

In this section we show that every two-connected CN-free graph G has an induced doubly dominating cycle or a good pair. Moreover, we present an efficient algorithm that, for a given two-connected CN-free graph G, finds either a good pair or an induced doubly dominating cycle. *Proofs of the following three lemmas are omitted.*

Lemma 3. *Every hole of a CN-free graph G dominates G.*

Corollary 3. *Let H be a hole of a CN-free graph G. Every vertex of $V \setminus H$ is adjacent to at least two vertices of H.*

A *subgraph G' of G (doubly) dominates G* if the vertex set of G' (doubly) dominates G.

Lemma 4. *Every induced subgraph of a CN-free graph G which is isomorphic to S_3 or S_3^- (see Figure 3) dominates G.*

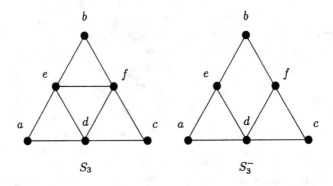

Fig. 3.

Lemma 5. *Let P be an induced path connecting vertices v and u of a CN-free graph G. Let s be a vertex of G such that $s \notin N[P]$ and $dist(v, s) \leq dist(v, u)$, $dist(u, s) \leq dist(v, u)$. Then G has an induced doubly dominating cycle and such a cycle can be found in linear time.*

Theorem 4. *There is a linear time algorithm that, for a given CN-free graph G, either finds an induced doubly dominating cycle or gives a dominating shortest path of G.*

Proof. Let G be a CN-free graph. One can construct an induced doubly dominating cycle or a dominating shortest path of G as follows (compare with the proof of Theorem 1). Take an arbitrary vertex v of G. Find a vertex u with the largest distance from v and a shortest path P connecting u with v. Check whether P dominates G. If so, we are done; P is a dominating shortest path of G. Assume now that the set $S = V \setminus N[P]$ is not empty. Find a vertex s in S with the largest distance from v and a shortest path P_v connecting v

with s. Create again a new path P^* by 'joining' shortest paths P and P_v as in the proof of Theorem 1. We have proven there that P^* dominates G. Let now P_u be a shortest path between s and u. If $dist(s, u) \leq dist(v, u)$ or both $dist(s, u) > dist(v, u)$ and $v \notin N[P_u]$, then Lemma 5 can be applied to get an induced doubly dominating cycle of G in linear time. Therefore, we may assume that $dist(s, u) > dist(v, u) \geq dist(v, s)$ and $v \in N[P_u]$. Now we show that the shortest path P_u dominates G. If v lies on the path P_u, then $P^* = P_u$ and we are done. Otherwise, let x be a neighbor of v in P_u. Note that $dist(v, s) > 1$ and so $x \neq s, u$. Since G is claw-free v is adjacent to a neighbor $y \in P_u$ of x. Assume, without loss of generality, that x is closer to s than y. If we show that $dist(v, s) = 1 + dist(x, s)$ and $dist(v, u) = 1 + dist(y, u)$, then again, by the proof of Theorem 1, the path P_u will dominate G (as a path obtained by 'joining' two shortest paths that connect v with u and v with s, respectively). By the triangle condition, we have $dist(u, s) < dist(v, u) + dist(v, s)$ (strict inequality because $v \notin P_u$) and $dist(v, s) \leq 1 + dist(x, s)$, $dist(u, v) \leq 1 + dist(y, u)$. Consequently, $dist(v, u) + dist(v, s) > dist(u, s) = dist(u, y) + 1 + dist(x, s) \geq dist(v, u) - 1 + 1 + dist(v, s) - 1 = dist(v, u) + dist(v, s) - 1$. That is, $dist(u, s) = dist(v, u) + dist(v, s) - 1$ and $dist(v, s) = 1 + dist(x, s)$, $dist(u, v) = 1 + dist(y, u)$. □

Corollary 4. *There is a linear time algorithm that, for a given (arbitrary) graph G, either finds an induced doubly dominating cycle, or gives a dominating shortest path, or outputs a claw or a net of G.*

Lemma 6. *Let $P = (v, x_2, \ldots, x_{k-1}, u)$ be a dominating shortest path of a graph G. Then $max\{ecc(v), ecc(u)\} \geq diam(G) - 1$.*

A pair of vertices u, v of G with $dist(u, v) = ecc(u) = ecc(v)$ is called a *pair of mutually furthest vertices*.

Corollary 5. *For a graph G with a given dominating shortest path, a pair of mutually furthest vertices can be found in linear time.*

In what follows we will use the fact, that in a two-connected graph, every pair of vertices is joint by two internally disjoint paths. In order to actually find such a pair of paths, one can use Tarjan's linear time DFS-algorithm for finding the blocks of a given graph. For the proof of Lemma 7 we refer to [18].

Lemma 7. *Let G be a two-connected graph and let x, y be two different nonadjacent vertices of G. Then one can construct in linear time two induced, internally disjoint paths, both joining x and y.*

Theorem 5. *There is a linear time algorithm that, for a given two-connected CN-free graph G, either finds an induced doubly dominating cycle or gives a good pair of G.*

Proof. By Theorem 4, we get either an induced doubly dominating cycle or a dominating shortest path of G in linear time. We show that, having a dominating shortest path of a two-connected graph G, one can find in linear time a good pair or an induced doubly dominating cycle. By Corollary 5, we may assume that a pair x, y of mutually furthest vertices of G is given. Let also P_1, P_2 be two induced internally disjoint paths connecting x and y in G. They exist and can be found in linear time by Lemma 7 (clearly, we may assume that $xy \notin E$, because otherwise $N[x] = V = N[y]$ and x, y together with a vertex $z \in V \setminus \{x, y\}$ will form a doubly dominating triangle). If one of these paths, say P_1, is not dominating, then there must be a vertex $s \in V \setminus N[P_1]$. Since x, y are mutually furthest vertices of G we have $dist(s, x) \leq dist(x, y)$, $dist(s, y) \leq dist(x, y)$. Hence, we are in the conditions of Lemma 5 and can find an induced doubly dominating cycle of G in linear time. □

Corollary 6. *There is a linear time algorithm that, for a given (arbitrary) two-connected graph G, either finds an induced doubly dominating cycle, or gives a good pair, or outputs a claw or a net of G.*

5   Hamiltonian Cycle

In this section we prove that, for claw-free graphs, the existence of an induced doubly dominating cycle or a good pair is sufficient for the existence of a Hamiltonian cycle. The proofs are also constructive and imply linear time algorithms for finding a Hamiltonian cycle.

Theorem 6. *Every claw-free graph G, that contains an induced doubly dominating cycle, has a Hamiltonian cycle. Moreover, given an induced doubly dominating cycle, a Hamiltonian cycle of G can be constructed in linear time.*

Proof is omitted.

Corollary 7. *Every claw-free graph, containing an induced dominating cycle of length at least 4, has a Hamiltonian cycle and, given that induced dominating cycle, one can construct a Hamiltonian cycle in linear time.*

Let $G = (V, E)$ be a graph and let $P = (x_1, \ldots, x_k)$ be an induced dominating path of G. P is called *enlargeable* path if there is some vertex v in $V \setminus P$ that is either adjacent to x_1 or to x_k but not to both of them and, additionally, to no other vertex in P. Consequently, an induced dominating path P is called *non-enlargeable* if such a vertex does not exist. Obviously, every graph G that has an induced dominating path, has a non-enlargeable induced dominating path as well. Furthermore, given an induced dominating path P, one can find in linear time a non-enlargeable induced dominating path P' by simply scanning the neighborhood of both x_1 and x_k. For the next theorem we will need an auxiliary result (*proof is omitted*).

Lemma 8. *Let G be a claw-free graph and $P = (x_1, x_2, \ldots, x_k)$ $(k > 2)$ be an induced non-enlargeable dominating path of G such that there is no vertex y in G with $N(y) \cap P = \{x_1, x_k\}$. Then there is a Hamiltonian path in G that starts in x_1 and ends in x_k and, given the path P, one can construct this Hamiltonian path in linear time.*

Theorem 7. *Let G be a two-connected claw-free graph with a good pair u, v. Then G has a Hamiltonian cycle and, given the corresponding induced dominating paths, one can construct a Hamiltonian cycle in linear time.*

Proof. Let $P_1 = (u = x_1, \ldots, v = x_k)$, $P_2 = (u = y_1, \ldots, v = y_l)$ be the induced dominating paths, corresponding to the good pair u, v. By the definition of a good pair both k and l are greater than 2. We may also assume that, for any induced dominating path $P = (a_1, \ldots, a_s)$ of G with $s > 2$, no vertex $y \in V \setminus P$ exists such that $N(y) \cap P = \{a_1, a_s\}$. Otherwise, P together with y would form an induced dominating cycle of length at least 4, and we can apply Corollary 7 to construct a Hamiltonian cycle of G in linear time.

Let A_1 be the set of vertices a_1 that are adjacent to x_1 but to no other vertex of P_1; let B_1 be the set of vertices b_1 that are adjacent to x_k but to no other vertex of P_1. A_2 and B_2 are defined accordingly for P_2. Of course, each of the sets A_1, A_2, B_1, B_2 forms a clique of G.

First we assume that one of these paths, say P_1, is non-enlargeable, i.e. $A_1 = \emptyset$, $B_1 = \emptyset$. In this case we do the following. We remove the inner vertices of P_2 from G and get the graph $G - (P_2)$, where (P_2) denotes the inner vertices of P_2. Then, using P_1, we create a Hamiltonian path in $G - (P_2)$, that starts at u and ends at v (Lemma 8) and add (P_2) to this path to create a Hamiltonian cycle of G.

We can use this method for creating a Hamiltonian cycle of G, whenever we have two internally disjoint paths P, P' of G both connecting u with v, such that one of them is an induced dominating and non-enlargeable path of the graph obtained from G by removing the inner vertices of the other path.

Now we suppose that both paths P_1, P_2 are enlargeable. Because of symmetry we have to consider the following three cases.

Case 1. *There exist a vertex $a_1 \in A_1 \setminus A_2$ and a vertex $b_1 \in B_1 \setminus B_2$.*
In this case there must be edges from a_1, b_1 to inner vertices y_i, y_j of P_2. Consequently, we can form a new path P_2' by starting in u and traversing through A_1, y_i, $\ldots$, y_j, B_1, v, where $(y_i, \ldots, y_j)$ is the subpath of P_2 between y_i and y_j. Evidently, P_2' contains all vertices of B_1, A_1 and is internally disjoint from P_1, which is non-enlargeable in $G - (P_2')$.

Case 2. *$B_1 = B_2$ and either $A_1 = A_2$ or there exists a vertex $a_1 \in A_1 \setminus A_2$.*
In this case none of the vertices of $B := B_1 = B_2$ (if $B \neq \emptyset$) has a neighbor in $P_1 \cup P_2$, other than v. As G is two-connected, for some vertex $b \in B$ there has to be a vertex $z \in V \setminus (P_1 \cup P_2 \cup B)$ with $zb \in E$. Since P_2 dominates G and $z \notin B$, vertex z must be adjacent to a vertex $y \in P_2 \setminus \{v\}$. If z is only adjacent to $y_1 = u$ but to no other vertex of P_2, then z necessarily belongs to A_2 and we can

form a new path P_1' by starting in u, using all vertices of A_2, B and ending in v. Again, P_1' is internally disjoint from P_2 and P_2 is non-enlargeable in $G - (P_1')$. If $N(z) \cap P_2 = \{u, v\}$ then we can apply Corollary 7.

Therefore, we may assume that z is adjacent to an inner vertex y of P_2. Now, if there exists a vertex $a_1 \in A_1 \setminus A_2$, then a_1 is adjacent to some vertex y' of (P_2) and we can construct a new path P_2' by using u, A_1, y', ..., y, z, B, v (if B was empty, then P_2' ends at $\ldots, y', \ldots, y_{l-1}, v$). This path is internally disjoint from P_1, which is non-enlargeable in $G - (P_2')$. If $A_1 = A_2$ then from the discussion above we may assume that either $A := A_1 = A_2$ is empty or there is a vertex $z' \in V \setminus (P_1 \cup P_2 \cup A)$ which is adjacent to a vertex of A and has a neighbor y' in (P_2). Hence, we can construct a path P_2' by using u, A, z', y', ..., y, z, B, v, which is internally disjoint from P_1 (if $z' = z$ then P_2' is constructed by using u, A, z, B, v).

Case 3. A_2 is strictly contained in A_1 and B_1 is strictly contained in B_2. Proof for this case is omitted.

It is not hard to see that the above described method can be implemented to run in linear time.   □

Theorem 8. *Every two-connected claw-free graph G that contains a dominating pair, has a Hamiltonian cycle and, given a dominating pair, a Hamiltonian cycle can be constructed in linear time.*

Proof is omitted.

Theorem 9. *Every two-connected CN-free graph G has a Hamiltonian cycle and such a cycle can be found in $O(n + m)$ time.*

Corollary 8. *There is a linear time algorithm that for a given (arbitrary) two-connected graph G either finds a Hamiltonian cycle or outputs a claw or a net of G.*

Corollary 9. *A Hamiltonian cycle of a two-connected (claw,AT)-free graph can be found in $O(n + m)$ time.*

Remark 1. Corollary 8 implies that every two-connected unit interval graph has a Hamiltonian cycle, which is, of course, well known (see [21,17]). The interesting difference of the above algorithm compared to the existing algorithms for this problem on unit interval graphs, is, that it does not require the creation of an interval model. It also follows from Corollaries 2 and 8 that both the Hamiltonian path problem and the Hamiltonian cycle problem are linear time solvable on proper circular arc graphs. Note that previously known algorithms for these problems had time bounds $O(m + n \log n)$ [16].

It should also be mentioned that Theorem 2 and Theorem 7 do cover a class of graphs, that is not contained in the class of CN-free graphs, Figure 4 shows a graph, that is claw-free, does contain a dominating/good pair, and, consequently, a dominating path but obviously, it is neither AT-free nor net-free.

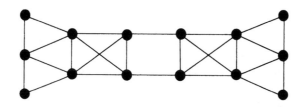

Fig. 4. Claw-free graph, containing a dominating pair and a net.

References

1. A.S. ASRATIAN, Every 3-connected, locally connected, claw-free graph is Hamilton-connected, *J. Graph Theory*, 23 (1996), 191–201.
2. B. BOLLOBÁS, O. RIORDAN, Z. RYJÁČEK, A. SAITO, and R.H. SCHELP, Closure and Hamiltonian-connectivity of claw-free graphs, *Discrete Math.*, 195 (1999) 67–80.
3. S. BRANDT, O. FAVARON and Z. RYJÁČEK, Closure and stable hamiltonian properties in claw–free graphs, *Tech. Rep. No 97*, University of West Bohemia, November, 1996.
4. J. BROUSEK, Z. RYJÁČEK and O. FAVARON, Forbidden Subgraphs, Hamiltonicity and Closure in Claw–Free Graphs, *Discrete Math.*, 196 (1999) 29–50.
5. D.G. CORNEIL, S. OLARIU and L. STEWART, A linear time algorithm to compute a dominating path in an AT-free graph, *Information Processing Letters*, 54 (1995), 253–257.
6. D.G. CORNEIL, S. OLARIU and L. STEWART, Asteroidal Triple–free Graphs, *SIAM J. Discrete Math.*, 10 (1997), 399–430.
7. D.G. CORNEIL, S. OLARIU and L. STEWART, Linear time algorithms for dominating pairs in asteroidal triple–free graphs, to appear in *SIAM J. Computing*.
8. P. DAMASCHKE, Hamiltonian-hereditary graphs, unpublished.
9. R. DIESTEL, Graph Theory, *Graduate Texts in Mathematics 173*, Springer, 1997.
10. D. DUFFUS, M.S. JACOBSON and R.J. GOULD, Forbidden Subgraphs and the Hamiltonian Theme, *Proceedings, 4th Int. Conference on the Theory and Applications of Graphs*, 1980, 297–316.
11. R.J. FAUDREE, E. FLANDRIN and Z. RYJÁČEK, Claw-free graphs – a survey, *Discrete Math.*, 164 (1997), 87–147.
12. R.J. FAUDREE and R.J. GOULD, Characterizing forbidden pairs for hamiltonian properties, *Discrete Math.*, 173 (1997), 45–60.
13. R.J. FAUDREE, Z. RYJÁČEK and I. SCHIERMEYER, Forbidden subgraphs and cycle extendability, *J. Combin. Math. Combin. Comput.*, 19 (1995), 109–128.
14. E. FLANDRIN, J.L. FOUQUET and H. LI, On Hamiltonian claw-free graphs, *Discrete Math.*, 111 (1993), 221–229.
15. M.C. GOLUMBIC, Algorithmic Graph Theory and Perfect Graphs, *Academic Press*, New York, 1980.
16. P. HELL, J. BANG-JENSEN and J. HUANG, Local Tournaments and Proper Circular Arc Graphs, *Algorithms*, Lecture Notes in Computer Science (T. Asano, T. Ibaraki, H. Imai, and T. Nishizeki, eds.), Vol. 450, Springer-Verlag, New-York (1990) pp. 101–108.

17. J. M. KEIL, Finding hamiltonian circuits in interval graphs, *Information Processing Letters*, 20 (1985), 201–206.
18. E. KÖHLER, Linear time algorithms for Hamiltonian problems in claw-free AT-free graphs, *manuscript*, 1999.
19. E. KÖHLER and M. KRIESELL, Edge-Dominating Trails in AT-free Graphs, *Tech. Rep. No 615*, Technical University Berlin, 1998.
20. M. LI, Hamiltonian cycles in 3-connected claw-free graphs, *J. Graph Theory*, 17 (1993), 303–313.
21. G. K. MANACHER, T. A. MANKUS and C. J. SMITH, An optimum $\Theta(n \log n)$ algorithm for finding a canonical hamiltonian path and a canonical hamiltonian circuit in a set of intervals, *Information Processing Letters*, 35 (1990), 205–211.
22. Z. RYJÁČEK, On a closure concept in claw–free graphs, *Journal of Combinatorial Theory, Series B*, 70 (1997), 217–224.
23. F.B. SHEPHERD, Hamiltonicity in Claw-Free Graphs, *Journal of Combinatorial Theory, Series B*, 53 (1991), 173–194.
24. F.B. SHEPHERD, Claws, *Master's Thesis*, University of Waterloo, 1987.
25. D. B. WEST, Introduction to Graph Theory, Prentice–Hall, 1996, (Problem 6.3.14, Chapter 6).

On Claw-Free Asteroidal Triple-Free Graphs

Harald Hempel[1] and Dieter Kratsch[1]

Fakultät für Mathematik und Informatik
Friedrich-Schiller-Universität
07740 Jena, Germany
{hempel,kratsch}@minet.uni-jena.de

Abstract. We present an $O(n^{2.376})$ algorithm for recognizing claw-free AT-free graphs and a linear-time algorithm for computing the set of *all* central vertices of a claw-free AT-free graph. In addition, we give efficient algorithms that solve the problems INDEPENDENT SET, DOMINATING SET, and COLORING. We argue that all running times achieved are optimal unless better algorithms for a number of famous graph problems such as triangle recognition and bipartite matching have been found. Our algorithms exploit the structure of 2LexBFS schemes.

1 Introduction

In this paper we study claw-free AT-free graphs and show that a large number of combinatorial problems can be efficiently solved. Moreover, we give strong arguments that the algorithms we present are optimal. In particular, we consider the following:

1. RECOGNITION: We present an $O(n^{2.376})$ recognition algorithm for claw-free AT-free graphs. We show that this is optimal unless one finds an algorithm for triangle recognition that runs faster than $O(n^{2.376})$.

2. RADIUS: We prove that there exists a linear-time algorithm for computing the radius of a claw-free AT-free graph.

3. ALL CENTRAL VERTICES: Though no linear-time algorithm for finding just *one* central vertex for AT-free graphs is known we give a linear-time algorithm that computes the set of *all* central vertices of a claw-free AT-free graph.

4. INDEPENDENT SET: A linear-time algorithm for claw-free AT-free graphs exists.

5. DOMINATING SET: There is a linear-time algorithm for claw-free AT-free graphs.

6. COLORING: We prove that computing an optimal coloring for claw-free AT-free graphs is related to both computing a maximum matching on bipartite graphs and on general graphs. There is an $O(\sqrt{n}m)$ coloring algorithm for claw-free AT-free graphs. Any improvement of this time bound implies the existence of an algorithm for computing a maximum matching of bipartite graphs that runs in time less than $O(n^2 + \sqrt{n}m)$.

Claw-free AT-free graphs, the class of all graphs neither containing a claw as an induced subgraph nor an asteroidal triple, form a subclass of AT-free graphs

Widmayer et al. (Eds.): WG'99, LNCS 1665, pp. 377–390, 1999.

and contain all complements of bipartite graphs. A *claw* is a graph on four vertices such that one of them is adjacent to the other three which themselves are pairwise non-adjacent. An *asteroidal triple* is a triple of vertices such that for every pair of two of those vertices there exists a path joining them that does not contain any vertex of the closed neighborhood of the third vertex.

With respect to the combinatorial problems listed above the following is known for AT-free graphs. The straightforward and currently the best known upper bound for recognizing (dense) AT-free graphs is $O(n^3)$. We mention in passing that $O(nm + n^{2.82})$ and $O(\overline{m}^{1.5} + n^2)^1$ algorithms can be obtained [4] that are faster than the straightforward algorithm on sparse graphs and complements of sparse graphs, respectively. The best (conditional) lower bound for recognizing AT-free graphs is closely related to the upper bound of triangle recognition which is currently $O(n^\alpha)$ [23], where $O(n^\alpha)$ is the time to multiply two binary $n \times n$ matrices. (At present $\alpha = 2.376...$) It has been shown that the diameter of AT-free graphs can be computed in linear time with an absolute error of 1 by 2LexBFS (see [6]). However, for both computing the diameter and the radius no linear-time algorithm is known for AT-free graphs. To the best of our knowledge there exists no algorithm for computing even *one* central vertex of an AT-free graph. COLORING is one of the most interesting open problems for AT-free graphs, still waiting for a polynomial-time algorithm or a proof of NP-completeness. The best algorithms for INDEPENDENT SET and DOMINATING SET for AT-free graphs have running times of $O(n^4)$ [5] and $O(n^6)$ [17], respectively.

Claw-free AT-free graphs have been considered in the literature before [16,20]. As a consequence of a theorem in [6], the diameter of a claw-free AT-free graph can be computed in linear time. HAMILTON PATH and HAMILTON CIRCUIT can be solved in linear time on claw-free AT-free graphs [4]. Kloks et al. have given a characterization of claw-free AT-free graphs. For each connected graph G holds: G is claw-free AT-free if and only if either $\alpha(G) \leq 2$ or G is a claw-free cocomparability graph [16].

Since that characterization does not offer a general tool for solving many of the combinatorial problems we address our approach is mainly based on structural properties of 2LexBFS schemes for claw-free AT-free graphs. 2LexBFS schemes have played a major role in showing that a dominating pair for AT-free graphs can be computed in linear time [8]. Among others we show that almost all levels of a 2LexBFS scheme of a claw-free AT-free graph are cliques, and that the union of any two consecutive levels does not contain three vertices forming an independent set.

---

[1] As is standard in graph theory, n and m denote the number of vertices and edges of the input graph, respectively. Similarly, $\overline{m}$ denotes the number of edges in the complement of the input graph.

2 Preliminaries

We consider only finite, undirected, simple, and *connected* graphs. For a graph $G = (V, E)$ and $W \subseteq V$, $G[W]$ denotes the subgraph of G induced by the vertices of W. For a vertex x of $G = (V, E)$, $N(x) = \{y \in V : \{x, y\} \in E\}$ is the neighborhood of x and $N[x] = N(x) \cup \{x\}$ is the closed neighborhood of x. We denote the maximum cardinality of an independent set of G by $\alpha(G)$. We refer the reader to [24] for standard graph theory notations and to [3,14] for definitions and properties of special graph classes not listed here.

Definition 1. *A* claw *is a graph on four vertices such that one of them, called the* center, *is adjacent to the other three vertices which themselves are pairwise non-adjacent. A graph G is called* claw-free *if it has no claw as an induced subgraph.*

A triple $\{x, y, z\}$ of vertices of a graph is an asteroidal triple *(AT) if for every two of these vertices there is a path between them avoiding the closed neighborhood of the third. A graph G is called* asteroidal triple-free *(AT-free) if it has no asteroidal triple.*

An AT-free graph is claw-free AT-free *if it does not contain a claw as an induced subgraph.*

Corneil, Olariu, and Stewart initiated and contributed substantially to the extensive research on structural and algorithmic properties of AT-free graphs [7,8]. One of their major discoveries is the relation between an old graph search procedure, called LEXICOGRAPHIC BREADTH-FIRST SEARCH (LexBFS), and so-called dominating pairs [8]. A pair (x, y) of vertices of a graph G is a *dominating pair* (short DP) if for every path P between x and y, the vertex set of P is a dominating set of G.

Theorem 1 ([7]). *Every connected AT-free graph has a dominating pair.*

In [8] a linear-time algorithm for computing a dominating pair of a given connected AT-free graph was established. The main idea of this algorithm is the use of LexBFS. LexBFS is a variant of BREADTH-FIRST SEARCH that has been introduced by Rose, Tarjan, and Lueker in 1976 [21]. It has been used as a subroutine in a variety of efficient (often linear-time) graph algorithms. We reproduce the details of the linear-time algorithms LexBFS and 2LexBFS from [21,8].

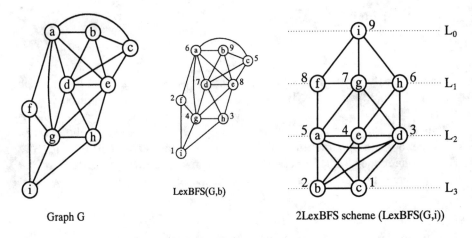

Graph G   2LexBFS scheme (LexBFS(G,i))

Fig. 1. A claw-free AT-free graph G and one of its 2LexBFS schemes.

Procedure LEXBFS(G, x);
Input: a connected graph $G = (V, E)$ and a distinguished vertex x of G
Output: a numbering σ of the vertices of G

begin
  label$(x) := |V|$;
  for each vertex v in $V \setminus \{x\}$ **do**
  label$(v) := \Lambda$;
  for $i := |V|$ **downto** 1 **do**
  begin
    pick an unnumbered vertex v with (lexicographically) largest label;
    $\sigma(v) := i$;   {assign to v number i}
    for each unnumbered vertex u in $N(v)$ **do**
      append i to label(u)
  end
  return σ
end; {LexBFS}

Procedure 2LEXBFS(G);
Input: a connected graph $G = (V, E)$
Output: a numbering σ_2 of the vertices of G

begin
  choose a vertex w of G;
  LexBFS(G, w);
  let σ_1 be the output of LexBFS(G, w) and let x be the vertex with $\sigma_1(x) = 1$;
  LexBFS(G, x);
  let σ_2 be the output of LexBFS(G, x);
  return σ_2
end; {2LexBFS}

    A vertex ordering $x_n, x_{n-1}, \ldots, x_1$ is called a *2LexBFS ordering* of G if some
2LEXBFS(G) returns σ_2 such that $\sigma_2(x_j) = j$ for all j. A 2LexBFS ordering
and the levels $L_0 = \{x_n\}, L_1 = N(x_n), \ldots, L_i = \{x_j : d(x_j, x_n) = i\}, \ldots, L_r$
are called a *2LexBFS scheme* of G. Observe that for all $0 \leq i \leq r$, $L_i = \{x_t, x_{t-1}, \ldots, x_{t'}\}$ for suitable $t \geq t'$. For convenience let the vertices of L_i be

linearly ordered on a horizontal line according to the 2LexBFS ordering such that x_t is the leftmost vertex (see Fig. 1).

Theorem 2 ([8]). *Every 2LexBFS ordering $x_n, x_{n-1}, \ldots, x_1$ of a connected AT-free graph has the* dominating pair-property *(DP-property), i.e., for all $i \in \{1, 2, \ldots, n\}$, (x_n, x_i) is a dominating pair of $G[\{x_i, x_{i+1}, \ldots, x_n\}]$.*

Consequently, for every 2LexBFS ordering $x_n, x_{n-1}, \ldots, x_1$ of an AT-free graph G, (x_1, x_n) is a DP. Thus a DP can be computed in linear time by 2LexBFS [8]. A 2LexBFS scheme of an AT-free graph and its DP-property has also been used in [16] to establish a bandwidth approximation algorithm for AT-free graphs. The following characterization of claw-free AT-free graphs is given in the same paper.

Theorem 3 ([16]). *Let $G = (V, E)$ be a connected graph. Then G is claw-free AT-free if and only if G is a claw-free cocomparability graph or $\alpha(G) \leq 2$.*

Note that *cocomparability graphs* (complements of comparability graphs) can be defined as those graphs that admit a *cocomparability ordering*, i.e., a vertex ordering $x_1, x_2, \ldots, x_n$ such that for all $i < j < k$: $\{x_i, x_k\} \in E$ implies $\{x_i, x_j\} \in E$ or $\{x_j, x_k\} \in E$ (see [3]).

3 Recognizing Claw-Free Asteroidal Triple-Free Graphs

The work of Corneil, Olariu, and Stewart on AT-free graphs led to a variety of interesting questions. A challenging one asks for an efficient recognition algorithm for AT-free graphs and it seemed natural to expect a dramatic improvement over the $O(n^3)$ running time of the straightforward algorithm [7]. However, recently Spinrad revealed a surprising connection to triangle recognition [23]. Triangle recognition is a well-studied fundamental graph algorithms problem and thus is sometimes even used as a lower bound for the complexity of other graph problems.

Theorem 4 ([23]). *If there is an $O(n^c)$ algorithm for recognizing AT-free graphs then there is an $O(n^{\max(2,c)})$ algorithm for recognizing triangle-free graphs.*

The best known algorithm for recognizing a triangle in dense graphs, here called TRIANGLE(G), uses matrix multiplication and has running time $O(n^\alpha)$ (see [1]). Thus any $O(n^c)$ algorithm for recognizing AT-free graphs with $c < \alpha$ would not only significantly improve the best known running time of a recognition algorithm for AT-free graphs, it would also improve the best known time bound for triangle recognition.

A deeper look into the reduction of Spinrad provides a (conditional) lower bound for the time to recognize claw-free AT-free graphs.

Theorem 5. *If there is an $O(n^c)$ algorithm for recognizing claw-free AT-free graphs (of diameter 2) then there is an $O(n^{\max(2,c)})$ algorithm for recognizing triangle-free graphs.*

Proof. We take the opportunity to present the nice construction of Spinrad [23].

Let $G = (V, E)$ be any graph. Assume $V = \{v_1, v_2, \ldots, v_n\}$. A new graph H_G is constructed as follows. Take the graph G, add vertices $v'_1, v'_2, \ldots, v'_n$, and edges $\{v'_i, v'_j\}$ for all $1 \leq i, j \leq n$ and $\{v'_k, v_\ell\}$ for all $1 \leq k, \ell \leq n$ such that $k \neq \ell$. Note that $\{v'_1, v'_2, \ldots, v'_n\}$ is a clique in H_G. (Observe that the diameter of H_G is 2.)

Claim: G has three pairwise non-adjacent vertices, called an independent triple, if and only if H_G has a claw or an AT.

Suppose $\{v_i, v_j, v_k\}$ is an independent triple in G. Then $\{v_i, v_j, v_k\}$ is an AT in H_G which can be seen by inspecting the paths (v_i, v'_k, v_j), (v_i, v'_j, v_k), and (v_j, v'_i, v_k) in H_G. For the inverse direction of the claim to be shown suppose H_G has an AT $\{x, y, z\}$ or a claw $\{c, x, y, z\}$ with center vertex c. In each case $\{x, y, z\}$ is an independent triple in H_G and by the construction of H_G it follows immediately that $\{x, y, z\}$ is also an independent triple in G.

Now suppose that we have a $t(n)$ algorithm for recognizing claw-free AT-free graphs. The following algorithm then recognizes triangle-free graphs in time $O(n^2) + t(2n)$: On input G compute $\overline{G}$ and $H_{\overline{G}}$ in time $O(n^2)$. Then test whether $H_{\overline{G}}$ is claw-free AT-free. By assumption this can be done in time $t(2n)$. The correctness of the algorithm follows immediately from the above proven claim. $\qquad\square$

In the following we present an $O(n^\alpha)$ algorithm for recognizing claw-free AT-free graphs. According to Theorem 5 this is the best possible running time of an algorithm for recognizing claw-free AT-free graphs without strong consequences, i.e., without an improvement over the best known running time for triangle recognition.

Lemma 1. *Let $G = (V, E)$ be a claw-free AT-free graph. Let $L_0 = \{x\}, L_1 = N(x), L_2, \ldots, L_i = \{w \in V : d(w, x) = i\}, \ldots, L_r$ be the levels of a 2LexBFS scheme of G. Then the following statements hold:*

1. *L_i is a clique for all $i = 0, 2, 3, \ldots, r$. (L_1 might not be a clique.)*
2. *$\alpha(G[L_1]) \leq 2$.*

Proof. Clearly L_0 is a clique. Consider L_i with $i \geq 2$. Suppose $u, v \in L_i$ and $\{u, v\} \notin E$. Then u and v have a common neighbor w in L_{i-1} due to the DP-property of any 2LexBFS ordering of an AT-free graph. Clearly, w has a neighbor $z \in L_{i-2}$ and hence $\{w, u, v, z\}$ induces a claw in G, a contradiction.

Suppose $\alpha(G[L_1]) \geq 3$. Let $\{a, b, c\}$ be an independent set of $G[L_1]$. Then the set $\{x, a, b, c\}$ induces a claw in G, a contradiction. $\qquad\square$

Our recognition algorithm for claw-free AT-free graphs works as follows.

Algorithm RECOGNITION

1. Let $G = (V, E)$ be the input graph and let $V = \{v_1, v_2, \ldots, v_n\}$. Compute the binary adjacency matrix $A = (a_{ij})_{i,j=1,\ldots,n}$ of G (i.e., $a_{ij} = 1$ if and only if $\{v_i, v_j\} \in E$, thus $a_{ii} = 0$ for all $i \in \{1, 2, \ldots, n\}$). Compute A^2.

2. If all off diagonal entries of A^2 are equal to 1 (i.e., if $diam(G) \leq 2$) call TRIANGLE($\overline{G}$). If the subroutine reports a triangle then reject G. Otherwise accept G. {This completes the test of input graphs G with $diam(G) \leq 2$.}

3. Let $a_{ij}^2 = 0$ for some i and j, $i \neq j$, so $diam(G) \geq 3$. Call 2LEXBFS(G) such that first LEXBFS(G, v_i) is performed. Let S be the resulting 2LexBFS scheme, $x_n, x_{n-1}, \ldots, x_1$ be the 2LexBFS ordering, and $L_0, L_1, \ldots, L_r$ be the levels of S.

4. Compute another 2LexBFS scheme S' by calling LEXBFS(G, x_1). Let $x_1 = y_n, y_{n-1}, \ldots, y_1$ be the resulting 2LexBFS ordering and $L_0', L_1', \ldots, L_{r'}'$ be the corresponding levels.

5. If some level L_i or L_i', $i \neq 1$, is not a clique then reject G.

6. If either TRIANGLE($\overline{G}[L_1]$) or TRIANGLE($\overline{G}[L_1']$) reports a triangle then reject G.

7. Compute the *gentle squares* G^S and $G^{S'}$ of G with respect to S and S' in the following way. Make L_1 (respectively L_1') a clique and add all those edges $\{u, w\} \in E(G^2) \backslash E(G)$ to G for which $u \in L_i$ and $w \in L_{i+2}$ (respectively $u \in L_i'$ and $w \in L_{i+2}'$) for some i.

8. If $x_n, x_{n-1}, \ldots, x_1$ is not a cocomparability ordering of G^S or $y_n, y_{n-1}, \ldots, y_1$ is not a cocomparability ordering of $G^{S'}$ then reject G.

9. Otherwise accept G.

Theorem 6. RECOGNITION *recognizes claw-free AT-free graphs in time* $O(n^\alpha)$.

Proof. First consider the correctness. Suppose $diam(G) \leq 2$. We will argue that then G has an independent triple if and only if G has an AT or a claw. Since both AT and claw contain independent triples one direction of that claim is trivial. For the other direction let $\{a, b, c\}$ be an independent set of G. If $N(a) \cap N(b) \cap N(c) \neq \emptyset$ then $\{w, a, b, c\}$ induces a claw for each vertex $w \in N(a) \cap N(b) \cap N(c)$. Otherwise there are pairwise different vertices u, v, w such that $u \in N(a) \cap N(b)$, $v \in N(a) \cap N(c)$, and $w \in N(b) \cap N(c)$ since G has diameter 2. Thus $\{a, b, c\}$ is an AT of G. This proves the correctness of step 2 of our algorithm.

Assume $diam(G) \geq 3$. Consider the two 2LexBFS schemes S and S' computed by the algorithm. Assume both have the properties of Lemma 1. Note that Lemma 1 establishes the correctness of steps 5 and 6. We claim that then G is AT-free and any claw in G is a *standard claw* in S or S'. A standard claw in S (resp. S') is a claw $\{w, a, b, c\}$ with center w such that $w, b \in L_i$, $a \in L_{i-1}$, $c \in L_{i+1}$ (resp. $w, b \in L_i'$, $a \in L_{i-1}'$, $c \in L_{i+1}'$) for some $i \in \{2, 3, \ldots, r-1\}$.

Suppose $\{a, b, c\}$ is an AT in G. Then the construction of S and S', in particular $r, r' \geq 3$, guarantees that either $|L_1 \cap \{a, b, c\}| \leq 1$ or $|L_1' \cap \{a, b, c\}| \leq 1$. Without loss of generality assume $|L_1 \cap \{a, b, c\}| \leq 1$. Thus no two vertices of $\{a, b, c\}$ are in one level of S. Hence, if $a \in L_i$, $b \in L_j$ and $c \in L_k$ with $i < j < k$ then there is no path between vertices a and c avoiding $N(b) \supseteq L_j$. Thus $\{a, b, c\}$ is no AT. Consequently G is AT-free. Now assume that $\{w, x, y, z\}$ induces a claw with center w in G. Notice that the construction of S and S' guarantees that either $|L_1 \cap \{x, y, z\}| \leq 1$ or $|L_1' \cap \{x, y, z\}| \leq 1$. Without loss of generality assume $|L_1 \cap \{x, y, z\}| \leq 1$. Then $\{w, x, y, z\}$ is a standard claw in S since every L_i, $i \neq 1$, is a clique.

The standard claw test is done in steps 7 and 8. A standard claw $\{w, a, b, c\}$ in S (S') with $w, b \in L_i$, $a \in L_{i-1}$ and $c \in L_{i+1}$ is transformed into a forbidden configuration for which $\{a, c\}$ is an edge in the gentle square of G while $\{a, b\}$ and $\{b, c\}$ remain non-edges, and b is sandwiched between a and c in the 2LexBFS ordering. If the 2LexBFS ordering is not a cocomparability ordering of the gentle square then any violating triple $\{a, b, c\}$ must be from pairwise different levels, since all levels of the gentle square are cliques. Hence $a \in L_{i-1}$, $b \in L_i$ and $c \in L_{i+1}$. The fact that $\{a, c\}$ is an edge of the gentle square implies the existence of a common neighbor w of a and c in L_i. Since $i \geq 2$ we have that $\{w, b\} \in E$ and hence $\{w, a, b, c\}$ induces a standard claw in G.

Finally, consider the running time. All parts of the algorithm can obviously be done in time $O(n^\alpha)$ using triangle recognition or matrix multiplication, except the test for standard claws. The gentle square of G can be computed in time $O(n^\alpha)$ by using A^2. The test whether a 2LexBFS ordering is a cocomparability ordering for the gentle square can be done in $O(n^\alpha)$ by checking whether the digraph obtained by orienting the edges of the gentle square according to the 2LexBFS ordering (i.e., orient the edges from the smaller to the larger vertex) is transitive. Notice that testing whether a digraph is transitive is equivalent to matrix multiplication [11]. □

4   All Central Vertices

In this section we consider distance properties of claw-free AT-free graphs. Distances in graphs and related graph theoretic parameters such as diameter and radius play an important role in the design and analysis of networks in a variety of networking environments like communication networks, electric power networks, and transportation networks. Until now no fast algorithms for computing the diameter of an arbitrary graph, avoiding the computation of the whole distance matrix, have been designed. There is a collection of work on distance problems for graphs of some special graph class. We refer to [9,10] for additional references.

Let $G = (V, E)$ be a graph. The *distance* $d(u, v)$ between vertices u and v is the length of a shortest path from u to v. The *eccentricity* $e(v)$ of a vertex v is defined as $e(v) := \max_{u \in V} d(u, v)$. The *radius* $r(G)$ and the *diameter* $diam(G)$ are defined as $r(G) := \min_{v \in V} e(v)$ and $diam(G) := \max\{d(u, v) : u, v \in V\}$, respectively. Thus $diam(G) = \max_{v \in V} e(v)$. Finally, a vertex v is a *central vertex* of G if $e(v) = r(G)$. We will show that all these parameters and the set of all central vertices can be computed in linear time.

We start with another property of 2LexBFS schemes of claw-free AT-free graphs.

Lemma 2. *Let $L_0 = \{x\}, L_1 = N(x), \ldots, L_r$ be the levels of a 2LexBFS scheme of a claw-free AT-free graph G. Then $\alpha(G[L_i \cup L_{i+1}]) \leq 2$ for all $i \geq 0$. Each vertex $v \in L_1$ has a neighbor in L_2 unless $L_1 \subseteq N[v]$.*

Proof. By Lemma 1, each level L_i, $i \neq 1$, is a clique of G and $\alpha(G[L_1]) \leq 2$. Thus we obtain $\alpha(G[L_i \cup L_{i+1}]) \leq 2$ for all $i \neq 1$ and $\alpha(G[L_1 \cup L_2]) \leq 3$.

Suppose that $G[L_1 \cup L_2]$ contains an independent set $\{a, b, c\}$. Since L_2 is a clique we get $|L_2 \cap \{a, b, c\}| = 1$. Suppose $a, b \in L_1$ and $c \in L_2$. Clearly c has a neighbor $w \in L_1$. Thus $\alpha(G[L_1]) \leq 2$ implies either $\{w, a\} \in E$ or $\{w, b\} \in E$ but not both, since otherwise $\{w, a, b, c\}$ would induce a claw. Suppose both a and b have a neighbor in L_2, called a' and b' (possibly $a' = b'$), respectively. Since L_2 is a clique we obtain that $G[\{a, b, c, x, w, a', b'\}]$ is a graph of diameter 2. But graphs of diameter 2 containing an independent triple, $\{a, b, c\}$ in our case, always contain either a claw or an AT, a contradiction.

Hence the only remaining case to consider is, without loss of generality, that a has no neighbor in L_2. Thus $a \in L_1$, $N[a] \subset L_1$ and $b \in L_1 \setminus N[a]$. Thus $N[a] \subset N[x]$. Therefore for all vertices w of G, each execution of LexBFS(G, w) either assigns a label to x before it assigns one to a, or after LexBFS(G, w) assigns a label to a the set label(x) is lexicographically larger than label(b). Hence in both cases x can not be the last vertex of the execution of LexBFS(G, w) and thus no 2LexBFS ordering σ_2 of G yields $\sigma_2(x) = n$. Hence $L_0 \neq \{x\}$, a contradiction.

□

From now on let G be a claw-free AT-free graph. Let $L_0 = \{x\}, L_1 = N(x), \ldots, L_r$ be the levels of a 2LexBFS scheme of G.

Definition 2. *We call v a* predecessor *of w if there is a path $v = u_i, u_{i+1}, \ldots, u_k = w$ such that $u_j \in L_j$ for $j = i, i+1, \ldots, k$ and $i < k$. For every vertex $w \in L_i$, $1 \leq i \leq r$, we define $N^\uparrow(w) = N(w) \cap L_{i-1}$. For every vertex $w \in L_i$, $0 \leq i \leq r - 1$, we define $N_\downarrow(w) = N(w) \cap L_{i+1}$. Consequently, $d^\uparrow(w) = |N^\uparrow(w)|$ and $d_\downarrow(w) = |N_\downarrow(w)|$. We call $v \in L_i$, $0 \leq i \leq r$, a* good vertex *if $d(v, u) = r - i$ for all $u \in L_r$. Note in particular, that x is a good vertex.*

Define $A = \{u \in L_{r-1} : d_\downarrow(u) = |L_r|\}$ and $B = L_{r-1} \setminus A$. Let B_{max} be the set of vertices v from B that have a maximal down-neighborhood $N_\downarrow(v)$ (with respect to set inclusion) among all vertices in B.

Before turning to the main results of this section we state some helpful lemmas.

Lemma 3. *Let $u, v \in L_i$, $i \in \{1, \ldots, r - 1\}$, such that $N_\downarrow(u)$ and $N_\downarrow(v)$ are incomparable (with respect to set inclusion). Then $N^\uparrow(u) = N^\uparrow(v)$.*

Lemma 4. *Let $u \in L_i$ and $v \in L_j$ with $i, j \in \{0, 1, \ldots, r\}$ and $i \neq j$. Then $d(u, v) \in \{|i - j|, |i - j| + 1\}$. Furthermore, $diam(G) = r$.*

We now turn to the good vertices in a 2LexBFS scheme. The notion of a good vertex will be a useful tool to characterize the central vertices in a claw-free AT-free graph.

Lemma 5. *Let $r \geq 2$. Let $u \in B_{max}$ and let $a \in L_{r-2}$ such that $a \notin N^\uparrow(u)$ and $a \notin N^\uparrow(v)$ for all $v \in A$. Then a is not a good vertex.*

The next theorem gives a characterization of all good vertices of a 2LexBFS scheme of G. We will later show that essentially only the good vertices from certain levels of a 2LexBFS scheme of a claw-free AT-free graph are central vertices.

Theorem 7. *A vertex $v \in L_i$, $0 \leq i \leq r$, is good if and only if one of the following holds:*

1. *$i = r$ and $L_r = \{v\}$.*
2. *$i = r - 1$ and $v \in A$.*
3. *$i = r - 2$ and $v \in N^\uparrow(u)$ for a vertex $u \in A$, or $i = r - 2$, $\bigcup_{u \in B_{max}} N_\downarrow(u) = L_r$, and $v \in N^\uparrow(u)$ for some $u \in B_{max}$.*
4. *$i \leq r - 3$ and v is predecessor of a good vertex $u \in L_{r-2}$.*

Proof. Let $v \in L_i$, $i \geq 0$. The claim is immediate for $i \in \{r - 1, r\}$. Suppose $i = r - 2$. It is obvious that every $v \in N^\uparrow(u)$ for a vertex $u \in A$ is a good vertex. Let $\bigcup_{u \in B_{max}} N_\downarrow(u) = L_r$ and $v \in N^\uparrow(u)$ for a vertex $u \in B_{max}$. Observe that in light of Lemma 3 it holds that for all vertices $u, u' \in B_{max}$, $N^\uparrow(u) = N^\uparrow(u')$. Thus v is not only a predecessor of u but of all $u' \in B_{max}$. Hence, if $\bigcup_{u \in B_{max}} N_\downarrow(u) = L_r$ then v is a good vertex. It remains to show that no other good vertices exist in L_{r-2}. So suppose that $a \in L_{r-2}$ is a good vertex and $a \notin N^\uparrow(u)$ for all $u \in A \cup B_{max}$. Hence a satisfies the assumptions of Lemma 5 and thus can not be a good vertex, a contradiction.

Suppose $i \leq r - 3$. Clearly, each predecessor of a good vertex in L_{r-2} is a good vertex. It remains to show that there are no other good vertices. Suppose that a is a good vertex in L_i, $i \leq r - 3$, and a is not a predecessor of a good vertex $u \in L_{r-2}$. There are two cases to be distinguished. The first case is that the only good vertices in L_{r-2} are predecessors of vertices in A. In that case there exists a vertex $b \in L_r$ satisfying $N^\uparrow(b) = A$. But then for any vertex $a \in L_i$, $d(a, b) = r - i$ if and only if a is a predecessor of a vertex $u \in A$. The second case is that there are good vertices in L_{r-2} not being predecessors of vertices in A. We have already shown that such good vertices in L_{r-2} have to be predecessors of some vertex $u \in B_{max}$. Let $u_{max} \in B_{max}$ be a vertex with maximal down-degree among all vertices in B_{max}. Since $B_{max} \subseteq B = L_{r-1} \setminus A$ there exists a vertex $b \in L_r$ such that $b \notin N_\downarrow(u_{max})$. Now suppose that a is a good vertex but a is not a predecessor of any $u \in A \cup B_{max}$. Hence a is not a predecessor of u_{max}, yet, since a is good, $d(a, b) = r - i$. It follows that there have to exist vertices $a' \in L_{r-2}$ and $a'' \in L_{r-1}$ such that $d(a, a') = r - i - 2$, $\{a', a''\}, \{a'', b\} \in E$, and neither a' nor a'' are good vertices (since otherwise a would be a predecessor of a good vertex from L_{r-2}). In particular, $\{a', u_{max}\} \notin E$. Hence $\{a'', a', u_{max}, b\}$ induces a claw with center vertex a'', a contradiction. $\square$

Corollary 1. *Let G be a connected claw-free AT-free graph.*
Then either $r(G) = diam(G) = 2$ or $r(G) = \lceil diam(G)/2 \rceil$. Hence the radius of a claw-free AT-free graph can be computed in linear-time by 2LexBFS.

Theorem 8. *There exists a linear-time algorithm that given a 2LexBFS scheme of a claw-free AT-free graph computes all good vertices.*

The algorithm exploits the characterization of good vertices given in Theorem 7. Now we are prepared to state a characterization of all central vertices in a claw-free AT-free graph.

Theorem 9. *Let G be a claw-free AT-free graph. Let $L_0 = \{x\}, L_1 = N(x), \ldots, L_r$ be the levels of a 2LexBFS scheme of G. Then v is a central vertex of G if and only if one of the following holds:*

1. *$r = 0$ or $r = 1$.*
2. *$r = 2$ and $d(v) = n - 1$, or $r = 2$ and $d(u) < n - 1$ for all $u \in V$.*
3. *$r = 2k$, for some integer $k \geq 2$, $v \in L_k$, and v is a good vertex.*
4. *$r = 2k + 1$, for some $k \geq 1$, $v \in L_k$ and v is a good vertex, or $v \in L_{k+1}$.*

The main idea of our linear-time algorithm for computing all central vertices is to determine all good vertices in those levels that potentially (see Theorem 9) contain central vertices.

Theorem 10. *There exists a linear-time algorithm for computing all central vertices of a claw-free AT-free graph.*

5   Independence, Domination, and Coloring

5.1   Independent Set

As mentioned earlier there exists an $O(n^4)$ algorithm for computing a maximum independent set for AT-free graphs [5]. There is also a polynomial-time algorithm for computing a maximum independent set for claw-free graphs [19]. Notice that the latter problem contains the well-known MATCHING problem as a subproblem.

Let $G = (V, E)$ be a claw-free AT-free graph and $L_0, L_1, \ldots, L_r$ be the levels of a 2LexBFS scheme of G. Let I be any independent set of G. Then Lemma 1 and Lemma 2 imply $|L_i \cap I| \leq 1$ if $i \neq 1$, $|L_1 \cap I| \leq 2$ and $|(L_{i-1} \cup L_i) \cap I| \leq 2$. Using this observation it can be shown that the search for a maximum independent set can be restricted to independent sets containing at most one vertex per level. This leads to a reachability type problem on an auxiliary graph $G^* = (V, E^*)$, where E^* is the set of all non-edges between any two consecutive levels in the 2LexBFS scheme of G, that can be solved in time linear in the size of G^*. Showing $|E^*| = O(|E|)$ we obtain

Theorem 11. *There is a linear-time algorithm for computing a maximum independent set for claw-free AT-free graphs.*

5.2   Dominating Set

It has been shown that a minimum dominating set in AT-free graphs can be computed in time $O(n^6)$ [17]. In contrast, computing a minimum dominating set in claw-free graphs is NP-complete since EDGE DOMINATING SET which is equivalent to DOMINATING SET on line graphs is NP-complete (see the comments to problem [GT2] in [13]).

Lemma 6. *Let $G = (V, E)$ be a claw-free AT-free graph and $L_0, L_1, \ldots, L_r$ be the levels of a 2LexBFS scheme of G. Let $u \in L_i$ and $v \in L_{i+2}$, $i \in \{0, 1, \ldots, r - 2\}$ such that u and v have a common neighbor $w \in L_{i+1}$. Then $L_{i+1} \subseteq (N[u] \cup N[v])$.*

Hence there is a dominating set of G containing one vertex of every level L_i, for i even or $i = r$. On the other hand, for every dominating set D of G and for every three consecutive levels L_i, L_{i+1}, and L_{i+2}, at least one of these levels must contain a vertex of D. The only thing left for the algorithm is to test whether for any two consecutive levels there is a dominating set without a vertex in these two levels.

Theorem 12. *There is a linear-time algorithm for computing a minimum dominating set and a minimum independent dominating set for claw-free AT-free graphs.*

5.3 Coloring

The COLORING problem is NP-complete for claw-free graphs since it remains NP-complete on line graphs which form a subclass of claw-free graphs [15]. In contrast, the algorithmic complexity of the COLORING problem for AT-free graphs is still an open problem. Let $t_{\text{MATCH}}(n, m)$ and $t_{\text{BIPMATCH}}(n, m)$ be the times needed to compute a maximum matching on general graphs and on bipartite graphs, respectively. Similarly, let $t_{\text{COLOR}}(n, m)$ denote the time to color a claw-free AT-free graph.

Theorem 13. [2] *There is an $O(n^2 + \sqrt{n}m)$ coloring algorithm for claw-free AT-free graphs. Furthermore, $t_{\text{COLOR}}(n, m) \leq \max\{t_{\text{MATCH}}(n, m),$ $t_{\text{BIPMATCH}}(2n, m)\} + O(n^2)$ and $t_{\text{BIPMATCH}}(n, m) \leq t_{\text{COLOR}}(n, m) + O(n^2)$.*

Proof. Let G be a claw-free AT-free graph. The following algorithm colors G with a minimum number of colors: Compute a 2LexBFS scheme of G. Let $L_0, L_1, \ldots, L_r$ be its levels. If $r \leq 2$ then by Lemma 2, $\alpha(G) \leq 2$ and we thus exploit the well-known fact that a minimum coloring of a graph G with $\alpha(G) \leq 2$ can be determined by computing a maximum matching M of $\overline{G}$ and by coloring G such that two vertices u and v of G obtain the same color if and only if $\{u, v\}$ is an edge of the maximum matching M. If $r > 2$ we either have that L_1 is a clique, in which case the 2LexBFS ordering of G is a cocomparability ordering of G since now all levels are cliques, or L_1 is not a clique implying that $\alpha(G) \geq 3$ and thus G is cocomparability due to Theorem 3. So we use an algorithm solving the COLORING problem for cocomparability graphs by a reduction to MATCHING on a bipartite auxiliary graph [12] (see also [22, pp. 232–238]). It is not hard to verify that the just outlined algorithm runs in time $\max\{t_{\text{MATCH}}(n, m), t_{\text{BIPMATCH}}(2n, m)\} + O(n^2)$ which proves the first inequality. Due to the fact that the current best algorithm for maximum matching on graphs has running time $O(\sqrt{n}m)$ [18] it follows that there exists a $O(\sqrt{n}m + n^2)$ coloring algorithm for claw-free AT-free graphs.

For the other inequality to be proven simply observe that the complement of a bipartite graph G satisfies $\alpha(\overline{G}) \leq 2$ and thus due to Theorem 3 $\overline{G}$ is a claw-free cocomparability graph. So an optimal coloring of $\overline{G}$ induces a maximum matching on G. □

---

[2] A polynomial-time coloring algorithm for claw-free AT-free graphs was independently found by E. Köhler.

Though the above theorem might give the impression that full complementation of the input graph, and thus an extra $O(n^2)$ in the running time, is necessary in order to apply matching algorithms we emphasize that this is not true. The authors have found an algorithm for coloring claw-free AT-free graphs that has running time $O(\sqrt{n}m)$.

References

1. N. Alon, R. Yuster, and U. Zwick, Finding and counting given length cycles, *Algorithmica* **17** (1997), 209–223.
2. H. Alt, N. Blum, K. Mehlhorn, and M. Paul, Computing a maximum matching in a bipartite graph in time $O(n^{1.5}\sqrt{m/\log n})$, *Inform. Proc. Lett.* **37** (1991), 237–240.
3. A. Brandstädt, Special graph classes — a survey, Schriftenreihe des Fachbereichs Mathematik, SM-DU-199, Universität Gesamthochschule Duisburg, 1993.
4. A. Brandstädt, F. Dragan, and E. Köhler, Efficient algorithms for Hamiltonian problems on (claw, net)-free graphs, *Proc. of WG'99*, To appear.
5. H. Broersma, T. Kloks, D. Kratsch, and H. Müller, Independent sets in asteroidal triple-free graphs, *Proceedings of ICALP'97*, Springer-Verlag, LNCS 1256, 1997, Berlin, 760–770.
6. D.G. Corneil, F.F. Dragan, M. Habib, and C. Paul, Diameter determination on restricted graph families, *Proc. of WG'98*, Springer-Verlag, LNCS 1517, 1998, 192–202.
7. D.G. Corneil, S. Olariu, and L. Stewart, Asteroidal triple-free graphs, *SIAM J. Dis. Math.* **10** (1997), 399–430.
8. D.G. Corneil, S. Olariu, and L. Stewart, A linear time algorithm for dominating pairs in asteroidal triple-free graphs, *Proc. of ICALP'95*, Springer-Verlag, LNCS 944, 1995, 292–302.
9. V. Chepoi and F. Dragan, A linear-time algorithm for finding a central vertex of a chordal graph, *Proc. of ESA'94*, Springer-Verlag, 1994, Berlin, LNCS 855, 159–170.
10. V. Chepoi and F. Dragan, Finding a central vertex in HHD-free graphs, Preprint, University of Rostock, 1998.
11. M.J. Fischer and A.R. Meyer, Boolean matrix multiplication and transitive closure, *Proc. of the Annual Symposium on Switching and Automata Theory*, 1971, 129–131.
12. D.R. Fulkerson, Note on Dilworth's decomposition theorem for partially ordered sets, *Proc. Amer. Math. Soc.* **7** (1956), 701–702.
13. R.M. Garey and D.S. Johnson, *Computers and Intractability: A Guide to the Theory of NP-completeness*, W.H. Freeman, San Francisco, 1979.
14. M.C. Golumbic, *Algorithmic Graph Theory and Perfect Graphs*, Academic Press, 1980.
15. I. Holyer, The NP-completeness of edge-coloring, *SIAM J. Comput.* **10** (1981), 718–720.
16. T. Kloks, D. Kratsch, and H. Müller, Approximating the bandwidth for AT-free graphs, *Proc. of ESA'95*, Springer-Verlag, LNCS 979, 1995, 434–447.
17. D. Kratsch, Domination and total domination on asteroidal triple-free graphs, *Forschungsergebnisse Math/Inf/96/25*, FSU Jena, Germany, 1996.
18. S. Micali, V.V. Vazirani, An $O(V^{1/2}E)$ algorithm for finding maximum matching in general graphs, *Proc. of FOCS'80*, New York, 1980, 17–25.

19. G.J. Minty, On maximal independent sets of vertices in claw-free graphs, *J. Comb. Theory, Ser. B* **28** (1980), 284–304.
20. A. Parra and P. Scheffler, Characterizations and algorithmic applications of chordal graph embeddings, *Disc. App. Math.* **79** (1997), 171–188.
21. D.J. Rose, R.E. Tarjan and G.S. Lueker, Algorithmic aspects of vertex elimination on graphs, *SIAM J. Comput.* **5** (1976), 266–283.
22. K. Simon, *Effiziente Algorithmen für perfekte Graphen*, B.G. Teubner, Stuttgart, 1992.
23. J. Spinrad, private communication.
24. D. West, *Introduction to graph theory*, Prentice Hall, Upper Saddle River, 1996.

Vertex Partitioning of Crown-Free Interval Graphs

Giuseppe Confessore[1], Paolo Dell'Olmo[2], and Stefano Giordani[1]

[1] Dipartimento di Informatica, Sistemi e Produzione,
Università di Roma "Tor Vergata"
Via di Tor Vergata 110, I-00133 Roma, Italy
{confessore, giordani}@disp.uniroma2.it
[2] Dipartimento di Probabilità, Statistica e Statistiche Applicate,
Università di Roma "La Sapienza"
Piazzale Aldo Moro 5, I-00185 Roma, Italy
dellolmo@iasi.rm.cnr.it

Abstract. We study the problem of finding an acyclic orientation of an undirected graph G such that each path is contained in a limited number of maximal cliques of G. In general, in an acyclic oriented graph, each path is contained in more than one maximal cliques. We focus our attention on crown-free interval graphs, and show how to find an acyclic orientation of such a graph, which guarantees that each path is contained in at most four maximal cliques. The proposed technique is used to find approximated solutions for a class of related optimization problems where a solution corresponds to an acyclic orientation of graphs.

Keywords: Dynamic storage allocation, interval graphs, interval coloring, acyclic orientation, approximation algorithm.

1 Introduction

We study the problem of finding an acyclic orientation of a undirected graph $G = (V, E)$ such that each path is contained in a limited number of maximal cliques of G. This problem is related to a wide class of optimization problems (i.e., scheduling, coloring, packing), in which each feasible solution corresponds to an acyclic orientation of a weighted graph, and the solution value is equal to the length of the longest weighted path of the oriented graph. Optimal solutions correspond to acyclic orientations having the length of the longest weighted path as small as possible; we call these orientations optimal. Any clique of the graph, once oriented, produces a path of length equal to the weight of the clique, therefore the weight of the maximum weighted clique is a valid lower bound on the optimal solution value. If G is a comparability graph, then it admits a transitive (and hence acyclic) orientation that can be found in polynomial time (Golumbic [13]), in which, by transitivity, each path is contained in a maximal clique of G, hence any such orientation is optimal. In all other cases, even for a perfect graph G, in any acyclic orientation there exists at least a path contained in more

Widmayer et al. (Eds.): WG'99, LNCS 1665, pp. 391–401, 1999.

than one maximal clique, and finding the optimal orientation is a strongly $\mathcal{NP}$-hard problem (Dell'Olmo et al. [8]). For such graphs, one can attempt to obtain in polynomial time paths whose length can be limited. This could be achieved, for example, by providing an orientation for which one can guarantee that any path is contained in a limited number k of maximal cliques. Therefore, the value of this orientation would be at most k times the optimum, for any vertex weight function. In particular, for the class of claw-free chordal graphs, Confessore et al. [4,6] provided an acyclic orienting algorithm which assures that any path is contained in at most 2 maximal cliques.

In this paper, we focus our attention on a subclass of interval graphs. Finding an optimal orientation of an interval graph is polynomially equivalent to find the optimal solution of a number of optimization problems such as interval coloring problem on interval graphs [13], ship building problem [13], bandwidth allocation problem (Habib [14]), and dynamic storage allocation problem (Coffman [2]). In such problems a feasible solution corresponds to an acyclic orientation of a weighted interval graph.

In particular, in the dynamic storage allocation problem it is given a set of items to be stored and, for each item, an arrival time, a departure time, and a size are known. Two items are in conflict if the arrival time of one of them is included between the arrival and departure times of the another one. Two in conflict items have to be stored in different storage intervals. The objective of the problem is to allocate all items in order to minimize the total used storage size. Any problem instance can be represented by a conflict graph, namely a weighted interval graph, in which the vertex set corresponds to the set of items to be stored, each vertex weight is equal to the size of the corresponding item, and there is an edge between two vertices if and only if the corresponding items are in conflict. The objective of the problem, represented by a weighted graph, is to find an acyclic orientation of the graph where the length of longest weighted oriented path is as small as possible.

In the general case in which the conflict graph is an interval graph, the dynamic storage allocation problem is known to be $\mathcal{NP}$-hard in strong sense (Garey and Johnson [10]). Approximation results for this problem have been given by Kierstead [15], with performance ratio 6, and by Gergov [11], with performance ratio 5. If the conflict graph is a proper interval graph, that is a $K_{1,3}$-free interval graph, the dynamic storage allocation problem is $\mathcal{NP}$-hard (Confessore et al. [3,7]). For this special case, Confessore et al. [3,7] and Naor et al. [17] independently gave 2-approximation algorithms. However, the performance ratio of the approximated solution in [3,7] holds for any weight function, that is the algorithm can be applied also (or especially) when weights are dynamic random variables or are completely unknown. This result is achieved by using a partitioning technique. The same technique has also been used by Confessore et al. [5] to provide an approximation algorithm for a periodic allocation problem.

In this paper we study the case in which the conflict graph is a $K_{1,1,3}$-free interval graph, that is a crown-free interval graph. This graph class properly contains the class of proper interval graphs, then, by restriction, the dynamic storage allocation problem in the case in which the conflict graph is a crown-free interval graph is $\mathcal{NP}$-hard.

The main contributions of our work are as follows. We show how to partition the vertex set of a crown-free interval graph into two disjoint subsets, such that each one induces a proper interval subgraph. Then, we partition the vertex set of each proper interval subgraph into cliques. Based on this partition, we show how to acyclically orient the whole graph, such that each path in the oriented graph is contained in at most four maximal cliques. This latter result allows us to state that there exists a 4-approximation algorithm for the interval coloring problem of crown-free interval graphs, and then for the related dynamic storage allocation problem. Moreover, with respect to previous results in [11,15,17], the performance ratio of our solution holds even for the case in which vertex weights are unknown.

2   Basic Notations and Definitions

A *graph* is a pair $G = (V, E)$, where V is a finite set of $n = |V|$ elements called *vertices*, and $E \subseteq \{(x, y) \in V \times V \mid x \neq y\}$ is a set of $m = |E|$ *unordered* vertex pairs called *edges*. For two distinct vertices x, y, we say that x is *adjacent* to y (or equivalently, y is adjacent to x) if $(x, y) \in E$; otherwise, they are *independent*.

Given $G = (V, E)$ and a subset $X \subseteq V$, the graph $G(X) = (X, E(X))$ is the *subgraph* of G *induced* by X if $E(X) = \{(x, y) \in E \mid x, y \in X\}$.

A set $V' \subseteq V$ of vertices is a *clique* of G if the subgraph $G(V')$ of G induced by V' is a complete graph. A *maximal clique* is a clique not properly contained in any other clique. A *maximum clique* is a (maximal) clique with largest number of vertices among all cliques.

A *weighted graph* G is a graph with an associated weight function $w : V \to N \cup \{0\}$ which assigns a non negative integer weight to each vertex of V. For vertex $x, w(x)$ is the weight of x. Vertices of zero weight are irrelevant in this context, but technically it is convenient to allow their presence in some cases. The weight $w(S)$ of a set $S \subset V$ is defined as $\sum_{x \in S} w(x)$. A *maximum weighted clique* of G is a clique with largest weight among all the (maximal) cliques; let $\omega_w(G)$ be the weight of the maximum weighted clique of G.

If we orient an edge $(x, y) \in E$ of $G = (V, E)$ from x to y, we write $(x \to y)$, and the oriented edge is called *arc*. If all edges in E are oriented, the resulting graph is called an *oriented* (or *directed*) graph.

A *path* in an oriented graph is a sequence of vertices $(x_1, x_2, \ldots, x_k)$ of V such that $(x_1 \to x_2), (x_2 \to x_3), \ldots, (x_{k-1} \to x_k)$.

An *orientation* of a graph $G = (V, E)$ is the set of arcs A (or the oriented graph $H = (V, A)$) such that $|A| = |E|$, and for each $(x, y) \in E$ either $(x \to y) \in A$, or $(y \to x) \in A$. An orientation $H = (V, A)$ of $G = (V, E)$ is *acyclic* if there is no path $(x_1, x_2, \ldots, x_k)$ in H such that $x_1 \equiv x_k$.

An orientation $H = (V, A)$ of $G = (V, E)$ is *transitive* if for each $(x \to y), (y \to z) \in A$ then $(x \to z) \in A$; hence, a transitive orientation is also acyclic. A graph $G = (V, E)$ which is transitive orientable is called a *comparability graph*.

A graph G is *chordal* if each cycle in G of length at least 4 has a chord.

The *intersection graph* of a finite family of non-empty sets is a graph where each vertex represents a set, and two vertices are joined by an edge if and only if the corresponding sets have a non-empty intersection. The intersection graph of a family of intervals of a linear ordered set (like the real line) is an *interval graph*. Gilmore and Hoffman [12] showed that an undirected graph G is an interval graph if and only if G is chordal and its complement $\bar{G}$ is a comparability graph.

Let $K_{1,3}$ be the *claw* graph, that is a graph $G = (\{a, b, c, d\}, E)$ with $E = \{(a, b), (a, c), (a, d)\}$. The vertex a is called the *center* of the claw.

A graph G is a *proper interval* graph if it is interval and *claw-free*, that is there is no subgraph of G isomorphic to the claw graph.

Let $K_{1,1,3}$ be the *crown* graph (see Faudree et al. [9]), that is the *complete multipartite* graph $G = (\{a\} \cup \{e\} \cup \{b, c, d\}, E)$ with $E = \{(a, b), (a, c), (a, d), (a, e), (e, b), (e, c), (e, d)\}$ (see Figure 1(a)).

A graph G is a *crown-free interval* graph if it is interval and there is no subgraph of G isomorphic to the crown graph.

Finally, let $S(K_{1,3})$ (see [9]) be the graph obtained from a claw $K_{1,3}$ by subdividing each edge (see Figure 1(b)).

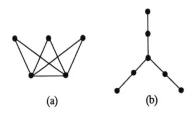

(a)   (b)

Fig. 1. The $K_{1,1,3}$ (or *crown*) graph (a), and the $S(K_{1,3})$ graph (b)

3   Partitioning Crown-Free Interval Graphs

In this section we study how to partition the vertex set V of a crown-free interval graph $G = (V, E)$ in two disjoint subsets V_1, V_2, such that the induced subgraphs $G(V_1), G(V_2)$ are proper interval graphs.

Let us start studying a property of an interval graph G.

Theorem 1. *Given a claw subgraph $G(\{a, b, c, d\})$, with center a, of an interval graph $G = (V, E)$, there exists at least one vertex $u \in \{b, c, d\}$, such that for each maximal clique $\check{C}$ of G, with $u \in C$, it also results that $a \in C$.*

Proof. Given a claw subgraph $G(\{a, b, c, d\})$, with center a, of an interval graph $G = (V, E)$, let us suppose, by contradiction, that there exist three maximal cliques C_b, C_c, C_d for which $b \in C_b$, $c \in C_c$, $d \in C_d$, and $a \notin C_b \cup C_c \cup C_d$. Since C_b, C_c, C_d are maximal cliques, there will be three vertices $x, y, z \in (V \backslash \{a, b, c, d\})$ such that $x \in (C_b \backslash (C_c \cup C_d))$, $y \in (C_c \backslash (C_b \cup C_d))$, $z \in (C_d \backslash (C_c \cup C_b))$, and $(a, x), (a, y), (a, z) \notin E$. In particular, x, y, z are distinct vertices; in fact if, by contradiction, $x \equiv y$, $G(\{a, b, c, x\})$ is a chordless cycle of length 4, and this is not possible because G is chordal (in the same way we prove that $x \not\equiv z$, $y \not\equiv z$). Moreover, since G is chordal, $(b, y), (b, z), (c, x), (c, z), (d, x), (d, y), (x, y), (x, z), (y, z) \notin E$ otherwise $G(\{a, b, c, d, x, y, z\})$ contains a chordless cycle of length at least 4. Then, the subgraph $G(\{a, b, c, d, x, y, z\})$ is a $S(K_{1,3})$ graph, but this is not possible because the $S(K_{1,3})$ graph is a forbidden structure for an interval graph G (Lekkerkerker and Boland, [16]). □

According to the previous theorem, let us give the following definition.

Definition 1. *Given a claw subgraph $G(\{a, b, c, d\})$, with center a, of an interval graph $G = (V, E)$, an embedded vertex of the claw is a vertex $u \in \{b, c, d\}$ such that, for each maximal clique C of G containing u, it results also that a belongs to C.*

Then, Theorem 1 states that, for each claw subgraph of an interval graph G, there exists at least one embedded vertex. By using this property, let us now consider the following algorithm that partitions the vertex set V of an interval graph G into two disjoint subsets V_1 and V_2.

Algorithm 1. *Partition*

```
Input: G = (V, E);
Output: V₁, V₂;
{
    let V₁ := V, and V₂ := ∅;
    while there exists a claw subgraph K of G(V₁);
    {
        let b ∈ V₁ be an embedded vertex of the claw subgraph K;
        let V₁ := V₁\{b} and V₂ := V₂ ∪ {b};
    }
}
```

By using Algorithm 1, it results that $G(V_1)$ is claw-free, and that, $\forall u \in V_2$, u is an embedded vertex of a claw of G. Note that, even if G is a connected graph, $G(V_1)$ and/or $G(V_2)$ could be disconnected. As for time complexity, the algorithm runs in polynomial time. Roughly speaking, a non optimized version of the algorithm, running in $O(n^5)$ time, can be implemented by using the procedure for recognizing claw-free graphs, which requires $O(n^4)$ time (see Alon and Boppana [1]), while, given a claw, an its embedded vertex can be found in linear time. A more efficient algorithm, that makes use of interval graph properties, might be designed, but this is out of the scope of the paper.

Lemma 1. *Given an interval graph $G = (V, E)$ and a claw subgraph $G(\{a, b, c, d\})$ of center a, if a is an embedded vertex of another claw $G(\{a', a, c', d'\})$ of center a', it results $a' \notin \{b, c, d\}$.*

Proof. Let $G(\{a', a, c', d'\})$ be a claw, where a is an embedded vertex and a' is the center. By contradiction, let us suppose that $a' \equiv b$. Since $G(\{a, b, c, d\})$ is a claw, we have that $c', d' \notin \{c, d\}$. Being $(a, c), (a, d) \in E$, and a an embedded vertex of the claw $G(\{b, a, c', d'\})$ of center b, by Theorem 1 it results that $(b, c), (b, d) \in E$; but this is not possible because $G(\{a, b, c, d\})$ is a claw. In the same way we prove that $a' \notin \{c, d\}$. □

Proposition 1. *Given an interval graph $G = (V, E)$, if there exists a claw $G(\{a, b, c, d\})$ of center a, and a claw $G(\{e, a, f, g\})$ of center e and embedded vertex a, then the subgraph $G = (\{a\} \cup \{e\} \cup \{b, c, d\})$ is a crown graph.*

Proof. Firstly, let us show that the vertices a, b, c, d, e, f are distinct. By Lemma 1, the center e of the second claw is not contained in $\{b, c, d\}$; moreover, since $G(\{a, b, c, d\})$ is a claw of center a, and hence $(a, b), (a, c), (a, d) \in E$, if there exists another claw $G(\{e, a, f, g\})$ of center e, it result that $g, f \notin \{b, c, d\}$.

By Theorem 1, since a is an embedded vertex of $G(\{e, a, f, g\})$, $(b, e), (c, e)$, $(d, e) \in E$. Since $(a, b), (a, c), (a, d)$ $(a, e), (e, b), (e, c), (e, d) \in E$, and (b, c), $(b, d), (c, d) \notin E$, it results that $G = (\{a\} \cup \{e\} \cup \{b, c, d\})$ is a crown graph. □

By Proposition 1 and by applying Algorithm 1, it is possible to prove that:

Theorem 2. *Given a crown-free interval graph $G = (V, E)$, it is possible to partition the vertex set V into two disjoint subsets V_1 and V_2, such that the subgraphs $G(V_1)$ and $G(V_2)$ are proper interval graphs.*

Proof. By using Algorithm 1, each vertex $b \in V_2$ is an embedded vertex of a claw K. Since there is no subgraph isomorphic to the crown graph in G, by Proposition 1, b cannot be the center of any other claw; therefore, there is no claw in $G(V_2)$. Moreover, since by construction $G(V_1)$ does not contains claws, we have that both $G(V_1)$ and $G(V_2)$ are claw free, i.e. proper interval graphs. □

3.1   Partitioning Proper Interval Graphs

In this subsection we analyze a proper interval graph $G_P = (V_P, E_P)$, and we show how to partition the vertex set V_P into cliques of G_P. Then, we study adjacency properties between partition elements.

It is known that:

Theorem 3. (Gilmore and Hoffman [12]) *An undirected graph G is an interval graph if and only if the maximal cliques of G can be linearly ordered such that, for every vertex x of G, the maximal cliques containing x occur consecutively.*

Obviously, this also holds for a proper interval graph G_P. Let $Q = (C_1, \ldots, C_q)$ be the linearly ordered list of all the maximal cliques of G_P, according to Theorem 3.

Lemma 2. *Let $G_P = (V_P, E_P)$ be an interval graph. G_P is proper if and only if, for any vertex $x \in V_P$ contained in $k + 1$ maximal cliques $(C_h, \ldots, C_{h+k})$ of G_P, with $k \geq 2$, any vertex $y \in C_j$, with $h < j < (h + k)$, belongs, at least, to C_h or to C_{h+k}.*

Proof. Let us consider a vertex x and let $(C_h, \ldots, C_{h+k})$, with $k \geq 2$, be the ordered list of all the maximal cliques containing it. Let us suppose, by contradiction, that there exists a vertex $y \in C_{h+t}$, with $0 < t < k$, such that $y \notin C_h \cup C_{h+k}$. Being $C_h, \ldots, C_{h+k}$ maximal cliques, there exists a vertex $u \in V_P$, such that $u \in C_h$ and $u \notin C_{h+1}$, and a vertex $v \in V_P$, such that $v \notin C_{h+k-1}$ and $v \in C_{h+k}$; note that $(u, v) \notin E_P$, because there is no maximal clique containing both u and v. For the same reason, $(u, y) \notin E_P$ and $(y, v) \notin E_P$, hence G_P contains a claw of vertices x, u, y and v, and this contradicts the hypothesis that G_P is proper.

Conversely, let us suppose that G_P is not proper, and then there exists in G_P a claw of vertices x, u, y, v. W.l.o.g., let x be the center of the claw, and let $(C_h, \ldots, C_{h+k})$ be the ordered list of all the maximal cliques containing x. As u, y, v are adjacent to x, there exist at least three maximal cliques C_{h+i}, C_{h+t}, C_{h+k-j}, containing x, which contain u, y, v, respectively. Moreover, since $\{u, y, v\}$ is a set of mutually independent vertices, without loss of generality we can consider $0 \leq i < t < k - j$, where $j \geq 0$ and $k \geq 2$; hence, as all the maximal cliques containing y are listed consecutively, $y \notin C_h \cup C_{h+k}$, contradicting the statement of the lemma. □

W.l.o.g., let $G_P = (V_P, E_P)$ a connected proper interval graph. Given $Q = (C_1, \ldots, C_q)$, the list of all the maximal cliques of G_P, rearranged according to Theorem 3, let $\mathcal{P} = (X_1, \ldots, X_r)$ be a vertex partition of V_P obtained as follows:

- $X_1 = C_{j_1}$, where $j_1 = 1$;
- $X_i = C_{j_i} \setminus C_{j_{i-1}}$, where $j_i = \max_{k > j_{i-1}} [k : \exists x \in (C_{j_{i-1}} \cap \cdots \cap C_k)]$, for $i = 2, \ldots, r - 1$;
- $X_r = C_{j_r} \setminus C_{j_{r-1}}$, where $j_r = q$.

Theorem 4. *If $G_P = (V_P, E_P)$ is a proper interval graph then $\mathcal{P} = (\mathcal{X}_\infty, \ldots, \mathcal{X}_\nabla)$ is a partition of V_P.*

Proof. Given $Q = (C_1, \ldots, C_q)$, by construction it follows that the elements of $\mathcal{P}$ are mutually disjoint. Therefore, we have only to show that $\mathcal{P}$ is a covering of V_P. Let us suppose that $\mathcal{P}$ is not a covering, that is there exists at least a vertex y not covered by $\mathcal{P}$. In this case, following the procedure used to obtain $\mathcal{P}$, there exists at least a maximal clique C_t containing y, such that $j_i < t < j_{i+1}$, and, being y not covered by $\mathcal{P}$, $y \notin (C_{j_i} \cup C_{j_{i+1}})$; moreover, by construction of $\mathcal{P}$, there exists a vertex $x \in (C_{j_i} \cap \ldots \cap C_{j_{i+1}})$. Hence, by Lemma 2, G_P is not proper. □

Moreover, we can show that:

Theorem 5. *Given the partition* $\mathcal{P} = (X_1, \ldots, X_r)$ *of* $G_P = (V_P, E_P)$, *there is no couple of partition elements* (X_i, X_{i+k}), *with* $k \geq 2$, *such that there exist two vertices* $x \in X_i$ *and* $y \in X_{i+k}$, *with* $(x, y) \in E_P$.

Proof. Given $Q = (C_1, \ldots, C_q)$, by construction, for $k \geq 2$, it results $X_{i+k} \cap C_{j_{i+k-1}} = \emptyset$, and it is simple to verify that $C_{j_i} \cap C_{j_{i+k}} = \emptyset$, hence $C_{j_i} \cap X_{i+k} = \emptyset$. By contradiction, suppose that there exist two vertices $x \in X_i$ and $y \in X_{i+k}$ such that $(x, y) \in E_P$. In this case, since the maximal cliques containing x occur consecutively, and knowing that both $X_{i+k} \cap C_{j_{i+k-1}}$ and $X_{i+k} \cap C_{j_i}$ are empty, there must be a maximal clique C_ρ containing both x and y, with $j_i < \rho < j_{i+k-1}$; but this implies that the maximal cliques containing y are not consecutively ordered. $\square$

Summarizing, given a crown-free interval graph $G = (V, E)$, it is possible to partition the vertex set V in two disjoint subsets V_1, V_2, such that $G(V_1)$ and $G(V_2)$ are proper interval graphs. Moreover, the subsets V_1, V_2 can be partitioned into disjoint cliques $X_1^1, \ldots, X_{r_1}^1$, and $X_1^2, \ldots, X_{r_2}^2$, respectively, such that X_i^1, X_j^1 (X_i^2, X_j^2) are *adjacent* —that is, there exists an edge $(u, v) \in E$, with $u \in X_i^1$ and $v \in X_j^1$— if and only if $j = i + 1$. However, X_i^1, X_j^2 could be adjacent for any i, j.

Let $G_0 = (V_0, E_0)$ be a graph, where $V_0 = \{x_1^1, \ldots, x_{r_1}^1, x_1^2, \ldots, x_{r_2}^2\}$, and $(x_i^h, x_j^k) \in E_0$ if and only if X_i^h and X_j^k are adjacent, with $h, k \in \{1, 2\}$, $i = 1, \ldots, r_1, j = 1, \ldots, r_2$.

Then, the vertex set V of a crown-free interval graph $G = (V, E)$ can be partitioned in such a way that the graph G_0 is a 4-partite graph.

4   On the Interval Coloring Problem of Crown-Free Interval Graphs

As application problem in which feasible solutions can be represented by acyclic orientations of a crown-free interval graph let us consider the interval coloring problem of a weighted crown-free interval graph. The problem is defined as follows. W.l.o.g., assuming that all vertex weights are positive integers, a (weighted) *interval coloring* (or *t-coloring*) of G is a function $c : V \to \{1, \ldots, t\}$ such that $c(x) + w(x) - 1 \leq t$ and, if $c(x) \leq c(y)$ and x is adjacent to y, then $c(x) + w(x) - 1 \leq c(y)$. An interval coloring of G can be viewed as the assignment of an interval $\{c(x), c(x) + 1, \ldots, c(x) + w(x) - 1\}$ of $w(x)$ colors to each vertex x, such that the intervals of colors assigned to two adjacent vertices do not overlap. The interval coloring problem is finding an interval coloring with minimum interval coloring number t, and the optimal solution value, i.e. the minimum interval coloring number, is called *interval chromatic number* $\chi_w(G)$ of G.

In general, given an undirected weighted graph $G = (V, E)$, a feasible interval coloring of G can be obtained by considering any acyclic orientation A of G, and visiting the oriented graph $H = (V, A)$ in $O(m)$ time by a breadth-first search, in the following way. Define the coloring function $c : V \to \{1, \ldots, t\}$ as follows. For each vertex $s \in V$ with no ingoing arc, let $c(s) = 1$. Proceeding recursively, for each other vertex $j \in V$ let $c(j) = \max_{(i,j) \in A}[c(i) + w(i)]$. Assign to each $j \in V$ the coloring interval $[c(j), \ldots, c(j) + w(j) - 1]$. Going back through the arcs by means of the coloring function above defined, we get a maximum weighted path π of $H = (V, A)$, whose length $l(\pi) = \sum_{i \in \pi} w(i)$ is equal to t. Vice-versa an interval coloring of $G = (V, E)$ defines an implicit acyclic orientation A of G, directing and edge toward the vertex whose coloring interval is to the right of the other on the real line. Hence, for an undirected weighted graph G, the minimum interval coloring number $\chi_w(G) = \min_A[\max_\pi l(\pi)]$ [13].

Now, we show how to orient the edge set of a crown-free interval graph G in order to obtain an acyclic orientation in which each path is contained in at most four maximal cliques of G.

Since, by Theorem 2, given a crown-free interval graph $G = (V, E)$, it is possible to partition the vertex set V in two disjoint subsets V_1 and V_2 such that subgraphs $G(V_1)$ and $G(V_2)$ are proper interval graphs, let us start to show how we can obtain an acyclic orientation $\mathcal{H}_P$ of a proper interval graph G_P.

Given $G_P = (V_P, E_P)$, we compute the partition $\mathcal{P}$ of V_P into cliques $(X_1, \ldots, X_r)$. Let the *odd* cliques and the *even* cliques be the cliques of $\mathcal{P}$ with odd and even indices, respectively. First, orient the edges of G_P from odd cliques to even cliques. More precisely, orient all the edges $(x, y) \in E_P$, from x to y with x and y belonging to an odd and to an even clique, respectively. Then, let us orient the edges between vertices of the same clique X_i, according to any one of its transitive orientation. In this way we have oriented all edges of G_P, the orientation so obtained is acyclic, and each (maximal) path in the oriented graph goes from a vertex of an odd clique to a vertex of an even clique and exactly contains all the vertices of these two cliques. Therefore, each path of the oriented graph obtained by orienting the proper interval graph G_P according to $\mathcal{H}_P$, is contained in at most two maximal cliques of G_P. Following this procedure, let $\mathcal{H}_1$ and $\mathcal{H}_2$ be the orientation of $G(V_1)$ and $G(V_2)$, respectively.

Now, in order to complete the acyclic orientation $\mathcal{H}$ of the whole crown-free interval graph $G = ((V_1 \cup V_2), E)$, we orient the remaining edges $(x, y) \in E$, from x to y with $x \in V_1$ and $y \in V_2$. In this way we have obtained an acyclic orientation $\mathcal{H}$ of G, where each (maximal) oriented path goes from a vertex of V_1 to a vertex of V_2, and is contained in at most four maximal cliques of G. Therefore, it follows that

Theorem 6. *Each path of the oriented graph obtained by orienting the crown-free interval graph G according to $\mathcal{H}$, is contained in at most four maximal cliques of G.*

Then, for any vertex weight function, it results that

Theorem 7. *There exists a 4-approximation algorithm for the interval coloring problem of crown-free interval graph.*

5 Conclusions

In this paper we investigated the problem of finding an acyclic orientation of a crown-free interval graph G, such that each (oriented) path is contained in at most four maximal cliques of G. We showed how this approach can be used to give a 4-approximated solution for $\mathcal{NP}$-hard optimization problems where each feasible solution can be represented as an acyclic orientation of a crown-free interval graph. Moreover, the performance ratio of the approximated solution holds for any weight function.

Acknowledgments

The authors owe thanks to anonymous referees for improvements over the manuscript.

References

1. N. Alon and R.B. Boppana, The Monotone Circuit Complexity of Boolean Functions, *Combinatorica*, 7, (1987) 1–22.
2. E.G. Coffman, An Introduction to Combinatorial Models of Dynamic Storage Allocation, *SIAM Review*, 25, (1983) 311–325.
3. G. Confessore, P. Dell'Olmo and S. Giordani, An Approximation Algorithm for Proper Dynamic Storage Allocation, Tech. Rep. 448, IASI-CNR, Rome, Italy, (January 1997), submitted to *Discrete Mathematics*.
4. G.Confessore, P.Dell'Olmo and S.Giordani, Decomposition Properties of Claw-free Chordal Graphs, Tech. Rep. RR-98.14, Dip. Informatica Sistemi e Produzione, Univ. of Rome "Tor Vergata", Rome, Italy (April 1998).
5. G. Confessore, P. Dell'Olmo and S. Giordani, An Approximation Result for a Periodic Allocation Problem, Tech. Rep. RR-98.17, Dip. Informatica Sistemi e Produzione, Univ. of Rome "Tor Vergata", Rome, Italy, (September 1998), submitted to *Discrete Applied Mathematics*.
6. G.Confessore, P.Dell'Olmo and S.Giordani, Partitioning Cliques of Claw-free Strongly Chordal Graphs, in: P.Kall and H.J.Lüthi (eds.), *Operations Research Proceedings 1998*, (Springer, 1999) 142–151.
7. G. Confessore, P. Dell'Olmo and S. Giordani, A Linear Time Approximation Algorithm for a Storage Allocation Problem, in: M.P. Polis et al. (eds.), *Proc. of 18th IFIP Conference on Syst. Model. and Opt. 1997*, CRC Research Notes in Math., (Chapman and Hall, 1999) 253–260.
8. P. Dell'Olmo, M.G Speranza and Zs. Tuza, Comparability Graph Augmentation for some Multiprocessor Scheduling Problems, *Discr. Appl. Math.*, 72, (1997) 71–94.
9. R. Faudree, E. Flandrin and Z. Ryjáček, Claw-free Graphs – a Survey, *Discr. Math.*, 164, (1997) 87–147.
10. M.R. Garey and D.S. Johnson, *Computers and Intractability: A Guide to the Theory of NP-Completeness*, (Freeman, San Francisco, CA, 1979).
11. J. Gergov, Approximation Algorithms for Dynamic Storage Allocation, in: *Proc. of the 4th Europ. Symp. on Algs.*, Lecture Notes in Computer Science, (Springer Verlag, 1996) 52–61.

12. P.C. Gilmore and A.J. Hoffman, A Characterization of Comparability Graphs and of Interval Graphs, *Canadian J. Math.*, 16, (1964) 539–548.

13. M.C. Golumbic, *Algorithmic Graph Theory and Perfect Graphs*, (Academic Press, New York, 1980).

14. I. Habib, Bandwidth Allocation in ATM Networks, *IEEE Communication Magazine*, (May 1997) 120–121.

15. H.A. Kierstead, A Polynomial Time Approximation Algorithm for Dynamic Storage Allocation, *Discrete Math.*, 88, (1991) 231–237.

16. C.G. Lekkerkerker and J.Ch. Boland, Representation of a Finite Graph by a Set of Intervals on the Real Line, *Fundam. Math.*, 51, (1962) 45–64.

17. J.S. Naor, A. Orda and Y. Petruschka, Dynamic Storage Allocation with Known Durations, in: R. Burkard and G. Woeginger (eds.), *Proc. of the 5th Europ. Symp. Algs.*, Lecture Notes in Computer Science 1284, (Springer Verlag, 1997) 378–387.

Triangulated Neighbourhoods in C_4-Free Berge Graphs

I. Parfenoff, F. Roussel, and I. Rusu

Université d'Orléans, L.I.F.O.,
B.P. 6759, 45067 Orléans Cedex 2, France

Abstract. We call a *T-vertex* of a graph $G = (V, E)$ a vertex z whose neighbourhood $N(z)$ in G induces a triangulated graph, and we show that every C_4-free Berge graph either is a clique or contains at least two non-adjacent T-vertices. An easy consequence of this result is that every C_4-free Berge graph admits a *T-elimination scheme*, *i.e.* an ordering $[v_1, v_2, \ldots, v_n]$ of its vertices such that v_i is a T-vertex in the subgraph induced by $v_i, \ldots, v_n$ in G.

Keywords: lexicographic breadth-first search, triangulated graph, perfectness

1 Introduction

A graph is called *perfect* if, for each of its induced subgraphs, the chromatic number equals the clique number. Since Berge [1] introduced perfect graphs, a lot of partial results have been obtained on the Strong Perfect Graph Conjecture (SPGC) which claims that a graph is perfect if and only if it is a Berge graph, that is, it contains no chordless cycle on odd number of vertices, and no complement of such a cycle. Among the particular classes of graphs which have been shown to satisfy the SPGC we can find every class of H-free graphs, for each graph H on four vertices but C_4 (the chordless cycle on four vertices) and its complement $2K_2$. It can be noticed that these two exceptions are in fact only one exception, since a graph is perfect if and only if its complement is (see [3]).

The perfection of C_4-free Berge graphs cannot be approached without a good knowledge of the structure of these graphs, and this knowledge is lacking. Our aim in this paper is to show that a special vertex exists in every C_4-free Berge graph; the immediate consequence of this is that C_4-free Berge graphs (which form an hereditary family) admit elimination schemes (obtained by considering a special vertex, removing it from the graph and repeating this operation on the remaining graph until an empty graph is found).

For us, a special vertex will be a *T-vertex*, that is, a vertex z whose neighbourhood is triangulated (it contains no chordless cycle on four vertices or more). To find it, we will use the lexicographic breadth-first search algorithm [4] (abbreviated LexBFS), which orders the vertices of the graph such that the last vertex in the ordering is the T-vertex we are looking for.

Widmayer et al. (Eds.): WG'99, LNCS 1665, pp. 402–412, 1999.

2   Graph Decomposition Using LexBFS

The algorithm LexBFS has been proposed by Rose, Tarjan and Lueker [4] in order to give a linear recognition algorithm for triangulated graphs. It turned out that for triangulated graphs, the last vertex in the order found by LexBFS is always a *simplicial* vertex, i.e. its neighbourhood induces a clique. Moreover, it has been shown [5] that for i-triangulated graphs (that is, graphs in which the vertices of every odd cycle induce a triangulated graph) the last vertex in the order given by LexBFS is always a multi-partite graph.

To describe LexBFS we attach to every vertex a label which consists of a set of numbers listed in decreasing order. The labels are compared using lexicographic order. The input graph G is assumed connected, although the algorithm works for non-connected graphs too. This is because our results on connected graphs are easily extended to non-connected graphs (by performing the same reasoning on each connected component). Then the algorithm is the following:

Algorithm LexBFS

> **Input:** a connected graph $G = (V, E)$ with n vertices
> **Output:** a function $\sigma : \{1, 2, \ldots, n\} \to V$ (an order on V)

> assign the label $\emptyset$ to each vertex;
> for $i := n$ down to 1 do
> > pick an unnumbered vertex v with a largest label (in lexicographic order);
> > $\sigma(i) := v$; {comment: this assigns to v the number i}
> > for each unnumbered vertex $t \in N(v)$ do
> > > add i to the label of t (at the end)

Notice that in the order σ obtained by this algorithm, the *last* vertex which has been chosen in the graph is $\sigma(1)$ (that we denote z all along the paper). This is the special vertex both in the case of triangulated graphs, and in the case of i-triangulated graphs. And this will be the special vertex in the case of C_4-free Berge graphs too.

As it can be easily seen, LexBFS can be applied to every graph. Based on the order given by LexBFS, a graph decomposition and structural results can be found, which are also valid for every graph. We give them below, in this section. In the next section, we will study the particular properties of C_4-free Berge graphs with regard to this decomposition. In Section 4, we will prove the existence of at least two non-adjacent T-vertices in every C_4-free Berge graph different from a clique.

Let $G = (V, E)$ be a graph with at least two vertices for which LexBFS was applied. Denote $w = \sigma(n)$. Define a partition $H_0, H_1, \ldots, H_p, H_p'$ of V as follows. Put $H_0 = \{w\}$ and notice that $H_0 \subseteq V - \{z\}$ (recall the notation $z = \sigma(1)$). By induction, assume that $H_i \subseteq V - \{z\}$ is already defined (for some $i \geq 0$) and set

$H_i' = \{h' \in V \mid \text{for all } h \in H_i, \sigma^{-1}(h) > \sigma^{-1}(h')\}.$

Furthermore, partition H_i into three subsets (the notation $N_{H_i'}(h)$ designates the set of neighbours of h which belong to H_i'):

$T_i = H_i \cap N(z)$
$U_i = \{h \in H_i - T_i \mid N_{H_i'}(h) \neq \emptyset\}$, and
$V_i = H_i - (T_i \cup U_i).$

Now, define $H_{i+1} = \cup_{u \in U_i} N_{H_i'}(u)$ (this last set will be denoted $N_{H_i'}(U_i)$). Assume this construction is repeated p times, so that we have the sets $H_0, H_1, \ldots, H_p$ (which are all non-empty) and H_p'. Let us say that a set $R \subseteq V$ is an interval if every $u \in V - R$ either satisfies $\sigma^{-1}(r) > \sigma^{-1}(u)$ for every $r \in R$, or satisfies $\sigma^{-1}(r) < \sigma^{-1}(u)$ for every $r \in R$. In [5], two of the authors proved that (see Fig. 1):

Claim 1. *For every graph $G = (V, E)$ we have (under the assumption that $H_{i+1} \neq \emptyset$):*

(P1) for all i $(0 \leq i \leq p - 1)$, the set H_{i+1} is an interval and $H_{i+1} \subseteq H_i'$ (so that $H_{i+1}' \subset H_i'$);

(P2) for all i $(0 \leq i \leq p - 1)$, the set $H_i \cup H_{i+1}$ is an interval (i.e. no $v \in V$ satisfies $\sigma^{-1}(v) > \sigma^{-1}(h')$ for every $h' \in H_{i+1}$, and $\sigma^{-1}(v) < \sigma^{-1}(h)$ for every $h \in H_i$);

(P3) for all i $(0 \leq i \leq p - 1)$, every $h' \in H_{i+1}'$ has $N(h') \subseteq H_{i+1}' \cup H_{i+1} \cup T_i \cup T_{i-1} \cup \ldots \cup T_0$.

An immediate consequence of (P1) and (P2) is that $H_0 \cup H_1 \cup \ldots \cup H_p \cup H_p'$ is a partition of V. Now, since $H_0 \cup H_1 \cup \ldots \cup H_p$ grows when p grows, after a finite number of steps we will have $H_{p+1} = \emptyset$, i.e. $U_p = \emptyset$ (notice that H_p' is never empty, since $z \in H_p'$). This is the value of p we will consider in the rest of the paper.

Now, calling a *module* of G an induced subgraph H of G such that every vertex in $G - H$ is either adjacent to all H or to no vertex in H, we prove that:

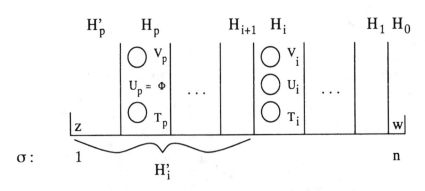

Fig. 1. The sets H_i, H_i' $(i = 0, 1, \ldots, p)$

Claim 2. *For every graph $G = (V, E)$ and every starting vertex w, the partition $H_0 \cup H_1 \cup \ldots \cup H_p \cup H'_p$ has the properties:*

(P4) for all i $(0 \le i \le p - 1)$, every vertex in T_i is adjacent to every vertex of H'_{i+1};

(P5) every vertex in T_p is adjacent to every vertex of H'_p;

(P6) H'_p is a module of G, and its neighbourhood in $G - H'_p$ is $T_0 \cup T_1 \cup \ldots \cup T_p$.

Proof. If (P4) is not true, then take a vertex $y \in H'_{i+1}$ non-adjacent to some $t \in T_i$ and recall that, during the execution of LexBFS, y has been chosen before z; that is, the label of y was larger (in lexicographic order) than the label of z at the precise moment when y was chosen. Now, by (P3), $N(y) \subseteq H'_{i+1} \cup H_{i+1} \cup T_i \cup T_{i-1} \cup \ldots \cup T_0$ which implies $N(y) - (H'_{i+1} \cup H_{i+1}) \subseteq T_i \cup T_{i-1} \cup \ldots \cup T_0 = N(z) - (H'_{i+1} \cup H_{i+1})$ and the inclusion is strict because of t. But then the label of y was smaller (in lexicographic order) than the label of z at the precise moment when y was chosen, because of t. So z should have been chosen instead of y, a contradiction.

If (P5) is not true, then take $y \in H'_p$ and notice that, since $U_p = \emptyset$, $N(y) \subseteq H'_p \cup T_p \cup T_{p-1} \cup \ldots \cup T_0$. As before, $N(y) - H'_p \subset T_p \cup T_{p-1} \cup \ldots \cup T_0 = N(z) - H'_p$, so z should have been chosen before y by LexBFS, a contradiction.

Statement (P6) is now easy to deduce: every vertex x in $G - H'_p$ is either in some T_i $(0 \le i \le p)$ or in no such set. In the first case, x is adjacent to all H'_p by (P4) and (P5). In the second one, $x \in (H_0 - T_0) \cup (H_1 - T_1) \cup \ldots \cup (H_{p-1} - T_{p-1}) \cup V_p$ (since $U_p = \emptyset$) and, by (P3) for every vertex in H'_p, x has no neighbour in H'_p.   □

Finally, we can add these two very useful properties:

Claim 3. *For every graph $G = (V, E)$ we have:*

(P7) if x and y are distinct vertices in H_i, there exists a chordless path joining x, y whose internal vertices are in $H_0 \cup H_1 \cup \ldots \cup H_{i-1} - N(H'_i)$ $(1 \le i \le p)$;

(P8) if G is a Berge graph and x, y are distinct vertices in H_i $(1 \le i \le p)$, then all the chordless paths of length at least three joining x, y in $G - xy$ (i.e. the graph obtained by removing, if it exists, the edge xy) and whose internal vertices are in $H_{i+1} \cup \ldots \cup H_p \cup H'_p$ have the same parity.

Proof. Statement (P7) is easy to prove by induction. Obviously, if $x, y \in H_1$ then either they are adjacent (and we have a path with only one edge and no internal vertices) or they are not adjacent and then the path of length 2 given by x, w, y is the one we need. Now, if we assume this is true for $i - 1$, then, as before, only the case where $xy \notin E$ is interesting. Each of $x, y \in H_i$ has a neighbour x_1, respectively y_1 in H_{i-1}. If $x_1 y \in E$ or $y_1 x \in E$, then we have a chordless path of length 2 joining x, y; otherwise, we apply induction hypothesis for $x_1, y_1 \in H_{i-1}$ to find a path joining them, and then add x, y to this path.

To deduce statement (P8), it is sufficient to notice that if two paths with the indicated property and different parities exist, then two chordless cycles with different parities and cardinality at least four may be found by putting together each of these two paths and the path given by (P7).   □

Remark 1. Obviously, if in (P8) the vertices x, y are non-adjacent, then the condition on the length of the paths (at least three) can be dropped.

In the next section, we investigate C_4-free Berge graphs using these general properties of the decomposition given by LexBFS.

3 Preliminary Results

Let G be a connected C_4-free Berge graph and assume we applied to G the decomposition above. The entire notation (for w and z included) is preserved.

In the aim to prove that z is a T-vertex, we notice that $N(z) - H'_p = T_0 \cup \ldots \cup T_p$ and recall that H'_p is a module. Our approach is the following one: in this section, we prove particular properties of the sets T_i $(0 \le i \le p)$, with regard to the P_4 (chordless paths on four vertices). Then, in the next section, we prove that z is a T-vertex.

An important role is played in our reasoning (see next section) by the chordless paths on four vertices. We show below that the distribution of their vertices in the non-empty sets T_i obeys to some constraints.

For that, we need some supplementary notation. In the proofs below, we will define sets A_1, B_1, C_1, D_1 which are always non-empty. The notation a_1 (respectively b_1, c_1, d_1) will then concern an arbitrary vertex in A_1 (respectively B_1, C_1, D_1). The notation $(ab)_1$ will concern an arbitrary vertex in $A_1 \cap B_1$, whenever this set is (or is supposed) non-empty (and similarly for the other intersections). The notation $(a.b)_1$ will concern an arbitrary vertex in $A_1 - B_1$, whenever this set is (or is supposed) non-empty (and similarly for the other set differences).

A path will be denoted by $[v_1, v_2, \ldots, v_k]$ and a cycle by $[v_1, v_2, \ldots, v_k, v_1]$. We use the classical notation P_k (respectively C_k) to denote a chordless path (respectively cycle) on $k \ge 4$ vertices.

Lemma 1. *For every i $(0 \le i \le p)$, T_i is P_4-free.*

Proof. Assuming the contrary, let T_i be a set which contains a P_4 denoted $[a, b, c, d]$. Since T_0 contains at most one vertex, we have $i \ge 1$. Define

$$A_1 = N_{U_{i-1}}(a),$$
$$B_1 = N_{U_{i-1}}(b),$$
$$C_1 = N_{U_{i-1}}(c) \text{ and}$$
$$D_1 = N_{U_{i-1}}(d).$$

These sets are non empty since $H_i = N_{H'_{i-1}}(U_{i-1})$ (by definition). We have that:

- $A_1 \cap C_1 = \emptyset$; otherwise $[a, (ac)_1, c, z, a]$ is a C_4.
- $A_1 \cap D_1 = \emptyset$; otherwise $[a, (ad)_1, d, z, a]$ is a C_4.
- $B_1 \cap D_1 = \emptyset$; otherwise $[b, (bd)_1, d, z, b]$ is a C_4.
- $A_1 \subseteq B_1$ or $D_1 \subseteq C_1$; otherwise the paths $[(a.b)_1, a, z, d, (d.c)_1]$ and $[(a.b)_1, a, b, c, d, (d.c)_1]$ contradict (P8).
- $A_1 \subseteq B_1$ can be assumed without loss of generality, because of the preceding affirmation (since edges ab, cd are symmetrical with respect to the P_4 $[a, b, c, d]$).

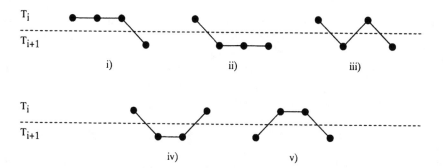

T_i

T_{i+1}

      i)   ii)   iii)

T_i

T_{i+1}

          iv)   v)

Fig. 2. Forbidden P_4.

- $C_1 \subseteq B_1$, otherwise $[a_1, a, z, c, (c.b)_1]$ and $[a_1, b, c, (c.b)_1]$ are paths which contradict (P8).
- $C_1 \cap D_1 = \emptyset$ because of $C_1 \subseteq B_1$ and $B_1 \cap D_1 = \emptyset$.

Then the chordless paths $[c_1, c, d, d_1]$ and $[c_1, b, z, d, d_1]$ contradict (P8). $\qquad \square$

Lemma 2. *For every i ($0 \le i \le p-1$), a P_4 $[a, b, c, d]$ with $a, b, c, d \in T_i \cup T_{i+1}$ satisfies none of the properties below (see Fig. 2):*
   i) $a, b, c \in T_i$ and $d \in T_{i+1}$;
   ii) $a \in T_i$ and $b, c, d \in T_{i+1}$;
   iii) $a, c \in T_i$ and $b, d \in T_{i+1}$;
   iv) $a, d \in T_i$ and $b, c \in T_{i+1}$;
   v) $b, c \in T_i$ and $a, d \in T_{i+1}$.

Proof. By contradiction, we assume that a P_4 satisfying i), ii), iii), iv) or v) exists, and we consider each case separately in the aim to obtain a contradiction. It is not difficult to see that in each case we have $i \ge 1$. As a general observation, recall that no vertex in H_{i+1} has neighbours in $H_{i-1} - T_{i-1}$ (because of (P3)). This observation will be often used in the sequel without any specification.

Case i) : $a, b, c \in T_i$ and $d \in T_{i+1}$.
   Because of $H_{i+1} = N_{H'_i}(U_i)$, there exists $d' \in U_i$ such that $d'd \in E$. Denote
$A_1 = N_{U_{i-1}}(a)$,
$B_1 = N_{U_{i-1}}(b)$,
$C_1 = N_{U_{i-1}}(c)$ and
$D_1 = N_{U_{i-1}}(d')$.
Because of $H_i = N_{H'_{i-1}}(U_{i-1})$, these sets are non-empty. We have:
- $d'b \notin E$; otherwise $[b, d', d, z, b]$ is a C_4.
- $d'a \notin E$; otherwise $[a, d', d, z, a]$ is a C_4.
- $A_1 \cap C_1 = \emptyset$; otherwise $[a, (ac)_1, c, z, a]$ is a C_4.
- $A_1 \cap D_1 = \emptyset$; otherwise $[a, (ad)_1, d', d, z, a]$ is a C_5.
- $B_1 \cap D_1 = \emptyset$; otherwise $[b, (bd)_1, d', d, z, b]$ is a C_5.
- $d'c \in E$, since in the contrary case we successively have: $D_1 \cap C_1 = \emptyset$ (otherwise $[c, (cd)_1, d', d, c]$ is a C_4), $A_1 \subseteq B_1$ (otherwise the paths $[(a.b)_1, a, z, d, d', d_1]$

and $[(a.b)_1, a, b, c, d, d', d_1]$ contradict (P8)), $C_1 \subseteq B_1$ (otherwise the paths $[a_1, b, c, (c.b)_1]$ and $[a_1, a, z, c, (c.b)_1]$ contradict (P8)) and consequently the paths $[c_1, b, z, d, d', d_1]$ and $[c_1, c, d, d', d_1]$ give again a contradiction with (P8).

- $D_1 \subseteq C_1$, since in the contrary case we successively have: $B_1 \subseteq C_1$ (otherwise the paths $[(d.c)_1, d', c, b, (b.c)_1]$ and $[(d.c)_1, d', d, z, b, (b.c)_1]$ contradict (P8)), $A_1 \cap B_1 = \emptyset$ (because of $B_1 \subseteq C_1$ and $A_1 \cap C_1 = \emptyset$) and consequently the paths $[a_1, a, b, b_1]$ and $[a_1, a, z, c, b_1]$ contradict (P8).
- $A_1 \subseteq B_1$; otherwise the paths $[(a.b)_1, a, b, c, d_1]$ and $[(a.b)_1, a, z, d, d', d_1]$ contradict (P8).

Then the chordless paths $[a_1, b, c, d_1]$ and $[a_1, a, z, c, d_1]$ contradict (P8).

Case ii) : $a \in T_i$ and $b, c, d \in T_{i+1}$.

Because of $H_{i+1} = N_{H_i'}(U_i)$, there exist $b', d' \in U_i$ such that $b'b, d'd \in E$ (so $b' \neq d'$ otherwise a C_4 is induced with b, z, d). Define

$A_1 = N_{U_{i-1}}(a)$,
$B_1 = N_{U_{i-1}}(b')$ and
$D_1 = N_{U_{i-1}}(d')$.

As usual, these sets are non-empty. We have:

- $d'b \notin E$; otherwise $[d', b, z, d, d']$ is a C_4.
- $d'c \in E$; otherwise the paths $[d', d, z, a]$ and $[d', d, c, b, a]$ contradict (P8).
- $d'a \notin E$; otherwise $[d', a, z, d, d']$ is a C_4.
- $b'd \notin E$; otherwise $[b, b', d, z, b]$ is a C_4.
- $b'd' \notin E$; otherwise $[b, b', d', d, z, b]$ is a C_5.
- $A_1 \cap D_1 = \emptyset$; otherwise $[d', (ad)_1, a, b, c, d']$ is a C_5.
- $B_1 \cap D_1 = \emptyset$; otherwise $[d', (bd)_1, b', b, c, d']$ is a C_5 or contains a C_4 (depending on whether $b'c \in E$ or not).
- $b'c \in E$; otherwise the paths $[b_1, b', b, c, d', d_1]$ and $[b_1, b', b, z, d', d_1]$ contradict (P8).
- $A_1 \cap B_1 = \emptyset$; otherwise the paths $[(ab)_1, b', c, d', d_1]$ and $[(ab)_1, a, z, d, d', d_1]$ contradict (P8).
- $ab' \in E$; otherwise the paths $[a_1, a, b, b', b_1]$ and $[a_1, a, z, c, b', b_1]$) contradict (P8).

Then $[b', a, z, c, b']$ is a C_4, a contradiction.

Case iii) : $a, c \in T_i$ and $b, d \in T_{i+1}$.

Because of $H_{i+1} = N_{H_i'}(U_i)$, there exist $b', d' \in U_i$ such that $b'b, d'd \in E$ (so $b' \neq d'$ otherwise a C_4 is induced with b, z, d). Define

$A_1 = N_{U_{i-1}}(a)$,
$B_1 = N_{U_{i-1}}(b')$,
$C_1 = N_{U_{i-1}}(c)$ and
$D_1 = N_{U_{i-1}}(d')$.

As usual, these sets are non-empty. Then we have:

- $d'a \notin E$; otherwise $[d', a, z, d, d']$ is a C_4.
- $d'b \notin E$; otherwise $[d', b, z, d, d']$ is a C_4.
- $b'd \notin E$; otherwise $[b', d, z, b, b']$ is a C_4.
- $b'd' \notin E$; otherwise $[b', d', d, z, b, b']$ is a C_5.

- $A_1 \cap C_1 = \emptyset$; otherwise $[c, (ac)_1, a, z, c]$ is a C_4.
- $A_1 \cap D_1 = \emptyset$; otherwise $[d', (ad)_1, a, z, d, d']$ is a C_5.
- $d'c \in E$, since in the contrary case we have $D_1 \cap C_1 = \emptyset$ (otherwise $[c, (dc)_1, d', d, c]$ is a C_4) and then the paths $[d_1, d', d, c, b, a, a_1]$ and $[d_1, d', d, z, a, a_1]$ contradict (P8).
- $C_1 \cap D_1 = \emptyset$; otherwise, $[(cd)_1, c, b, a, a_1]$ and $[(cd)_1, d', d, z, a, a_1]$ contradict (P8).
- $b'c \in E$, since in the contrary case we have $B_1 \cap C_1 = \emptyset$ (otherwise $[c, (bc)_1, b', b, c]$ is a C_4) and then the paths $[b_1, b', b, c, d', d_1]$ and $[b_1, b', b, z, d, d', d_1]$ contradict (P8).
- $b'a \notin E$; otherwise $[b', a, z, c, b']$ is a C_4.
- $A_1 \cap B_1 = \emptyset$; otherwise $[a, (ab)_1, b', b, a]$ is a C_4.
- $B_1 \cap D_1 = \emptyset$; otherwise the paths $[(bd)_1, b', b, a, a_1]$ and $[(bd)_1, d', d, z, a, a_1]$ contradict (P8).
- $B_1 \cap C_1 = \emptyset$; otherwise the paths $[(bc)_1, c, d', d_1]$ and $[(bc)_1, b', b, z, d, d', d_1]$ contradict (P8).

Then the paths $[a_1, a, b, b', b_1]$ and $[a_1, a, z, c, b', b_1]$ contradict (P8).

Case iv) : $a, d \in T_i$ and $b, c \in T_{i+1}$.

Denote $A_1 = N_{U_{i-1}}(a)$ and $D_1 = N_{U_{i-1}}(d)$ (non-empty sets of vertices). Then $A_1 \cap D_1 = \emptyset$ (otherwise $[d, (ad)_1, a, b, c, d]$ is a C_5) and the paths $[a_1, a, b, c, d, d_1]$ and $[a_1, a, z, d, d_1]$ contradict (P8).

Case v) : $b, c \in T_i$ and $a, d \in T_{i+1}$.

Because of $H_{i+1} = N_{H'_i}(U_i)$, there exist $a', d' \in U_i$ such that $a'a, d'd \in E$ (so $a' \neq d'$ otherwise a C_4 is induced with a, z, d). Define
$$A_1 = N_{U_{i-1}}(a'),$$
$$B_1 = N_{U_{i-1}}(b),$$
$$C_1 = N_{U_{i-1}}(c) \text{ and}$$
$$D_1 = N_{U_{i-1}}(d').$$
As usual, these sets are non-empty and we have:
- $a'd \notin E$; otherwise $[a', a, z, d, a']$ is a C_4.
- $d'a \notin E$; otherwise $[d', a, z, d, d']$ is a C_4.
- $a'd' \notin E$; otherwise $[a', a, z, d, d', a']$ is a C_5.
- $a'c \notin E$; otherwise $[a', a, z, c, a']$ is a C_4.
- $d'b \notin E$; otherwise $[d', d, z, b, d']$ is a C_4.
- $A_1 \cap C_1 = \emptyset$; otherwise $[c, (ac)_1, a', a, z, c]$ is a C_5.
- $B_1 \cap D_1 = \emptyset$; otherwise $[(bd)_1, d', d, z, b, (bd)_1])$ is a C_5.
- $a'b \in E$ or $d'c \in E$, since in the contrary case we successively have: $A_1 \cap B_1 = \emptyset$ (otherwise $[b, (ab)_1, a', a, b]$ is a C_4), $C_1 \cap D_1 = \emptyset$ (otherwise $[c, (cd)_1, d', d, c]$ induces a C_4) and then the paths $[a_1, a', a, z, d, d', d_1]$ and $[a_1, a', a, b, c, d, d', d_1]$ contradict (P8).
- $d'c \in E$ can be assumed without loss of generality, because of the preceding affirmation (since edges ab, cd are symmetrical with respect to the P_4 $[a, b, c, d]$).
- $a'b \notin E$, since in the contrary case we successively have: $a_1b \in E$ or $d_1c \in E$ (otherwise the paths $[a_1, a', b, c, d', d_1]$ and $[a_1, a', a, z, d, d', d_1]$ contradict (P8)),

and we assume without loss of generality (for symmetry reasons) that $a_1 b \in E$; $d_1 c \notin E$ (otherwise the paths $[a_1, b, c, d_1]$ and $[a_1, a', a, z, d', d, d_1]$ contradict (P8)) and then the paths $[a_1, b, c, d', d_1]$ and $[a_1, b, z, d, d', d_1]$ contradict (P8).

- $A_1 \cap B_1 = \emptyset$; otherwise $[b, (ab)_1, a', a, b]$ is a C_4.
- $D_1 \cap C_1 = \emptyset$ since in the contrary case we have $(cd)_1 a' \notin E$ (because of $A_1 \cap C_1 = \emptyset$) and then the paths $[(cd)_1, d', d, z, a, a', a_1]$ and $[(cd)_1, c, b, a, a', a_1]$ contradict (P8).
- $b_1 c \in E$; otherwise the paths $[b_1, b, c, d', d_1]$ and $[b_1, b, z, d, d', d_1]$ contradict (P8).

Then the paths $[b_1, c, z, a, a', a_1]$ and $[b_1, b, a, a', a_1]$ contradict (P8).   □

4   Main Result

This section is devoted to the result we announced, that we state here:

Theorem 1. *Let G be a C_4-free Berge graph and z be the last vertex numbered by LexBFS. Then z is a T-vertex.*

Proof. We use induction on the number of vertices of the graph. The theorem is obviously true for a graph with two vertices. Now, assume the theorem is true for all graphs with strictly less than n vertices, and let us prove it for $G = (V, E)$ which has n vertices. We can assume G is connected, otherwise the conclusion follows by induction (for the last connected component which is numbered by LexBFS). Recall that H'_p is a module of G, whose neighbourhood in $G - H'_p$ is $T_0 \cup T_1 \cup \ldots \cup T_p$.

Claim 4. *The theorem is true if H'_p has at least two vertices.*

Proof. Consider the precise moment where the first vertex in H'_p (i.e. the vertex y with $\sigma^{-1}(y) = max\{\sigma^{-1}(x) \,|\, x \in H'_p\}$) has been chosen by the algorithm LexBFS. As H'_p is a module, and $T_0 \cup \ldots \cup T_p$ contains only vertices that have been numbered *before* every vertex in H'_p, all the vertices in H'_p have the same label as y at the indicated moment. Therefore, the execution of LexBFS restricted to the subgraph $G[H'_p]$ (induced by H'_p in G) is exactly the one we would have if $G[H'_p]$ was considered as a separate, entire graph and LexBFS was applied to it. Then, we can apply the induction hypothesis to deduce that the neighbourhood of z in $G[H'_p]$ is triangulated.

We consider $V' = V - H'_p \cup \{z\}$ and observe that the execution of LexBFS on G restricted to the subgraph $G[V']$ is identical to the one we would have if $G[V']$ was considered as a separate, entire graph and LexBFS was applied to it (since the vertices of $H'_p - \{z\}$ are chosen at the end of the execution on G, so they do not influence the numbering of the vertices in $G - H'_p$). Moreover, since H'_p has at least two vertices, V' has at most $n - 1$ vertices and we can apply the induction hypothesis for $G[V']$ to deduce that $T_0 \cup \ldots \cup T_p$ (i.e. $N(z) - H'_p$) is triangulated. As $H'_p \cap N(z)$ is triangulated (see above) and $T_0 \cup \ldots \cup T_p$ is triangulated too, $N(z)$ (which is the complete join of these two sets) is triangulated too (no cycle is created, except possibly for some C_4; but G contains no C_4). In fact, we can notice that at least one of the two sets is a clique, otherwise a C_4 can be found by taking a pair of non-adjacent vertices in each set.   □

Claim 5. *The theorem is true if $H'_p = \{z\}$.*

Proof. Suppose by contradiction that z is not a T-vertex. Then there exists a chordless cycle $C = [v_1, v_2, ..., v_k, v_1]$ in $N(z)$ with at least four vertices. Moreover, the fact that G is a C_4-free Berge graph implies that C contains at least six vertices. Now, call i the minimum index such that C has at least one vertex in T_i, and assume without loss of generality that $v_1 \in T_i$. So, for every $v \in C$, we have $v \in T_j$ with $j \geq i$.

By (P4), if $t_j \in T_j, t_k \in T_k$ with $|j - k| \geq 2$, then $t_j t_k \in E$. Now, since v_3, v_4 and v_5 are non-adjacent to v_1, these three vertices belong necessarily to $T_i \cup T_{i+1}$. The vertex v_2 is in $T_i \cup T_{i+1} \cup T_{i+2}$, since it is non-adjacent to v_4 and v_5. Now we distinguish three cases :

1. v_2 belongs to T_{i+2}.
   In this case, v_4 and v_5 must be in T_{i+1} (in the contrary case they are adjacent to v_2). Thus, $v_3 \in T_i$, otherwise the P_4 $[v_2, v_3, v_4, v_5]$ verifies the property i) in Lemma 2, a contradiction. Now, v_6 is not adjacent to v_3 and v_2, and must be (by (P4)) in T_{i+1}. But $[v_3, v_4, v_5, v_6]$ is a P_4 which verifies the property ii) in Lemma 2, a contradiction.
2. v_2 belongs to T_{i+1}.
   Since the P_4 $[v_1, v_2, v_3, v_4]$ cannot verify the property ii), iii) or iv) in Lemma 2, the two vertices v_3 and v_4 must be in T_i. If $v_5 \in T_i$ (resp. $v_5 \in T_{i+1}$) then the P_4 $[v_2, v_3, v_4, v_5]$ verifies the property ii) (respectively the property v)) in Lemma 2, a contradiction.
3. v_2 belongs to T_i.
   One of v_3 or v_4 belongs to T_{i+1}, otherwise $[v_1, v_2, v_3, v_4]$ is a P_4 of T_i, a contradiction with Lemma 1. As $[v_1, v_2, v_3, v_4]$ cannot verify the property i) in Lemma 2, v_3 must be in T_{i+1}. Now, we can perform a change of notation as follows: for every $1 < i \leq p$, $v'_{i-1} = v_i$; moreover, $v'_k = v_1$. Then for v'_1 and v'_2 we have the situation in case 2 (with v'_1 instead of v_1 and v'_2 instead of v_2), so we can conclude as in case 2.   □

Claim 4 and Claim 5 prove the theorem.   □

Theorem 1 has the following easy corollaries:

Corollary 1. *Every C_4-free Berge graph G which is not a clique has at least two non-adjacent T-vertices.*

Proof. Since G is not a clique, there exists at least one non-universal vertex w in G. Perform LexBFS by choosing firstly the vertex w (so that $\sigma(n) = w$). Then $z = \sigma(1)$ is a T-vertex, by Theorem 1. Moreover, z is not universal, since $zw \notin E$. Perform again LexBFS, this time by choosing z as the first vertex (i.e. $\sigma'(n) = z$) and denote $x = \sigma'(1)$. Then z, x are two non-adjacent T-vertices.   □

Corollary 2. *The order* $[\sigma(1), \sigma(2), \ldots, \sigma(n)]$ *given by LexBFS is a T-elimination scheme, i.e. for every* $j \in \{1, 2, \ldots, n\}$, *the vertex* $\sigma(j)$ *is a T-vertex in the graph* $G[\sigma(j), \sigma(j+1), \ldots, \sigma(n)]$.

Proof. It is sufficient to apply Theorem 1 for $G[\sigma(j), \sigma(j+1), \ldots, \sigma(n)]$.   □

5   Conclusion

In Theorem 1 and its corollaries, we presented a structure of Berge C_4-free graphs which could be a starting point for the proof of their perfection, and we confirmed Rose, Tarjan and Lueker's feeling that LexBFS can be a good algorithm for many other classes than triangulated graphs.

However, the question arises on whether Theorem 1 is the best result of this kind we can obtain, or the class which is involved (of triangulated graphs) can be reduced, in order to have more constraints on the neighbourhood of the special vertex.

References

1. C. Berge - Färbung von Graphen, deren sämtliche bzw. deren ungerade Kreise starr sind, Wiss. Z. Martin Luther Univ., Halle Wittenberg (1961), 114–115.
2. M. C. Golumbic, "Algorithmic graph theory and perfect graphs", Academic Press (1980).
3. L. Lovász - A characterization of perfect graphs, *J. Combin. Theory Ser. B* **13** (1972), 95–98.
4. D. J. Rose, R. E. Tarjan, G. S. Lueker - Algorithmic aspects of vertex elimination on graphs, *SIAM J. Comput.* 5 (1976), 266–283.
5. F. Roussel, I. Rusu - A linear algorithm to color i-triangulated graphs, *Information Processing Letters* **70** (1999), 57–62.

Author Index

Lecture Notes in Computer Science

For information about Vols. 1–1652
please contact your bookseller or Springer-Verlag

Vol. 1689: F. Solina, A. Leonardis (Eds.), Computer Analysis of Images and Patterns. Proceedings, 1999. XIV, 650 pages. 1999.

Vol. 1690: Y. Bertot, G. Dowek, A. Hirschowitz, C. Paulin, L. Théry (Eds.), Theorem Proving in Higher Order Logics. Proceedings, 1999. VIII, 359 pages. 1999.

Vol. 1691: J. Eder, I. Rozman, T. Welzer (Eds.), Advances in Databases and Information Systems. Proceedings, 1999. XIII, 383 pages. 1999.

Vol. 1692: V. Matoušek, P. Mautner, J. Ocelíková, P. Sojka (Eds.), Text, Speech and Dialogue. Proceedings, 1999. XI, 396 pages. 1999. (Subseries LNAI).

Vol. 1693: P. Jayanti (Ed.), Distributed Computing. Proceedings, 1999. X, 357 pages. 1999.

Vol. 1694: A. Cortesi, G. Filé (Eds.), Static Analysis. Proceedings, 1999. VIII, 357 pages. 1999.

Vol. 1695: P. Barahona, J.J. Alferes (Eds.), Progress in Artificial Intelligence. Proceedings, 1999. XI, 385 pages. 1999. (Subseries LNAI).

Vol. 1696: S. Abiteboul, A.-M. Vercoustre (Eds.), Research and Advanced Technology for Digital Libraries. Proceedings, 1999. XII, 497 pages. 1999.

Vol. 1697: J. Dongarra, E. Luque, T. Margalef (Eds.), Recent Advances in Parallel Virtual Machine and Message Passing Interface. Proceedings, 1999. XVII, 551 pages. 1999.

Vol. 1698: M. Felici, K. Kanoun, A. Pasquini (Eds.), Computer Safety, Reliability and Security. Proceedings, 1999. XVIII, 482 pages. 1999.

Vol. 1699: S. Albayrak (Ed.), Intelligent Agents for Telecommunication Applications. Proceedings, 1999. IX, 191 pages. 1999. (Subseries LNAI).

Vol. 1700: R. Stadler, B. Stiller (Eds.), Active Technologies for Network and Service Management. Proceedings, 1999. XII, 299 pages. 1999.

Vol. 1701: W. Burgard, T. Christaller, A.B. Cremers (Eds.), KI-99: Advances in Artificial Intelligence. Proceedings, 1999. XI, 311 pages. 1999. (Subseries LNAI).

Vol. 1702: G. Nadathur (Ed.), Principles and Practice of Declarative Programming. Proceedings, 1999. X, 434 pages. 1999.

Vol. 1703: L. Pierre, T. Kropf (Eds.), Correct Hardware Design and Verification Methods. Proceedings, 1999. XI, 366 pages. 1999.

Vol. 1704: Jan M. Żytkow, J. Rauch (Eds.), Principles of Data Mining and Knowledge Discovery. Proceedings, 1999. XIV, 593 pages. 1999. (Subseries LNAI).

Vol. 1705: H. Ganzinger, D. McAllester, A. Voronkov (Eds.), Logic for Programming and Automated Reasoning. Proceedings, 1999. XII, 397 pages. 1999. (Subseries LNAI).

Vol. 1706: J. Hatcliff, T. Æ. Mogensen, P. Thiemann (Eds.), Lectures on Partial Evaluation. Proceedings, 1998. IX, 433 pages. 1999. (Subseries LNAI).

Vol. 1707: H.-W. Gellersen (Ed.), Handheld and Ubiquitous Computing. Proceedings, 1999. XII, 390 pages. 1999.

Vol. 1708: J.M. Wing, J. Woodcock, J. Davies (Eds.), FM'99 – Formal Methods. Proceedings Vol. I, 1999. XVIII, 937 pages. 1999.

Vol. 1709: J.M. Wing, J. Woodcock, J. Davies (Eds.), FM'99 – Formal Methods. Proceedings Vol. II, 1999. XVIII, 937 pages. 1999.

Vol. 1710: E.-R. Olderog, B. Steffen (Eds.), Correct System Design. XIV, 417 pages. 1999.

Vol. 1711: N. Zhong, A. Skowron, S. Ohsuga (Eds.), New Directions in Rough Sets, Data Mining, and Granular-Soft Computing. Proceedings, 1999. XIV, 558 pages. 1999. (Subseries LNAI).

Vol. 1712: H. Boley, A Tight, Practical Integration of Relations and Functions. XI, 169 pages. 1999. (Subseries LNAI).

Vol. 1713: J. Jaffar (Ed.), Principles and Practice of Constraint Programming – CP'99. Proceedings, 1999. XII, 493 pages. 1999.

Vol. 1714: M.T. Pazienza (Eds.), Information Extraction. IX, 165 pages. 1999. (Subseries LNAI).

Vol. 1715: P. Perner, M. Petrou (Eds.), Machine Learning and Data Mining in Pattern Recognition. Proceedings, 1999. VIII, 217 pages. 1999. (Subseries LNAI).

Vol. 1716: K.Y. Lam, E. Okamoto, C. Xing (Eds.), Advances in Cryptology – ASIACRYPT'99. Proceedings, 1999. XI, 414 pages. 1999.

Vol. 1717: Ç. K. Koç, C. Paar (Eds.), Cryptographic Hardware and Embedded Systems. Proceedings, 1999. XI, 353 pages. 1999.

Vol. 1718: M. Diaz, P. Owezarski, P. Sénac (Eds.), Interactive Distributed Multimedia Systems and Telecommunication Services. Proceedings, 1999. XI, 386 pages. 1999.

Vol. 1719: M. Fossorier, H. Imai, S. Lin, A. Poli (Eds.), Applied Algebra, Algebraic Algorithms and Error-Correcting Codes. Proceedings. 1999. XIII, 510 pages. 1999.

Vol. 1721: S. Arikawa, K. Furukawa (Eds.), Discovery Science. Proceedings, 1999. XI, 374 pages. 1999. (Subseries LNAI).

Vol. 1722: A. Middeldorp, T. Sato (Eds.), Functional and Logic Programming. Proceedings, 1999. X. 369 pages. 1999.

Vol. 1723: R. France, B. Rumpe (Eds.), UML'99 – The Unified Modeling Language99. XVII, 724 pages. 1999.

Vol. 1726: V. Varadharajan, Y. Mu (Eds.), Information and Communication Security. Proceedings, 1999. XI, 325 pages. 1999.

Vol. 1727: P.P. Chen, D.W. Embley, J. Kouloumdjian, S.W. Liddle, J.F. Roddick (Eds.), Advances in Conceptual Modeling. Proceedings, 1999. XI, 389 pages. 1999.

Vol. 1728: J. Akoka, M. Bouzeghoub, I. Comyn-Wattiau, E. Métais (Eds.), Conceptual Modeling – ER '99. Proceedings, 1999. XIV, 540 pages. 1999.

Vol. 1729: M. Mambo, Y. Zheng (Eds.), Information Security. Proceedings, 1999. IX, 277 pages. 1999.

Vol. 1734: H. Hellwagner, A. Reinefeld (Eds.), SCI: Scalable Coherent Interface. XXI, 490 pages. 1999.

Vol. 1564: M. Vazirgiannis, Interactive Multimedia Documents. XIII, 161 pages. 1999.

Vol. 1591: D.J. Duke, I. Herman, M.S. Marshall, PREMO: A Framework for Multimedia Middleware. XII, 254 pages. 1999.